D1614656

are provided at [illegible]
mechanics, and electromagnetism which underl[illegible]
related to real measurements and to the experimental techniq[illegible] used by physicists in academe and industry. Books in this series are written as course books, and include ample tutorial material, examples, illustrations, revision points, and problem sets. They can likewise be used as preparation for students starting a doctorate in physics and related fields, or for recent graduates starting research in one of these fields in industry.

CONDENSED MATTER PHYSICS

1. M. T. Dove: *Structure and dynamics: an atomic view of materials*
2. J. Singleton: *Band theory and electronic properties of solids*
3. A. M. Fox: *Optical properties of solids*
4. S. J. Blundell: *Magnetism in condensed matter*
5. J. F. Annett: *Superconductivity*
6. R. A. L. Jones: *Soft condensed matter*

ATOMIC, OPTICAL, AND LASER PHYSICS

7. C. J. Foot: *Atomic physics*
8. G. A. Brooker: *Modern classical optics*
9. S. M. Hooker, C. E. Webb: *Laser physics*

PARTICLE PHYSICS, ASTROPHYSICS, AND COSMOLOGY

10. D. H. Perkins: *Particle astrophysics*
11. T. P. Cheng: *Relativity and cosmology*

Modern Classical Optics

GEOFFREY BROOKER

Department of Physics
University of Oxford

UNIVERSITY PRESS

OXFORD
UNIVERSITY PRESS

Great Clarendon Street, Oxford OX2 6DP

Oxford University Press is a department of the University of Oxford.
It furthers the University's objective of excellence in research, scholarship,
and education by publishing worldwide in

Oxford New York

Auckland Bangkok Buenos Aires Cape Town Chennai
Dar es Salaam Delhi Hong Kong Istanbul Karachi Kolkata
Kuala Lumpur Madrid Melbourne Mexico City Mumbai Nairobi
São Paulo Shanghai Taipei Tokyo Toronto

Oxford is a registered trade mark of Oxford University Press
in the UK and in certain other countries

Published in the United States
by Oxford University Press Inc., New York

First published 2003

A catalogue record for this title
is available from the British Library

Library of Congress Cataloging in Publication Data
(Data available)

ISBN 0 19 859965 X (Pbk)
ISBN 0 19 859964 1 (Hbk)

10 9 8 7 6 5 4 3 2 1

Typeset by the author using LaTeX

Printed in Great Britain
on acid-free paper by The Bath Press, Avon

To: Amie, Lois, Oona, Phoebe, Alfie and Blaise.

Preface

The level of treatment in this book is that of the fourth year of an M. Phys. undergraduate course in the UK. However, I have tried to give descriptions that are simple enough to be followed by someone at an earlier stage who seeks a 'different' account of the more basic material. And graduates may find that some 'well-understood' ideas offer unexpected challenges. The topics included here are more than could be covered in the time available in any one undergraduate course, but different courses will, quite properly, make different selections of material.

I concentrate on 'physical' optics (light as a wave), and describe only as much geometrical optics as is really necessary. A thick lens is mentioned only three times. Lens design and optical aberrations are hardly mentioned at all, and then in terms of an optical transfer function rather than Seidel sums. I justify this exclusion on the ground that lens design is now wholly done by computer-aided optimization, description of which would require a very different style of presentation.

This book might better have been called 'semi-classical optics', since the photon nature of light is not ignored. Indeed, photon emission and detection are inherently quantum-mechanical. However, our main concern, the passage of light between emission and detection, can usually be treated classically. Those phenomena, such as entanglement or antibunching, that require 'quantum optics' proper lie outside our remit. Even so, I have tried, in Chapter 10, to explain where the interface lies between the (semi-)classical and quantum regimes.

In a book of this length, some selection of topics is unavoidable, even within physical optics. In particular I regret the omission of interference microscopes (too large a digression) and of adaptive optics applied to Earth-bound astronomical telescopes (too computational).

A book is a linear structure: from beginning to end. Understanding is not like that. It's achieved by reading interactively: checking calculations; cross-linking new information with old; asking 'what if'; thinking of implications and possible objections. Why is μ_r always assumed to be 1 at optical frequencies? Why is an electromagnetic wave always discussed in terms of its $\boldsymbol{E}$-field when $\boldsymbol{B}$ is equally significant in the Maxwell equations? A Fabry–Perot and a thin film are very similar structures; why then are the methods of analysis so different? Can we trust the Kirchhoff-assumption boundary conditions used in diffraction, and how could we find out? Why are the fields inside a laser cavity mathematically similar to the wave functions for a simple harmonic oscillator? The bright student will want answers to such questions, and if a full answer can't be given at this level then at least s/he wants to be given reassurance that the questions are not regarded as troublemaking. The less imaginative should be encouraged by example to see that these are the kind of questions that ought to be asked.

Critical questions shouldn't be allowed to clutter the exegesis, yet a place must be found for them. I have therefore put much exploratory material into 'problems'. A 'problem' is not necessarily an exercise set (though it may be) by the author or at his suggestion by a teacher; it is often an opportunity to make statements (to be checked) about special cases, to point out a link to a similar idea elsewhere, or to encourage discrimination by challenging the 'well-known'. In short, problems are meant to be read, as a commentary integral to the material; the reader can decide which ought to be worked through in detail as well. Most problems are referenced at an appropriate point in the text, to indicate that additional information or critical discussion is available. Solutions are given where mathematical steps are not all obvious or where the physics needs (even more!) discussion.

Problems are graded on a three-point scale from the most basic (a), usually easy, to the most challenging (c). A few 'entertainment' problems, labelled (e), are opportunities to take a wider view. Whatever its classification, each problem is intended to draw attention to some insight, technique or order of magnitude.

This book has grown out of lecture courses that I have given, over a number of years and at different levels. (The author sometimes complains that the Oxford Physics Department has type-cast him as an optician.) The content and attitude have been strongly influenced by frequent interaction with students, one or two at a time, in tutorial discussions that range over most areas of degree-level physics. To those students who asked searching questions or insisted on clearer explanations, I owe a great debt.

Several colleagues have contributed to this book, not always knowingly, by reading it in draft or via a number of discussions. I am particularly indebted to Professor D. N. Stacey and Professor C. E. Webb for their patience in the discussion of a number of tricky points. Dr S. J. Blundell is responsible for an improvement in the clarity of Chapter 5. Mr C. Goodwin very kindly supplied the practical detail on thin-film deposition in Chapter 6. Dr J. Halliday was most helpful in providing information on CD and DVD technology. Dr R. A. Taylor gave valuable help with the description of LEDs in Chapter 11.

I thank Springer-Verlag for permission to reproduce Figs 3.8, 3.9 and 3.12 from the *Atlas of Optical Phenomena.* I thank the IEEE and Dr H. Kogelnik for permission to reproduce Fig. 8.3. I thank Lucent Technology and Drs Fox, Li, Boyd and Gordon for permission to base Fig. 8.7 on a diagram that originally appeared in the *Bell System Technical Journal.* All other diagrams were prepared by the author.

Finally, I thank Marlene, who in her last three years shared less time with me than we both would have wished because this book was in preparation.

Contents

Electromagnetism and basic optics

1

1.1 Introduction

In this chapter we revise electromagnetism and a rather minimal amount of optics. Although the material is all elementary, we wish to set up a notation and to have a number of statements on record so that they can be referred to throughout the remainder of the book.

1.2 The Maxwell equations

The Maxwell equations relate the electric field $\boldsymbol{E}$ and the magnetic field[1] $\boldsymbol{B}$ to the density of charge ρ and the density of current $\boldsymbol{J}$. In the first instance, they are written in their 'microscopic' form:

$$\operatorname{div} \boldsymbol{B} = 0, \qquad \operatorname{div}(\varepsilon_0 \boldsymbol{E}) = \rho, \tag{1.1a}$$

$$\operatorname{curl} \boldsymbol{E} = -\partial \boldsymbol{B}/\partial t, \qquad \operatorname{curl}(\boldsymbol{B}/\mu_0) = \boldsymbol{J} + \varepsilon_0 \partial \boldsymbol{E}/\partial t. \tag{1.1b}$$

Here ρ is the charge density per unit volume, and it counts each electron (or proton or quark) as a separate piece of charge. A similar statement applies to the current density $\boldsymbol{J}$. This way of handling charges is what is meant when Maxwell's equations are said to be given, as here, in their 'microscopic' version or 'in vacuum': there is no completely empty vacuum, or there would be no charges; but charges have nowhere been grouped into a dipole moment density or treated as zero on the average for an uncharged body. There is no 'medium' present, and no room for the equations to contain an ε_r or a μ_r.

Maxwell's equations will rarely be written henceforth in their microscopic form for reasons that are easy to see: all quantities (ρ, $\boldsymbol{E}$, ...) vary wildly with position, and do so over distances that may well be much smaller than those of interest to us. Look at problem 1.1 at the end of this chapter.

When we are not interested in looking at field quantities (meaning ρ, $\boldsymbol{J}$, $\boldsymbol{E}$ and $\boldsymbol{B}$) except in a 'coarse-grained' way, we average them over a scale of distances that is small compared to distances of interest but still large enough to achieve useful smoothing. If our concern is with visible-frequency light, distances of interest are of the order of the wavelength, several thousand times the atomic spacing for a solid or liquid, so a distance scale for averaging is easily devised. (Of course, there are contexts

[1] *Comment*: Standard nomenclature makes $\boldsymbol{H}$ 'magnetic field', while $\boldsymbol{B}$ is called 'magnetic induction' or 'magnetic flux density'. This is a left-over from the time when $\boldsymbol{H}$ was thought of as the field quantity generated by magnetic poles and acting on those poles. Since we now think of magnetism as caused by currents, it is $\boldsymbol{B}$ that is the fundamental field quantity. I do my small bit here for reform by contentiously referring to $\boldsymbol{B}$ as 'field'.

in which it may not be possible to find a scale of distances to average over; see problem 1.2.) When averaging has been done, Maxwell's equations are given in their 'macroscopic' version.

It is not necessary to specify the precise distance scale used in taking an average over the microscopic fields, provided that one exists. We must, however, imagine that all quantities are being averaged in the same way over the same scale.[2] Averaging yields smoothed values $\langle \rho \rangle$ for ρ and similarly for the other quantities. Furthermore, the charges on atoms and molecules (and within solids and liquids) are often conveniently represented as being partly grouped into electric or magnetic dipoles. After averaging we represent[3] the averaged dipole moments by an electric dipole moment density $\boldsymbol{P}$ and a magnetic dipole moment density $\boldsymbol{M}$. Then the results of averaging are:

$$\langle \rho \rangle = \rho_{\text{accessible}} + (-\operatorname{div} \boldsymbol{P}), \tag{1.2a}$$

$$\langle \boldsymbol{J} \rangle = \boldsymbol{J}_{\text{accessible}} + \partial \boldsymbol{P}/\partial t + \operatorname{curl} \boldsymbol{M}. \tag{1.2b}$$

The charge density $\rho_{\text{accessible}}$ is the space average of those charges that we choose not to pair off into dipoles; and a similar interpretation applies to the current density $\boldsymbol{J}_{\text{accessible}}$. The description 'accessible' arises because these charges are not (necessarily) bound inside atoms or molecules; a conduction current inside a metal is the commonest example of an 'accessible' current.

The reader will be familiar (if not, work problem 1.3) with the substitution of these statements into Maxwell's (averaged) equations to yield the standard macroscopic Maxwell equations

$$\operatorname{div} \boldsymbol{B} = 0, \qquad \operatorname{div} \boldsymbol{D} = \rho_{\text{accessible}}, \tag{1.3a}$$

$$\operatorname{curl} \boldsymbol{E} = -\partial \boldsymbol{B}/\partial t, \qquad \operatorname{curl} \boldsymbol{H} = \boldsymbol{J}_{\text{accessible}} + \partial \boldsymbol{D}/\partial t. \tag{1.3b}$$

Here the electric displacement $\boldsymbol{D}$ and corresponding magnetic quantity $\boldsymbol{H}$ have been defined by[4]

$$\boldsymbol{D} = \varepsilon_0 \boldsymbol{E} + \boldsymbol{P}, \qquad \boldsymbol{H} = \boldsymbol{B}/\mu_0 - \boldsymbol{M}. \tag{1.4}$$

The fields $\boldsymbol{E}$ and $\boldsymbol{B}$ in eqns (1.3) and (1.4) should really be written as $\langle \boldsymbol{E} \rangle$ and $\langle \boldsymbol{B} \rangle$ to distinguish them[5] from the unaveraged quantities in eqns (1.1); however, 'microscopic' fields will not be used again here. Similarly, ρ and $\boldsymbol{J}$ will now represent the 'accessible' charge density and current density, and the label 'accessible' will be dropped.

[2] *Aside*: The reason why averaging yields a charge density $(-\operatorname{div} \boldsymbol{P})$ (known as the **polarization charge density**) in the presence of atomic dipoles with space-varying magnitude is nicely explained in Robinson (1973*a*). The formal recipe for undertaking the averaging process in a respectable way is detailed in Robinson (1973*b*). We could describe the (respectable) averaging as 'looking at the charges and fields through Mr Abbe's microscope' (see Chapter 12).

[3] *Comment*: These averaged dipole-moment densities are understood in the obvious way: as the total of the dipole moments in some small volume, divided by that volume. Such an average can make sense only if the volume is large enough to contain several dipoles. So the presence of $\boldsymbol{P}$ or $\boldsymbol{M}$ in a formula is a reminder of the averaging and its scale of distances. The same signal is conveyed by the use of symbols $\boldsymbol{D}$ or $\boldsymbol{H}$ which—recommendation—are written only when it is correct for that signal to be given.

[4] *Question*: Has any assumption been used in this, perhaps about a relation linking $\boldsymbol{P}$ to $\boldsymbol{E}$ or linking $\boldsymbol{M}$ to $\boldsymbol{B}$?

[5] There is no question of attaching averaging brackets $\langle \rangle$ to $\boldsymbol{P}$, $\boldsymbol{M}$, $\boldsymbol{D}$ or $\boldsymbol{H}$, because these quantities are defined from the beginning to be averages, or to be built up from averages.

1.3 Linear isotropic media

For much (but not all) of this book we shall think about electromagnetic fields within a medium that is *linear*, *isotropic* and *homogeneous*, often called an LIH medium. Such a medium can be described by giving its relative permittivity ε_r and its relative permeability μ_r defined by the **constitutive relations**

$$\boldsymbol{D} = \varepsilon_r \varepsilon_0 \boldsymbol{E}, \qquad \boldsymbol{B} = \mu_r \mu_0 \boldsymbol{H}. \tag{1.5}$$

The relative permittivity ε_r and magnetic permeability μ_r are in general dependent upon the frequency ν at which the fields undergo changes, in which case the medium is said to be **dispersive**. However, dispersion is a complication that we shall ignore in sections 1.3 to 1.6. Inevitably, some things we say will be specific to a non-dispersive medium. We shall return to the more general case in §1.7.

With eqns (1.5) the Maxwell equations (1.3) can be manipulated straightforwardly to yield, for a medium containing no accessible charge or current, the wave equations[6]

$$\nabla^2 \boldsymbol{E} = \frac{1}{v^2}\frac{\partial^2 \boldsymbol{E}}{\partial t^2}, \qquad \nabla^2 \boldsymbol{H} = \frac{1}{v^2}\frac{\partial^2 \boldsymbol{H}}{\partial t^2}. \tag{1.6}$$

Here

$$v = \text{phase velocity} = (\mu_r\,\mu_0\,\varepsilon_r\,\varepsilon_0)^{-1/2} = c(\mu_r\,\varepsilon_r)^{-1/2} \tag{1.7}$$

and the **refractive index** n is defined by[7]

$$n^2 = (c/v)^2 = \mu_r\,\varepsilon_r, \tag{1.8}$$

which for a non-magnetic medium (problem 1.4) reduces to

$$n^2 = \varepsilon_r.$$

[6] *Comment*: Given the insistence on $\boldsymbol{E}$ and $\boldsymbol{B}$ as the fundamental fields in sidenotes 1 and 3, it may seem surprising that I have switched from $\boldsymbol{B}$ to $\boldsymbol{H}$ here. No point of principle is involved, only one of convenience. It is $\boldsymbol{H}$ that appears in the Poynting vector of eqn (1.19); and it is $\boldsymbol{H}_{\text{tangential}}$ that is continuous at boundaries such as those in Figs. 6.2 and 6.3.

[7] This definition makes no reference to Snell's law of refraction. However, it requires only a straightforward argument involving optical paths to show that the n introduced here is equivalent to the Snell-law refractive index.

1.4 Plane electromagnetic waves

One possible solution of the wave equations (1.6) is a plane wave

$$\boldsymbol{E}(x, y, z, t) = \boldsymbol{E}_0 \cos(\boldsymbol{k}\cdot\boldsymbol{r} - \omega t - \alpha), \tag{1.9}$$

or, using a complex-exponential notation

$$\boldsymbol{E}(x, y, z, t) = \boldsymbol{E}_0\, \mathrm{e}^{\mathrm{i}(\boldsymbol{k}\cdot\boldsymbol{r} - \omega t - \alpha)}, \tag{1.10}$$

where in eqn (1.10) it is understood that the actual field $\boldsymbol{E}$ is obtained by taking the real part.[8] Here:

$\boldsymbol{r}$ = position vector = $x\hat{\imath} + y\hat{\jmath} + z\hat{\boldsymbol{k}}$ where $\hat{\imath}$, $\hat{\jmath}$, $\hat{\boldsymbol{k}}$ are unit vectors
ω = angular frequency = $2\pi\nu = 2\pi/(\text{period})$
ν = frequency, whose SI unit is the hertz
$\boldsymbol{k}$ = wave vector, direction of $\boldsymbol{k}$ is the direction of the phase velocity
magnitude of $\boldsymbol{k} = 2\pi/(\text{wavelength in medium})$
α = a phase *lead* (as opposed to a lag) measured in radians.

[8] We have perpetrated a notational duplication here by using the same symbol $\boldsymbol{E}$ to represent both the real field and its complex representation. Nevertheless the practice is common and doesn't normally cause confusion; which meaning is intended will be clear from the context.

When eqn (1.9) or eqn (1.10) is substituted into the Maxwell equations (1.3) for an uncharged non-conducting medium we find (problem 1.5):

$$\boldsymbol{k}\cdot\boldsymbol{B} = 0, \qquad \boldsymbol{k}\cdot\boldsymbol{D} = 0, \qquad \omega\boldsymbol{B} = \boldsymbol{k}\times\boldsymbol{E}, \qquad -\omega\boldsymbol{D} = \boldsymbol{k}\times\boldsymbol{H}. \tag{1.11}$$

From these vector relations,

$\boldsymbol{B}$ is perpendicular to $\boldsymbol{k}$ and $\boldsymbol{E}$; $\boldsymbol{D}$ is perpendicular to $\boldsymbol{k}$ and $\boldsymbol{H}$.

For an isotropic medium we find that $\boldsymbol{E}$, $\boldsymbol{H}$ and $\boldsymbol{k}$ are mutually perpendicular.[9]

For the rest of this section it's convenient to choose the z-axis along the direction of $\boldsymbol{k}$ (give or take a sign). Then eqns (1.11) are consistent provided that $k = k_z$ satisfies[10]

$$\frac{\omega}{k} = \frac{c}{n}, \tag{1.12}$$

where n is given by eqn (1.8).

Backtracking to eqns (1.11) we find further that our plane wave has field components of the form

$$E_x(z,t) = E_0 \cos(kz - \omega t - \alpha) \tag{1.13}$$

and these field components are related to each other by[11]

$$\frac{E_x}{H_y} = \frac{E_y}{-H_x} = Z \times \operatorname{sign}(k), \tag{1.14}$$

where the **impedance** Z is defined by

$$Z \equiv \left(\frac{\mu_r \mu_0}{\varepsilon_r \varepsilon_0}\right)^{1/2} = Z_0 \left(\frac{\mu_r}{\varepsilon_r}\right)^{1/2} \tag{1.15}$$

and

$$Z_0 \equiv (\mu_0/\varepsilon_0)^{1/2} = \text{impedance of free space} \approx 377\,\Omega.$$

For any non-magnetic medium ($\mu_r = 1$) then, we have

$$Z = Z_0/\sqrt{\varepsilon_r} = Z_0/n, \tag{1.16}$$

and, so long as the medium is carrying a wave travelling in one direction only,

$$E_x/H_y = (Z_0/n) \times \operatorname{sign}(k). \tag{1.17}$$

The $\operatorname{sign}(k)$ in eqns (1.14) and (1.17) is intended to be a surprise; it can be such a habit to write $\omega/k = v$ that we risk getting in a tangle in cases where k may have either sign.

Finally, we mention that travel through a distance Δz implies a phase change—a lag—of[12]

$$k\,\Delta z = (\omega/c)(n\,\Delta z), \tag{1.18}$$

in which $(n\,\Delta z)$ is called the **optical path difference**.

1.5 Energy flow

A standard result in electromagnetic theory tells us that the energy flow vector, representing the power crossing unit area, is the

$$\text{Poynting vector} = \boldsymbol{S} = \boldsymbol{E} \times \boldsymbol{H}. \tag{1.19}$$

The **intensity** is defined as the *time-averaged* power crossing unit area.[13]

[9] *Comment*: The most direct way of obtaining the property 'electromagnetic waves are transverse' from Maxwell's equations is via $\operatorname{div}\boldsymbol{D} = 0$ and $\operatorname{div}\boldsymbol{B} = 0$, because these yield $\boldsymbol{k}\cdot\boldsymbol{E} = 0$ and $\boldsymbol{k}\cdot\boldsymbol{B} = 0$. Is this trying to tell us something? There is a nice precedent in the polarization of sound waves (displacement $\boldsymbol{u}$) in an isotropic solid: the mathematical property that identifies transverse waves is $\operatorname{div}\boldsymbol{u} = 0$, and what identifies longitudinal waves is $\operatorname{curl}\boldsymbol{u} = 0$. (A good reference is Landau and Lifshitz (1986), beginning of Chapter 3.) Knowing this precedent then: if you suspect some new wave of being transverse, look at its divergence (and if you suspect it of being longitudinal look at its curl). Your favourite proof that electromagnetic waves are transverse fits very well into the wider scene of wave motions in physics.

[10] Usually, n is taken as positive while k may have either sign, and then eqn (1.12) should read $\omega/|k| = c/n$. But sometimes (examples in problem 1.14(9) and in §§7.4 and 7.8) we shall treat n as negative for a wave whose k_z is negative. The omission of modulus signs in eqn (1.12) is therefore not just a piece of carelessness.

[11] Expression $\operatorname{sign}(k)$ is defined in the obvious way as

$$\operatorname{sign}(k) = \begin{cases} +1 & \text{for } k > 0 \\ -1 & \text{for } k < 0. \end{cases}$$

[12] Notice that $(n\,\Delta z)$ is a valid measure of phase difference only for waves possessing a single ω.

[13] The 'official' name for this quantity, in an optical context, is **irradiance**. The justification for introducing 'irradiance' is to have a consistent nomenclature for energy quantities, distinguishing those where all frequencies are treated equally, as here, from those that are weighted according to the sensitivity to frequency of the human eye (where the corresponding quantity is **illuminance**). There is, however, no sign that 'irradiance' will replace 'intensity' for waves of all kinds across the whole of physics. The present author therefore sticks with 'intensity', regarding it as unnecessary that areas within physics should acquire 'local' jargon.

It is obtained from the Poynting vector by:

$$\text{intensity } \boldsymbol{I} = (\text{time average of } \boldsymbol{S}) = \langle \boldsymbol{S} \rangle \equiv \frac{1}{T}\int_{-T/2}^{T/2} \boldsymbol{S}(t)\,\mathrm{d}t, \tag{1.20}$$

where T = the averaging time ($\gg 2\pi/\omega$) for our observing apparatus.

Consider the energy carried by the monochromatic plane wave of eqn (1.13), with k positive and with $\boldsymbol{E}$ in the x-direction only. Then

$$\begin{aligned} I_z &= \frac{1}{T}\int_{-T/2}^{T/2} E_x(z,t)\, H_y(z,t)\,\mathrm{d}t \\ &= \frac{1}{Z}\frac{1}{T}\int_{-T/2}^{T/2} E_x(z,t)^2\,\mathrm{d}t \qquad (1.21) \\ &= \frac{1}{Z}\frac{1}{T}\int_{-T/2}^{T/2} E_0^2\cos^2(kz-\omega t-\alpha)\,\mathrm{d}t \qquad (1.22) \\ &= E_0^2/(2Z) + \text{a left-over that} \to 0 \text{ as } T\to\infty. \end{aligned}$$

For the case of a non-magnetic material, we have by eqn (1.16)

$$I_z = \frac{E_0^2}{2Z_0}\,n. \tag{1.23}$$

Comment 1 Note the n: the intensity is no longer proportional only to (amplitude)^2 when the wave can move between media of different refractive index n.

Comment 2 The $\cos^2$ in eqn (1.22) reminds us that the energy flow in a travelling electromagnetic wave is 'lumpy',[14] because E and H have their maxima together and their minima together. Drawings displaying this lumpy structure are usually given emphasis in introductory courses on electromagnetic waves. It is because of the 'lumpiness' that it is necessary to define the intensity as an average over time: the aim is to discard the lumpy variations and keep just the average value.

[14] *Question*: For visible-frequency light, what is the typical duration of one 'lump'? This shows that the averaging time T need not be longer than a few optical periods. 'Intensity' has not had *all* dependence on time defined away; indeed, we can think of intensity as varying with time on all 'ordinary' timescales. This point will be important when we think about fluctuations of intensity in Chapter 10.

1.6 Scalar wave amplitudes

When only one Cartesian component of each field ($\boldsymbol{E}$ or $\boldsymbol{H}$) is significant, eqns (1.6) reduce to a *scalar* wave equation

$$\nabla^2 V = \frac{1}{v^2}\frac{\partial^2 V}{\partial t^2}, \tag{1.24}$$

where V is a scalar amplitude that represents an appropriate choice from E_x, E_y, E_z, H_x, H_y, H_z.

For a large part of our work on optics (roughly speaking, everything except polarization) it will be sufficient to use a scalar V to represent an appropriate field amplitude. Problem 1.6 asks you to investigate the conditions within which we may use a single V to give adequate information about both $\boldsymbol{E}$ and $\boldsymbol{H}$.

It is necessary to decide whether V is to represent E or H. The reason is to be found in eqn (1.23) and in problem 1.6(2): The 'E' choice makes the intensity proportional to $\langle V^2 \rangle n$, while the 'H' choice makes it proportional to $\langle V^2 \rangle / n$: both are correct on their own terms, but we have to select one (and stick to it) or we'll be in a muddle. The convention used here will be that V stands for the *electric* field. The reason for this choice does not lie in electromagnetism, which treats E and H (or B) more or less symmetrically, but in atomic physics: atoms are usually much more interested in the E-field of light than in the B-field, because their strongest transitions are electric dipole transitions.

The scalar wave equation permits us to introduce a second wave solution: the spherical wave

$$V(r, t) = V_0 \frac{\cos(kr - \omega t - \alpha)}{r}. \tag{1.25}$$

This represents a wave propagating outwards (if k is positive) from the origin, with r as the distance from the origin.

Although eqns (1.9) and (1.25) are in principle special-case solutions of their wave equations, they will suffice for much of this book. The reason is to be found in the device of Huygens secondary waves, used, above all, in the discussion of diffraction in Chapter 3: waves of other shapes (however complicated) can be constructed mathematically as superpositions of spherical waves (or of plane waves, but these are a special case of spherical waves).

1.7 Dispersive media

At the beginning of §1.3 we mentioned that the development started there was restricted to the case of a non-dispersive medium. Some of the results obtained remain valid for a dispersive medium; some do not. The issue is investigated in problems 1.7 and 1.8.

1.8 Electrical transmission lines

Many pieces of physics involve one-dimensional wave motion in the presence of boundaries. Examples are: quantum particles encountering a potential step; electromagnetic waves incident on a boundary between two media; voltage–current waves on an electrical transmission line encountering a change of impedance; sound waves incident on a boundary between two media. The wave–boundary problem that is conventionally used as the paradigm for all of these is the case of joined electrical transmission lines. The reason for this choice is that it naturally leads to a physical understanding and a terminology involving characteristic impedance, impedance mismatch and impedance matching. We, therefore, briefly review the description of waves on transmission lines; there will be a payoff in Chapter 6 when we find similar physics in the behaviour of light in thin films.

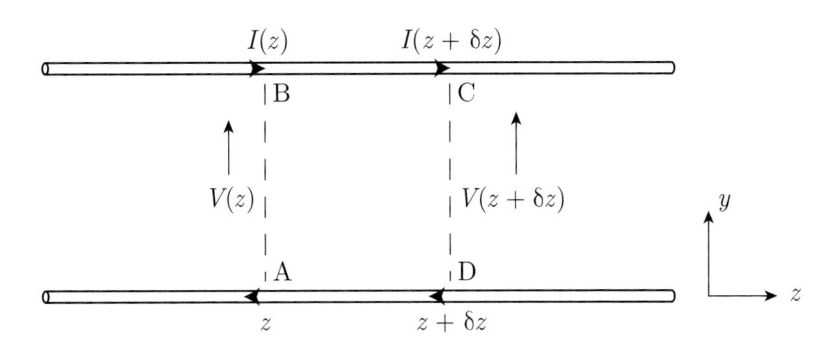

Fig. 1.1 A representative transmission line. The two conductors are drawn in the form of two parallel wires, but other geometries are commonly used. In particular, the transmission line most frequently encountered is a coaxial line, such as connects a television aerial to a television receiver.

The 'voltage difference' and current at location z and time t are $V(z,t)$ and $I(z,t)$; the symbol t is omitted from the diagram to avoid clutter. A sign convention must be defined for V and for I; the positive direction for I and the direction of increasing V are indicated by arrows.

A section of transmission line is shown in Fig. 1.1. Positions along the line are identified by coordinate z. The 'top' conductor carries current $I(z,t)$ to the right at position z and time t. An equal current flows to the left at the same z, t along the 'bottom' conductor.

A 'voltage difference' $V(z,t)$ between the conductors at z, t is defined by[15] $V(z,t) = -\int E_y(z,t)\,dy$, where the integral is taken 'straight across' the line from A to B.[16]

Section δz of line has capacitance $C\,\delta z$ between the conductors, where C is the capacitance per unit length.

Take the geometrical path ABCDA and consider the magnetic flux threading it—into the paper to be consistent with vector-analysis conventions and the directions of the arrows attached to $I(z,t)$. Magnetic flux $(L\,\delta z)I$ threads this area, where L is the inductance per unit length of the structure.[17]

As V changes with time, charge accumulates on the element of conductor BC (and equal and opposite charge on AD), charging up the capacitance. This charge is accounted for by a difference between the current $I(z,t)$ entering that element and the current $I(z+\delta z,t)$ leaving, so $I(z) - I(z+\delta z) = (C\,\delta z)\partial V/\partial t$, which can be reshaped as

$$\frac{\partial I}{\partial z} = -C\frac{\partial V}{\partial t}. \tag{1.26}$$

As I changes with time, so does the flux threading area ABCD, and an e.m.f. (defined as $\oint \boldsymbol{E}\cdot d\boldsymbol{l}$) is set up round the perimeter path ABCDA. The e.m.f. is $-V(z)+V(z+\delta z)$, so $-V(z)+V(z+\delta z) = -(L\,\delta z)\partial I/\partial t$, which reshapes to

$$\frac{\partial V}{\partial z} = -L\frac{\partial I}{\partial t}. \tag{1.27}$$

It is not hard now (problem 1.9) to eliminate between eqns (1.26) and (1.27) to obtain wave equations for V and I, with wave speed v given by

$$v = \frac{1}{\sqrt{LC}}. \tag{1.28}$$

[15] This V is not to be confused with the scalar-wave amplitude V of eqn (1.24).

[16] *Comment*: Taking the path of integration 'straight across' should make good sense: $V(z,t)$ is most clearly a function of z if this one z only is explored by the integration path.

[17] (Flux) = (inductance) × (current) is the fundamental definition of inductance.

Knowing that the one-dimensional wave equation applies, we can assume a solution[18] $V = V_0\,\mathrm{e}^{\mathrm{j}(\omega t - kz)}$; this wave may travel towards $+z$ if k is positive, or towards $-z$ if k is negative. Substituting this expression into eqn (1.26) yields

$$\frac{V}{I} = \frac{k}{\omega C} = \frac{\sqrt{LC}\,\mathrm{sign}(k)}{C} = \sqrt{\frac{L}{C}}\,\mathrm{sign}(k) = Z \times \mathrm{sign}(k), \tag{1.29}$$

where $Z = \sqrt{L/C}$ is the **characteristic impedance** of the transmission line. The similarity of shape between (1.29) and (1.14) is one instance among many that establishes an isomorphism between electromagnetic plane waves and voltage–current waves on a transmission line. See problem 1.10 for more on this isomorphism and others.

It is not our task here to usurp the functions of a textbook of electromagnetism, so we quote only one other result from transmission-line theory. Let a wave $V = V_{\mathrm{inc}}\,\mathrm{e}^{\mathrm{j}(\omega t - kz)}$ (k positive now) be incident from within a transmission line of impedance Z_1 onto a join to another line[19] of impedance Z_2. There is a reflected wave $V = V_{\mathrm{ref}}\,\mathrm{e}^{\mathrm{j}(\omega t + kz)}$ that travels back from the join. The amplitude ratio is

$$\frac{V_{\mathrm{ref}}}{V_{\mathrm{inc}}} = \frac{Z_2 - Z_1}{Z_2 + Z_1}. \tag{1.30}$$

We shall encounter an exactly similar expression[20] in eqn (6.9) when electromagnetic plane waves encounter a change of impedance at a boundary between two media.

[18] As with the $\boldsymbol{E}$ of eqn (1.10), this is written with the understanding that the actual physical voltage difference is the real part of the complex expression.

[19] This 'second' line is assumed to carry no wave returning from $+z$, so it is terminated by its characteristic impedance, or is infinitely long, or at least is long enough that no signal has yet returned from its far end.

[20] When $Z_2 \neq Z_1$, the impedances of the two transmission lines are **mismatched**, and we use this description to explain to ourselves why a part of the incident wave is reflected. Conversely, if we arrange that $Z_2 = Z_1$ then we achieve zero reflection because the impedances are **matched**.

1.9 Elementary (ray) optics

1.9.1 The thin lens

Figure 1.2 shows a thin lens in air forming an image of an object. Points F, F′ are the **focal points** of the lens, with F′L = LF = f, the **focal length**. Three rays are drawn from object to image, each illustrating one of the three most basic rules for the correct construction of rays.

In elementary work, it is usual to specify the locations of object and image using a 'real is positive' convention:[21] O is located at distance u = OL from the lens, positive when O is to the left of L, and I is located[22] at distance v = LI, positive when I is to the right of L. In effect, a coordinate system is used, as drawn in Fig. 1.2. With this coordinate system,

$$\frac{1}{v} + \frac{1}{u} = \frac{1}{f} = P \qquad \text{and} \qquad \frac{h_{\mathrm{i}}}{h_{\mathrm{o}}} = \frac{v}{u}. \tag{1.31}$$

Here P is the **power** of the lens.[23] The SI unit of power is the dioptre (metre^{-1}).

The ray-drawing rules exhibited in Fig. 1.2 can usefully be supplemented by the rule displayed in Fig. 1.3. The incident ray has nothing special about it: it is not parallel to the axis, and doesn't pass through a focal point or the lens centre. How is this ray refracted by the lens?

[21] The name arises because a real object lies to the left of the lens, at some positive value of u; a virtual object (one towards which rays travel but encounter the lens before reaching it) lies at negative u. Similarly, a real image lies at positive v, a virtual image at a negative value of v.

[22] There should be no confusion between this v and the $v = c/|n|$ representing the speed of electromagnetic waves in a medium of refractive index $|n|$.

[23] *Comment*: 'Refracting power' might seem a nicer name, since 'power' has other meanings. But a spherical mirror has an optical power, defined in a similar way to that of a refracting surface; so this kind of power is not always refracting.

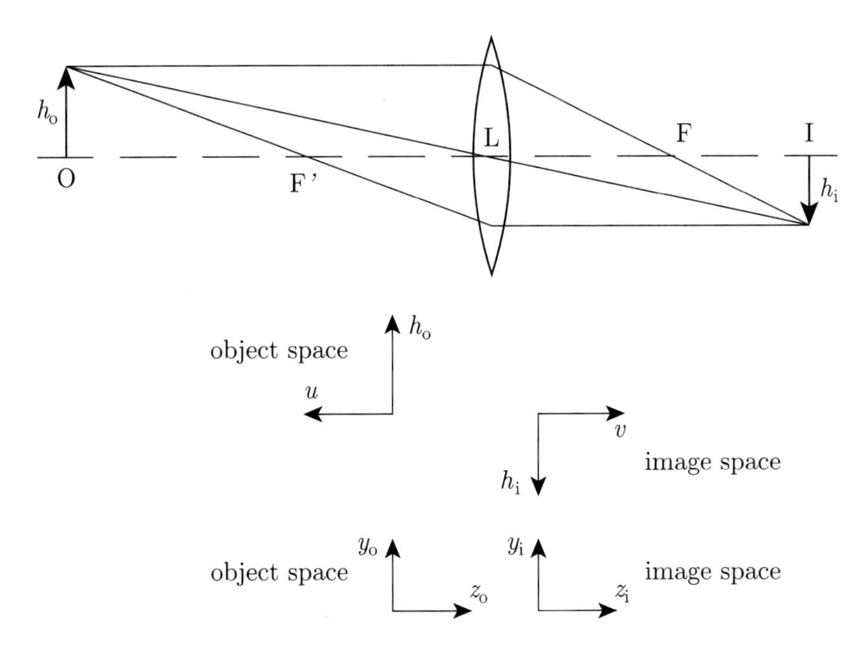

Fig. 1.2 The formation of an image by a thin lens. The three rays traced through the lens are chosen to illustrate the three most basic ray-tracing rules: a ray parallel to the axis is refracted through F; a ray through F′ is refracted to exit parallel to the axis; a ray through L is undeviated.

Two sets of coordinate axes are shown. The upper set is used with the 'real is positive' sign convention. The lower set, preferred in this book, uses Cartesian axes with x, y, z always facing the same way. The object location is specified by giving coordinates in the **object space** using (h_o, u) or (x_o, y_o, z_o) axes, while the image is located in the **image space** (h_i, v) or (x_i, y_i, z_i). For a thin lens, all axis systems have a common origin at L (the drawing is 'exploded'), but Fig. 1.5 shows the more general case that obtains with a thick lens.

Answer: Draw line LP, parallel to the incoming ray, from lens centre to the **focal plane**[24] through F; it meets that plane at P. The refracted ray passes through P. There is a way of looking at Fig. 1.3 that makes this statement obvious. Think of lens and focal plane as a camera 'focused for infinity', that is with the film lying in the focal plane. All light rays travelling in the direction shown, coming from infinity, meet at a focus, a single point in the focal plane; that is what 'focused for infinity' means. To reason out where this point lies, use one of the rays whose behaviour is known: a ray passing through L is undeviated, so where that ray meets the focal plane identifies where all rays go; and that point is P.[25]

The geometry shown in Fig. 1.3 is more useful than it looks. Since all light travelling in the direction shown ends up at P, the lens *maps directions onto positions*: each direction of incident light is brought to its own unique place in the focal plane. A device with this behaviour is often needed.[26]

[24] The lens's focal planes are planes normal to the z-axis and drawn through the focal points F and F′.

[25] Notice also that the three rules exemplified in Fig. 1.2 can all be derived as special cases of the new one. Remembering the one rule of Fig. 1.3 can therefore save a certain amount of routine learning.

[26] The rule of Fig. 1.3 may be seen in action in Fig. 3.4, for example.

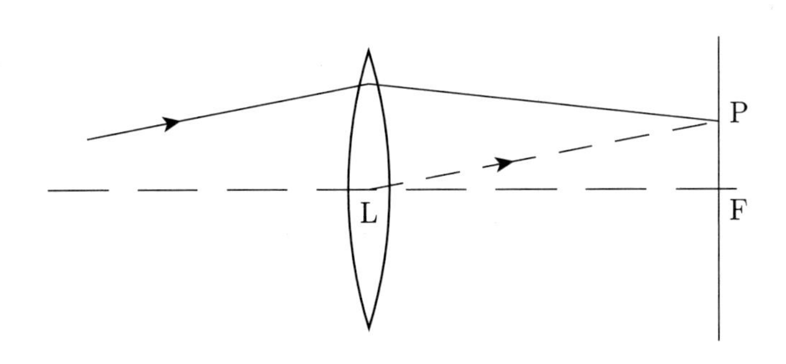

Fig. 1.3 The most important characteristic of refraction at a thin lens. The incoming ray is none of the special cases drawn in Fig. 1.2. To find how it is refracted at the lens, draw line LP parallel to the incoming ray, from lens centre to the focal plane through F. Construction line LP identifies the location of point P; the ray is refracted to pass through P. Any other ray parallel to the first uses the same construction and is refracted through the same point P.

1.9.2 Sign conventions

A number of different imaging configurations can be drawn after the fashion of Fig. 1.2, depending on whether O is to the left of F′, between F′ and L, or to the right of L, and depending on whether the lens is converging ($f > 0$) or diverging ($f < 0$). All such arrangements can be investigated using the same three ray-drawing rules, though if $f < 0$, F and F′ are on reversed sides of the lens. And eqns (1.31) continue to hold. The price we pay for having eqns (1.31) hold always is that we must obey rigorously the rules for attaching signs to u, v, f, etc. Students (and almost everyone else too!) find the assignment and interpretation of signs something of a pain.

As soon as we have to pursue rays through two or more lenses, the 'real is positive' convention rapidly becomes a muddle.[27] In this book, therefore, we invite you to switch (if need be) to the more rational Cartesian convention. The pain isn't entirely removed, but it's easier to keep a clear head.[28]

[27] There are differences between books as to whether the h_i-axis lies parallel or antiparallel to the h_o-axis. We have arranged that a real object with positive u and h_o gives (if $u > f$) a real image with positive v and h_i, or (if $u < f$) a virtual image with both negative—real is positive literally.

[28] As we move from object space to image space, or from one optical component to another, the z-origin changes, but at least coordinate axes don't keep changing direction.

From now on, we use the lower sets of coordinates in Fig. 1.2. The object lies at z-coordinate z_o, which is numerically negative for the case drawn because the object is to the left of the object-space origin; and the image lies at image-space z-coordinate z_i, which as drawn is positive. Equations (1.31) now become

$$\frac{1}{z_i} - \frac{1}{z_o} = \frac{1}{f} \quad \text{and} \quad \frac{y_i}{y_o} = \frac{z_i}{z_o}. \tag{1.32}$$

Again, these equations cover all cases. The price paid for having 'one size fits all' equations is that the new sign convention must be applied with all the rigour of the old.

To complete the specification of our sign convention, we must add one more statement: a spherically curved surface has radius of curvature R_s, which is taken as a positive quantity if the surface is 'convex to the left', as drawn in Fig. 1.4,[29] so light arriving from the left meets the surface to the left of the surface's centre. The spherical surface is given a negative value of R_s (unchanged in absolute magnitude) if it is concave to the left. These statements are independent of whether the surface tends to converge or diverge the light; such matters are dealt with 'mechanically' when refractive indices are inserted into formulae.

[29] *Comment*: An opposite convention, with appropriate modifications to the equations, would of course work just as well. In fact, the reversed convention would be somewhat more convenient when working with Gaussian beams, as will appear in Chapter 7. Nevertheless, the choice made here is entrenched in geometrical optics, and it would cause even more confusion if this book were not to go along with the majority.

1.9.3 Refraction at a spherical surface

With the algebraic signs assigned as above we now have the following statements about reflection and refraction at spherical surfaces to add to eqns (1.32):

- The object–image relations for a single spherical refracting surface, as drawn in Fig. 1.4, are[30]

$$\frac{n_2}{z_i} - \frac{n_1}{z_o} = P \quad \text{and} \quad \frac{y_i}{y_o} = \frac{n_1}{n_2}\frac{z_i}{z_o}, \tag{1.33}$$

[30] Although Fig. 1.4 is drawn with dimensions u, v marked in preparation for problem 1.14, we write equations here using the Cartesian sign convention.

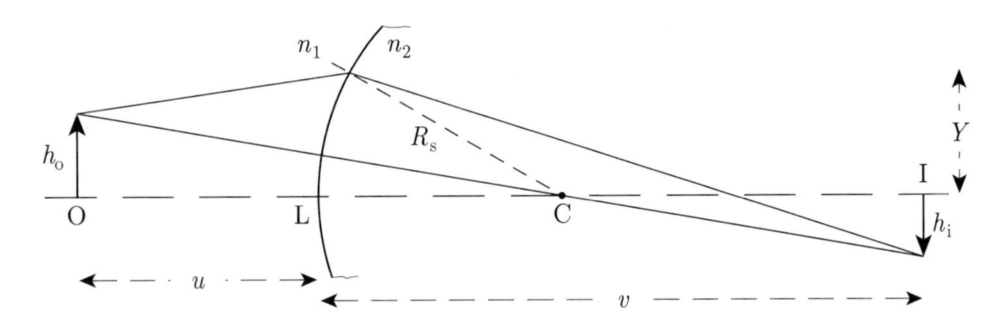

Fig. 1.4 A spherical refracting surface separates media with refractive indices n_1 (on the left) and n_2 (on the right). Coordinate axes for u, v, h_o, h_i, y, z are not drawn but have the same orientations as in Fig. 1.2; their origins all lie at L. An object of height $y_o = h_o$ is located at distance $u = \text{OL} = -z_o$ to the left of L. It is imaged to height $y_i = -h_i$ at distance $z_i = v = \text{LI}$. The Cartesian coordinates are used in eqn (1.33); the u, v, h, Y dimensions are used in problem 1.14.

where n_1 and n_2 are the refractive indices of the media separated by the surface (n_2 on the right),[31] $z_o = -u$ is the object distance from L and $z_i = v$ is the image distance. The quantity

$$P = (n_2 - n_1)/R_s \tag{1.34}$$

is the power of the surface.

- A thin lens of refractive index n_g, embedded in air of refractive index n_a, has power P and focal length f given by[32]

$$P = \frac{1}{f} = \frac{(n_g - n_a)}{n_a}\left(\frac{1}{R_1} - \frac{1}{R_2}\right), \tag{1.35}$$

where R_1 and R_2 are the radii of curvature of the two lens surfaces (numbered in the order in which the light reaches them, signs allocated by the rule given).
- A 'converging' lens, or a concave mirror, has $f > 0$, $P > 0$, a property that emerges 'mechanically' from application of the rules already given.[33]
- Equations (1.33) apply to reflection at a spherical surface as well as to refraction if we make the convention that $n_2 = -n_1$. This apparently bizarre procedure is explored in problem 1.14.

Comment on approximations: We should mention that equations like eqn (1.31) or eqn (1.33) are approximate: good to first order in a ray's distance from the axis and to first order[34] in every angle (see in particular problem 1.14). Rays for which such an approximation is adequate are said to be **paraxial**,[35] and the body of equations and rules that results from considering paraxial rays is often called paraxial (or Gaussian) optics.

The approximations contained within paraxial optics are often not as damaging as they might seem. The art of designing a lens (in particular) consists in making the lens behave as if paraxial optics applied, by arranging for errors to be cancelled.

[31] Although eqns (1.33) are quoted here in connection with refraction at a single spherical surface, their shape is more general. A similar pair of equations (though expressed there in the Newton form) is obtained in problem 7.3(7) where a lens separates two media with different refractive indices. More generally, similar equations describe the object–image transformation that must apply to any reasonable imaging system, as is shown in problem 13.8.

[32] This equation for f is often called the lensmaker's formula.

[33] This property holds in both sign-convention systems.

[34] Since $\sin\theta \approx \theta - \theta^3/3!$, corrections to the paraxial-optics treatment are of third order in angles and distances from the axis.

[35] Meaning 'near to the axis'.

1.9.4 The thick lens

A 'thick lens' may be a single piece of glass whose thickness is not negligible compared with its focal length, but it may also be any composite system composed of several pieces of glass (possibly mirrors also). The ray-drawing rules for a thick lens are very similar to those for a thin lens, the only change being the introduction of two planes, the **principal planes** H_1, H_2 indicated in Fig. 1.5. The rules for drawing rays are just the same as those in Figs. 1.2 and 1.3, except that 'a gap has been pulled'[36,37] between H_1 and H_2.

[36]When making a drawing of a thick lens, imagine the paper pleated so as bring H_2 on top of H_1; if the diagram is brought to look like Fig. 1.2, with no breaks in the rays and no kink in the ray through L, the drawing has been made correctly.

[37]In this book we shall make almost no use of the thick lens. However, an understanding of the principal planes is needed in problem 13.8.

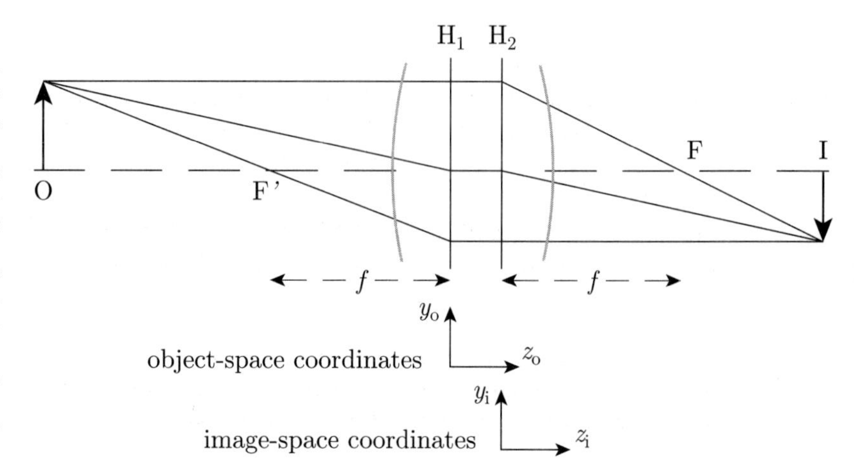

Fig. 1.5 A schematic thick lens. Details of the glass structure are only hinted at, as any reasonable structure is being catered for. Planes H_1 and H_2 are the principal planes, and the points where they cross the axis are their **principal points**. The object space has its origin at the principal point where H_1 crosses the axis, while the image space relates in a similar way to H_2. Representative rays are drawn, corresponding to those in Fig. 1.2. The 'middle' ray has two sections that are parallel to each other. If the gap between H_1 and H_2 were closed these two sections would become a single straight line.

1.10 Rays and waves

There is a special meaning attached to a **ray** in optics: it's a *representative normal to a wavefront.*[38] That is, we should always be 'thinking waves' even when drawing the most earthy of ray diagrams. And the wave being represented is broad unless something indicates otherwise.[39]

The reason for drawing rays at all is simply this: if we were to draw the wavefronts instead we'd end up with extremely cluttered diagrams. So we draw the normals to wavefronts and ask the reader to supply the wavefronts for himself in the mind's eye.[40]

If only one ray is drawn, as for example in Fig. 1.3, we are given no information about the 'shape' of the wavefronts: the wave might be a plane wave or a spherical wave; perhaps it doesn't matter. Sometimes a single ray is drawn but the context demands that the wave be plane; such is the case in Fig. 4.3, for example, because a plane diffraction grating is always[41] illuminated with collimated light. Figure 1.2 shows a more informatively drawn case. Three rays are shown radiating out from a point on the object, and it's implied that these are three among many. To be normal to all these rays, wavefronts must be spherical: we are to read the diagram as telling us that a spherical wave propagates outwards from this point on the object. Similar reasoning shows that a spherical

[38]Exception: In double refraction, the extraordinary ray is by convention drawn in the direction of the group velocity, while it is the phase velocity that is normal to surfaces of constant phase. This exception need not concern us until Chapter 15.

[39]The common misunderstanding that's being attacked here is that a ray represents any kind of 'thin pencil' of light.

[40]*Aside*: The rays are the 'duals' of the wavefronts in much the same sense that 'dual' is used in tensor analysis to indicate the relation of one tensor field to another.

[41]Well, almost always.

wave converges[42] towards the corresponding point on the image.

The summary given in sections 1.9 and 1.10 is a bare minimum of ray optics that we need in preparation for later chapters. We shall feel free to make use of any of these concepts without comment. The reader who is insecure with anything stated here should consult a book on basic optics, such as that by Hecht (2002). Problems 1.13–1.17 provide a rather short work-out.

[42] More properly: a portion of a spherical wave.

Problems

Problem 1.1 (a) Orders of magnitude

(1) Calculate the magnitude of the electric field inside a hydrogen atom, at a distance of one Bohr radius from the proton. It is this field that holds the electron in its orbit. (This field magnitude is needed when we work out what size of $\boldsymbol{E}$-field is needed to elicit a non-linear response from a material: the beginnings of the large subject of non-linear optics.)

(2) Estimate the order of magnitude of the electric field inside a hydrogen atom, just outside the surface of the nucleus.

Problem 1.2 (a) Scales of distance for averaging fields

Suggest a suitable averaging distance (if there is one) for constructing macroscopic-average fields, or explain why there is no such distance, in the case of:

(1) visible-light waves travelling in glass

(2) visible-light waves travelling in a crystal such as sodium chloride

(3) X-rays Bragg-reflected from a crystal such a sodium chloride

(4) light reflected from an opal[43]

(5) light diffracted from a hologram

(6) light propagating in air at STP

(7) light propagating in the upper atmosphere at a height where the air has a pressure of 10^{-6} atm.

[43] An opal contains a more-or-less regular structure like tiny spheres packed together and crushed somewhat. The period is comparable to an optical wavelength and the mineral exhibits colours caused by Bragg scattering.

Problem 1.3 (a) Macroscopic charges and currents

(1) In eqns (1.2), $-\operatorname{div}\boldsymbol{P}$ is called **polarization charge density**, $\partial\boldsymbol{P}/\partial t$ is **polarization current density** and $\operatorname{curl}\boldsymbol{M}$ is **magnetization current density**. Show that the terms $-\operatorname{div}\boldsymbol{P}$ and $\partial\boldsymbol{P}/\partial t$, taken together, conserve charge. Show that the same is true of current $\operatorname{curl}\boldsymbol{M}$ and its non-existent partner. Show that in consequence, $\rho_{\text{accessible}}$ and $\boldsymbol{J}_{\text{accessible}}$ conserve the accessible part of the charge. The accountancy in §1.2 has been done in a thoroughly rational way.

(2) Substitute expressions (1.2) for $\langle\rho\rangle$ and $\langle\boldsymbol{J}\rangle$ into averaged versions of eqns (1.1) to derive the macroscopic Maxwell equations (1.3).

(3) Obtain wave equations (1.6) from eqns (1.3) and (1.5).

(4) Show that the refractive index n defined in eqn (1.8) is identical to that defined by Snell's law of refraction.[44]

[44] *Hint*: Draw a diagram showing a plane wave, incident at an oblique angle on an interface between two media. Show how the wave undergoes refraction, and find two $\boldsymbol{k}\cdot\boldsymbol{r}$ quantities to equate.

Problem 1.4 (a) Materials are non-magnetic at optical frequencies Why is the magnetic permeability μ_r always taken to be 1 at optical frequencies? Equivalently, why can an applied $\boldsymbol{B}$-field not induce a magnetization $\boldsymbol{M}$ on the timescale of an optical period?[45]

[45] *Hint*: A free magnetic dipole obeys the Bloch equation $d\boldsymbol{M}/dt = \gamma \boldsymbol{M} \times \boldsymbol{B}$ where γ is the magnetogyric ratio, so the component of $\boldsymbol{M}$ parallel to $\boldsymbol{B}$ has no means of changing with time. Look up the full form of the Bloch equation that I have quoted incompletely; find how $\boldsymbol{M}$ does in fact manage to change; find the order of magnitude of the relaxation time T_1; and compare T_1 with a typical optical period. See, e.g., Kittel (1996).

Problem 1.5 (b) Wave basics

(1) Let $\boldsymbol{E} = \boldsymbol{E}_0 \, e^{i(\boldsymbol{k} \cdot \boldsymbol{r} - \omega t - \alpha)}$ represent a plane wave as in eqn (1.10). Show from Maxwell's equations that in an uncharged and non-conducting dielectric medium this plane wave satisfies the four relations (1.11).

(2) The four equations (1.11) look like *exact* consequences of Maxwell's equations. Are they, or did you sneak in some assumptions about the dielectric, like linearity or isotropy?[46]

[46] *Answer*: Of the three conditions mentioned in §1.3, you need all except isotropy. But why are those two needed?

(3) Suppose our dielectric medium is not isotropic. We can't now derive wave equations (1.6); try it. Nevertheless, plane waves *can* propagate—double refraction. How can we have found a wave when it's a solution of something other than a standard wave equation? Analysis isn't trivial (see problem 15.2); but make yourself comfortable with the idea by thinking of another example in physics where it happens.

(4) This problem started with the statement that the medium is uncharged. It's not practical for us to intervene to ensure that every point in some medium is uncharged; whether the medium is charged or not depends on what *it* wants to do. Show that any medium (assumed LIH and non-dispersive now) with a non-zero conductivity is uncharged in this sense: any initial charge density ρ decays exponentially, and in a manner unaffected by—and therefore not affecting—any wave that may also be present.

Problem 1.6 (a) Scalar representation of waves

(1) Suppose that we set up an optical arrangement resembling Young's slits. Waves originating from the two slits are combined at places downstream from the slits. We know that the wave amplitudes are combined by adding them, because (in a linear medium) the Maxwell equations are linear. Strictly, what we should add are the vector fields $\boldsymbol{E}$ and $\boldsymbol{H}$. In what circumstances does it suffice

(a) to add fields $\boldsymbol{E}$ only and let $\boldsymbol{H}$ look after itself?

(b) to add the scalar magnitudes of fields $\boldsymbol{E}$ (perhaps representing them by our standard symbol V) and trusting that the result will correctly give the magnitude of the vector total $\boldsymbol{E}$?

(2) Look at eqn (1.23). If V stands for the electric-field amplitude whose magnitude is $E_0 \cos(kz - \omega t - \alpha)$, the intensity I_z is proportional to (the time average of) nV^2. Show that if instead V stands for the magnetic $\boldsymbol{H}$-field whose magnitude is $H_0 \cos(kz - \omega t - \alpha)$ then the intensity I_z is proportional to the time average of V^2/n.

(3) Show that the spherical wave given by eqn (1.25) is a solution of the scalar wave equation (1.24). Explain physically why the wave amplitude *must* include the factor $1/r$.

Problem 1.7 (b) The effect of having a dispersive medium
Investigate the statements made in sections 1.3–1.6. Decide which remain correct in the presence of dispersion, and give the modifications required for those that do not.

Problem 1.8 (c) A paradox
Here are three statements about the electromagnetic properties of a material that is linear, isotropic and homogeneous, but dispersive:

(a) The energy density is $U = \frac{1}{2}\boldsymbol{D}\cdot\boldsymbol{E} + \frac{1}{2}\boldsymbol{B}\cdot\boldsymbol{H}$.
(b) The energy flow is (energy density) $\times$ (group velocity) $= U\,\mathrm{d}\omega/\mathrm{d}k$.
(c) The energy flow is given by the Poynting vector $\boldsymbol{E}\times\boldsymbol{H}$.

These three statements are incompatible, because the first two taken together make the energy flow depend on $\mathrm{d}n/\mathrm{d}\omega$, while the third does not. So at least one of the statements must be wrong. Which?[47]

[47] The answer is very unexpected. See, e.g., Landau *et al.* (1993), §80.

Problem 1.9 (a) Transmission line basics
(1) In §1.8 it is stated that the currents in the two conductors at z, t are equal and opposite. Why are they?
(2) In §1.8, $V(z,t)$ is called a 'voltage difference'. Why must it *not* be called a 'potential difference'?
(3) The conductors are assumed resistanceless. What then is the E-field along the top conductor from B to C? What is the E-field along the lower conductor from D to A? How then can $V(z+\delta z)$ be different from $V(z)$? Can either conductor be called an equipotential? Is one conductor being thought of as a 'ground plane' with all the activity taking place in the other? (Read again my definition of L: it's important for a lot of things that were *not* said.)
(4) Fill in the missing steps in the derivation of eqns (1.26) and (1.27). Take particular care with the signs in eqn (1.27).
(5) Eliminate I between eqns (1.26) and 1.27) to obtain a one-dimensional wave equation for V, and confirm expression (1.28) for the speed of wave travel v.
(6) Work the algebra that leads to eqn (1.29).
(7) Two transmission lines are joined end-to-end, as described in the run-up to eqn (1.30). What boundary conditions on V and I apply at the join? Use these boundary conditions to derive eqn (1.30). Show that a similar expression applies to the ratio of currents $I_{\mathrm{ref}}/I_{\mathrm{inc}}$, but it is different by a sign.

Problem 1.10 (b) The transmission-line analogy
It is stated in §1.8 that all one-dimensional wave–boundary problems are isomorphic to the corresponding problem with joined transmission lines. Here we investigate how this is demonstrated, and how a correspondence can be set up.

Waves in some new context will resemble those propagating along transmission lines if they have the following properties:

- A wave has two 'amplitude' quantities, which we may call V and I. If it's possible we'll use I to describe some sort of flow, and then V may look like some sort of force acting on that flow.
- When a wave travels towards $+z$, V and I are related by $V/I = Z$, and when a wave travels towards $-z$, V and I are related by $V/I = -Z$, where these equations define an impedance-quantity Z.
- The definitions of V and I will be so chosen that Z is real[48] and positive when the wave travels without attenuation.
- When the wave encounters a boundary, there are boundary conditions: V and I are both continuous.
- There is a conserved quantity, most usually energy, whose flow is given by $\text{Re}(V) \times \text{Re}(I)$. Equivalently, the flow has time average $\frac{1}{4}(VI^* + V^*I)$.

(1) Show that the field amplitudes E_x and H_y for a plane (and linearly polarized) electromagnetic wave have all the properties listed if we identify E_x with V and H_y with I.

(2) A sound wave can be described by a pressure change p and the displacement ξ of a parcel of material. Show that p and ξ are not a suitable pair of variables to identify with V and I because the associated impedance quantity is imaginary for an unattenuated wave. Show that p and $\partial\xi/\partial t$ do form a satisfactory pair of amplitudes, and that the 'acoustic impedance' is then $Z = \rho c$, where ρ is the density and c is the speed of sound.

(3) A quantum particle of energy E moves along the z-axis with wave function $\Psi \propto \mathrm{e}^{\mathrm{i}(kz-\omega t)}$ and encounters a potential step of height $V_0 < E$. The wave function is partly reflected and partly transmitted, with conservation of the probability current. Show that if Ψ is chosen to correspond to V and $(-\mathrm{i}\partial\Psi/\partial z)$ to correspond to I then properties in line with those above are obtained.[49,50]

Problem 1.11 (e) Spherical waves?

Show that an electromagnetic wave (or any other wave with transverse character) cannot form a spherical wave over the full solid angle of 4π steradians, and with non-zero intensity in all directions.

How then do we justify using spherical waves in discussing phenomena like diffraction? Spherical waves are visible in eqn (3.4), for example.[51]

Problem 1.12 (b) Polarization

A light wave is represented by

$$\left.\begin{aligned} E_x &= A\cos(\omega t - kz), \\ E_y &= B\cos(\omega t - kz + \alpha). \end{aligned}\right\} \tag{1.36}$$

The wave is *truly monochromatic*, possessing just a single ω.

(1) Show that 'truly monochromatic' means there can be no change with time of A, B, ω, k, or α. Therefore, the polarization state of the wave is fixed for all time.[52,53]

[48] This is of course on the understanding that V and I are represented by complex-exponential expressions similar to that in eqn (1.10).

[49] *Comment*: The choices suggested here for V and I make $(VI^* + V^*I)/4$ proportional to the probability current, but not equal to it, and an extra factor $2\hbar/m$ needs to be included. It would be possible to incorporate the extra factor into the definition of V or of I or to distribute it between them; but the algebra would be cluttered for no benefit in understanding.

[50] *Comment*: Students are sometimes puzzled as to what is being recommended by drawing attention to the above similarities. Should we translate the quantum-mechanics problem into the transmission-line equivalent, solve that problem, and then translate back again? My answer is **no**. Each problem should be set up in its own language. To do two translations would add to the labour for no benefit. But it helps to be aware of the *shape* of the problem: if you can solve one you can solve all the others, and you know good ways to proceed. You also know what algebraic shapes are to be expected; in particular, if the amplitude reflection doesn't resemble eqn (1.30) something has gone wrong.

[51] *Hint*: There is a theorem that you cannot comb the hair on a hairy ball all over without having to introduce singularities. The key word to look up is **disclinations**.

[52] *Hint*: Remember the mathematics of 'carrier wave plus sidebands' from electronics: any attempt at changing amplitude, frequency or phase is a modulation ....

[53] *Comment*: Therefore 'unpolarized' or 'randomly polarized' light is something that can exist only because it contains a range of frequencies.

(2) Now allow the wave to be 'nearly monochromatic', so it contains angular frequencies occupying a range $\Delta\omega$ ($\ll \omega$). Argue that such a wave possesses a polarization state at any instant, but that state changes over a timescale[54] of order $1/\Delta\omega$.

[54] There is a link here with coherence time, discussed in Chapter 9.

Problem 1.13 (a) Basic lens optics

(1) Choose two appropriate pairs of similar triangles on Fig. 1.2 and use them to derive eqns (1.31).

(2) In a setup like that of Fig. 1.2 the lens has focal length $f = 0.50\,\text{m}$. The object lies at a distance of 0.80 m from the lens, measured along the axis, and is 10 mm away from the axis. Draw a ray diagram carefully on graph paper, using the three rules exemplified in Fig. 1.2, and find by drawing where the image is located.

(3) Assign values, with signs, to u and h_o for the object of part (2). Use eqns (1.31) to calculate v and h_i for the image, and interpret those values to state where the image lies. Check for compatibility with the drawing of part (2).

(4) Assign values, with signs, to z_o and y_o for the object of part (2). Use eqns (1.32) to calculate z_i and y_i. Again interpret these values and check against parts (2) and (3).

(5) Draw a diagram in which an object is imaged by a diverging lens. Derive eqns (1.31) for this case, showing that they continue to apply when signs are chosen according to the real-is-positive rules given in §1.9.2. Repeat the steps in parts (2)–(4) for the case of a lens whose focal length is $f = -0.50\,\text{m}$.

(6) A lens is made from glass of refractive index 1.50. It is meniscus-shaped, meaning that one surface is concave and the other convex. The concave surface has radius of curvature 2.00 m, and the convex surface has radius 1.50 m (both of these requiring to be given signs according to the sign convention in §1.9.2). Imagine the lens is placed on the z-axis with its concave surface facing left. Assign values, with conventional signs, to the radii of curvature and use them to calculate the focal length of the lens. Reverse the lens and do it again. Does the focal length come out the same?

Problem 1.14 (a) Spherical surfaces and mirrors

Use the geometry of Fig. 1.4 on p. 11. Light from an object at O is refracted at a spherical surface separating media with refractive indices n_1, n_2 and forms an image at I. Two rays have been drawn from a point on the object; after refraction, they meet again at the image, which is how we know that that point *is* the image. One ray passes through the centre C of the surface; therefore, it passes normally through the surface and is undeviated. The other is incident at an angle and is deviated by refraction. Let the second ray meet the refracting surface at a point distant Y from the axis. Small-angle approximations may be made throughout.

(1) Show that Snell's law applied to the deviated ray leads to

$$n_1\left(\frac{Y}{R_s}+\frac{Y-h_o}{u}\right)=n_2\left(\frac{Y}{R_s}-\frac{Y+h_i}{v}\right).$$

(2) Argue that object and image have the property that v and h_i are independent of Y. Deduce that

$$\frac{n_2}{v}+\frac{n_1}{u}=\frac{(n_2-n_1)}{R_s}=P \qquad \text{and} \qquad \frac{h_i}{h_o}=\frac{n_1}{n_2}\frac{v}{u}.$$

(3) Change to a Cartesian coordinate system[55] similar to the lower set of axes drawn in Fig. 1.2. Show that the object–image relations end up identical to eqns (1.33).

[55] *Suggestion*: Write $u = -z_o$, etc., and substitute mechanically.

(4) Obtain eqns (1.33) by a quicker route. Draw a ray from the top of the object at O to meet the refracting surface at L. This ray is refracted to pass through the tip of the image at I. Apply Snell's law to this ray and obtain the second of eqns (1.33) at once. Use the geometry of this ray, and of the ray passing through the centre of curvature C, to obtain[56] the other equation relating u and v.

[56] This route to eqns (1.33) is simpler than that of parts (1) and (2), and is also more conventional. Nevertheless, we had a point in doing things the harder way. The fact that v and h_i work out independent of Y is the only demonstration we have given that refraction leads to the formation of a 'stigmatic' image (point object yielding a point image), even within small-angle approximations. Once this demonstration has been given, it can of course be extended to a case such as that of a lens where there are several refracting surfaces.

(5) Figure 1.4 has been drawn for the case $n_2 > n_1$, $R_s > 0$ and $z_o < 0$. Draw the other cases, and show that eqns (1.33) apply always when the algebraic signs are assigned according to the Cartesian rules of §1.9.2.

(6) Consider imaging by a spherical mirror as shown in Fig. 1.6. Light passing through the mirror's centre C meets the mirror normally and is reflected back along its original path. Light arriving parallel to the z-axis is reflected to pass through F. Show that, in a small-angle approximation, CF = FL so $f = -R_s/2$. Yes, a minus sign, because the mirror as drawn has negative R_s while f is positive for this 'converging' mirror.

(7) Use the geometry of the two rays drawn to find the image location

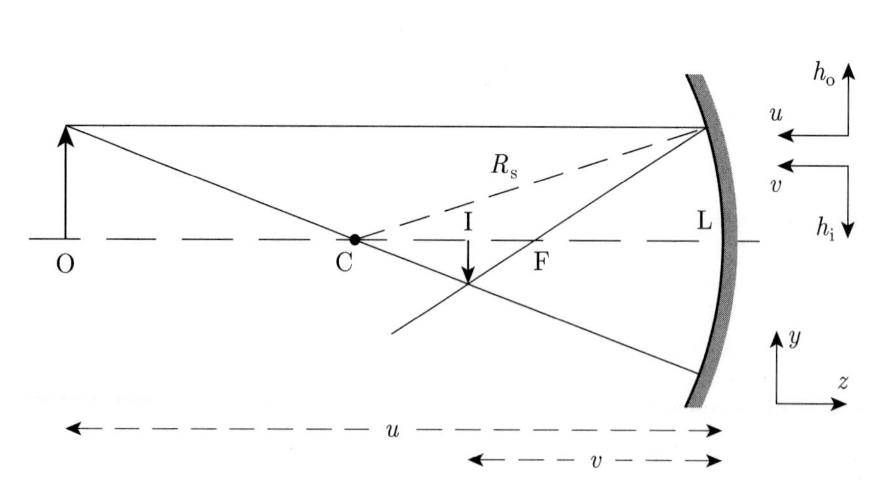

Fig. 1.6 Image formation by a concave mirror. Axes drawn at top right define the 'real is positive' convention; Cartesian axes are drawn at lower right. As with Fig. 1.2, the axes have been drawn displaced and they should all have their origins at L.

$v = \mathrm{IL}$ and the height h_i of the image. Show that

$$\frac{1}{v} + \frac{1}{u} = \frac{1}{f} = P = \frac{-2}{R_s} \qquad \text{and} \qquad \frac{h_i}{h_o} = \frac{v}{u}.$$

(8) In Fig. 1.2 three rays left the object and met again at the image. Figure 1.6 shows only two such rays. Draw the third.

(9) Change to the Cartesian axis system drawn at bottom right of Fig. 1.6. Make one other change as well.[57] Define $n_2 = -n_1$, and distribute n_1 and n_2 judiciously through the equations, to yield

$$\frac{n_2}{z_i} - \frac{n_1}{z_o} = \frac{(n_2 - n_1)}{R_s} \qquad \text{and} \qquad \frac{y_i}{y_o} = \frac{n_1\, z_i}{n_2\, z_o}.$$

Here we attach n_2 to dimensions in the 'image space' of the mirror, mimicking the presence of n_2 in similar places in the refraction case. After these contortions, the equations just derived are the same as eqns (1.33): reflection has been made mathematically identical to refraction, so the two phenomena have one unified mathematical description.

[57] *Comment*: The manoeuvres undertaken in part (9) must look somewhere between weird and capricious. Nevertheless, we shall find in Chapters 7 and 8 that rays and waves travelling to the left are neatly dealt with by attaching to them a negative refractive index, and then it will look rather tidy to have eqns (1.33) applying to both refractions and reflections. The convention introduced here was anticipated in eqn (1.8) where the sign of n was not made explicit, and in eqn (1.12) which continues to hold for left-going waves when both k and n are made negative.

Problem 1.15 (a) Newton's lens formula

Define distance X in the object space by $u = f + X$, so that X is measured leftwards but from focal point F$'$ instead of from L (notation as in Fig. 1.2). Similarly define distance Y in the image space by $v = f + Y$, so that Y is measured rightwards from F. Show that the object and image locations are given by the equations (simpler than in eqn (1.31)), known as Newton's lens formula:[58,59]

$$XY = f^2, \qquad \frac{h_i}{h_o} = \frac{Y}{f} = \frac{f}{X}. \tag{1.37}$$

Examine each possible case (f positive or negative, object more distant than $|f|$ from the lens, object less distant than $|f|$ ...) and check whether eqn (1.37) applies to all of them, given sign allocations as explained. Is any adaptation required when a mirror (convex or concave) is used instead of the lens?

[58] *Comment*: Newton's lens formula will be useful when we deal with the propagation of Gaussian beams through an optical system involving a lens or mirror, as in problem 7.11.

[59] A more general version of eqn (1.37) is derived in problem 7.3(7), where we consider the case of a lens that has different media on either side of it.

Problem 1.16 (a) A simple magnifier or eyepiece

A simple magnifier is shown in Fig. 1.7. It forms an upright virtual image at a distance X that is comfortable for an eye. The same arrangement describes the eyepiece of a telescope or microscope.

The magnification M is defined as the angle ratio $(H/X)/(h/D_{\mathrm{ldv}})$. Here H/X is the angular size of the image, and h/D_{ldv} is the angular size of the object held at the 'least distance of distinct vision' D_{ldv}, conventionally taken to be 250 mm.

(1) Show that $M = \dfrac{D_{\mathrm{ldv}}}{f}\left(1 + \dfrac{f - w}{X}\right)$.

(2) Show that in the limit $X \to \infty$, the magnification $M \to D_{\mathrm{ldv}}/f$.

(3) Show that all choices of X and w yield $M < D_{\mathrm{ldv}}/f + 1$.

(4) Show that $M > D_{\mathrm{ldv}}/f - 1$ so long as $w < f(1 + X/D_{\mathrm{ldv}})$. Argue that the user will almost always wish to meet this condition on w.

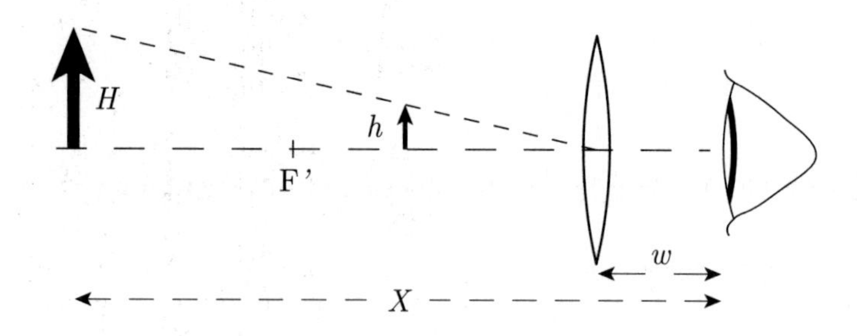

Fig. 1.7 A simple magnifier (or eyepiece) takes an object of height h and magnifies it to an image of height H, located at a distance X that is far enough from the eye to be focused upon comfortably.

Parts (3) and (4) show that $M = D_{\rm ldv}/f$, the value conventionally assigned to the magnification, is correct to a good approximation, for all reasonable choices of adjustment.[60]

[60] The knowledgeable user should adjust the magnifier to give an image at $X = \infty$, so that the eye can be in its relaxed state. Such an adjustment gives a magnification close to the greatest possible, and so exacts no penalty.

Problem 1.17 (b) A single-lens reflex camera

Figure 1.8 shows a single-lens reflex camera in cross section. 'Single-lens' indicates that the one objective lens is used for two functions: viewfinding in preparation for taking a photograph; and the actual photographic recording. In the left-hand diagram, the photographer sees the scene imaged onto the ground-glass screen at the top of the camera. The image is right-way-up, but is inverted left–right, the result of forming the image via reflection in a mirror. A more comfortable arrangement is shown at the right, where additional reflections take place in a 'pentaprism'. The image no longer suffers from inversion.

But something is wrong! The diagram[61] shows three reflections in all, and to make the image non-inverted we ought to have an even number of reflections. How does the diagram misrepresent things?

[61] Three-dimensional versions of Fig. 1.8, displaying the incorrect optical arrangement, are often to be found in print, usually in diagrams supplied by manufacturers.

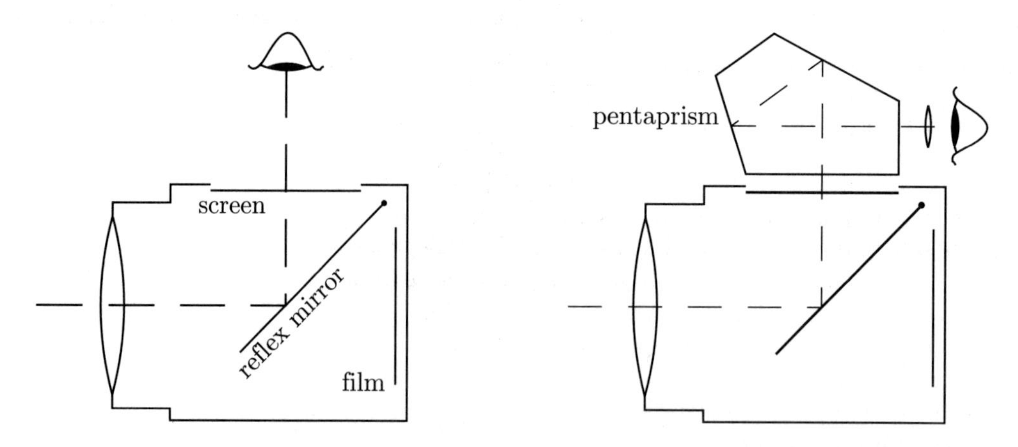

Fig. 1.8 A single-lens reflex camera is shown in cross section. While the photographer is 'viewfinding', light enters through the lens and is reflected from the reflex mirror onto a screen at the top of the camera. Some cameras permit the screen to be viewed 'downwards' as in the left-hand diagram, but the scene then appears reversed left–right. More commonly the screen is viewed through a 'pentaprism' through which the scene appears as normal. When a photograph is to be taken, the mirror swings upwards out of the way of light passing from lens to film, and is arranged to seal the camera against light entering via the viewfinder. The shutter (not shown) is just in front of the film.

The diagram is schematic. In particular, the lens normally has to be built from six or seven pieces of glass if it is to give acceptable image quality; a zoom lens is even more complicated.

Fourier series and Fourier transforms

2

2.1 Introduction

A familiarity with Fourier mathematics will be needed on several occasions throughout this book. We take the opportunity in this chapter to set out those mathematical properties that will be needed later.

It is assumed that the reader has encountered both Fourier series and Fourier transforms before, but may not be entirely confident with transforms, and may not have a fully developed insight into the meaning of it all. The present chapter therefore introduces afresh the Fourier transform, using the Fourier series as a jumping-off point.

We have of course already met sinusoidal waves, represented by a cosine in eqn (1.9) and by a complex exponential in eqn (1.10). Fourier methods can be thought of as just a way of handling waves that are composed of more than one such sinewave.[1]

[1] In this book **sinewave** (one word) will describe a wave like $\sin(\omega t+\alpha)$ with any value of α: anything sinusoidal. 'Sine wave' (two words) will indicate a sine, as opposed to a cosine, so $\sin(\omega t)$ or similar.

Applications of Fourier ideas range widely: from Fraunhofer diffraction (§§3.8–3.12) to the uncertainty principle (problem 2.25).

2.2 Fourier series: spectrum of a periodic waveform

Consider a 'signal', meaning some function $V(t)$ that consists of an infinite sequence of repetitions of a 'building block' waveform repeated at intervals τ. Such a function is illustrated in Fig. 2.1. The independent variable t may be thought of as time, but this is not necessary as the

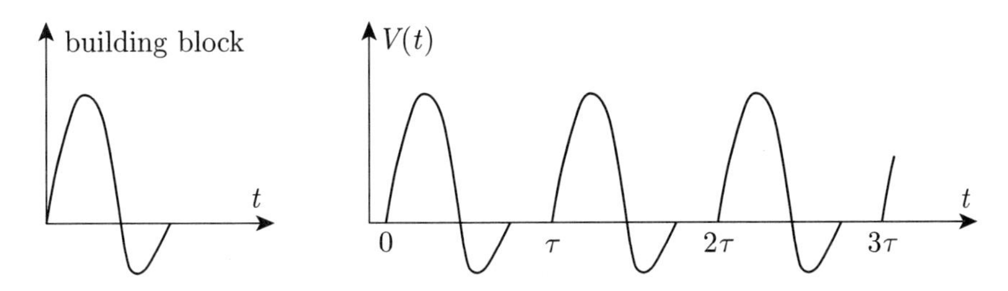

Fig. 2.1 A periodic waveform is made up by repeating a building block at regular intervals τ. The repetitions continue indefinitely many times, to $t \to -\infty$ as well as to $t \to \infty$.

discussion is mathematically general. The signal can be represented by a **Fourier series**:

$$V(t) = c_0 + \sum_{p=1}^{\infty} c_p \cos(p\omega_0 t) + \sum_{p=1}^{\infty} s_p \sin(p\omega_0 t), \tag{2.1}$$

or, equivalently, as

$$V(t) = \sum_{p=0}^{\infty} a_p \cos(\alpha_p + p\omega_0 t), \tag{2.2}$$

where $\omega_0 = 2\pi/\tau$ is the fundamental angular frequency.[2] The amplitude of the component (the pth harmonic) at angular frequency $p\omega_0$ is a_p, which is of course related to c_p and s_p since

$$c_p = a_p \cos\alpha_p, \qquad s_p = -a_p \sin\alpha_p.$$

[2] In this book we use p as a general-purpose integer, since n has already been committed to representing refractive index.

The coefficients of the sines and cosines may be found by multiplying $V(t)$ by $\cos(m\omega_0 t)$ or by $\sin(m\omega_0 t)$ and integrating over a period; the expressions that result are:[3]

$$s_p = \frac{2}{\tau}\int_0^{\tau} V(t)\sin(p\omega_0 t)\,\mathrm{d}t \qquad \text{for all } p, \tag{2.3a}$$

$$c_p = \frac{2}{\tau}\int_0^{\tau} V(t)\cos(p\omega_0 t)\,\mathrm{d}t \qquad \text{for } p \neq 0, \tag{2.3b}$$

$$c_0 = \frac{1}{\tau}\int_0^{\tau} V(t)\,\mathrm{d}t. \tag{2.3c}$$

[3] The integrations described yield values for the 'picked-out' coefficients c_m and s_m. But since m is thereafter treated as a general variable we may revert to using p in writing eqns (2.3).

The manipulation leading to eqns (2.3) is rehearsed in problem 2.1.

2.3 Fourier series: a mathematical reshape

We now rearrange the mathematics outlined above, to make it more economical, and to make its shape anticipate that of the Fourier transform.

It's unnecessarily complicated to work out three separate integrals for the c_p, s_p as given in eqns (2.3). Instead, we can work out

$$A_p = \frac{1}{\tau}\int_0^{\tau} V(t)\exp(\mathrm{i}p\omega_0 t)\,\mathrm{d}t, \tag{2.4}$$

so defined for all integer values of p, positive, negative or zero. The new coefficients A_p are related to the old ones by

$$\left.\begin{aligned} A_p &= \tfrac{1}{2}a_p \exp(-\mathrm{i}\alpha_p) = \tfrac{1}{2}(c_p + \mathrm{i}s_p) && \text{for } p > 0,\\ A_{-p} &= \tfrac{1}{2}a_p \exp(\mathrm{i}\alpha_p) \;\;= \tfrac{1}{2}(c_p - \mathrm{i}s_p) && \text{for } p > 0,\\ \text{but} \quad A_0 &= a_0 \qquad\qquad\quad\;\; = c_0 && \text{for } p = 0. \end{aligned}\right\} \tag{2.5}$$

Conversely,

$$c_p = A_p + A_{-p}, \qquad s_p = (A_p - A_{-p})/\mathrm{i}, \qquad c_0 = A_0. \tag{2.6}$$

We see that the complex amplitude A_p contains all the information we need about the Fourier coefficients. Moreover, finding the c_p and s_p by adding and subtracting (or taking real and imaginary parts) is much quicker than working out a second or third integral.

When $V(t)$ is real, as it usually will be, the c_p and s_p are all real, so that

$$c_p = 2\,\mathrm{Re}\,A_p, \qquad s_p = 2\,\mathrm{Im}\,A_p, \qquad a_p = 2|A_p|, \qquad \alpha_p = -\arg A_p \quad (2.7)$$

(but c_0 and a_0 are given by half of these expressions). And the coefficients A_p have the symmetry property

$$\text{if } V(t) \text{ is real:} \qquad A_{-p} = A_p^*. \quad (2.8)$$

Not only is A_p convenient for working out the coefficients c_p, s_p, but it leads to a simplified form of expansion, correct for all p, even $p = 0$:

$$V(t) = \sum_{p=-\infty}^{\infty} A_p\,\mathrm{e}^{-\mathrm{i}p\omega_0 t}, \qquad A_p = \frac{1}{\tau}\int_0^{\tau} V(t)\,\mathrm{e}^{\mathrm{i}p\omega_0 t}\,\mathrm{d}t. \quad (2.9)$$

This seems to be trying to tell us something: the complex exponential forms, though perhaps less concrete than sines and cosines, are the easy things to work with. Problem 2.2 fills out the last statement, showing that the simplification is more widespread than might at first appear.

Comment: Equations (2.9) are beginning to look rather abstract, so it may be worth spelling out again what they are doing for us. The left-hand statement says that function $V(t)$, whatever function it happens to be, can be assembled—Fourier **synthesis**—by adding up cosines and sines. The cosines and sines are slightly disguised in the exponentials $\mathrm{e}^{-\mathrm{i}p\omega_0 t}$, but they should still be clear to see. The right-hand statement tells us[4] how to obtain—Fourier **analysis**—the coefficients A_p for a given $V(t)$. As with all complex-exponential waveforms, each coefficient A_p is itself (probably) complex; its magnitude tells us 'how much' while its argument tells us 'with what phase' the pth complex exponential contributes to the sum. To see confirmation of these statements, look again at eqns (2.7), and see how a_p and α_p are linked to A_p.

Problem 2.3 discusses one way in which the behaviour of $V(t)$ affects the values of the A_p.

[4] A cross-reference: In quantum mechanics we often expand some wave function $\psi(x)$ in terms of a set of expansion functions $\phi_p(x)$. The coefficient of ϕ_p is obtained by multiplying through by ϕ_p^* and integrating. We can see $\mathrm{e}^{-\mathrm{i}p\omega_0 t}$ as an (unnormalized) expansion function in the sum on the left of eqn (2.9). And we can see its conjugate $(\mathrm{e}^{-\mathrm{i}p\omega_0 t})^*$ in the integral on the right of eqn (2.9) doing precisely the same job for us that ϕ_p^* does in quantum mechanics.

2.4 The Fourier transform: spectrum of a non-periodic waveform

Let $V(t)$ be a function, probably non-periodic now, of (perhaps time) t. We shall represent $V(t)$ as a sum over sinewaves (compare eqn (2.2)):

$$V(t) = \int_0^{\infty} a(\omega)\cos\{\alpha(\omega) + \omega t\}\,\mathrm{d}\omega/2\pi. \quad (2.10)$$

Here angular frequency ω takes a continuous range of values, where in eqn (2.2) we had discrete angular frequencies $p\omega_0$; hence the replacement of a discrete sum by an integral. In the first instance, the old $p\omega_0$ and the new ω range over positive values only.

We see that $a(\omega)\,\mathrm{d}\omega/2\pi$ (replacing a_p) is the magnitude of the sinewave belonging to the frequency range $\mathrm{d}\nu = \mathrm{d}\omega/2\pi$, while $\alpha(\omega)$ (replacing α_p) gives the phase of that sinewave.

It is usually more convenient to work with a complex amplitude, for the reasons found in §2.3. So we rearrange the integral:[5]

[5]Notice that ω starts out being defined for positive values only, but acquires a definition covering negative values during the working. Check with problem 2.6.

$$\begin{aligned} V(t) &= \int_0^\infty a(\omega)\left(\frac{\mathrm{e}^{-\mathrm{i}\{\alpha(\omega)+\omega t\}} + \mathrm{e}^{\mathrm{i}\{\alpha(\omega)+\omega t\}}}{2}\right)\frac{\mathrm{d}\omega}{2\pi} \\ &= \int_0^\infty \left(\frac{a(\omega)\,\mathrm{e}^{-\mathrm{i}\alpha(\omega)}}{2}\right)\mathrm{e}^{-\mathrm{i}\omega t}\frac{\mathrm{d}\omega}{2\pi} + \int_{-\infty}^0 \left(\frac{a(-\omega)\,\mathrm{e}^{\mathrm{i}\alpha(-\omega)}}{2}\right)\mathrm{e}^{-\mathrm{i}\omega t}\frac{\mathrm{d}\omega}{2\pi} \\ &= \int_{-\infty}^\infty \widetilde{V}(\omega)\,\mathrm{e}^{-\mathrm{i}\omega t}\frac{\mathrm{d}\omega}{2\pi}, \end{aligned}$$

where

$$\left.\begin{aligned} \widetilde{V}(\omega) &\equiv \tfrac{1}{2}\,a(\omega)\,\mathrm{e}^{-\mathrm{i}\alpha(\omega)} \\ \text{and}\qquad \widetilde{V}(-\omega) &\equiv \tfrac{1}{2}\,a(\omega)\,\mathrm{e}^{\mathrm{i}\alpha(\omega)}, \end{aligned}\right\}\ \text{both for positive } \omega. \tag{2.11}$$

The function $\widetilde{V}(\omega)$ is called the **Fourier transform** of $V(t)$.

All this is useful only if we can find $\widetilde{V}(\omega)$. Fortunately, the Fourier transform has well-known mathematical properties, so the inverse transform can be quoted at once:

$$V(t) = \int_{-\infty}^\infty \widetilde{V}(\omega)\,\mathrm{e}^{-\mathrm{i}\omega t}\frac{\mathrm{d}\omega}{2\pi}, \quad \text{where} \quad \widetilde{V}(\omega) = \int_{-\infty}^\infty V(t)\,\mathrm{e}^{\mathrm{i}\omega t}\,\mathrm{d}t. \tag{2.12}$$

(A proof—of sorts—of this is rehearsed in problem 2.8.) Thus, we can find $\widetilde{V}(\omega)$ by evaluating an integral over $V(t)$. As with the coefficients A_p of Fourier series, function $\widetilde{V}(\omega)$ contains all the information[6] we need about the frequency components of $V(t)$.

[6]Extract the meaning from eqns (2.12): $V(t)$ can be reconstructed—**synthesized**—by building it up from sinewaves. The ideas are exactly the same as those discussed in the comment after eqns (2.9). The transform $\widetilde{V}(\omega)$ is the coefficient that tells us 'how much' and 'with what phase' each sinewave $\mathrm{e}^{-\mathrm{i}\omega t}$ contributes to building up $V(t)$. The introduction on Fourier series was provided to make it easy to see all these similarities and thereby to make the Fourier transform seem less forbidding.

If $V(t)$ is real, as it usually will be, the quantities $a(\omega)$ and $\alpha(\omega)$ can be evaluated from

$$a(\omega) = 2|\widetilde{V}(\omega)|, \qquad \alpha(\omega) = -\arg\{\widetilde{V}(\omega)\}. \tag{2.13}$$

Moreover, a consequence of eqn (2.11) closely resembles eqn (2.8):

$$\text{if } V(t) \text{ is real:} \qquad \widetilde{V}(-\omega) = \widetilde{V}(\omega)^*. \tag{2.14}$$

In fact, eqns (2.10)–(2.14) all have obvious correspondences to earlier equations. In particular, the (τA_p) of eqn (2.9) resembles $\widetilde{V}(\omega)$; and problem 2.4 shows the series of eqn (2.9) merging into the integral of eqn (2.12) when a limit is taken.

Additional basic properties of $\widetilde{V}(\omega)$ are derived in problem 2.5. The behaviour of $\widetilde{V}(\omega)$ for large ω is investigated in problem 2.17, with results resembling those already encountered in problem 2.3.

2.5 The analytic signal

Approach §2.4 another way. Henceforth, $V(t)$ will be restricted to being real. Repeating eqn (2.10),

$$V(t) = \int_0^\infty a(\omega)\cos\{\alpha(\omega) + \omega t\}\frac{d\omega}{2\pi}.$$

Here $V(t)$ is the known thing, but Fourier's theorem fixes $a(\omega)$ and $\alpha(\omega)$ so we can think of them as known—or at least knowable—too. Define an associated signal $V^{\mathrm{I}}(t)$, also real, given by

$$V^{\mathrm{I}}(t) = -\int_0^\infty a(\omega)\sin\{\alpha(\omega) + \omega t\}\frac{d\omega}{2\pi}.$$

From these we can define a complex amplitude $U(t)$, called the **analytic signal**,

$$\begin{aligned} U(t) &= V(t) + \mathrm{i}V^{\mathrm{I}}(t) \\ &= \int_0^\infty a(\omega)\exp\{-\mathrm{i}\alpha(\omega) - \mathrm{i}\omega t\}\frac{d\omega}{2\pi} = 2\int_0^\infty \widetilde{V}(\omega)\,\mathrm{e}^{-\mathrm{i}\omega t}\frac{d\omega}{2\pi}. \end{aligned} \quad (2.15)$$

Then by construction of U, the real optical signal is

$$V(t) = \text{the real part of } U(t). \quad (2.16)$$

Comment: The analytic signal $U(t)$ is familiar, though it is not usually introduced in quite this way.[7] We often use a complex amplitude, e.g. in electric circuit theory where voltages and currents are made proportional to $\mathrm{e}^{\mathrm{j}\omega t}$, and we agree to take the real part at the end of the problem. That means we adopt by implication the following procedure:

(a) We note that the circuit equations are linear.
(b) We break each voltage or current into its constituent sinewaves, and we concentrate for the moment on just one of those constituents.
(c) We use a complex representation (containing $\mathrm{e}^{\mathrm{j}\omega t}$ or $\mathrm{e}^{-\mathrm{i}\omega t}$) of that sinewave for mathematical convenience.
(d) We work out a solution for the one sinewave, taking the real part at the end.[8]
(e) We assemble the whole solution by adding up all the contributing sinewaves.

Procedure (a)–(e) is rarely carried through all five stages, as (e) can usually be dispensed with. However, it is always available, if only as an 'in desperation' method, for any linear problem.

The definition (2.15) of $U(t)$ reminds us (by explicitly writing down a sum over them) that all the sinewave components are (or may be) present together, while it retains the simplifying advantages of complex arithmetic.

[7] At the beginning of this chapter, we said that t did not have to represent time, but could be any variable of interest. In the present discussion, and in much of the rest of the chapter, t does represent time.

[8] Alternatively, we may instead find the magnitude and phase from the modulus of the solution and from its argument. This operation is routine and familiar. Yet the description of an exactly similar way of understanding $\widetilde{V}(\omega)$ (eqns (2.13) and sidenote 6 opposite) usually causes the student's eyes to glaze over.

Notice that the last step in eqn (2.15) can be cross-referenced to the working written out before eqns (2.11). One way of describing $U(t)$ is to say that it is 'built up out of the positive-frequency part' of $V(t)$. Positive frequency? Do problem 2.6.

Some subtleties that follow from the definition of U are explored in problem 2.10.

2.6 The Dirac δ-function

The δ-function is defined by

$$\delta(x) \equiv \begin{cases} \infty & \text{for } x = 0 \\ 0 & \text{for } x \neq 0, \end{cases} \qquad \text{such that} \qquad \int_{(\text{anything}<0)}^{(\text{anything}>0)} \delta(x)\,\mathrm{d}x = 1. \tag{2.17}$$

For any continuous function $f(x)$,

$$\int_{(\text{anything}<0)}^{(\text{anything}>0)} f(x)\,\delta(x)\,\mathrm{d}x = f(0),$$

and by extension,

$$\int_{-\infty}^{\infty} f(x)\,\delta(x - x_0)\,\mathrm{d}x = f(x_0). \tag{2.18}$$

We sometimes express this by saying that $\delta(x - x_0)$ is a 'picking-out function'.

There are several ways of building up a 'spiky' function to represent $\delta(x)$. We'll use

$$\delta(x) = \lim_{N\to\infty} \frac{1}{2\pi} \int_{-N}^{N} \mathrm{e}^{-\mathrm{i}xy}\,\mathrm{d}y = \frac{1}{2\pi} \int_{-\infty}^{\infty} \mathrm{e}^{-\mathrm{i}xy}\,\mathrm{d}y, \tag{2.19}$$

where the last version is sloppy mathematics but is often written. Problem 2.7 shows that the function defined by eqn (2.19) does indeed have the properties required for representing $\delta(x)$.

A number of uses of the δ-function will appear in this chapter, and later. Now is a good time to work problem 2.8, in which the δ-function is used in a plausibility argument to make you comfortable both with it and with the Fourier relations (2.12).

2.7 Frequency and angular frequency

Sometimes it helps to think about Fourier transforms (and series) in a slightly different way. Define $F(\nu) \equiv \widetilde{V}(2\pi\nu) = \widetilde{V}(\omega)$. Then

$$F(\nu) = \widetilde{V}(\omega) = \int_{-\infty}^{\infty} V(t)\,\mathrm{e}^{\mathrm{i}2\pi\nu t}\,\mathrm{d}t \quad \text{and} \quad V(t) = \int_{-\infty}^{\infty} F(\nu)\,\mathrm{e}^{-\mathrm{i}2\pi\nu t}\,\mathrm{d}\nu.$$

Everything is now expressed[9] in terms of frequency ν, rather than angular frequency ω.

[9] *Comment*: It is usually not a good idea to present the mathematics in terms of ν, because the 2π clutters the exponentials $\mathrm{e}^{\pm\mathrm{i}2\pi\nu t}$, and clutter here exacts a higher price than having the 2π elsewhere. I have tried to minimize clutter in this chapter by using ω, yet clumping $\mathrm{d}\omega/2\pi$ together in order to emphasize that the clump is an element of 'real frequency' $\mathrm{d}\nu$.

Mathematicians usually deal with the untidy 2π another way, by splitting the 2π in eqn (2.12) so that $1/\sqrt{2\pi}$ is placed in front of each integral. This makes for a greater symmetry between the two statements in eqn (2.12).

However, in the present chapter we are thinking of the 'forward' and 'backward' transforms as doing rather different jobs: the first finds the coefficients of sinewaves (Fourier analysis); the second assembles those sinewaves into a replica of the original function (Fourier synthesis). Thought of this way, it is almost an accident that there is any close mathematical resemblance in shape between the 'forward' and 'backward' integrals. Indeed, the similarity of mathematical shape in the two integrals of eqn (2.12)—much the same operation except for a sign in the exponent and possibly a factor 2π—seems itself to contribute to a feeling of insecurity in a beginner's mind.

2.8 The power spectrum

The *intensity* of an electromagnetic wave is defined in §1.5 and shown to be given by eqn (1.21):

$$I = \frac{1}{T}\int_{-T/2}^{T/2} E(t)\,H(t)\,\mathrm{d}t = \frac{n}{Z_0}\frac{1}{T}\int_{-T/2}^{T/2} E(t)^2\,\mathrm{d}t.$$

In this statement, E and H are *real* field amplitudes.[10] To express things in terms of frequency, we need to make contact with the notation of §2.4, so we define $V(t)$, also real, to represent $E(t)$ during the observation time from $-\frac{1}{2}T$ to $\frac{1}{2}T$, but $V(t) = 0$ for all other t. Then

$$I = \frac{n}{Z_0}\frac{1}{T}\int_{-\infty}^{\infty} V(t)^2\,\mathrm{d}t.$$

[10] Subscripts indicating Cartesian components are now omitted, it being taken that E and H are appropriately defined fields for a plane or near-plane wave.

A consequence of the Fourier relations (2.12) is the **Parseval theorem**:[11]

$$\text{for real } V(t), \qquad \int_{-\infty}^{\infty} V(t)^2\,\mathrm{d}t = \int_{-\infty}^{\infty} \left|\widetilde{V}(\omega)\right|^2 \frac{\mathrm{d}\omega}{2\pi}. \tag{2.20}$$

[11] You are invited to construct a proof of the Parseval theorem in problem 2.9.

An obvious substitution now leads us to an expression for intensity as an integral over frequencies:[12]

$$I = \frac{n}{Z_0}\frac{1}{T}\int_{-\infty}^{\infty} \left|\widetilde{V}(\omega)\right|^2 \frac{\mathrm{d}\omega}{2\pi} = \frac{2n}{Z_0}\frac{1}{T}\int_{0}^{\infty} \left|\widetilde{V}(\omega)\right|^2 \frac{\mathrm{d}\omega}{2\pi}, \tag{2.21}$$

[12] Another application of the Parseval theorem may be found in problem 3.11.

where the last step makes use of a property derived in problem 2.5(2). Thus, the intensity (mean power per unit area) attributable to frequency range $\mathrm{d}\nu = \mathrm{d}\omega/2\pi$ is $P(\omega)\,\mathrm{d}\omega/2\pi$, where[13]

$$P(\omega) = \frac{2n}{Z_0}\frac{\left|\widetilde{V}(\omega)\right|^2}{T} \propto \begin{cases} & \widetilde{V}(\omega)^*\,\widetilde{V}(\omega) \\ \text{or} & F(\nu)^*\,F(\nu) \\ \text{or} & a(\omega)^2, \end{cases} \tag{2.22}$$

[13] The quantity $P(\omega)$ is known as the **power spectrum**.

where $a(\omega)$ is defined in §2.4.[14] We may notice that expressions (2.22) give electromagnetic respectability to an idea[15,16] introduced in elementary courses on waves: that (intensity) $\propto$ (amplitude)^2.

[14] *Comment*: Statement (2.21) exhibits the intensity as a sum (or rather an integral) over the intensities contributed by the individual frequencies that are present. Different frequencies don't give rise to any kind of interference, such as 'beats'—at least for now ....

[15] *Comment*: Although the main emphasis here has been given to $|\widetilde{V}(\omega)|^2$, notice again the presence of the refractive index n, carried forward from eqn (1.23).

[16] *Question*: Incidentally, for the whole of §2.8, an assumption has been sneaked in: that n is frequency independent. Can you see where? And what would you have to do if that assumption wasn't adequate? [Help in problem 2.22.]

2.9 Examples of Fourier transforms

2.9.1 A single rectangular pulse

The pulse is shown in Fig. 2.2(a), and has the algebraic definition

$$V(t) = \begin{cases} V_0 & \text{for } 0 < t < \tau_{\mathrm{w}} \\ 0 & \text{for all other } t. \end{cases}$$

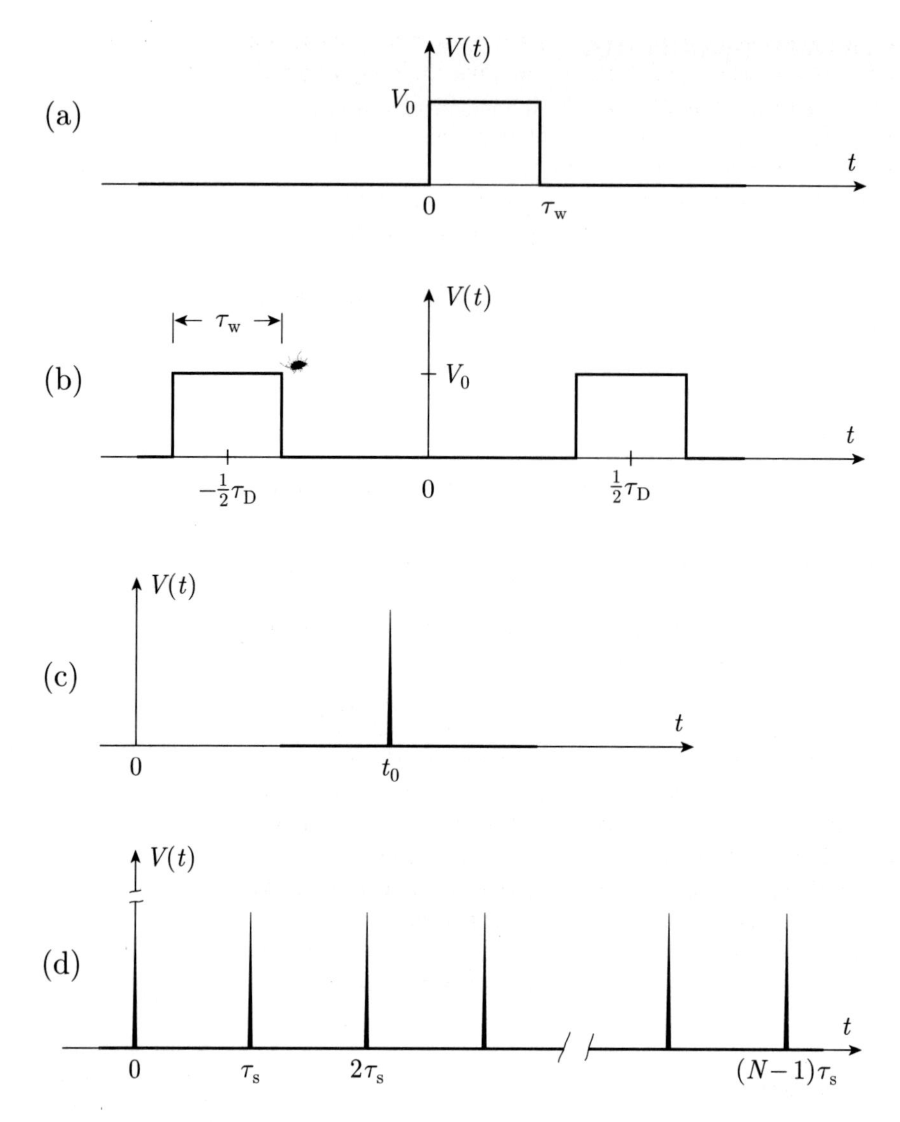

Fig. 2.2 Functions $V(t)$ whose Fourier transforms are given in §2.9. (a) A rectangular pulse of duration τ_{W}. (b) A double pulse: two pulses of duration τ_{W}, separated (centre-to-centre) by τ_{D}. (c) A δ-function pulse located at $t = t_0$. (d) An array of N δ-function pulses separated by τ_{s}.

The Fourier transform $\widetilde{V}(\omega)$ is easily worked out to be

$$\begin{aligned}\widetilde{V}(\omega) &= \int_{-\infty}^{\infty} V(t)\,\mathrm{e}^{\mathrm{i}\omega t}\,\mathrm{d}t = V_0 \int_0^{\tau_{\mathrm{W}}} \mathrm{e}^{\mathrm{i}\omega t}\,\mathrm{d}t = V_0 \left(\frac{\exp(\mathrm{i}\omega\tau_{\mathrm{W}}) - 1}{\mathrm{i}\omega} \right) \\ &= V_0 \exp(\tfrac{1}{2}\mathrm{i}\omega\tau_{\mathrm{W}}) \left(\frac{\exp(\frac{1}{2}\mathrm{i}\omega\tau_{\mathrm{W}}) - \exp(-\frac{1}{2}\mathrm{i}\omega\tau_{\mathrm{W}})}{\mathrm{i}\omega} \right) \\ &= V_0\tau_{\mathrm{W}} \exp(\tfrac{1}{2}\mathrm{i}\omega\tau_{\mathrm{W}}) \left(\frac{\sin(\frac{1}{2}\omega\tau_{\mathrm{W}})}{(\frac{1}{2}\omega\tau_{\mathrm{W}})} \right).\end{aligned} \tag{2.23}$$

Although the notation suggests a pulse dependent upon time (perhaps a voltage pulse from an electronic pulse generator), we could equally well be thinking about Fraunhofer diffraction from a single slit.

The exponential factor in expression (2.23) may be unexpected. It exemplifies a 'shifting theorem'; see problem 2.11.

Even though our first concrete example of a Fourier transform is rather simple, there are some surprisingly useful things we can do with it. Some of them are explored in problem 2.12.

You are invited to confirm the mathematics for all the examples in this section by working problem 2.13.

2.9.2 The double pulse

The two pulses are shown in Fig. 2.2(b). The transform works out to be

$$\widetilde{V}(\omega) = (2V_0\tau_{\mathrm{w}}) \times \underbrace{\left(\frac{\sin(\frac{1}{2}\omega\tau_{\mathrm{w}})}{(\frac{1}{2}\omega\tau_{\mathrm{w}})}\right)}_{\text{envelope from shape of } each \text{ single pulse}} \times \underbrace{\cos(\tfrac{1}{2}\omega\tau_{\mathrm{D}})}_{\text{modulation caused by } double \text{ pulse}}. \tag{2.24}$$

There is a 'shape' to this expression, whose significance will be clearer when we look back after understanding the convolution theorem. The 'envelope' factor $\sin(\frac{1}{2}\omega\tau_{\mathrm{w}})/(\frac{1}{2}\omega\tau_{\mathrm{w}})$ is identical to the transform of the single pulse: it originates from the shape of *each* single pulse within the pair. The final factor $\cos(\frac{1}{2}\omega\tau_{\mathrm{D}})$ has oscillatory behaviour caused by the fact that the pulse is repeated. (Obvious relevance to Fraunhofer diffraction from two slits; equivalently to the Young slits.)

The double-pulse $\widetilde{V}(\omega)$ can be checked in several common-sense ways that don't just repeat the algebra. For confirmation see problem 2.14.

2.9.3 A δ-function pulse

The pulse is shown in Fig. 2.2(c). We have[17]

$$\begin{aligned} V(t) &= V_0\,\delta(t - t_0). \\ \widetilde{V}(\omega) &= \int_{-\infty}^{\infty} V_0\,\delta(t - t_0)\,\mathrm{e}^{\mathrm{i}\omega t}\,\mathrm{d}t \\ &= V_0\,\mathrm{e}^{\mathrm{i}\omega t_0}. \end{aligned} \tag{2.25}$$

Then

$$\widetilde{V}(\omega)^*\,\widetilde{V}(\omega) = V_0^2,$$

which shows the important characteristic of the δ-function in this context: it is such an extreme function that its power spectrum is flat.[18]

[17] *Comment*: Notice in the exponential of eqn (2.25) a manifestation of the 'shifting theorem' already seen in eqn (2.23).

[18] *Aside*: Take the case where the δ-function is located at $t = 0$. Then if $V(t) = \delta(t)$, $\widetilde{V}(\omega) = 1$. Write down the reverse Fourier transform, carrying us back to $V(t)$ again, and you will see where eqn (2.19) came from.

2.9.4 A regular array of δ-functions

Figure 2.2(d) shows the array defining $V(t)$. The transform $\widetilde{V}(\omega)$ is calculated as follows:

$$V(t) = \sum_{p=0}^{N-1} \delta(t - t_p) = \sum_{p=0}^{N-1} \delta(t - p\tau_s).$$

$$\begin{aligned}\widetilde{V}(\omega) &= \int_{-\infty}^{\infty} \sum_{p=0}^{N-1} \delta(t - p\tau_s)\, e^{i\omega t}\, dt = \sum_{p=0}^{N-1} \int_{-\infty}^{\infty} \delta(t - p\tau_s)\, e^{i\omega t}\, dt \\ &= \sum_{p=0}^{N-1} e^{i\omega p\tau_s} = \sum_{p=0}^{N-1} \{e^{i\omega\tau_s}\}^p \\ &= 1 + e^{i\omega\tau_s} + e^{i2\omega\tau_s} + e^{i3\omega\tau_s} + \cdots + e^{i(N-1)\omega\tau_s} = \frac{e^{iN\omega\tau_s} - 1}{e^{i\omega\tau_s} - 1} \\ &= e^{i\alpha}\,\frac{\sin(\frac{1}{2}N\omega\tau_s)}{\sin(\frac{1}{2}\quad\omega\tau_s)}. \end{aligned} \tag{2.26}$$

Here[19] $\alpha = \frac{1}{2}(N-1)\omega\tau_s$; the $e^{i\alpha}$ is merely the consequence of choosing the origin of t at one end of the array (check using problem 2.13). The physics interest lies in the last factor, which should be recognizable ....

If we have an infinite array of δ-functions, we are back to thinking about periodic functions and Fourier series. The implications for the Fourier transform (which still exists) are explored in problem 2.24.

[19] The blank space in the denominator of eqn (2.26) is left to emphasize the shape of the expression; I won't do this often.

2.9.5 A random array of δ-functions

Let N identical δ-function pulses occur within a time interval T:

$$V(t) = \sum_p \delta(t - t_p),$$

where the t_p are random times within the range $-\frac{1}{2}T < t_p < \frac{1}{2}T$. We now have

$$\widetilde{V}(\omega) = \sum_{p=0}^{N-1} e^{i\omega t_p}.$$

For any one randomly chosen set of pulses, this sum might add up to anything. (Well, not quite anything, since of course $|\widetilde{V}(\omega)| \le N$.) We can, however, calculate statistical averages over an *ensemble* of such pulse sets (shades of statistical mechanics). The ensemble averages are

$$\langle \widetilde{V}(\omega) \rangle = 0, \qquad \langle \widetilde{V}(\omega)^* \, \widetilde{V}(\omega) \rangle = N. \tag{2.27}$$

These results will be helpful when thinking about coherence. (For a derivation, see problem 2.15.)

2.9.6 An infinite sinewave

A sinewave 'going on for ever' of the form[20]

$$V(t) = e^{-i\omega_0 t} \qquad \text{has transform} \qquad \widetilde{V}(\omega) = 2\pi\,\delta(\omega - \omega_0). \tag{2.28}$$

[20] Equation (2.28) is proved in problem 2.16. It will be insightful in problem 3.19 and also in the reasoning of §12.5.2.

2.10 Convolution and the convolution theorem

It often happens that we can express an optical signal $V(t)$ as the **convolution** of two simpler functions $\phi(t)$ and $\psi(t)$ thus:[21]

$$V(t) = \phi(t) \circledast \psi(t) \equiv \int_{-\infty}^{\infty} \phi(t')\,\psi(t-t')\,\mathrm{d}t'. \tag{2.29}$$

The **convolution theorem** states that the Fourier transform of a convolution is the product of the individual transforms,[22] that is

$$\widetilde{\phi(t) \circledast \psi(t)} = \widetilde{\phi}(\omega) \times \widetilde{\psi}(\omega),$$

or, equivalently,[23]

$$\widetilde{V}(\omega) = \int_{-\infty}^{\infty} \phi(t)\,\mathrm{e}^{\mathrm{i}\omega t}\,\mathrm{d}t \times \int_{-\infty}^{\infty} \psi(t)\,\mathrm{e}^{\mathrm{i}\omega t}\,\mathrm{d}t. \tag{2.30}$$

[21] *Comment*: Note that the 'dummy' variable of integration t' occurs with opposite signs in the two functions in the integrand; had the integrand been $\phi(t')\,\psi(t'-t)$ the integral would have been a *correlation*.

[22] The theorem is quite easy to prove: see problem 2.18.

[23] The notation of eqn (2.30) is careless, because the two different dummy variables should not both be represented by the same t; but I thought a switch of variables might be more distracting than an easily corrected lapse. By contrast, the ωs in the two integrals really are the same.

2.11 Examples of convolution

Figure 2.3 shows a building-block function $\phi(t)$ that is repeated N times at regular intervals τ_s to form a total function $V(t)$. A function of this shape might, for example, be the electric field of light that has just been reflected from a diffraction grating.

An easy and insightful way of working out the Fourier transform is to rearrange this $V(t)$ into the form of a convolution:

$$V(t) = \sum_{p=0}^{N-1} \phi(t - p\tau_\mathrm{s}) = \int_{-\infty}^{\infty} \phi(t-t')\,.\,\sum_{p=0}^{N-1} \delta(t' - p\tau_\mathrm{s})\,\mathrm{d}t'. \tag{2.31}$$

The Fourier transform may now be assembled from the transforms of the two convolved functions (ϕ and the sum) using eqn (2.26):

$$\begin{aligned}\widetilde{V}(\omega) &= \int_{-\infty}^{\infty} \phi(t)\,\mathrm{e}^{\mathrm{i}\omega t}\,\mathrm{d}t \qquad \times \int_{-\infty}^{\infty} \sum_{p=0}^{N-1} \delta(t' - p\tau_\mathrm{s})\,\mathrm{e}^{\mathrm{i}\omega t'}\,\mathrm{d}t' \\ &= \big\{\text{transform } \widetilde{\phi}(\omega) \text{ of } \phi(t)\big\} \times \mathrm{e}^{\mathrm{i}\alpha} \left\{ \frac{\sin(\frac{1}{2}N\omega\tau_\mathrm{s})}{\sin(\frac{1}{2}\omega\tau_\mathrm{s})} \right\}. \end{aligned} \tag{2.32}$$

This result has obvious application to diffraction at a grating.

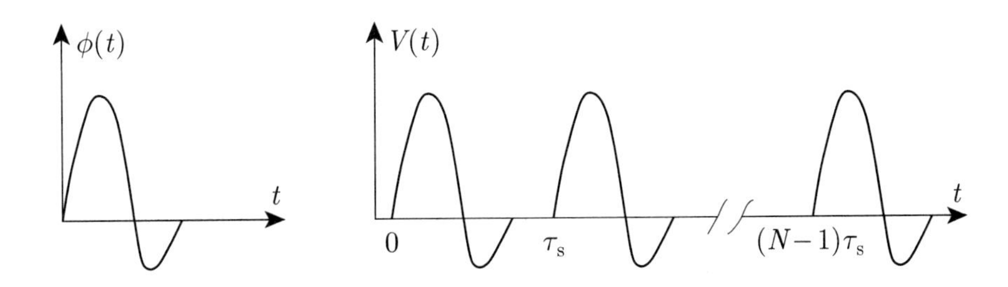

Fig. 2.3 A regular array of N building-block waveforms. In contrast to Fig. 2.1, these repetitions do not continue indefinitely.

Physics is full of convolutions.

- The example just given shows that the amplitude of light leaving a diffraction grating is a convolution: of one function describing the action of a *single* grating element; and a second function that represents the *repetitions* of that element.[24] (The δ-function array of Fig. 2.2(d) begins to look rather useful now .... Indeed, δ-functions as a whole are looking rather less contrived.)
- The double pulse is a special case of such a 'grating' in which there are only two repeated units; work problem 2.19.[25]
- The 'shifting theorem' of problem 2.11 is easily seen to be the result of convolving a building-block function with the shifted δ-function of eqn (2.25).
- Two-dimensional examples are equally plentiful. Standard things to mention are: a 'tiling pattern' (in the mathematician's sense) where a tile, or a small group of tiles, is repeated to cover a plane;[26] a papered wall which has a building-block piece of pattern repeated in a similar way (sometimes the δ-functions lie on a non-rectangular array, if the wallpaper pattern has a 'drop').
- The most important application of all is in three dimensions: crystal structure. A crystal can be described as made up of regular repetitions of a building block: a convolution of a single building block (the *basis*) with a regular array of δ-functions (the *lattice*). When X-rays are scattered from a crystal (under Fraunhofer conditions) the scattered amplitude factorizes: one factor, the *structure factor*, originates from scattering by the building block and puts an 'envelope' on the scattered amplitude; the other factor originates from the repetitions, and puts δ-function spikes at each location on the reciprocal lattice. Just like eqns (2.31) and (2.32).
- Convolutions are even more widespread. Think of the convolution theorem from its 'other side'. What about looking for a product? An equation such as $\boldsymbol{D}(\omega) = \varepsilon_0\, \varepsilon_{\mathrm{r}}(\omega)\boldsymbol{E}(\omega)$ contains a product of two functions of ω; so $\boldsymbol{D}$ ought to contain a convolution of two functions of time ...? This idea is explored, through a model example, in problem 2.22.

An idea of the usefulness of the convolution theorem can be gained from looking at problems 2.19–2.24, all of which employ the convolution theorem, either to assist calculation or to convey an insight.

[24] This factorization property is what we need to understand the blazing of a diffraction grating in §4.5.

[25] It may be clear now why so much excited attention is paid (by teachers and authors) to the factorization property on show in eqn (2.24). The product form is no accident, but the result of a convolution. The Fourier transform consists of one factor coming entirely from the shape of a single building-block pulse, and one factor coming entirely from the fact that the pulse is repeated. Algebraic signals confirm this: the pulse width τ_{w} appears only in the building-block factor; the pulse separation τ_{D} only in the 'repetition' factor.

[26] Penrose tiling is an exception because it does not repeat itself.

2.12 Sign choices with Fourier transforms

In eqns (2.12), and indeed as far back as eqn (2.4), an arbitrary choice was made, in putting $e^{-i\omega t}$ into the 'synthesis' integral, which then forces $e^{+i\omega t}$ into the 'analysis' integral. The same happened in eqn (2.15), where the 'analytic signal' was defined as built from the 'positive frequency part' of $\widetilde{V}(\omega)$.

Two conventions are used for the complex representation of waves in physics:

- amplitudes $\propto \mathrm{e}^{-\mathrm{i}\omega t}$, e.g. in quantum mechanics where a stationary state has
$$\Psi(\boldsymbol{r}, t) = \psi(\boldsymbol{r})\,\mathrm{e}^{-\mathrm{i}Et/\hbar}$$
- amplitudes $\propto \mathrm{e}^{\mathrm{j}\omega t}$, e.g. in electric circuit theory.

There is no prospect that either of these conventions will ever oust the other from its home territory, so we have to live with both and make things as tidy as possible when working 'in the middle'. In this book,[27] I use $\mathrm{e}^{-\mathrm{i}\omega t}$ for preference because it's convenient to have e^{+ikz} accounting for the effect on a wave of travelling through distance z. Exceptions are made when visiting electric circuits as, for example, in §1.8. More importantly, I use the two given conventions only[28] (never $+\mathrm{i}\omega t$ or $-\mathrm{j}\omega t$). And if it's necessary to switch between conventions, the conversion is made mechanical by replacing j with $-$i or the reverse.

The advantages of following the choice recommended here are:

- You can use whichever convention is more convenient for the problem in hand, and if necessary (as in problem 2.22) switch to the other by a routine substitution $\mathrm{j} = -\mathrm{i}$, so confusions of sign are easily avoided.
- The presence of i or j in a formula is a reminder of the convention within which it was derived, so you don't have to remember separately what that convention was.
- And if a formula contains neither i nor j then it's true within either convention.

[27] Any attempt at enforcing a single convention ends up making either electromagnetism or quantum mechanics unrecognizable. The author finds the solution recommended here—of using both $-\mathrm{i}\omega t$ and $+\mathrm{j}\omega t$ and saying $\mathrm{j} = -\mathrm{i}$ —so eminently sensible that he can't understand why it hasn't become universal.

[28] *Beware*: Amplitudes proportional to $\mathrm{e}^{+\mathrm{i}\omega t}$ are frequently found in textbooks (though rarely anywhere else!), especially books on optics.

Comment: In this book a Fourier transform (as in Fraunhofer diffraction) will sometimes look exactly like that defining $\widetilde{V}(\omega)$, sometimes it will have the opposite sign in the exponent. To avoid untidy sentences, I'll refer to both versions indiscriminately as Fourier transforms, leaving it to the reader to be aware and tidy up the description.

A phase *shift* will be taken to be positive when it's a *lead*. Within the convention used more often here ($\mathrm{e}^{-\mathrm{i}\omega t}$), $\mathrm{e}^{-\mathrm{i}\alpha}$ represents a phase lead when α is positive.

Problems

Problem 2.1 (a) Orthogonality for sines and cosines
Given that $\omega_0 = 2\pi/\tau$, prove that

$$\int_0^\tau \cos(m\omega_0 t)\cos(p\omega_0 t)\,\mathrm{d}t = \begin{cases} 0 & \text{for } m \neq p \\ \tfrac{1}{2}\tau & \text{for } m = p \neq 0 \\ \tau & \text{for } m = p = 0, \end{cases}$$

$$\int_0^\tau \sin(m\omega_0 t)\sin(p\omega_0 t)\,\mathrm{d}t = \begin{cases} 0 & \text{for } m \neq p \\ \tfrac{1}{2}\tau & \text{for } m = p \neq 0 \\ 0 & \text{for } m = p = 0, \end{cases}$$

and
$$\int_0^\tau \sin(m\omega_0 t)\cos(p\omega_0 t)\,\mathrm{d}t = 0 \quad \text{for all } m,\ p.$$

Use these orthogonality relations to show that in the Fourier series (2.1) the coefficients are given by eqns (2.3) given on p. 22.

Problem 2.2 (a) Fourier series basics

(1) Verify the working in §2.3, and obtain statements (2.9) by substituting for s_p and c_p in terms of A_p.

(2) Equations (2.6) are set up to give c_p and s_p correctly, whether or not $V(t)$ is real. Obtain expressions for a_p and α_p that likewise hold even when $V(t)$ is complex.

(3) Show that the complex exponentials have their own orthogonality relation:

$$\int_0^\tau \exp(im\omega_0 t) \times \exp(ip\omega_0 t)\, dt = \begin{cases} 0 & \text{for } m \neq -p \\ \tau & \text{for } m = -p, \end{cases} \tag{2.33}$$

true for all integer values of m and p, positive, negative or zero.

(4) Use the orthogonality relation (2.33) to derive[29,30] eqns (2.9) directly.

[29] *Hint*: One equation of the pair (2.9) can be taken as a definition of A_p, say the first. Then the equations are consistent if the second equation can be derived from the first. Multiply the first equation by $\exp(im\omega_0 t)$ and integrate.

[30] *Comment*: The route in part (4) is considerably simpler than that used in problem 2.1, but it might be hard to grasp the motivation for it if you hadn't seen the longer route first. Even now there is a tidier way of looking at the expression for A_p in statements (2.9); see if you can spot it.

Problem 2.3 (b) Behaviour of Fourier-series coefficients for large p

The following are 'bookwork' properties of the coefficients A_p that give the Fourier-series expansion of $V(t)$:

- if the function $V(t)$ possesses vertical steps, the coefficients A_p behave like (something)/p as $p \to \infty$;
- if function $V(t)$ has kinks but not steps (i.e. discontinuities in its first derivative but not in the function itself), the coefficients behave like (something)/p^2 as $p \to \infty$;
- if $V(t)$ has a discontinuity in its second derivative (but none in the first or in the function itself), the coefficients behave like (something)/p^3,

and so on.

A proof will not be asked for here in the Fourier-series case because we'll construct one in problem 2.17 for the more important case of Fourier transforms.

Think about the way in which the sinewaves of eqn (2.1) *rebuild* a $V(t)$ with a step singularity. Interpret the statements in this question in terms like 'you need high frequencies to rebuild the step'.

Problem 2.4 (b) The Fourier integral as the limit of a Fourier series

Let $V(t)$ represent a periodic waveform like that drawn in Fig. 2.1, i.e. built up by repeating a 'building block' at regular intervals with period τ. Then we can represent $V(t)$ by building it up out of a Fourier series. Now keep the shape and size of each building block unaltered, but stretch the spaces between the blocks to let $\tau \to \infty$. In this limit, $A_p \to 0$, while (τA_p) remains finite, and $\Delta\omega = (\omega_p - \omega_{p-1}) \to 0$. In words: as the periodic time τ between repetitions of the 'building block' increases, the amplitude at any fixed frequency tends to zero, while the discrete frequency components of the spectrum merge towards a continuous distribution.

(1) Show that at all stages in the above procedure, even at the beginning, the expression for (τA_p) is best thought of as identical to the continuous function $\widetilde{V}(\omega)$ (whose form is independent of τ), just 'sampled' at the discrete places $\omega = p\omega_0$.

(2) Show that as $\tau \to \infty$ the sampling points $p\omega_0$ crowd closer together until the Fourier-series sum in eqn (2.9) goes over into an integral, which is the Fourier (reconstructing = synthesis) integral of eqn (2.12).

Problem 2.5 (a) Basic mathematical properties of the Fourier transform

(1) Show that when $V(t)$ is real, $\widetilde{V}(\omega)$ is in general complex, but is restricted by the symmetry property (2.14): $\widetilde{V}(-\omega) = \widetilde{V}(\omega)^*$.

(2) Show that when $V(t)$ is real $|\widetilde{V}(\omega)|^2$ is an even function of ω.

(3) Show that when $V(t)$ is an even function of t, then $\widetilde{V}(\omega)$ is an even function of ω; and that if additionally $V(t)$ is real then so is $\widetilde{V}(\omega)$.

(4) Show that when $V(t)$ is an odd function of t, then $\widetilde{V}(\omega)$ is an odd function of ω; and that if additionally $V(t)$ is real then $\widetilde{V}(\omega)$ is pure imaginary.

(5) Link parts (1)–(4) back to corresponding properties of the Fourier-series coefficients A_p: eqns (2.5) and (2.8).

Problem 2.6 (a) Negative frequencies?

What was this in §2.5 about 'building up U from the positive-frequency part of V'? Are there such things as 'negative frequencies'? Certainly, ω is allowed to be negative in eqn (2.12) ....[31]

[31] *Hint*: It's a well known fact (sometimes flippantly called the Second Law of Thermodynamics) that you can't get something out of the end of a piece of mathematics that didn't go in at the beginning. Statement (2.10) contained positive ω only. And the same was true of eqn (2.2).

Problem 2.7 (a) Basic properties of the δ-function

(1) Define $\delta_N(x) \equiv \dfrac{1}{2\pi}\displaystyle\int_{-N}^{N} \mathrm{e}^{-\mathrm{i}xy}\,\mathrm{d}y$. Show that $\delta_N(x) = \dfrac{1}{\pi}\dfrac{\sin Nx}{x}$.

(2) Show that $\delta_N(0) = N/\pi$, so it tends to ∞ as $N \to \infty$.

(3) Show that $\int_{-\infty}^{\infty} \delta_N(x)\,\mathrm{d}x = 1$ independent of N. You may assume that $\int_{-\infty}^{\infty} \{(\sin x)/x\}\,\mathrm{d}x = \pi$, though if you have learned how to do contour integration you may like to construct a proof of this too.

(4) Use the properties obtained in parts (2) and (3) to construct sketch graphs of $\delta_N(x)$, as functions of x for increasing values of N. Argue from these graphs that $\delta_N(x)$ more and more resembles the δ-function $\delta(x)$ as N is made to approach infinity.[32]

[32] *Comment*: It is not trivial to deal with the 'wiggles' of $(\sin Nx)/x$. We really need to go back to eqn (2.18) and think of what would happen if we integrated the wiggly function multiplied by a smooth $f(x)$. As with the 'proof' of Fourier's theorem in problem 2.8, it's not the intention here to insist on mathematical rigour (or at least not when it's difficult), but rather to provide a way of feeling comfortable with what is being presented.

(5) Show that the $\delta(x)$ defined by eqn (2.19) is real. Therefore it does not matter whether the integrand contains $\mathrm{e}^{-\mathrm{i}xy}$ or $\mathrm{e}^{\mathrm{i}xy}$.

Problem 2.8 (b) The Fourier transform: there and back

In this problem I shall lead you by the hand to obtaining a 'proof' of the Fourier transform relations. The word 'proof' is in quotation marks, because it is not being claimed that a rigorous mathematical proof is in prospect. That is not the point. We *know* that proofs of Fourier's theorem exist, and we can trust our mathematician colleagues to have got them right. The intention is the reverse: it is to present some working that uses the δ-function and leads to a known correct result; and thereby to build confidence in the δ-function and in manipulations involving it.

We have two statements in eqns (2.12). We can use one of them to define $\widetilde{V}(\omega)$ without begging any questions. What we must prove is that the *other* equation correctly follows. We are to show then that

$$V(t) = \int_{-\infty}^{\infty} \widetilde{V}(\omega)\, \mathrm{e}^{-\mathrm{i}\omega t} \frac{\mathrm{d}\omega}{2\pi} \qquad \text{implies} \qquad \widetilde{V}(\omega) = \int_{-\infty}^{\infty} V(t)\, \mathrm{e}^{\mathrm{i}\omega t}\, \mathrm{d}t.$$

It's important here to keep a clear distinction between the ω in the first statement, which is a 'dummy' variable being integrated over, and the ω in the second statement, which is an independent variable upon which the answer depends. The first essential for keeping a clear head is therefore to invent a new symbol, say s, for the dummy variable:

$$V(t) = \int_{-\infty}^{\infty} \widetilde{V}(s)\, \mathrm{e}^{-\mathrm{i}st} \frac{\mathrm{d}s}{2\pi}.$$

Multiply through by $\mathrm{e}^{\mathrm{i}\omega t}$, where ω is some constant (so far as s and t are concerned):

$$V(t)\, \mathrm{e}^{\mathrm{i}\omega t} = \int_{-\infty}^{\infty} \widetilde{V}(s)\, \mathrm{e}^{-\mathrm{i}(s-\omega)t} \frac{\mathrm{d}s}{2\pi} = \int_{-\infty}^{\infty} \widetilde{V}(\omega + r)\, \mathrm{e}^{-\mathrm{i}rt} \frac{\mathrm{d}r}{2\pi},$$

where $r = s - \omega$. Check the correctness of the following:

$$\begin{aligned}
\int_{-\infty}^{\infty} V(t)\, \mathrm{e}^{\mathrm{i}\omega t}\, \mathrm{d}t &= \int_{-\infty}^{\infty} \int_{-\infty}^{\infty} \widetilde{V}(\omega + r)\, \mathrm{e}^{-\mathrm{i}rt} \frac{\mathrm{d}r}{2\pi}\, \mathrm{d}t \\
&= \int_{-\infty}^{\infty} \widetilde{V}(\omega + r) \left(\frac{1}{2\pi} \int_{-\infty}^{\infty} \mathrm{e}^{-\mathrm{i}rt}\, \mathrm{d}t \right) \mathrm{d}r \\
&= \int_{-\infty}^{\infty} \widetilde{V}(\omega + r)\, \delta(r)\, \mathrm{d}r \\
&= \widetilde{V}(\omega).
\end{aligned}$$

This is the required result, and it does something to fill in the steps missing from the discussion leading up to eqns (2.12).

Run the above argument in reverse. Define $\widetilde{V}(\omega)$ by the second equation of (2.12) and use it to derive the first equation.

Problem 2.9 (a) The Parseval theorem

(1) Prove the Parseval theorem, starting by substituting for *one* of the factors $V(t)$ on the left-hand side of eqn (2.20), and leaving the other alone. The calculation should take about three lines.

(2) [Harder] As an exercise in the use of the δ-function, prove the Parseval theorem by a longer route. Substitute for both factors $V(t)$ on the left-hand side of eqn (2.20), in terms of integrals over $\widetilde{V}(\omega)$. Remember that the two integrals over angular frequency must be written with different 'dummy variables'.

(3) The version of the Parseval theorem quoted in eqn (2.20) is one in which it is acceptable to take $V(t)$ as real. Guess the shape of a more general Parseval theorem that applies to the case where $V(t)$ is permitted to be complex. Set up a proof, following the techniques used in this problem, to see if you are right.

Problem 2.10 (a) Differences between U and V
There are some not-very-obvious consequences of the definitions made in §2.5. This problem explores some of these, which can be surprisingly muddling if attention is not drawn to them.

(1) Show from definition (2.15) that

$$\widetilde{U}(\omega) = \begin{cases} 2\widetilde{V}(\omega) & \text{for } \omega > 0, \\ 0 & \text{for } \omega < 0. \end{cases} \tag{2.34}$$

(2) Show that $\langle U(t)^2 \rangle = 0$ and $\langle U(t)\,U(t+\tau)\rangle = 0$, where $\langle\,\rangle$ may represent either a time average over a long time T or an ensemble average over many random wavetrains. (Similar results are *not* true for V.)

(3) Show that the intensity of an electromagnetic wave whose electric field is represented by the analytic signal $U(t)$ is[33]

$$I = \frac{n}{Z_0}\frac{1}{T}\int_{-T/2}^{T/2} \frac{|U(t)|^2}{2}\,\mathrm{d}t = \frac{n}{Z_0}\frac{|U(t)|^2}{2}. \tag{2.35}$$

This expression may be contrasted with that at the beginning of §2.8; notice the difference by a factor 2.

(4) Show that the power spectrum $P(\omega)$ is given in terms of U by[34]

$$P(\omega) = \frac{2n}{Z_0}\frac{1}{T}\frac{|\widetilde{U}(\omega)|^2}{4}.$$

Problem 2.11 (b) Shifting theorem
You might be surprised to see the factor $\exp(\frac{1}{2}\mathrm{i}\omega\tau_\mathrm{w})$ appearing in the $\widetilde{V}(\omega)$ of eqn (2.23). It happened because the origin of t was chosen to lie at one end of the pulse instead of in the middle. A general result immediately suggests itself. Show that if $V(t)$ has Fourier transform $\widetilde{V}(\omega)$, then the 'shifted function' $V(t-t_0)$ has transform $\exp(\mathrm{i}\omega t_0)\,\widetilde{V}(\omega)$.

Explain why the last step of eqn (2.19) is sloppy mathematics.

Problem 2.12 (b) The Fourier transform has unexpected uses

(1) Consider again the single rectangular pulse of §2.9.1. Use the reverse Fourier transform for the special time $t = \frac{1}{2}\tau_\mathrm{w}$ to show that[35]

$$\int_{-\infty}^{\infty} \frac{\sin x}{x}\,\mathrm{d}x = \pi.$$

(2) Prove that $\int_{-\infty}^{\infty}\{(\sin x)/x\}^2\,\mathrm{d}x = \pi$, using three different methods:[36]
(a) from the result of part (1) of this problem, integrating by parts

[33] The averaging over T is taken over a time interval that is only just long enough to remove terms in $(U^* + U)^2$ containing $\mathrm{e}^{\pm 2\mathrm{i}\omega t}$, in accordance with the understanding in sidenote 14 on p. 5. It is therefore correct that the final expression here contains $|U(t)|^2$, shown as still dependent upon t.

[34] *Comment*: Compare this expression for $P(\omega)$ with eqn (2.22). There is a factor 4 that comes from the difference in the definitions of V and U. But there is a more confusing difference also. The two expressions for $P(\omega)$ agree when ω is positive, as they should. However, the $P(\omega)$ of eqn (2.22) is formally an even function of ω (see problem 2.5(2)), although, of course, we attach a physical interpretation to the 'positive-frequency part' only. But the $P(\omega)$ given in part (4) of problem 2.10 is zero for $\omega < 0$. Mathematics involving $U(t)$ and $V(t)$ can sometimes give one $P(\omega)$, sometimes the other; and although the difference ought to be trivial it can cause muddle.

[35] It looks as though a circular argument has been made, since in problem 2.7(3) this integral was used to show that the area under $\delta_N(x)$ is 1, and that result was in turn used in problem 2.8 to 'prove' the Fourier theorem. However, we know that the Fourier theorem can be proved by much more rigorous mathematical methods than those given here, and that knowledge breaks the circle.

[36] *Comment*: Yet another route will be encountered in problem 2.20.

(b) by using the Parseval theorem (2.20) on the single rectangular pulse

(c) by finding the Fourier transform of the 'triangle pulse' defined by

$$V(t) = \begin{cases} V_0(1 - |t|/\tau) & \text{for } -\tau < t < \tau \\ 0 & \text{for all other } t, \end{cases}$$

and then transforming back for the special case $t = 0$.

Problem 2.13 (b) Checking the examples

(1) Work all the algebra in §2.9.

(2) Use the 'shifting theorem' of problem 2.11 to show that the factor $e^{i\alpha}$ disappears from eqn (2.26) if the origin of t is moved to the middle of the array.

(3) Consider the case of a double pulse with $\tau_W \ll \tau_D$: the pulses are narrow in relation to their spacing. Show that in this limit the single-pulse factor within eqn (2.24) is so wide that it encompasses many oscillations of the cosine term; the description 'envelope' is highly appropriate for the single-pulse factor in this case.

(4) Consider the case of a double pulse where $\tau_D = 2\tau_W$. Show that two periods of the cosine still fit within the central maximum of the envelope. Even in this case, the envelope is a slower-varying function of ω than is the cosine.

(5) Make an analysis of what happens if the two pulses of a double pulse are made so wide that the gap between them is just on the point of disappearing.

Problem 2.14 (a) Fourier analysis by inspection

Look at the single pulse of Fig. 2.2(a) and its transform (eqn 2.23). A conspicuous feature of the transform is the fact that it has zeros where $\sin(\frac{1}{2}\omega\tau_W) = 0$ (except at $\omega = 0$). These frequencies are—it appears—absent from the original $V(t)$. There ought to be a way of 'seeing' a result as dramatic as this. For the whole of this problem, you should *not* repeat the algebra done previously, but use sketch graphs.

(1) Draw the top-hat function $V(t)$; it is easiest to think about if the pulse is centred on the origin. Also draw (underneath and on the same horizontal scale) a sine wave $\sin(\omega t)$ with $\frac{1}{2}\omega\tau_W = \pi$. Think about whether this sine wave would be useful when adding up sine waves in doing a *synthesis* aimed at reconstructing the original $V(t)$. Convince yourself that this particular sine wave 'hinders as much as it helps'.

(2) Draw a $\cos(\omega t)$ wave with the same frequency as that of part (1), and use it in a similar argument.

(3) Use the sketch graphs of parts (1) and (2) to show, by picture-drawing the integrand, not by algebra, that the integral defining the Fourier transform $\widetilde{V}(\omega)$ is zero for this special ω. Things are consistent.

(4) Draw similar sketch graphs of sines and cosines with $\frac{1}{2}\omega\tau_W = 2\pi$, 3π. Use these graphs to show that $\widetilde{V}(\omega) = 0$ for these values of ω.

(5) Draw sketch graphs of sine and cosine waves for $\frac{1}{2}\omega\tau_{\rm W} = \pi/2$ and show that the cosine *does* make a useful contribution to the rebuilding of $V(t)$ from its constituent sinewaves.

(6) Extend these ideas to locate all the zeros of the Fourier transform of the double pulse (eqn 2.24). You know from optics that a double-slit diffraction pattern has the form of a sinc (meaning a $(\sin x)/x$ function) multiplied by a cosine, so sketch that roughly. Use a procedure modelled on parts (1)–(5) to locate the zeros of the cosine and of the sinc function. Hence, construct a sketch graph of $\widetilde{V}(\omega)$ with all significant points identified and marked on the horizontal axis—without redoing any integrals or looking anything up.

(7) Return to part (3) and work out the connection between it and the 'non-calculus method' for finding the first diffraction zero of light passing through a single slit. (By the non-calculus method is meant pairing Huygens secondary sources that cancel, rather than working out an integral—see problem 3.6.)

Problem 2.15 (b) A random array of δ-functions
Prove the properties (2.27) of a random array of δ-functions. You should have seen what is effectively a proof of this in an elementary course when it was shown that 'if the phases are random, add intensities'. If not, try to construct a proof now, using this cross-link as a hint.[37]

Statements (2.27) are actually not quite correct, because they go wrong for 'sufficiently small' ω. Think about why.[38]

Problem 2.16 (a) Transform of a sinewave
Prove eqn (2.28).[39]

Derive the 'shifting theorem' for the inverse transform.

Problem 2.17 (b) The Fourier transform for large ω
Look at the examples of Fourier transforms in §2.9. The first two examples show a $V(t)$ with a 'cornery' behaviour; the Fourier transforms $\widetilde{V}(\omega)$ behave like $1/\omega$ for large ω. We can understand that $\widetilde{V}(\omega)$ falls so slowly because the steps in $V(t)$ are 'difficult' to reproduce when $V(t)$ is reconstructed by adding up sinewaves. Delta-functions are even more difficult. Conversely, functions $V(t)$ with friendlier behaviour than steps are 'easier' to reconstruct. Let's now make this idea formal.

(1) Obtain $\widetilde{V}(\omega)$ again for the rectangular pulse: integrate by parts in such a way that the pulse gets differentiated. Its differential consists of a δ-function at each end and nothing in the middle. Notice that a further integration by parts can't be attempted because you can't (sensibly) differentiate a δ-function. Notice that a $1/\omega$ appeared necessarily in the course of the one integration by parts. Evaluate $\widetilde{V}(\omega)$ this way.

(2) Take a general $V(t)$ possessing a step discontinuity at $t = t_0$ (and no worse behaviour), and show that its $\widetilde{V}(\omega)$ falls like $1/\omega$ for large ω.

(3) Show that if $V(t)$ is continuous everywhere, but has a kink at $t = t_0$ (i.e. a step discontinuity in its first derivative), then $\widetilde{V}(\omega)$ falls like $1/\omega^2$

[37] *Comment*: There's another similarity: to a random walk. Each δ-function contributes a term like $e^{i\omega t_0}$ to the transform, and there is a random distribution of the t_0s (compare eqn (2.25)). You can represent these complex numbers as 'vectors' (the correct name is 'phasors') in the complex plane. All phasors here have unit length but their directions are random. If you were to draw the phasors end to end, you'd have a drawing of a two-dimensional random walk, all steps of equal length but taken in random directions. This makes us think about other random walks, such as the movement of a molecule in gas kinetic theory. The distance ξ travelled in time t (in the x-direction by a molecule) is a random variable obeying $\langle\xi^2\rangle = 2Dt$, where D is the diffusion coefficient and t is time. The mean square displacement is proportional the number of steps taken. The similarity is strong.

[38] The equivalent of this in Fraunhofer diffraction turns up in problem 4.16(2).

[39] *Hint*: It's easiest to work from right to left.

for large ω.

(4) Show that if $V(t)$ has a discontinuity somewhere in its pth derivative (and no worse singularity), $\widetilde{V}(\omega)$ falls like ω^{-p-1} for large ω.

(5) Argue that if $V(t)$ has no discontinuity in *any* of its derivatives, then transform $\widetilde{V}(\omega)$ must fall for large ω faster than any power of ω^{-1}. Think of an example of such a function and check that it has the property argued for.

(6) To complete the list, think finally about singularities worse than steps. A δ-function has transform given by eqn (2.25), which does not fall at all with increasing ω. The same of course applies to an array of two or more δ-functions as in eqn (2.26).

Problem 2.18 (a) The convolution theorem

(1) Prove the convolution theorem.

(2) Show that the integral defining the convolution in eqn (2.29) can equally well contain $\psi(t')\,\phi(t-t')$; it does not matter which way round ϕ and ψ are written.

Problem 2.19 (b) The double pulse as a convolution
Derive the Fourier transform of a double pulse:

(1) by direct integration over the two pulses;

(2) by considering first a single pulse centred on the origin, then applying the result of problem 2.11 to find the transform of a single pulse shifted to be centred on $-\frac{1}{2}\tau_{\mathrm{D}}$, next doing the same to find the transform of a single pulse centred on $+\frac{1}{2}\tau_{\mathrm{D}}$, and finally adding the contributions of the two shifted pulses;

(3) by expressing the double pulse as the convolution of a single pulse with a function consisting of two δ-functions, and applying the convolution theorem.

Problem 2.20 (b) The triangle pulse as a convolution
Show that the 'triangular pulse' of problem 2.12(2)(c) can be constructed from the convolution of two identical rectangular pulses. Use this to obtain the transform of the triangular pulse from that of the rectangular pulse, and check you agree with the direct calculation in problem 2.12.

Problem 2.21 (b) More convolutions
Functions $f(t)$, $g(t)$, which need not be real, possess Fourier transforms $F(\omega)$, $G(\omega)$.

(1) Show that:

(a) $F(-\omega)$ is the transform of $f(-t)$

(b) $F(\omega)\,G(\omega)$ is the transform of convolution $\int f(t')\,g(t-t')\,\mathrm{d}t'$

(c) $F(\omega)\,G(-\omega)$ is the transform of correlation $\int f(t')\,g(t'-t)\,\mathrm{d}t'$.

(2) If these results are applied now to real functions of t, all functions of ω found here should obey eqn (2.14). Do they?

(3) Apply part (1)(c) to the special case where functions $f(t)$ and $g(t)$ are the same real function. Show that the correlation integral (it is now

called an autocorrelation function) is an even function of t and that its transform is $|F(\omega)|^2$. We have just proved the Wiener–Khintchine theorem (§10.6).

(4) Take the special case $t = 0$ in part (3), and you should be able to make a new proof of the Parseval theorem.

Problem 2.22 (e) Frequency dependence implies a short-term memory

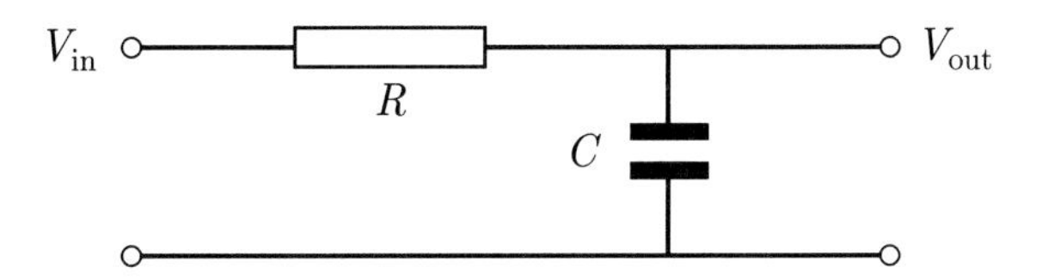

Fig. 2.4 The electrical RC circuit discussed in problem 2.22. This circuit illustrates the 'memory' possessed by anything that has a frequency dependence.

Consider the electrical filter shown in Fig. 2.4; it is perhaps the simplest physical system we can invent that introduces a frequency dependence. If we think about the circuit in the usual way, we shall assume that voltages v and currents i are proportional to $e^{j\omega t}$, and then it is easy to show that

$$v_{out}(\omega) = \frac{1}{1 + j\omega RC}\, v_{in}(\omega).$$

So $v_{out}(\omega)$ is a product of two functions of ω. By the convolution theorem, used backwards, $V_{out}(t)$ ought to be a convolution of two functions of time. Really ...?

(1) Go back to circuit-theory first principles and avoid using $e^{j\omega t}$. Show that the circuit equation is

$$RC\frac{dV_{out}}{dt} + V_{out}(t) = V_{in}(t).$$

(2) This differential equation can be solved without mention of $e^{j\omega t}$ by using the integrating-factor method; the integrating factor is $e^{t/RC}$. Use this to obtain

$$V_{out}(t) = \int_{-\infty}^{\infty} K(t - t')\, V_{in}(t')\, dt',$$

where
$$K(t) = \begin{cases} (1/RC)\, e^{-t/RC} & \text{for } t > 0 \\ 0 & \text{for } t < 0. \end{cases}$$

(3) Interpret the physical meaning[40] of this dependence of $V_{out}(t)$ on $V_{in}(t)$; consider separately the contributions from $t' < t$ and $t' > t$.

(4) Use the convolution theorem to find the Fourier transform $\widetilde{V}_{out}(\omega)$ of $V_{out}(t)$, and show that[41]

$$\widetilde{V}_{out}(\omega) = \frac{1}{1 - i\omega RC}\, \widetilde{V}_{in}(\omega). \tag{2.36}$$

This equation is almost identical—in form—to the equation first written in this problem. The main difference is the replacement of $+j$ by $-i$ in the denominator. Think about the sign conventions associated with the use of i and j, as recommended in §2.12, and reconcile the two equations being compared.

[40] Something like: 'a weighted average of recent inputs', or 'a short-term memory of past inputs'—and no knowledge of the future!

[41] *Comment*: You have been investigating the physics behind a relation of the form

$$v_{out}(\omega) = G(\omega)\, v_{in}(\omega).$$

A system with such a behaviour has—we now see—a 'memory' of recent inputs. You are invited to believe that this idea carries over to all other cases that involve a *linear input–output relation* with a frequency-dependent coefficient. For example, this interpretation should apply to

$$\boldsymbol{P}(\omega) = \varepsilon_0\, \chi(\omega)\, \boldsymbol{E}(\omega),$$

linking the electric dipole moment density $\boldsymbol{P}$ to the electric field $\boldsymbol{E}$ in a dispersive medium.

Returning to the first equation of this comment, the $G(\omega)$ might represent the gain of some amplifier as a function of frequency, or it might be the transmission function of some passive filter. So linear input–output relations are extremely widespread. Moreover, once we have our attention drawn to the inability of $K(t - t')$ to give knowledge of the future, Kramers–Kronig relations are not very far away (look them up).

(5) Was it necessary[42] to make a notational distinction between $v_{\text{out}}(\omega)$ and $V_{\text{out}}(\omega)$?

[42] *Hint*: Problem 2.10.

(6) [Much harder] Reverse the procedure of part (4), and find $K(t)$ from its transform. This may require a contour integration.

(7) [Also harder] We might have a lurking suspicion that everything found in this problem is something of a fluke, rather than offering a deep insight. To show that it isn't a fluke, make your own analysis of an *LRC* circuit, and show that it too has a $K(t-t')$ with physically reasonable properties.

Problem 2.23 (b) The convolution theorem for the reverse transform
Show that the convolution theorem that carries us 'backwards' is[43]

$$\widetilde{V}(\omega) = \int_{-\infty}^{\infty} \widetilde{\phi}(\omega')\,\widetilde{\psi}(\omega-\omega')\frac{\mathrm{d}\omega'}{2\pi} \quad \text{implies} \quad V(t) = \phi(t)\times\psi(t). \quad (2.37)$$

[43] *Comment*: It is a bit irritating that this differs in shape from eqn (2.30) by a factor 2π. This has happened because the Fourier relations given here contain an asymmetry, with $1/2\pi$ in the reverse transform and none in the forward transform. See sidenote 9 on p. 26.

Problem 2.24 (c) The Fourier transform for a periodic function
Return to the periodic waveform $V(t)$ of Fig. 2.1, where the building block is called $\phi(t)$.

(1) We are not forbidden from working out a Fourier transform just because $V(t)$ is periodic. What do you think the transform $\widetilde{V}(\omega)$ looks like? Now continue this problem to see if you were right.

(2) Show that $V(t)$ can be expressed in the form

$$V(t) = \sum_{p=-\infty}^{\infty} \phi(t-p\tau) = \int_{-\infty}^{\infty} \phi(t-t')\sum_{p=-\infty}^{\infty}\delta(t'-p\tau)\,\mathrm{d}t'.$$

Attention is now thrown onto the infinite sum over δ-functions.

(3) Show[44] that the transform of the sum over δ-functions is

$$\frac{2\pi}{\tau}\sum_{m=-\infty}^{\infty}\delta\left(\omega - m\frac{2\pi}{\tau}\right).$$

[44] The reasoning is not trivial, hence the (c) grading of this problem. I suggest you start by taking a sum over δ-functions from $p=-N$ to $p=N$, rather as in eqn (2.32), and then take the limit $N\to\infty$. You may find help in problems 4.2 and 2.7.

Thus, the transform consists of an infinite array of δ-functions at angular frequency $\omega_0 = 2\pi/\tau$ and its harmonics.[45]

[45] Given a 'reasonable' building block $\phi(t)$, there will be an 'envelope' $\widetilde{\phi}(\omega)$ that tapers off the δ-functions and makes it unnecessary for us to work with an infinity of them.

Problem 2.25 (e) The uncertainty principle
The uncertainty principle of quantum mechanics is rooted in Fourier mathematics.

Find a textbook derivation[46] of the uncertainty principle that proves

$$\Delta x \times \Delta p_x \geq \tfrac{1}{2}\hbar.$$

[46] A possible reference is Landau and Lifshitz (1977) §16. The Landau–Lifshitz reasoning assumes that $\psi(x)$ describes a particle with $\overline{x}=0$ and $\overline{p_x}=0$. A more versatile calculation is given by Schiff (1968), pp. 60–61.

Here Δx is the standard deviation for measurements of position, defined by

$$(\Delta x)^2 = \int_{-\infty}^{\infty} \psi^*(x)\,(x-\overline{x})^2\,\psi(x)\,\mathrm{d}x,$$

and Δp_x is defined similarly. What do you think is the definition of $\overline{x}$?

Rework the book's calculation to remove all mention of Planck's constant, and show that

$$\Delta x \times \Delta k_x \geq \tfrac{1}{2}, \tag{2.38}$$

where k_x is the Fourier-transform variable accompanying x. Equivalently, if you work with t and ω instead of x and k_x,

$$\Delta t \times \Delta\omega \geq \tfrac{1}{2}. \tag{2.39}$$

Statement (2.39) sets a least value of the product $(\Delta t)(\Delta\omega)$, and it is shaped to look like an uncertainty principle in the quantum-mechanical mould. Nevertheless it is *not* quantum mechanical: there is no mention of Planck's constant in either the result or (if you did what was intended) in its derivation. The same applies to eqn (2.38).

Comment: The uncertainty principle is often presented as one of the most fundamental statements in quantum mechanics. Well it is. At the same time we can take a slightly off-centre view. De Broglie showed that a wavelength λ is associated with a momentum, via $p = h/\lambda$ or equivalently $k_x = 2\pi/\lambda = p_x/\hbar$. There must be a Fourier-transform relationship between any function of k_x and its transform-function of x, and then $\Delta x \times \Delta k_x \geq \frac{1}{2}$ and $\Delta x \times \Delta p_x \geq \frac{1}{2}\hbar$. Such Fourier-transform statements are inescapable as soon as it's accepted that we are dealing with a wave. Perhaps we should think of the uncertainty principle as a consequence of having waves, rather than as a stand-alone founding principle?

3 Diffraction

3.1 Introduction

In this chapter we introduce diffraction using the concept of Huygens secondary sources. At the same time, we take the attitude that the whole subject must be grounded on Maxwell's equations. We do what we can at our level of treatment to make the connection, and look critically at the customary assumptions.

The whole subject of diffraction is, of course, itself founded on an understanding of simple cases of interference, such as the Young slits. It is assumed that the reader has acquired this foundation from a more elementary treatment of optics.

3.2 Monochromatic spherical wave

In §1.6 we introduced an expression for the scalar amplitude $V(\boldsymbol{r}, t)$ of a monochromatic spherical wave:

$$V(\boldsymbol{r}, t) = V_0 \frac{\cos(kr - \omega t - \alpha)}{r}.$$

Amplitude $V(\boldsymbol{r}, t)$ is a solution of the scalar wave equation, and represents a wave travelling radially outwards[1] from an origin at $r = 0$. As written, it is monochromatic, with a single (angular) frequency ω; it may, of course, be one Fourier component of something more complicated. As is almost always the case, it is mathematically more convenient to express the amplitude by means of the analytic signal $U(\boldsymbol{r}, t)$:

$$U(\boldsymbol{r}, t) = V_0 \frac{\mathrm{e}^{\mathrm{i}(kr - \omega t - \alpha)}}{r} = U_0 \frac{\mathrm{e}^{\mathrm{i}(kr - \omega t)}}{r}, \qquad \text{where} \qquad U_0 = V_0\, \mathrm{e}^{-\mathrm{i}\alpha}. \tag{3.1}$$

[1] The wave vector k is here being taken to be positive.

In §§3.3–3.5 it will be convenient to drop the factor $\mathrm{e}^{-\mathrm{i}\omega t}$ to save writing, so we further define a space-dependent part ψ, also complex, by

$$U(\boldsymbol{r}, t) = \psi(\boldsymbol{r})\, \mathrm{e}^{-\mathrm{i}\omega t}.$$

3.3 The Kirchhoff diffraction integral

An arrangement for exhibiting diffraction is shown in Fig. 3.1. In elementary discussions of diffraction, it is assumed that an element $\mathrm{d}S$ of area, illuminated by the incident wave, acts as a **Huygens secondary**

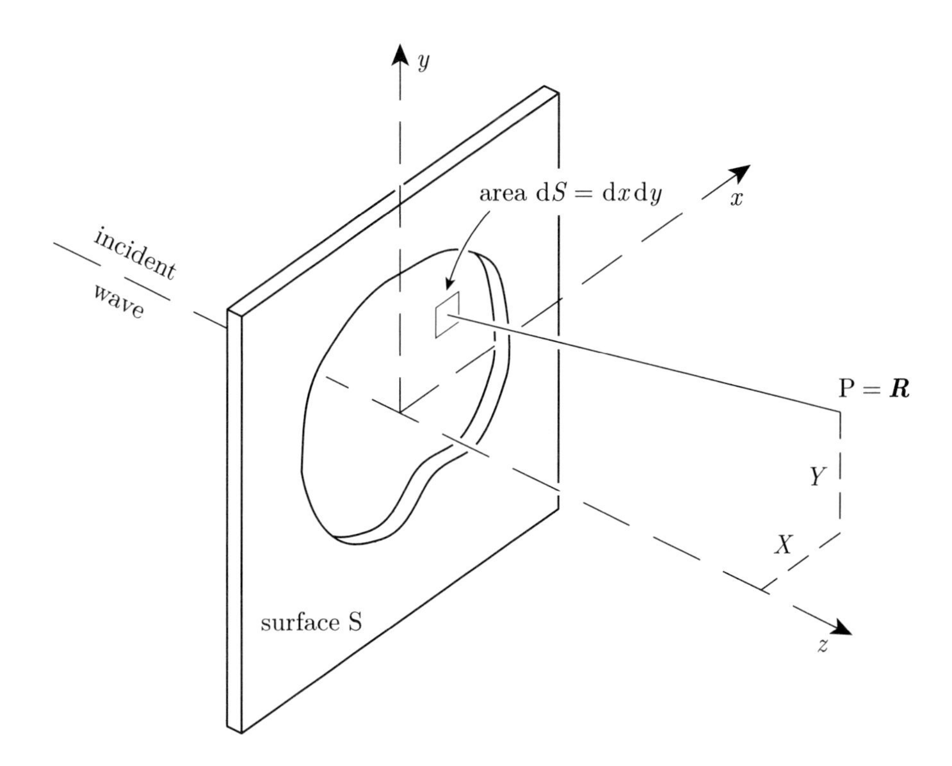

Fig. 3.1 An electromagnetic wave arriving from top left illuminates a surface S, part of which may be obscured by the opaque part of an aperture. An element of area $\mathrm{d}S$ at $\boldsymbol{r}$ acts as a Huygens secondary source that radiates spherical waves. Point P at $\boldsymbol{R}$ is a representative point at which secondary waves are summed to obtain the diffracted amplitude. For convenience of drawing and discussion, (x, y, z) coordinates are used and the secondary sources are taken to lie on the xy plane so $\boldsymbol{r} = (x, y, 0)$, though this is not a necessary restriction.

source and radiates **Huygens secondary waves**. Plausibility arguments give

$$\psi_{\text{dif}}(\boldsymbol{R}) \propto \int \psi_{\text{trans}}(\boldsymbol{r})\, \mathrm{e}^{\mathrm{i}k|\boldsymbol{R}-\boldsymbol{r}|}\, \mathrm{d}S, \tag{3.2}$$

where $\psi_{\text{trans}}(\boldsymbol{r})$ is the field amplitude at $\boldsymbol{r}$ on the secondary-source surface, and $\psi_{\text{dif}}(\boldsymbol{R})$ is the amplitude diffracted to a representative 'field point' P at location $\boldsymbol{R}$. The integration is taken over the whole area that contributes a non-zero $\psi_{\text{trans}}(\boldsymbol{r})$. At our present level of treatment, we should ask what happens if we try to derive eqn (3.2), or something like it, starting from the Maxwell equations.

The Maxwell equations are known (§1.3) to lead to wave equations for the Cartesian components E_x, E_y, E_z of field $\boldsymbol{E}$, and likewise for the components of $\boldsymbol{H}$. We also know (problem 1.6) that, when polarization isn't important and waves arrive from only a small range of directions, it is adequate to represent the electric-field amplitude by a scalar $V(\boldsymbol{r}, t)$; more conveniently by a complex scalar (analytic signal) $U(\boldsymbol{r}, t)$. The amplitudes V and U both obey the scalar wave equation (1.24).

It can be shown that an exact solution of the scalar wave equation is

$$\psi_{\text{dif}}(\boldsymbol{R}) = \frac{1}{4\pi} \int \Big(G(\boldsymbol{R}-\boldsymbol{r})\, \nabla\psi_{\text{trans}} - \psi_{\text{trans}}(\boldsymbol{r})\, \nabla G(\boldsymbol{R}-\boldsymbol{r}) \Big) \cdot \mathrm{d}\boldsymbol{S}. \tag{3.3}$$

Here $\mathrm{d}\boldsymbol{S}$ is an element at $\boldsymbol{r}$ of a closed surface surrounding point $\boldsymbol{R}$, with its positive direction taken 'outwards' according to the usual vector conventions; and $G(\boldsymbol{r}) = \exp(\mathrm{i}k|\boldsymbol{r}|)/|\boldsymbol{r}|$. The operator ∇ takes the gradient with respect to $\boldsymbol{r}$ holding $\boldsymbol{R}$ fixed.

Equation (3.3) shows that the electromagnetic field at $\boldsymbol{R}$ can be thought of as caused by waves arriving from all points $\boldsymbol{r}$ on the bounding surface. So the above diffraction integral

- makes respectable the idea of Huygens secondary sources originating secondary waves
- gives a full mathematical expression relating the diffracted field to the field at the aperture—at least within the limitations of a scalar-amplitude theory.

Problems 3.1 and 3.2 ask you to think about the mathematical statements given above, to see why the shape of the integral in eqn (3.3) is reasonable, and to do what you can to derive it.

3.4 The Kirchhoff boundary conditions

If we are to evaluate the integral in eqn (3.3), we need to know the value of $\psi_{\mathrm{trans}}(\boldsymbol{r})$ on the surface of the aperture. We usually assume[2] (assumptions known as the Kirchhoff boundary conditions or as the St Venant hypothesis) that

$$\psi_{\mathrm{trans}}(\boldsymbol{r}) = \begin{cases} \text{incident amplitude} & \text{for } \boldsymbol{r} \text{ in transparent area of aperture} \\ 0 & \text{for } \boldsymbol{r} \text{ on opaque part of aperture.} \end{cases}$$

These assumptions lead to a very good quantitative treatment of diffraction, and we'll use them with confidence. Nevertheless, the Kirchhoff conditions can't be the whole story; see §§3.18 and 4.10.

[2]The formulation given here takes the 'aperture' either to be opaque or to be fully transparent and non-phase-shifting: the $T(\boldsymbol{r})$ of §3.10 is 0 or 1. This is a temporary restriction, used for definiteness only. When we consider a blazed grating in Chapter 4 (for example) we shall still be using Kirchhoff boundary conditions in an obvious way, but with a $T(\boldsymbol{r})$ that is far from having the simple properties assumed here.

3.5 Simplifying the Kirchhoff integral

Step 1 Use only the active part of the surface
In the special case drawn in Fig. 3.1, the important part of the closed surface S surrounding $\boldsymbol{R}$ is the illuminated part of the xy plane,[3] and on that plane $\nabla\psi_{\mathrm{trans}} \cdot \mathrm{d}\boldsymbol{S} = -(\partial\psi_{\mathrm{trans}}/\partial z)\,\mathrm{d}x\,\mathrm{d}y$. The rest of the closed surface is the unilluminated part of the xy plane on which ψ is taken to be zero, together with a large hemisphere surrounding $\boldsymbol{R}$, whose contribution to the field at $\boldsymbol{R}$ is also zero (any contribution we might worry about tends to zero as we increase the radius of the hemisphere).

Step 2 Point source
Take another special case: a point source radiating a spherical wave $\psi(\boldsymbol{r}) = \psi_0 \exp(\mathrm{i}kr_{\mathrm{s}})/r_{\mathrm{s}}$ from the left as shown in Fig. 3.2. The diffracted amplitude $\psi_{\mathrm{dif}}(\boldsymbol{R})$ now reduces (problem 3.2(1)) to

[3]For convenience we talk as if the mathematical surface (that part of $\boldsymbol{S}$ that is illuminated) has to be plane. As mentioned in the caption to Fig. 3.1, the surface may have any reasonable shape and may have a coordinate system, not necessarily xy, appropriate to that shape.

We also talk of an 'aperture' in the xy plane, which may suggest that some obstacle obstructs part of the xy plane while allowing light to pass through some other part. It will be emphasized in §3.10 that the process of diffraction does need any obstacle to be present. So the 'aperture' may sometimes be a completely unobstructed mathematical surface.

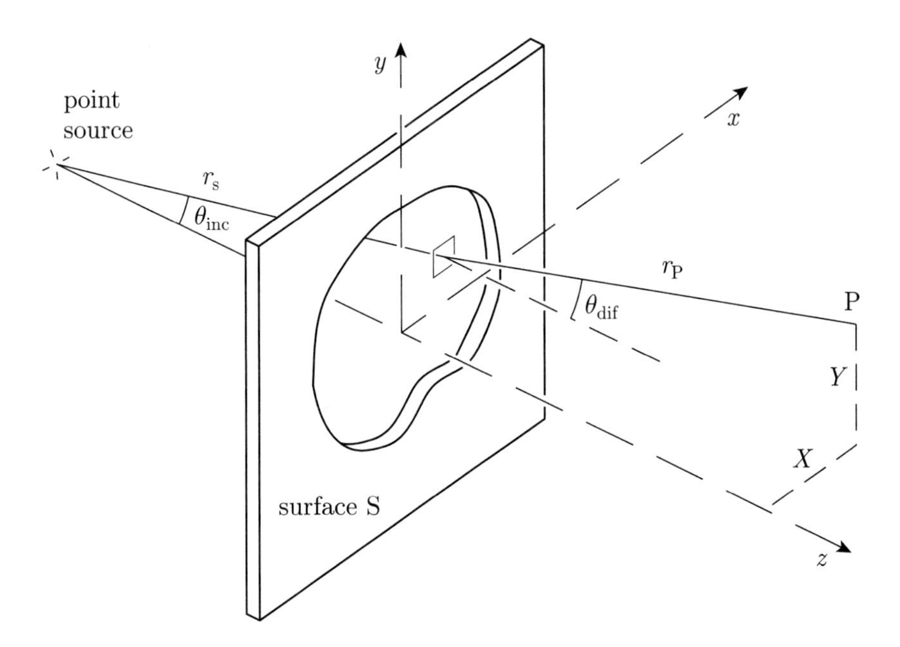

Fig. 3.2 The geometry for Step 2 of §3.5. Light from a point source illuminates area element dS of the secondary-source surface S, arriving at angle θ_{inc} to the normal. Secondary waves from dS arrive at field point P, having left dS at angle θ_{dif} to the normal. As with Fig. 3.1, it is not necessary for the secondary-source surface to be plane, but a simple case has been taken.

$$\psi_{dif}(\boldsymbol{R}) = \frac{1}{i\lambda} \int_{\text{open area}} \left(\frac{\cos\theta_{inc} + \cos\theta_{dif}}{2} \right) \psi_0 \, \frac{\exp(ikr_s)}{r_s} \, \frac{\exp(ikr_P)}{r_P} \, dS. \tag{3.4}$$

It is assumed here that $r_s \gg \lambda$ and $r_P \gg \lambda$, where λ is the wavelength of the light. The quantity $\frac{1}{2}(\cos\theta_{inc} + \cos\theta_{dif})$ is called the **obliquity factor**.

Usually, we are interested in cases where light passing through an aperture continues in the same general direction. In such cases, θ_{inc} and θ_{dif} are small angles; then $\cos\theta_{inc} \approx \cos\theta_{dif} \approx 1$ and the obliquity factor ≈ 1. We also keep the mathematics from getting too abstract by continuing to take the special case where the surface being integrated over is the xy plane. Then

$$\begin{aligned} \psi_{dif}(\boldsymbol{R}) &\approx \frac{1}{i\lambda} \int_{\text{open area}} \psi_0 \, \frac{\exp(ikr_s)}{r_s} \, \frac{\exp(ikr_P)}{r_P} \, dx \, dy \\ &= \frac{1}{i\lambda} \int \psi_{trans}(x, y) \, \frac{\exp(ikr_P)}{r_P} \, dx \, dy. \end{aligned} \tag{3.5}$$

Step 3 General incident wave
Approximation (3.5) expresses the diffracted amplitude in terms of

(a) $\psi_{trans}(x, y)$, the complex amplitude transmitted at (x, y) in the open area of the aperture and driving a secondary source there; and
(b) the geometry of optical paths r_P between each secondary source and the observation point P.

All oddment factors relating to the source side of the aperture have disappeared. Therefore, we may superpose the effects of any number of point sources, and treat eqn (3.5) as applying to any 'reasonable' incident wave. Restoring $e^{-i\omega t}$ we now have

$$U_{\text{dif}}(\boldsymbol{R}) = \frac{1}{i\lambda}\int U_{\text{trans}}(x,y)\,\frac{\exp(ikr_{\text{P}})}{r_{\text{P}}}\,dx\,dy, \tag{3.6}$$

where $U_{\text{trans}}(x,y)$ represents the amplitude transmitted at (x,y), however it originated.

Step 4 (Optional) Ignore constants and near-constants
We are often interested only in relative values of $U_{\text{dif}}(\boldsymbol{R})$ as a function of $\boldsymbol{R}$, and then we can ignore both the magnitude and phase information in the factor $1/i\lambda$. Usually also, $1/r_{\text{P}}$ varies by a negligible percentage, so we treat it as constant and ignore it too.[4] Finally, then we can justify the 'elementary-level' Kirchhoff integral

$$U_{\text{dif}}(\boldsymbol{R}) \propto \int U_{\text{trans}}(x,y)\exp(ikr_{\text{P}})\,dx\,dy, \tag{3.7}$$

which is identical in meaning to the statement (3.2) postulated in §3.3. Often in this book we shall wish to keep the extra factor $1/(i\lambda r_{\text{P}})$, so eqn (3.6) rather than eqn (3.7) will be the preferred statement of the Kirchhoff integral.

[4] This point is investigated in problem 3.7(1).

Comment 1 All the above is based on a *scalar* wave equation. This is fine when all the $\boldsymbol{E}$-field amplitudes being added at location $\boldsymbol{R}$ are near-parallel, so it doesn't matter whether we add them as vectors or as scalars. But it's easy to think of cases where that wouldn't be so. Treatments of diffraction (complicated!) exist which start from Maxwell's equations, and handle fields correctly as vectors; they don't add any insights useful to us for now.[5]

[5] But Fig. 3.20 is calculated from the full Maxwell equations so as to investigate the limits of reliability of the present treatment.

Comment 2 There is a reason in the history of diffraction theory for taking the obliquity factor rather seriously. People were anxious: how 'real' are the secondary sources? If they were real sources of radiation (like electric dipole moments induced on atoms by the incident field) they would radiate back towards the source just as strongly as they do in the forward direction. To make the difficulty as stark as possible, consider Huygens sources on a surface 'in the middle of nowhere', with no aperture to distract us. We get a forward wave downstream from the secondary-source surface because secondary waves interfere constructively;[6] a similar constructive interference should then happen for back-travelling waves. But we know there isn't any backward wave: a wave doesn't suddenly come to a lurching halt and start going backwards. Is this a fatal objection to the idea of secondary sources? and if not, how do we dispose respectably of the backward wave? The introduction of the obliquity factor (Fresnel) removed the backward wave, and with it the anxiety.

[6] An oversimplification. Problem 3.25 shows that there is considerable cancellation of secondary waves, even in the forward direction. What should be said is that cancellation is total in the backward direction, less than total forwards.

In the 'backward' direction, $\theta_{\text{dif}} \approx \pi$ and $\cos\theta_{\text{dif}} \approx -1$, making the obliquity factor close to zero. Problems 3.1 and 3.3 ask you to think round what has just been said.

Hindsight (infallible guide that it is) suggests that a rather different attitude is appropriate for us now. The mathematics, as far back as eqn (3.3), has told us that the secondary-source idea is respectable: the field at $\boldsymbol{R}$ *can* be thought of as a sum over secondary waves, spherical waves radiating out from each element of area $\mathrm{d}S$ of the secondary-source surface. Given that, a solution to the 'backward-wave problem' *had* to come out of the wave equation. So the interesting question should be not 'whether?' but 'how?'.

In this book I shall be cavalier with the obliquity factor. Usually we shall consider diffraction into near-forward directions only,[7] where the obliquity factor is close to 1. Indeed, if we were to consider angles so large that $\frac{1}{2}(\cos\theta_{\text{inc}}+\cos\theta_{\text{dif}})$ differed significantly from 1, then the fields arriving at P would need to have their $\boldsymbol{E}$ and $\boldsymbol{H}$ fields added vectorially, and we should have no business using a scalar theory.[8,9] The attitude then is: 'if it isn't 1 it's wrong'.

Comment 3 Even when we've satisfied ourselves on matters such as the scalar wave equation and the obliquity factor, there is (or should be) a niggling mistrust of the Kirchhoff boundary conditions ....

3.6 Complementary screens: the Babinet principle

Section 3.5 carries us through a sequence of approximations: from the Maxwell equations to a scalar theory; to the full version of the Kirchhoff integral (eqn 3.3); to approximations that rely on the distances r_{S}, r_{P} being much greater than a wavelength; to discarding the obliquity factor. And above all, there were the assumptions of §3.4 about the behaviour of $\psi_{\text{trans}}(\boldsymbol{r})$ on the surface spanning the aperture.

In looking back over those approximations, the step we ought to feel least happy about is the introduction of the Kirchhoff boundary conditions. When there is a physical aperture present, its open part may be only a few optical wavelengths across, and a disturbance to the electromagnetic fields generated by the presence of nearby opaque material surely can be expected to extend well into the open area. The theoretical aspects of this objection are discussed in §3.18. In the present section we take a different approach: can we create a simple testable prediction from the boundary conditions; and what does experiment have to say?

A test can be constructed by using two apertures whose opaque and transparent areas are interchanged. Such apertures are said to constitute a pair of **complementary screens.**[10] An example of such a pair of screens might be: a slit in an otherwise opaque screen; and a wire (with the same width as the slit) in an otherwise open area. The Kirchhoff boundary conditions lead directly to a testable prediction about the diffraction from complementary screens.[11]

Figure 3.3 shows in cross-section two complementary screens and the amplitudes $U_{\text{a}}(\boldsymbol{r})$, $U_{\text{b}}(\boldsymbol{r})$ assumed to occur just downstream from them.

[7] Figure 3.20 is an exception, where we exhibit the diffracted amplitude at angles up to 90° to the incident direction. But the amplitude there is calculated directly from Maxwell's equations and not from the equations of this section.

[8] We might think that light departing from a grating constitutes an exception, as it may well be diffracted through a large angle. However, waves diffracted by different grating elements can have their $\boldsymbol{E}$-fields added as scalars because they all travel in the same direction and have their $\boldsymbol{E}$-fields parallel to each other. The obliquity factor could affect the intensity of different grating orders, relative to each other and to the incident intensity. But the physics of real gratings introduces larger corrections, so the obliquity factor is not worth bothering with, even here.

[9] *Comment*: The statements here are given to highlight a danger signal: if the angles are large, we cannot expect to trust a scalar theory. Nevertheless, there are a few diffraction problems in which only one Cartesian component of $\boldsymbol{E}$ or of $\boldsymbol{H}$ is non-zero. A scalar treatment then happens to be valid for that component even though the angles are not all small. Examples may be found in Born and Wolf (1999), Chapter 11. The single slit of Fig. 3.20 is also such a happy case. Incidentally, the diffraction pattern of Fig. 3.20, calculated exactly from the full Maxwell equations, agrees very closely with Fig. 3.7, and the agreement is not made better by 'improving' Fig. 3.7 by patching in an obliquity factor.

[10] 'Complementary apertures' would be a better name, but 'screens' is standard jargon.

[11] *Comment*: It has to be admitted that the test proposed here is not in fact quite as searching as it may appear. There exists a rigorous derivation from the Maxwell equations of a Babinet principle for the case where the complementary screens are made from ideal electrical conductors. See, e.g., Born and Wolf (1999) §11.3. This result would permit both diffraction patterns to differ from those we calculate here while maintaining the Babinet relation between them.

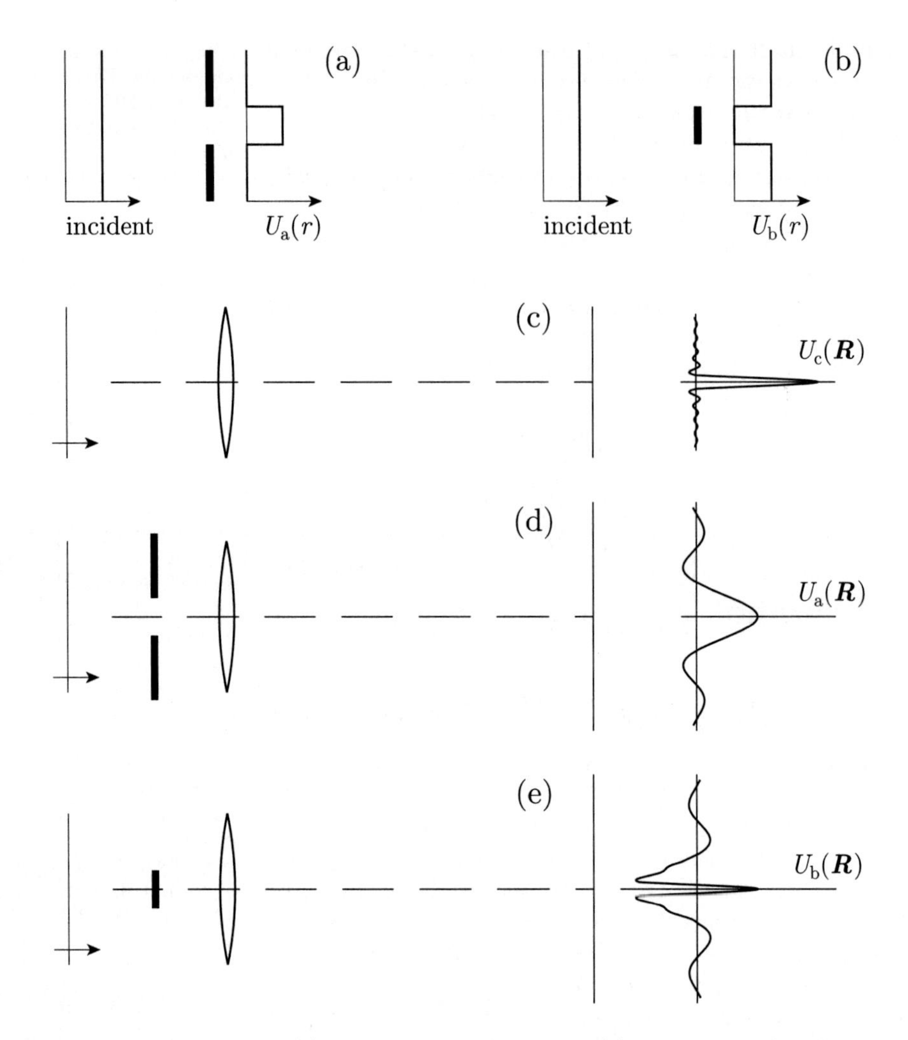

Fig. 3.3 Illustrating the Babinet principle. (a) A slit and the amplitude $U_a(\boldsymbol{r})$ it transmits when illuminated normally with a uniform plane wave. (b) A strip, complementary to the slit, and the amplitude $U_b(\boldsymbol{r})$ transmitted round it under similar conditions. (c) The Fraunhofer diffraction pattern (amplitude) $U_c(\boldsymbol{R})$ produced by a plane wave that is uninterrupted other than by the outline of the lens. In the diagram, the central peak has had both its height and the surrounding 'wiggles' de-emphasized (by different factors) to avoid clutter. (d) The amplitude $U_a(\boldsymbol{R})$ Fraunhofer-diffracted by passage through the slit of (a). (e) The amplitude $U_b(\boldsymbol{R})$ Fraunhofer-diffracted by passage past the strip of (b). The amplitude is drawn to conform to the Babinet predictions (except for distortions, similar to those of (c), in the proportions of the central feature).

Aperture (a) might be a slit whose length is normal to the paper, and aperture (b) might be a wire or strip, again with its length normal to the paper. Now $U_a(\boldsymbol{r}) + U_b(\boldsymbol{r})$ is the same as the incident amplitude $U_c(\boldsymbol{r})$. So the prediction we make, known as the Babinet principle, is this: At any place downstream from the two complementary apertures (standing in the same relation to both), *the sum of the two diffracted amplitudes will be the same as the amplitude diffracted from the unobstructed wavefront.*

The Babinet principle applies to any geometrical layout of apparatus, but its predictions are most dramatic, and most easily checked, if the optical system is set up with Fraunhofer conditions. Then the unobstructed wave is brought to a focus by a lens and concentrates all its energy into a small spot there; its amplitude $U_c(\boldsymbol{R})$ is shown in Fig. 3.3(c). Figures 3.3(d) and (e) show the amplitude $U_a(\boldsymbol{R})$ diffracted from the slit (known from elementary work) and the amplitude $U_b(\boldsymbol{R})$ diffracted from the wire, constructed to conform to the Babinet prediction. Away from the central spike, U_c is zero, $U_b(\boldsymbol{R}) = -U_a(\boldsymbol{R})$ and $|U_b(\boldsymbol{R})|^2 = |U_a(\boldsymbol{R})|^2$: *the Fraunhofer-diffracted intensities from complementary screens should be the same (except at the central spike).*[12]

The prediction just enunciated has been subjected to experimental tests in a wide range of conditions. It is confirmed to a surprisingly high accuracy. There is, therefore, a strong experimental reason for believing that the Kirchhoff boundary conditions do not seriously mislead us. Problem 3.4 invites you to check the reasonableness of the above argument by considering a numerical example.

[12] *Comment*: A somewhat surprising condition has to be imposed if $U_c(\boldsymbol{R})$ is to be negligible, even away from the centre. See problem 3.4(2) and the notes on its solution.

3.7 The Fraunhofer condition I: provisional

Most of the diffraction phenomena discussed in this chapter will be those obtaining under the Fraunhofer condition. A full explanation of what this means is deferred until §3.14. Here we give a provisional recipe that imposes stricter conditions on the optical setup than are really necessary.[13]

We agree for now to look at cases possessing the restrictions:

(1) A plane wave, travelling in the $+z$ direction, is incident normally upon an aperture in the xy plane.[14] Then, over the interesting part of the xy plane (possibly the open part of some aperture), $U_{\text{incident}}(x, y) = U_0\,\mathrm{e}^{-\mathrm{i}\omega t}$, independent of x and y. By the Kirchhoff assumptions, $U_{\text{transmitted}}$ takes the same value, $U_0\,\mathrm{e}^{-\mathrm{i}\omega t}$, over the same area.[15]

(2) The observation plane (representative point P) is at a large distance d from the centre of the illuminated part of the xy plane (or is in the focal plane of a lens as shown for example in Fig. 3.4), so paths leaving every (x, y) for P are parallel (remember Fig. 1.3).

[13] These conditions assume that we are using the second of the two viewpoints introduced in §3.10.

[14] The source of light may be a point at a large distance, or radiation from a point at a finite distance may have been collimated by a lens.

[15] We continue here to think of an 'aperture' having the restricted properties introduced in sidenote 2 on p. 46. As before, this special case is taken to avoid complicated wordings; it will not prevent us from dealing with other cases later.

3.8 Fraunhofer diffraction in 'one dimension'

Diffraction is said to be observed in 'one dimension' if light is passed through an aperture that is long and uniform in the y-direction, so that its interesting variation is with x only, and if the Y-dependence[16] of the diffraction pattern is discarded by some means. Such can be made the case with a single slit, a double slit or a diffraction grating—the examples that the reader is likely to have encountered in an elementary course in physical optics. Here we give only enough detail to relate

[16] Coordinates X, Y refer to the plane of observation, as in Fig. 3.4.

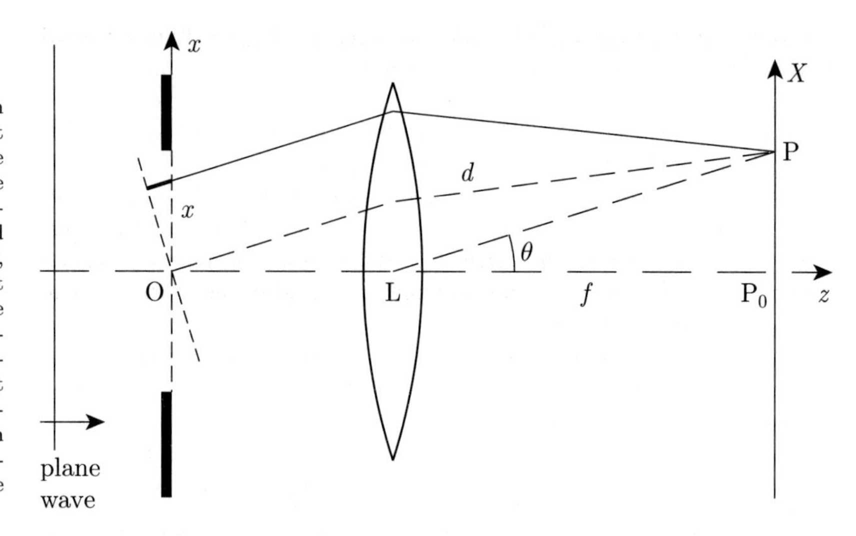

Fig. 3.4 Fraunhofer diffraction in 'one dimension'. The aperture is a slit whose long dimension is normal to the paper, or it has some other shape (like a double slit or a grating) that is likewise uniform with y. The diffracted amplitude is sought at points like P, on the X-axis passing through point P_0, (which in the present simple case is) the focal point of the lens. Secondary sources are, of course, not restricted to the plane of the diagram, but lie along the whole length of the aperture. Nevertheless, the optical path from (x, y) to P is $d - x \sin\theta$, independent of y. With these arrangements the amplitude at P obeys eqn (3.9).

these elementary examples to the present treatment and to the 'two-dimensional' cases discussed below. Problem 3.5 asks you to confirm that the simplifications from two dimensions to one are legitimate.

The conditions assumed (additional to the restrictions itemized in §3.7) are illustrated in Fig. 3.4:

(a) The amplitude transmitted through the diffracting aperture has an interesting dependence upon x only, which is arranged by making the aperture long and uniform in the y-direction.

(b) In the first instance, observation is restricted to points P in the plane of the diagram (on the X-axis in the lens's focal plane if we use a lens). Then the ray-optics rules for refraction at a lens (revise them if in any doubt) say that the optical path from (x, y) to P is

$$r_{\mathrm{P}} = d - x \sin\theta, \tag{3.8}$$

where d is the optical path from O to P along the ray path shown. The important 'shape' in expression (3.8) is the term linear in x. It is this linear dependence on x (and, later, on x and y) that has been engineered by imposing the conditions of §3.7.

A consequence of (a) and (b) is that integration over the y-coordinate of the aperture yields only a constant factor.

When these conditions are met,

$$\begin{aligned} U_{\mathrm{P}} &= \text{constant} \times \int_{-\infty}^{\infty} U_{\mathrm{trans}}(x)\, \exp(\mathrm{i}k r_{\mathrm{P}})\, \mathrm{d}x \\ &\propto \mathrm{e}^{\mathrm{i}kd} \int_{-\infty}^{\infty} U_{\mathrm{trans}}(x)\, \mathrm{e}^{-\mathrm{i}kx\sin\theta}\, \mathrm{d}x. \end{aligned} \tag{3.9}$$

For a workout on elementary applications of eqns (3.6) and (3.9) see problem 3.7.

It's often useful to identify the observation point by $\beta = k\sin\theta$ instead of θ, and then

$$U_{\mathrm{P}} = U_{\mathrm{dif}}(\beta) \propto \int_{-\infty}^{\infty} U_{\mathrm{trans}}(x)\,\mathrm{e}^{-\mathrm{i}\beta x}\,\mathrm{d}x, \tag{3.10}$$

where the factor $\mathrm{e}^{\mathrm{i}kd}$ has been dropped; it does not affect the intensity because $|\mathrm{e}^{\mathrm{i}kd}|^2 = 1$. The integral in eqn (3.10) should be instantly recognizable: it is a Fourier transform. We have a link between diffraction, within the Fraunhofer condition, and the insights and mathematical tools developed in Chapter 2.

Examples of one-dimensional Fraunhofer diffraction are given in Figs. 3.7 and 3.9 below, though we delay discussion of them until after dealing with diffraction in two dimensions.

Comment on signs: Comparing eqn (3.10) with eqn (2.12) shows that $U_{\mathrm{dif}}(\beta)$ is not quite the Fourier transform $\widetilde{U_{\mathrm{trans}}}(\beta)$ of $U_{\mathrm{trans}}(x)$, even ignoring the proportionality constant, because of the negative sign in the exponent. Instead, $U_{\mathrm{dif}}(\beta)$ is proportional to $\widetilde{U_{\mathrm{trans}}}(-\beta)$. As forecast in the comment in §2.12, I'm not going to be precise over this, as it would only clutter the English and the mathematics for no profit. Keep an eye on the signs for yourself and deal with them appropriately.[17]

Comment: Look back at eqn (2.14) and problem 2.5(1). The property $\widetilde{U_{\mathrm{dif}}}(-\beta) = \widetilde{U_{\mathrm{dif}}}(\beta)^*$ does *not* apply here because $U_{\mathrm{trans}}(x)$ is complex.

[17] *Comment*: The identification of $U_{\mathrm{dif}}(\beta)$ as the Fourier transform of $U_{\mathrm{trans}}(x)$ is not quite correct for another reason. Problem 3.8 investigates, and shows that the Fourier relationship holds strictly only if a lens is used and if distance OL in Fig. 3.4 (as well as $\mathrm{L\,P_0}$) is made equal to the focal length of the lens. The complication comes from the factor $\mathrm{e}^{\mathrm{i}kd}$, in which d, though independent of x, is not independent of β. For elementary work this detail is usually swept under the carpet because it doesn't affect the intensity of the diffraction pattern. But there will be cases in this book where we have to be more careful.

3.9 Fraunhofer diffraction in 'two dimensions'

The 'one-dimensional' conditions considered in the last section are very restrictive, and are really idealizations used for keeping things simple when teaching basic concepts. Most realistic cases of diffraction involve more general shapes of aperture. The secondary-source amplitude $U_{\mathrm{trans}}(x, y)$ can be expected to depend in a significant way (magnitude and/or phase) on both x and y, and $U_{\mathrm{dif}}(X, Y)$ may be required over more of the XY plane than just the X-axis.

Figures 3.5 and 3.6 show the geometry of optical paths from positions (x, y) in the aperture to positions $\mathrm{P}(X, Y)$ in the plane of observation. The optical path between these is

$$r_{\mathrm{P}} = d - x\sin\theta - y\sin\phi, \tag{3.11}$$

where point P lies at an orientation tilted through θ, ϕ about axes in the y, $-x$ directions. Distance d is the optical path along a ray following a ray-optics path from O to P. Then

$$U_{\mathrm{P}} \propto \int U_{\mathrm{trans}}(x, y)\,\mathrm{e}^{\mathrm{i}k(d - x\sin\theta - y\sin\phi)}\,\mathrm{d}x\,\mathrm{d}y. \tag{3.12}$$

Problem 3.9 asks you to derive eqn (3.11). As was noted in §3.8, the optical path r_{P} is a linear function of the coordinates x, y in the aperture.

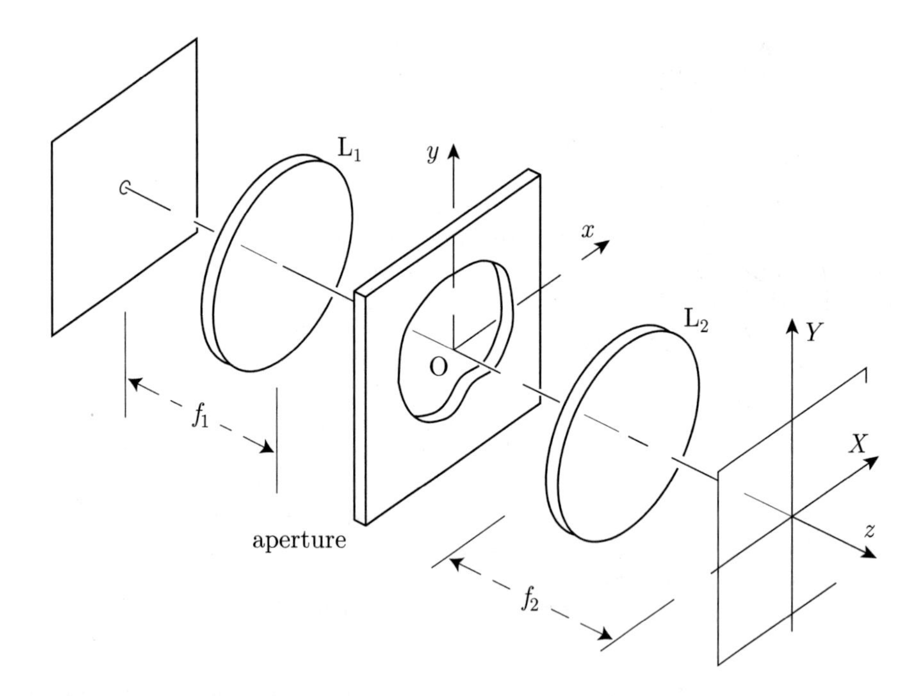

Fig. 3.5 An experimental arrangement for exhibiting Fraunhofer diffraction. A lamp, not shown, supplies light that passes through the pinhole at top left, thereby making that pinhole act as a point source. Light then passes through the aperture and two lenses to the XY plane. This arrangement conforms to conditions (1) and (2) of §3.7: lens L_1 is distant from the point source by its focal length, so that collimated light is produced to fall on the aperture; and the diffracted light is observed in the focal plane of lens L_2.

In the most important case, a Fraunhofer diffraction pattern is formed at a focus produced by a lens (rather than the alternative: looking at a large distance downstream). For that case, problem 3.10 shows that the constants of proportionality from eqn (3.6) can be reinserted to give the more complete form

$$U_\mathrm{P} = U_\mathrm{dif}(X,Y) = \frac{\mathrm{e}^{\mathrm{i}kd}}{\mathrm{i}\lambda f_2}\int U_\mathrm{trans}(x,y)\,\mathrm{e}^{-\mathrm{i}k(x\sin\theta + y\sin\phi)}\,\mathrm{d}x\,\mathrm{d}y, \quad (3.13)$$

where $X = f_2\tan\theta$ and $Y = f_2\tan\phi$. As was stated in connection with eqn (3.6), we shall often find it preferable to include the constants in front of the integral here. A consistency check on the magnitude of the coefficient is made in problem 3.11.

Statement (3.13) is the extension to two dimensions (meaning x and y) of eqn (3.10), this time with a more complete set of multiplying factors. It shows that there is a two-dimensional Fourier-transform relationship[18] between $U_\mathrm{dif}(X,Y)$ and the incident amplitude $U_\mathrm{trans}(x,y)$.

[18] *Comment*: A statement should be made here for honesty, paralleling that in sidenote 17 (previous page). Distance d is not independent of X and Y unless we locate lens L_2 one focal length away from the xy plane. If this condition is not met, a Fourier relationship exists, but it is between $U_\mathrm{trans}(x,y)$ and $\exp\{-\mathrm{i}k\,d(X,Y)\}\,U_\mathrm{dif}(X,Y)$, which is not very tidy.

Comment: A possible misunderstanding. There isn't a Fourier transform relationship, in a literal sense, between $U_\mathrm{dif}(X,Y)$ and $U_\mathrm{trans}(x,y)$, because that would imply an integral containing exponential $\mathrm{e}^{-\mathrm{i}(xX+yY)}$, which is of course dimensionally impossible. When talking of a Fourier-transform relationship in diffraction we always mean to say that the

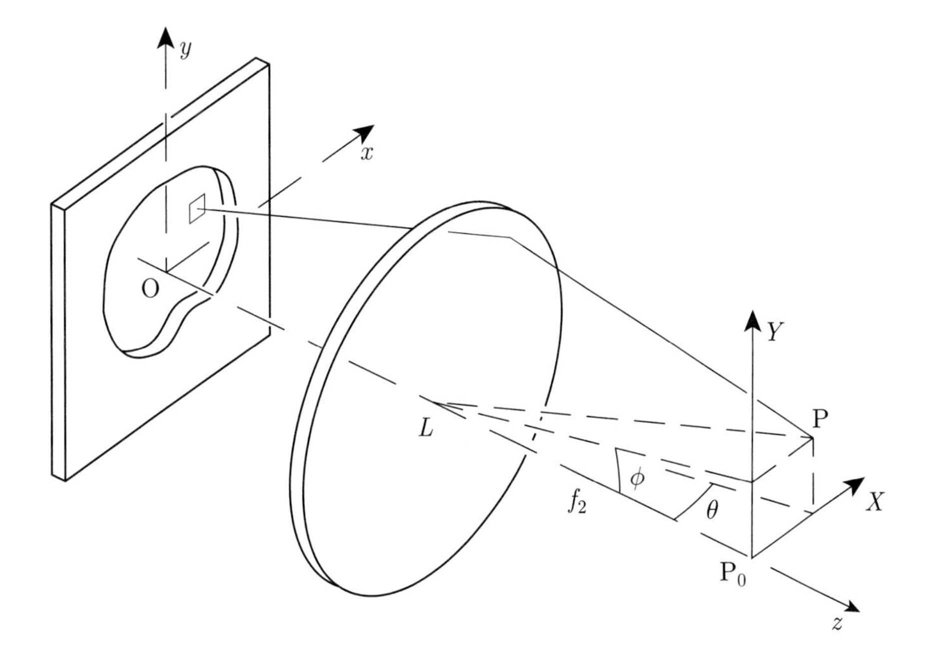

Fig. 3.6 Geometrical detail of part of Fig. 3.5 showing the optical paths from a Huygens source at (x, y) to point $P = (X, Y)$. These coordinates relate to angles θ and ϕ by $X = f_2 \tan\theta$, $Y = f_2 \tan\phi$. Problem 3.13 invites investigation of the ray drawn from the area element to P.

transform of $U_{\text{trans}}(x, y)$ is on display in the XY plane, but with scale factors (k/f_2 for small angles) linking the transform variables $\beta = k\sin\theta$, $\gamma = k\sin\phi$ to distances measured along the X- and Y-axes. A similar comment applies to the multiplying factor $e^{ikd}/(i\lambda f_2)$ which, even if constant, makes $U_{\text{dif}}(X, Y)$ differ in magnitude from the exact Fourier transform.

3.10 Two ways of looking at diffraction

Statements like eqns (3.6) and (3.13) are used in two ways, which differ rather little in the basic physics, but lead us through rather different thought processes.

First: we may think about the relationship between a *field distribution* $U_{\text{trans}}(\boldsymbol{r})$ existing on one surface and the diffracted field $U_{\text{dif}}(\boldsymbol{R})$ that it maps to on some other surface after propagating from the first surface to the second. It does not matter how the original $U_{\text{trans}}(\boldsymbol{r})$ is set up, or whether its dependence on $\boldsymbol{r}$ has been imposed by some aperture. Section 3.17 discusses examples of diffraction from this point of view, and Chapters 7 and 8 contain further developments along these lines.

The second of the possible viewpoints takes a special case, albeit an important one. We arrange for a uniform plane or spherical wave with amplitude $U_{\text{inc}}(\boldsymbol{r})$ to be incident on an aperture[19] that has amplitude transmission coefficient $T(\boldsymbol{r})$, in this context often called an **aperture**

[19] To avoid convoluted sentences, most statements here are made for the case of an aperture that can be spanned by a plane, and coordinates (x, y) refer to that plane. It is not hard to invent more general-purpose statements. By contrast, it is not a simplifying restriction that the diffraction pattern is formed in a plane; this will be the case if the focusing lens is free from aberrations.

function. The aperture may introduce phase shifts as well as changes of $|U|$, so function $T(\boldsymbol{r})$ may be complex. A special case of such an arrangement is shown in Fig. 3.5. To write a Kirchhoff integral, we must draw a mathematical surface spanning the aperture, on which we take there to be Huygens secondary sources. These secondary sources are driven by an electric field amplitude

$$U_{\text{trans}}(\boldsymbol{r}) = T(\boldsymbol{r})\, U_{\text{inc}}(\boldsymbol{r}).$$

Different choices of spanning surface are possible (bounded by the same outline), and they yield different values for $U_{\text{inc}}(\boldsymbol{r})$ and $U_{\text{trans}}(\boldsymbol{r})$, but the diffracted amplitude is independent of these choices.[20] In this second viewpoint, the transmission coefficient $T(\boldsymbol{r})$ is seen as all-important in controlling the development of the diffraction pattern downstream from the aperture.[21] Discussions of the quality of an image, or the resolution of an optical instrument, take this second viewpoint.

The diffraction pattern is, in both viewpoints, worked out by putting $U_{\text{trans}}(\boldsymbol{r})$ into a Kirchhoff diffraction integral, so it may seem hardly worth distinguishing two viewpoints at all. However, we shall see that Fraunhofer diffraction (in particular) is described rather differently according to which viewpoint we take.

We consider next the way in which the two viewpoints affect our thinking about the Fraunhofer case of diffraction. In the first (field-distribution) viewpoint, light propagates from one specific surface to another. The pair of surfaces might, e.g., be those shown in Fig. 3.19(a) or (b). Conditions are Fraunhofer if the optical path from (x, y) to (X, Y) is a linear function of x and y, and this is just a matter of geometry between the two specific surfaces.

Within the second viewpoint, the Fraunhofer case of diffraction is described by the following statements, to be proved later, which supersede the simplifying restrictions made in §3.7. The notation is that of Fig. 3.5.

(1) The diffraction pattern is observed in a plane at which the optical system forms an image of the light source, that image having a diffraction spread because a uniform beam on its way has had its width restricted[22] by an aperture with transmission function $T(x, y)$.
(2) The optical path from a point on the light source, via a point at (x, y) in the plane of the aperture,[23] to a point $\mathrm{P}(X, Y)$ in the diffraction pattern, is a linear function of x and y.

It will be shown that these statements are equivalent to each other: any arrangement of apparatus that guarantees one of them will guarantee the other.

A consequence of (1) and (2) is that a Fourier-transform relationship (with suitably scaled variables) exists between $T(x, y)$ and the amplitude $U_{\text{dif}}(X, Y)$ appearing in the image plane.

Elementary discussions of diffraction use our second picture without comment, and that was the case, for example, throughout §§3.7–3.9 as well as in problems 3.5–3.17. The slit (or double slit or whatever)

[20] *Comment*: This raises a question. The physics must make the diffracted field independent of the surface chosen; but does the mathematics agree? Here is a consistency check that ought to be explored. Fortunately, all is well, though to confirm this we have to go back to eqn (3.4) or even eqn (3.3) to get 'behind' some of our customary approximations. Different choices of surface have the effect of transferring pieces of optical path from r_{S} to r_{P} or the reverse (notation of Fig. 3.2). A discussion, including the accompanying changes of obliquity factor, can be found in Lipson *et al.* (1995), §7.2.5.

[21] The language we use often seems to lock in this way of thinking, when we say 'diffraction *at* a slit'. Diffraction occurs downstream from the slit, not at it; and diffraction needs no slit to make it happen.

[22] This statement has its heart in the right place, but it isn't quite correct. See problem 3.12.

[23] The wording 'via a point at (x, y)' is intended to emphasize that phase shifts and optical paths on the upstream side of the aperture are just as significant as those on the downstream side. In the notation of Fig. 3.2, it is $(r_{\mathrm{S}} + r_{\mathrm{P}})$ that must be a linear function of x and y.

was spoken of as the significant participant in controlling the diffraction process.

In §§3.11–3.16 we continue to use the 'aperture function' view of diffraction. We return to the 'field distribution' viewpoint in §3.17.

3.11 Examples of Fraunhofer diffraction

We consider here only a few standard examples of Fraunhofer diffraction.

- Figure 3.7 shows graphically the amplitude and intensity that result when a uniform beam of light is passed through a slit or a rectangular aperture. Problems 3.7 and 3.14 ask you to work out the mathematics for this case. Problem 3.6 gives a way 'seeing why' there are zeros at $\sin\theta = p\lambda/a$.

  The 'width'[24] of the diffraction pattern is conventionally given as the distance from the centre to the first zero.[25] From Fig. 3.7 we see that this corresponds to $\rho = \pi$. It is easy to work back through the definitions to obtain the familiar result

$$\sin\theta = \lambda/a, \qquad \text{width of diffraction pattern} = f\lambda/a, \qquad (3.14)$$

  where f is the focal length of the focusing lens, the f_2 of Fig. 3.5.

[24] *Comment*: Statement (3.14) is given for the special case where the optical layout conforms to that of Fig. 3.5: with 'parallel' light focused into the focal plane of lens L_2. A more general arrangement, like that of Fig. 3.16, requires an obvious change: the pattern width is now $\lambda(d_P/a)$. This draws our attention to the fact that the pattern width is really controlled by the range of angles (a/d_P when the angle is small) within which light arrives at the diffraction pattern.

[25] In other contexts we would normally define the width of a peak by giving its FWHM (full width at half maximum height). Here the FWHM is not very different from the distance between centre and first zero, so we use the quantity that is easier to calculate. Problem 3.15 investigates.

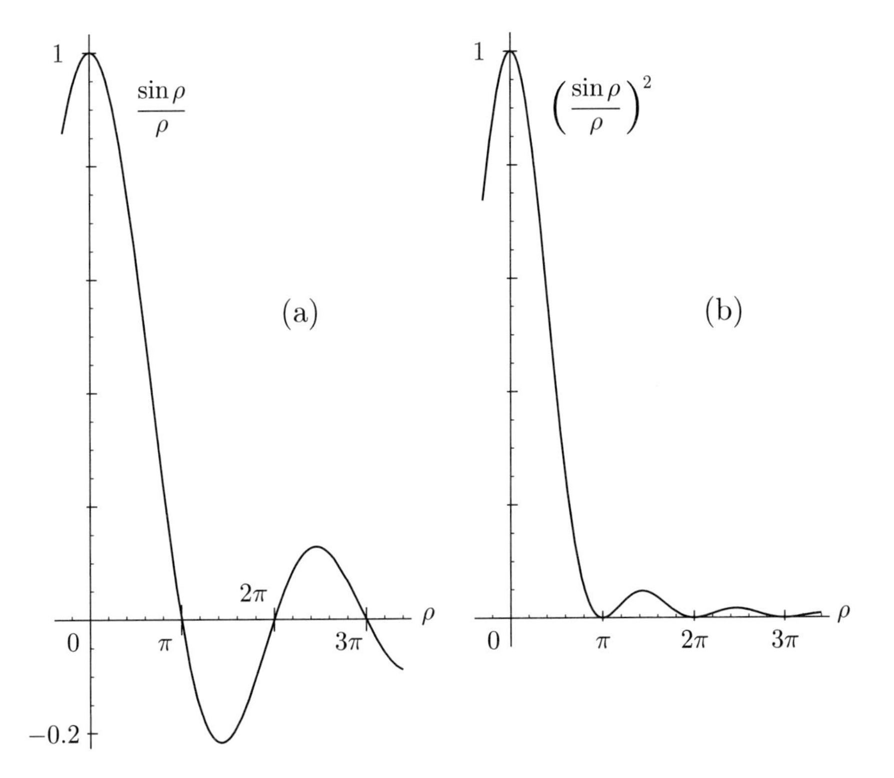

Fig. 3.7 The dependence of (a) amplitude and (b) intensity upon $\rho = \frac{1}{2}ka\sin\theta$ for Fraunhofer diffraction from a slit of width a. Diffraction via a rectangular aperture results in a product of two such functions (problem 3.14 and Fig. 3.8).

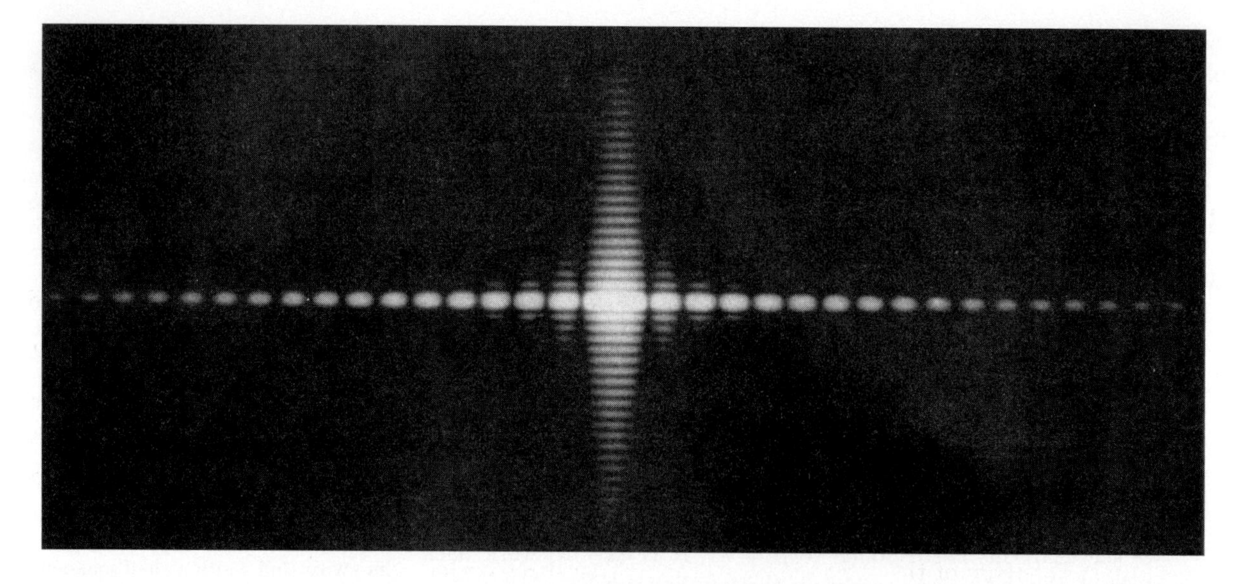

Fig. 3.8 Fraunhofer diffraction from a rectangular aperture, taller than it is wide.

- Figure 3.8 shows a beautiful photograph of the pattern for the case where the aperture is rectangular.[26]
- Figure 3.9 shows two 'one-dimensional' Fraunhofer diffraction patterns, produced by one slit and by two similar slits. The photographs exhibit the factorization property of eqn (2.24), by showing that the envelope of the two-slit pattern coincides with the profile of the single-slit pattern.

  The single-slit pattern of Fig. 3.9 resembles the rectangular-aperture pattern of Fig. 3.8, but it has been smeared in the vertical direction. Such smearing is conventional in the display of one-dimensional diffraction patterns, and is arranged by elongating the source to a slit as shown in Fig. 3.10. In effect, the smearing removes the dependence of the diffracted intensity on the vertical coordinate in Fig. 3.8, treating that dependence as irrelevant.[27]
- A light beam is often restricted to lie within a circular outline because it is passed through a circular lens. That restricted width causes a diffraction spread and sets a limit on the resolution obtainable if the lens is being used to form an image. Figures 3.11 and 3.12 show the amplitude, the intensity, and a photographic reproduction after the fashion of Figs 3.7 and 3.8. The mathematics of the circular aperture is rehearsed in problem 3.16.

  As with the case of a slit, we assign a conventional 'width' to the diffraction pattern by giving the distance from centre to first zero. From Fig. 3.11, the first zero is at $\rho = 1.22\pi$, equivalent to

$$\sin\theta = 1.22\lambda/D, \qquad \text{width of pattern} = 1.22 f\lambda/D. \qquad (3.15)$$

The effect of making the aperture circular has been to increase the width of the diffraction pattern by 22%, compared with that from

[26] *Comment*: Figure 3.8 is reproduced from Cagnet *et al.* (1962). It should be pointed out that the diffracted intensity is extremely weak at the extremities of the pattern and 'in the angles' of the cross-shape. It is a triumph of photographic technique to exhibit these weak maxima—at all. The reader must not be misled into thinking that these maxima are anything like as intense as they look. A similar comment applies to the very weak outer rings in Fig. 3.12. It may help to remind yourself: the first side maximum for the single slit has less than 5% of the intensity of the central maximum; and the first maximum 'in the angle' is only 5% of 5%. Likewise, the first outside ring for a circular aperture has only 2% of the intensity of the central maximum.

[27] *Comment*: It should be noted that the effect of an elongated source slit is to allow light to arrive at the diffracting aperture from a range of directions, and this means the illuminating amplitude is no longer independent of y. Rather, there is (usually, unless we arrange otherwise) incoherence of the illumination along the length of the aperture, while coherence across the aperture's width is maintained.

Fig. 3.9 Fraunhofer diffraction from a single slit (upper photo) and from two slits (lower photo). All three slits have the same width. The similarity in the 'envelope' is clearly displayed. The lower pattern is crossed by Young-type fringes resulting from interference of light from the two slits.

If the optical arrangement had been the same as that of Fig. 3.5, the two patterns would have been confined to thin lines running horizontally through the middle of each photograph. The patterns have been 'smudged' in the vertical direction to make them easier to appreciate. The optical arrangement by which the smudging is effected is shown in Fig. 3.10.

a square aperture of the same diameter. The physical reason for this is explored in problem 3.16(6).

Even though these few cases are the only ones given a detailed treatment here, there will be plenty of other diffraction examples later in the book. A test of understanding is provided in problem 3.17.

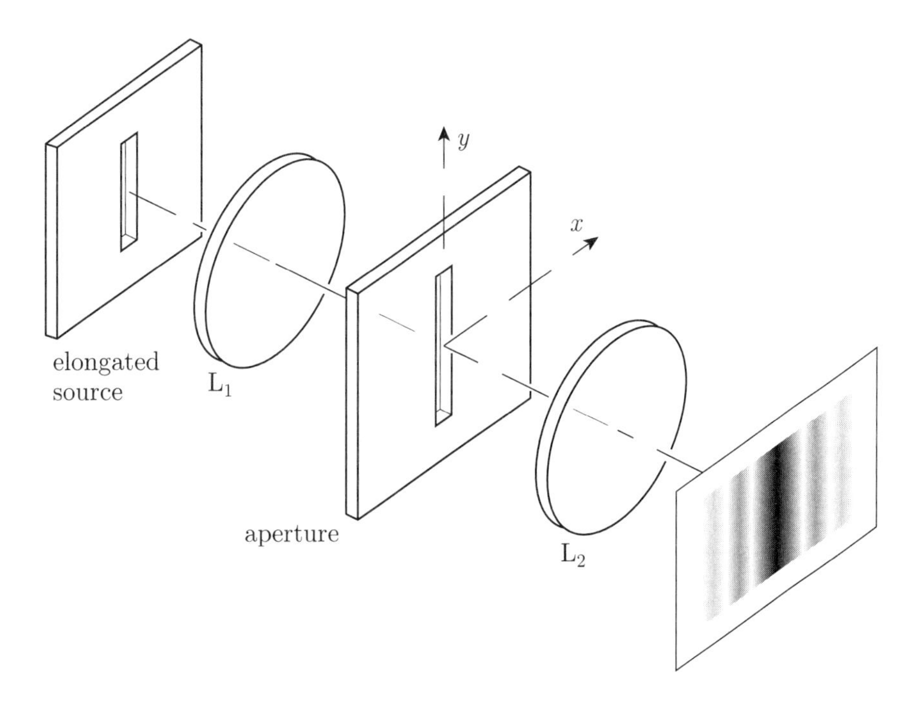

Fig. 3.10 The optical arrangement by which the 'smudged' diffraction patterns like those of Fig. 3.9 are obtained. The diagram should be contrasted with Fig. 3.5. The significant change in the equipment is the elongation of the source from a pinhole into a slit, with consequent vertical smearing of the diffraction pattern. The source slit is, of course, set parallel to the diffracting slit (or slits), so that the information in the X-dependence of the diffraction pattern is not degraded.

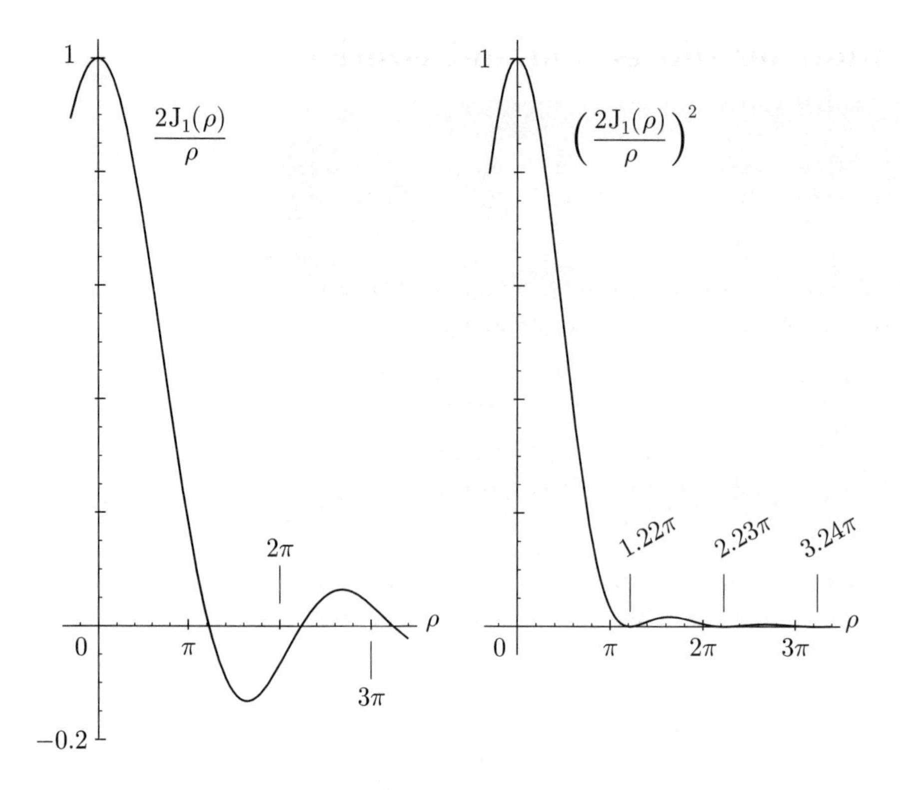

Fig. 3.11 The amplitude (a) and the intensity (b) for light diffracted via a circular aperture, as functions of $\rho = \frac{1}{2}kD\sin\phi$ where D is the diameter of the circle.

A beautiful compilation of two-dimensional Fraunhofer diffraction patterns, useful for understanding how one can proceed from the simple to the complicated, is given in Harburn *et al.* (1975).

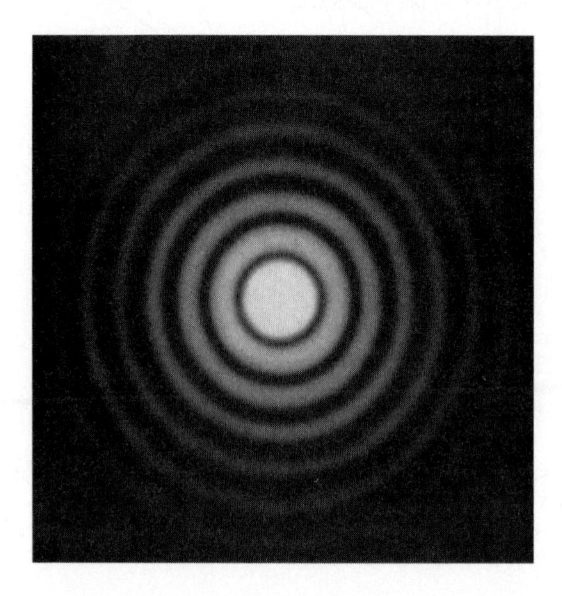

Fig. 3.12 The Fraunhofer diffraction pattern for light that has been transmitted through a circular aperture. The central bright patch is often called the **Airy disc**. The intensity in the outer rings has been exaggerated relative to that of the Airy disc, in order to demonstrate the correctness of the ring structure predicted by the calculations leading to Fig. 3.11.

3.12 Fraunhofer diffraction and Fourier transforms

The Fourier-transform relationship between the amplitude $U_{\text{trans}}(x, y)$ existing on one surface and the amplitude $U_{\text{dif}}(X, Y)$ Fraunhofer-diffracted to another suitable surface has been on display since eqns (3.10) and (3.13). Nevertheless, it helps understanding to see these statements brought to life by specific examples. Example 3.1 and problems 3.19 and 3.20 pursue the Fourier property in a rather thoroughgoing way.

Example 3.1 A grating that gives only three orders

A special diffraction grating has width (x-direction) D, and within that width has amplitude transmission factor $T(x) = \frac{1}{2}(1+\cos\beta_0 x)$ with $\beta_0 D \gg 1$. It is illuminated normally by a uniform monochromatic plane wave with amplitude $U_0\,\mathrm{e}^{-\mathrm{i}\omega t}$, and the transmitted diffraction pattern is observed under the Fraunhofer condition. We are to show that there are only three grating orders.

Solution: The aperture is correctly described as a grating because its transmission is periodic in x with period $d = 2\pi/\beta_0$. The number of periods is $D/d = \beta_0 D/2\pi$, which we are given is large.

It helps to write $T(x) = \frac{1}{4}(2 + \mathrm{e}^{\mathrm{i}\beta_0 x} + \mathrm{e}^{-\mathrm{i}\beta_0 x})$. Then the diffracted amplitude U_{dif} is

$$U_{\text{dif}} \propto \frac{1}{2}\int_{-D/2}^{D/2} \mathrm{e}^{-\mathrm{i}kx\sin\theta}\,\mathrm{d}x + \frac{1}{4}\int_{-D/2}^{D/2} \mathrm{e}^{-\mathrm{i}kx(\sin\theta-\beta_0/k)}\,\mathrm{d}x$$

plus a third integral with exponent containing $(\sin\theta + \beta_0/k)$.

The first integral is proportional to

$$\tfrac{1}{2}F(\sin\theta) \equiv \tfrac{1}{2}\frac{\sin(\frac{1}{2}kD\sin\theta)}{\frac{1}{2}kD\sin\theta},$$

which is a single-slit diffraction pattern with D as the slit width. The function $F(\sin\theta)$ has a peak where its argument (the thing in brackets) is zero. The 'width' of the peak, measured in the usual way from centre to first zero, is $\Delta(\sin\theta) = 2\pi/(kD) = \lambda/D$.

The second integral $\propto \frac{1}{4}F(\sin\theta - \beta_0/k)$. This is another single-slit diffraction pattern, this time centred on $\sin\theta = \beta_0/k = (1)\lambda/d$; we recognize the condition for a diffraction-grating maximum of order $p = 1$.

The third integral $\propto \frac{1}{4}F(\sin\theta + \beta_0/k)$: yet another single-slit pattern peaking at $\sin\theta = (-1)\lambda/d$ and representing grating order $p = -1$.

The first term can be said to have its peak at $\sin\theta = (0)\lambda/d$, so it represents the zeroth grating order. Thus, the grating yields the three promised orders: orders $0, \pm 1$.

The three grating-order peaks are well separated from each other: the ratio (separation)/(width) $= D/d$.

This example is a good vehicle for understanding the Fourier transform. When we wrote $T(x)$ as a sum of three complex exponentials, we came very close to performing a Fourier decomposition 'by inspection'. This decomposition would be exact if the grating had infinite width, and then the diffraction pattern would consist of three δ-functions. As it is, the three diffraction orders have the structure that is required by the fact that the grating starts and stops at $x = \pm D/2$.

Further insights as to 'why' an amplitude U_{trans} containing $e^{i\beta_0 x}$ diffracts to a peak centred on $\sin\theta = \beta_0/k$ are revealed in problem 3.19.

3.13 The Fraunhofer condition II: Rayleigh distance and Fresnel number

In §§3.7–3.9 we identified the Fraunhofer case of diffraction, and its associated Fourier-transform property, with a rather restrictive set of experimental conditions. We also associated it with the presence in eqns (3.8) and (3.11) of terms *linear* in the coordinates (x, y) of a secondary-source surface, probably the surface spanning an 'aperture'. These were provisional explanations and they raise several questions:

- Just what is it that has to be a linear function of x and y?
- What experimental arrangement is needed (as opposed to being sufficient) for achieving the required condition?
- Since nothing is ever exact, how are those experimental conditions toleranced?
- What happens if we significantly fail to meet the Fraunhofer condition?

We start by taking a special case, but a different one from that considered previously. As in §3.7, we supply a plane wave parallel to the z-axis and pass it through a slit of width a. But now we have no lens downstream of the slit, and we investigate how the diffraction pattern changes as we go in the downstream direction towards $+z$.

In the 'far field' we know from earlier work that we have a Fraunhofer diffraction pattern whose angular width is $\sim\lambda/a$. In the 'near field', most of the intensity lies within the geometric width (the shadow) of the aperture. Figure 3.13 shows these two limiting outlines for the diffracted beam. We can guess that the actual width of the beam is something like the larger of the two. A changeover between the two regimes occurs at distances around the **Rayleigh distance** d_R, where $d_R(\lambda/a) = a$, so

$$\text{Rayleigh distance } d_R = a^2/\lambda. \tag{3.16}$$

We have Fraunhofer diffraction—as an approximation, to which we shall have to attach a quantitative tolerance—at distances $\gtrsim d_R$.

We next consider how to construct such a quantitative tolerance. We continue for now to take the special case of normal incidence. The

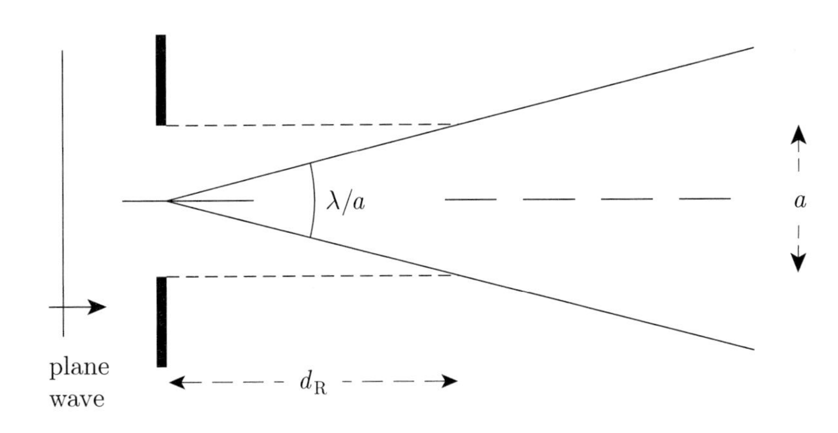

Fig. 3.13 The limiting outlines for light, originally collimated, that has passed through a slit of width a. Close to the slit, the beam spreads little and has an outline roughly at the shadow-edge of the slit. (A more careful description would mention the fringes of a Fresnel diffraction pattern in this region.) At large distances, the light spreads out in a Fraunhofer pattern whose angular width is of order λ/a. The changeover between the near-field (Fresnel) and far-field (Fraunhofer) regimes takes place at distances of the order of the Rayleigh distance $d_R = a^2/\lambda$.

optical path r_P from a secondary source at $(x, y, 0)$ to a point of observation P at (X, Y, Z) contains 'linear terms' involving x, y; 'quadratic terms' involving x^2, y^2; and higher powers. To investigate the limits of Fraunhofer conditions, we need to investigate the nuisance value of the quadratic terms: in what circumstances are these terms negligible?[28]

The geometry of optical paths is shown in Fig. 3.14. Problem 3.21(1) shows that for finding the quadratic contributions to optical paths it suffices to work with paths to point P_0 on the axis of the aperture. It is convenient to look back from P_0 and to divide the aperture into **Fresnel half-period zones** by drawing on it a set of circles centred on the axis. The first zone is a disc-shaped area defined by

$$0 < (r - Z) < \lambda/2,$$

the next zone is an annulus defined by

$$\lambda/2 < (r - Z) < \lambda,$$

and so on by steps of a half wavelength. Because of the added half-wavelengths, each zone makes a contribution to the amplitude at P_0 that is 180° out of phase with those of its immediate neighbours. The **Fresnel number** is the number of zones that are 'exposed', i.e., not obstructed by the opaque part of the aperture. In the case of a circular aperture of diameter a (with a plane wave incident normally), the Fresnel number N is

$$N = \frac{r_{\max} - Z}{\lambda/2} \approx \frac{(a/2)^2}{Z\lambda} = \frac{1}{4}\frac{d_R}{Z}, \tag{3.17}$$

a result that you are asked to show in problem 3.21.

The optical path differences involved in assigning Fresnel zones are the quadratic-dependence pieces of path that we have to design out when aiming for the Fraunhofer condition. So a way of imposing the Fraunhofer condition is to require that the Fresnel number $N \ll 1$.

In Fig. 3.13,[29] the Fresnel number is $\frac{1}{4}$ when the downstream distance Z from the aperture is the Rayleigh distance d_R, and the greatest

[28] *Comment*: When quadratic terms really can't be neglected, we have a Fresnel case of diffraction. The Fresnel case is of course just diffraction and is described by eqn (3.6), but it is mathematically more complicated than Fraunhofer—just because it requires handling the quadratic terms.

[29] We are now treating Fig. 3.13 as representing a circular aperture of diameter a, rather than a slit of width a.

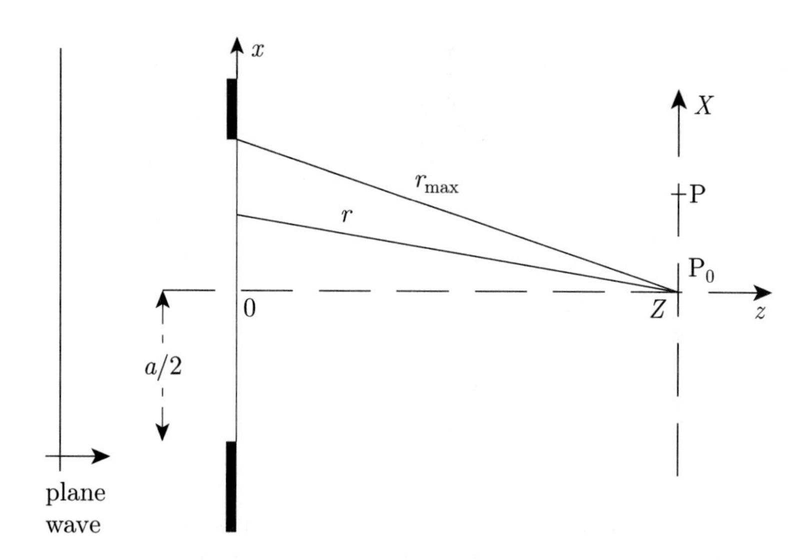

Fig. 3.14 The geometry used for calculating the Fresnel number when a collimated beam of light passes through a circular aperture of diameter a and is observed in a plane at distance Z downstream. A half-period zone is an annular (or for the first zone circular) area of the xy plane defined by giving distance r a range of $\lambda/2$.

Point P is referred to in problem 3.21.

quadratic contribution to the optical path is then $\lambda/8$; equivalently, secondary waves are added in the 'wrong' phase with a phase error that is at worst $\pi/4$. This is often regarded as the greatest acceptable error when aiming for the Fraunhofer condition; but it is, of course, easily scaled if one wishes to be more or less demanding. Summarizing, we have:

$$\begin{aligned}
&\text{Fraunhofer case when } N \ll \tfrac{1}{4} \quad &&(\text{quadratic path differences } \ll \lambda/8)\\
&\text{Fresnel case when } N \gtrsim \tfrac{1}{4} \quad &&(\text{quadratic path differences } \gtrsim \lambda/8).
\end{aligned}$$

3.14 The Fraunhofer condition III: object and image

Finally, we'll relax both conditions (1) and (2) of §3.7, and place both source and observation plane at finite distances from an aperture. A condition will have to be met if we are to have Fraunhofer diffraction, but that condition is different from what we used provisionally.

Let spherical waves be incident on an aperture of width a, from a point source S at distance d_s as shown in Fig. 3.15. From eqn (3.4) we see that it is $k(r_s + r_P)$ that appears in the exponential in the Kirchhoff integral. To have Fraunhofer diffraction (meaning the same mathematical and physical behaviour as we have investigated before), we shall want the optical path $(r_s + r_P)$ from S to P via (x, y) to be a linear function of x, y. There is no hope of this unless the path drawn in Fig. 3.15 from S to the central point P_0 is free from significant quadratic contributions (problem 3.21(5) shows that the quadratic contributions in the paths to P are the same as those to P_0). Thus,

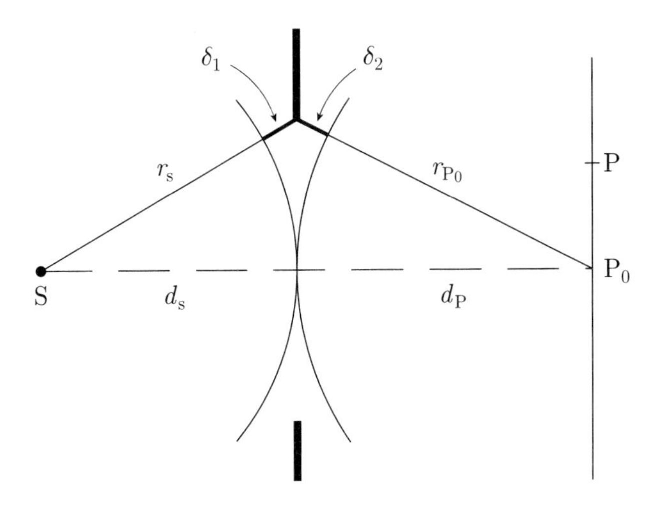

Fig. 3.15 Spherical waves radiate from point source S and pass through an aperture of width a; the resulting diffraction pattern is observed at points such as P in a plane through P_0. The curved surfaces are spheres centred on S and on P_0. To see whether the conditions approximate to Fraunhofer, we need to identify 'quadratic' contributions to the optical paths. The quadratic contributions to the paths from S to P_0 are indicated, and at greatest are $(\delta_1 + \delta_2)$.

$$\begin{aligned}(\text{longest path SP}_0) - (\text{shortest path SP}_0)& \\ &= \delta_1 + \delta_2 \\ &\approx \frac{1}{2}\left(\frac{a}{2}\right)^2 \left(\frac{1}{d_s} + \frac{1}{d_P}\right). \qquad (3.18)\end{aligned}$$

The largest phase error is $k(\delta_1 + \delta_2)$; we require this to be no more than, say, $\frac{1}{4}\pi$. Then

$$\delta_1 + \delta_2 \le \lambda/8 \qquad \text{and so} \qquad \frac{1}{d_s} + \frac{1}{d_P} \le \frac{\lambda}{4}\left(\frac{2}{a}\right)^2 = \frac{1}{d_R}, \qquad (3.19)$$

where d_R is the Rayleigh distance.[30]

How can we make creative use of this? Aha! We can use a lens to form an image of the source, and place that image profitably. Let's place the image right *at* P_0; the arrangement we've just invented is that of Fig. 3.16. Then $d_s = -d_P$, which makes the phase error zero for light reaching P_0: the Fraunhofer condition applies exactly.[31]

There is an inevitability to the result just obtained, which may be appreciated from Fig. 3.17. We can imagine that two thin lenses are

[30] Equation (3.18) can be stated equivalently as

$$\text{Fresnel number} = \frac{1}{4}\left(\frac{d_R}{d_s} + \frac{d_R}{d_P}\right);$$

see problem 3.21.

[31] A brute-force confirmation of this result is obtained in problem 3.23.

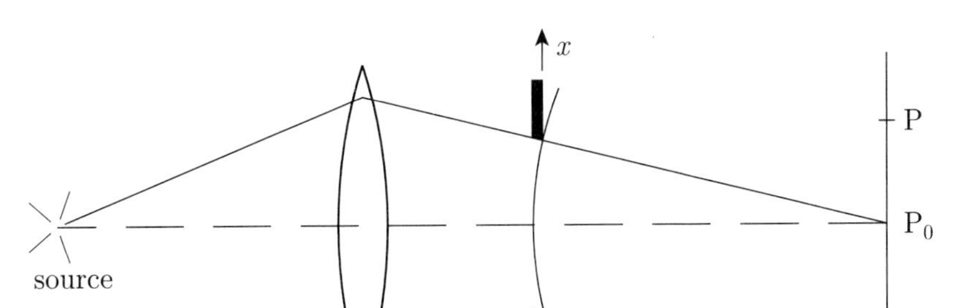

Fig. 3.16 The arrangement that removes 'quadratic' contributions to the optical paths from source to P_0. Both of the spherical surfaces drawn in Fig. 3.15 now coincide with the sphere drawn, centred on P_0. Since this arrangement eliminates the quadratic contributions to optical paths, leaving only linear contributions, it results in a Fraunhofer diffraction pattern in the plane through P_0. Rays show that this configuration places the plane of observation at the ray-optics image of the source.

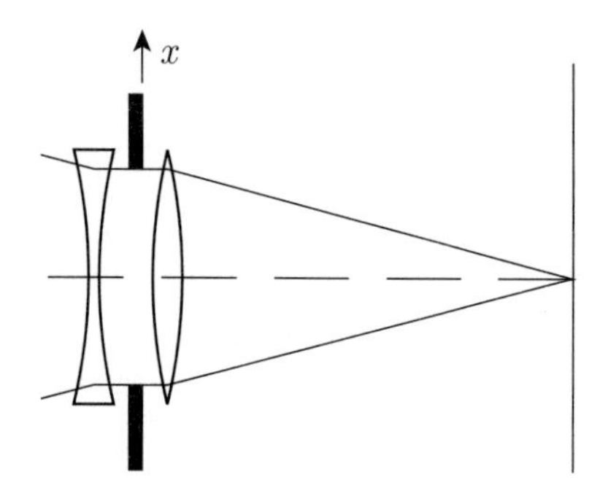

Fig. 3.17 This diagram adds two cancelling lenses to the arrangement of Fig. 3.16, in such a way that light travels as a parallel beam through the aperture. By this trick, the optical system of Fig. 3.16 is made optically identical to that of Fig. 3.5, and the Fraunhofer condition is satisfied in the form stated in §3.7. The result $d_s = -d_P$ (notation of Fig. 3.15) *had* to happen. The author is indebted to Professor D. N. Stacey for this insight.

added to the optical system of Fig. 3.16, in such a way that they cancel so far as the image formation is concerned. Yet, between these lenses, the light travels parallel to the axis, and the optics involving the aperture is identical to that in Fig. 3.4. This trick is sometimes used in elementary discussions of Fraunhofer diffraction to avoid the need for the mathematics of eqns (3.18) and (3.19).

Summarizing then:

- *Fraunhofer diffraction*—within the second of the viewpoints enunciated in §3.10—*is observed in a plane where there is a ray-optics image of the source.*

There is no need for collimated light to pass through the aperture so long as this object–image relationship holds. The tolerance obtained above (Fresnel number $\leq \frac{1}{4}$; quadratic contribution to the optical path from source to image $\leq \lambda/8$) can be used to find the tolerable departure from exact focus; see problem 3.22(3).

As forecast, the two conditions itemized in §3.10 constitute a single requirement on the configuration of the optical system. We have shown that meeting (2) requires meeting (1); it should be obvious that the converse is also true. Condition (2) ensures that the diffraction pattern is identical, physically and mathematically, to what we would calculate from eqn (3.9) or eqn (3.12), so it is appropriate to apply the description 'Fraunhofer' to diffraction obtained under the extended conditions.

People often parrot-learn the bulleted statement above and immediately misapply it. The image formed is of the light *source*, not of the diffracting aperture. We can see why. The beam of light travelling towards the image is being narrowed by the aperture, and it is that narrowing that causes and controls the diffraction spread. *Information*[32] is carried by light on its way to forming an image, and the narrowing removes some of that information; a degraded image (spread out by diffraction) is the result.

We now have on display the chief reason why physicists are so concerned with the Fraunhofer case. Optical systems almost always form a ray-optics image of something, and we are concerned to know how good that image can be.[33] Fraunhofer diffraction controls the quality of the image (at least it sets the fundamental limit when aberrations have been reduced), so it is the physics needed for understanding the resolution.

Problem 3.23 expresses the geometrical reasoning of this section in a different way.

[32]In the form of Fourier components (laid out across the aperture) representing high spatial frequencies present in the $\boldsymbol{E}$-field that left the source. A narrow aperture filters away these high spatial frequencies—high β in the notation of eqn (3.10)—and the image is built from what remains. These ideas are explored further in Chapter 12.

[33]Even a spectrometer forms an image: each spectral 'line' is an image of the entrance slit formed by light of a particular wavelength.

3.15 The Fresnel case of diffraction

We can understand the Fresnel case of diffraction by saying that it is the inverse of the Fraunhofer case: it deals with what happens away from an image. In particular, if a light beam is partly obstructed by some aperture, Fresnel diffraction describes the way light behaves in the 'shadow' (meaning both the light and dark parts) behind the aperture.

In Fig. 3.13 the 'near field' region, closer than the Rayleigh distance to the aperture, is the region where Fresnel diffraction is observed and where its intensity distribution is distinctly different from that for the Fraunhofer case.

In the history of optics, Fresnel diffraction was important in answering the question: if light is a wave, how can we account for the fact that it 'travels in straight lines' (unless we look closely enough to see details of its diffraction)? You can observe how light propagates beyond a straight edge (for example) and find that it really spreads very little into the geometrical shadow—even though it is a wave. (The smallness of the spread, of course, is because the secondary waves interfere destructively, combined with the fact that the wavelength of light is small compared with human-scale dimensions.)

In this book we shall take the attitude that 'light travels in straight lines' is a solved problem, and not rehearse the Fresnel-diffraction arguments that justify it.

There exist elegant methods, algebraic and graphical, for discussing Fresnel diffraction from slits or circular apertures (Fresnel integrals, Cornu's spiral, ...). These methods do their best to tame mathematics that can be rather unpleasant. A number of books give good and detailed accounts, and the reader who is interested can consult, e.g., Rossi (1965). In the present book we shall not pursue these 'traditional' methods. Fresnel-diffraction cases will, of course, be encountered and require analysis, e.g. in the propagation of laser beams (problem 7.9) and in holography. But fortunately our discussions will not need to be mathematically complicated. We give just two problems involving Fresnel diffraction at the end of this chapter: problem 3.24 investigates the Fresnel number for the light beam inside a laser cavity; and problem 3.25 investigates the intensity on the axis of a circular aperture as a function of aperture diameter.

The arguments presented here have avoided details of Fresnel diffraction, but even so they are not totally deficient in regard to 'light travels in straight lines'. The Fraunhofer diffraction pattern surrounding the ideal location of a point image is really quite small, for cases where we do not go out of our way to make it large. For collimated light passing through a lens with an f-number of 10, the diffraction spread is only 10 wavelengths (see eqn (3.21)). Ray optics would put all the light at the focal point, and diffraction concentrates it quite closely round that same point. 'Light travels in straight lines' can be seen to work pretty well in the (admittedly special) cases that we've investigated in detail.

Before we leave the subject, it should be mentioned that the insights of Fresnel diffraction carry over to other areas of physics. Fresnel-diffraction arguments have established that 'light travels in straight lines when the wavelength is too small to matter'. Exactly similar ideas are needed in quantum mechanics to show that classical particle mechanics is recovered from Ψ-waves when 'the de Broglie wavelength is too small to matter'. Four applications of diffraction ideas to non-optical physics are explored in problem 3.26.

3.16 Fraunhofer diffraction and optical resolution

Because it is concerned with the spread of light around an image, Fraunhofer diffraction is responsible for setting the fundamental limit to the performance of optical instruments that form images, such as telescopes and spectrometers.

We deal here only with the angular resolution obtainable from a telescope,[34] meaning any lens focused for infinity whose periphery is the aperture limiting the width of the transmitted light beam. Figure 3.18(a) shows the lens and the Fraunhofer-diffraction pattern formed by light from a point source at infinity on the lens axis. The width of the diffraction pattern is of order $f\lambda/D$, where f is the lens's focal length and D is its diameter. More usefully here, the *angular* width α is of order λ/D: λ/D for a lens with a square outline; $1.22\lambda/D$ for the more likely case of a circular lens.

[34] The resolution obtainable from a grating spectrometer is derived in §4.7.

Figure 3.18(b) shows light propagating through the lens from an object lying at an angle ϕ to the axis. It too produces a Fraunhofer diffraction pattern, this time displaced through ϕ from the axis. By convention (the **Rayleigh criterion**), the optical system is considered to resolve the two objects if $\phi \geq \alpha$. By 'resolve' we mean that a person looking at the images is able to discern with reasonable confidence that there are two objects. (Reasonable confidence? Well, yes, we are setting a criterion for the two objects to be 'just' resolved, so the observer should be only just sure.) Summarizing:

$$\text{rectangular aperture:} \qquad (\text{angular resolution}) = \lambda/D \tag{3.20a}$$

$$\text{circular aperture:} \qquad (\text{angular resolution}) = 1.22\lambda/D. \tag{3.20b}$$

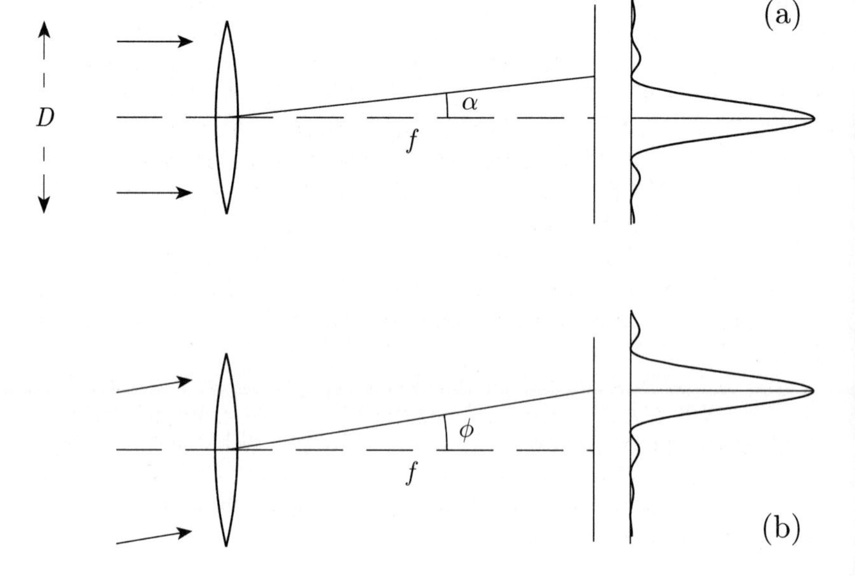

Fig. 3.18 The angular resolution of a telescope. Light from a point source at a large distance on the telescope axis is brought to a focus where it forms a Fraunhofer diffraction pattern with intensity having the form sketched in (a). The angular width, from centre to first zero, is $\alpha = 1.22\lambda/D$ if the lens is circular. A second source, displaced in direction from the first by ϕ, produces the shifted diffraction pattern shown in (b). According to the Rayleigh criterion, the telescope 'resolves' two such sources as distinct, meaning that the superposed images are clearly double, provided that $\phi \geq \alpha$.

The finite wavelength of light, of course, controls the (fundamental limit on the) resolution of any kind of optical measurement. We shall meet the resolution limit $\delta\bar{\nu}$ of a grating spectrometer in Chapter 4; the reasoning builds very directly on statement (3.20a). Implications for the image-forming performance of a microscope are discussed in Chapter 12. And it is diffraction in the reading optics that limits the density with which digital information can be stored on a compact disc, holding music (CD) or computer information (CD-ROM): problem 3.27 and Chapter 16.

In this context a restatement of the standard Fraunhofer-diffraction result (eqn 3.14) is often useful. Suppose collimated light is imaged by a lens of diameter D and focal length f. The image has width (centre to first zero) of order $f\lambda/D$ (perhaps 1.22 times this, but this isn't a dramatic factor). The ratio f/D is known as the f-number of the lens, so we have

$$\text{width of diffraction pattern} \sim \lambda \times f\text{-number}. \tag{3.21}$$

This equation acts as a useful reminder that the image size is of the order of the wavelength. But more importantly it gives the coefficient in a simple form, easily related to the shape of the equipment being used.

3.17 Surfaces whose fields are related by a Fourier transform

Almost all sections from §3.6 onwards have treated diffraction as the outcome of placing some aperture, with transmission function $T(\boldsymbol{r})$, in the path of a light beam, usually a light beam on its way to forming an image. But we have seen in §3.10 that there is another and more fundamental way of looking at diffraction, even within the Fraunhofer case. This is to think about a *field distribution* on a 'secondary-source surface' S_1, giving rise to Huygens secondary waves that propagate to some other surface S_2. From this viewpoint, there is no need to mention any aperture, even though an aperture may have shaped the field distribution on S_1. Indeed, the difference between the two viewpoints ($T(\boldsymbol{r})$ and field distribution) has already been somewhat blurred: problem 3.19 investigates diffraction from a field distribution where there's little participation by an aperture; and example 3.1 (p. 61) reminds us how far the meaning of 'aperture' can be stretched.

The 'field distribution' viewpoint prompts an obvious question: can we find pairs of surfaces on which the electromagnetic fields are related by Fraunhofer–Fourier relations; and how can such pairs of surfaces be identified?

Figure 3.19 shows two pairs of surfaces that meet the requirement just stated. Problem 3.28 investigates the Fraunhofer–Fourier properties of the surfaces and shows how we may reason our way towards recognizing the surfaces chosen as being appropriate. The case of two concave surfaces will reappear when we discuss cavities and the confocal Fabry–Perot in Chapter 8.

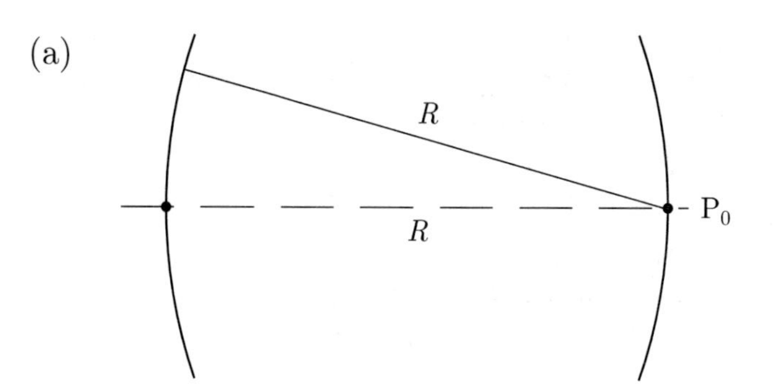

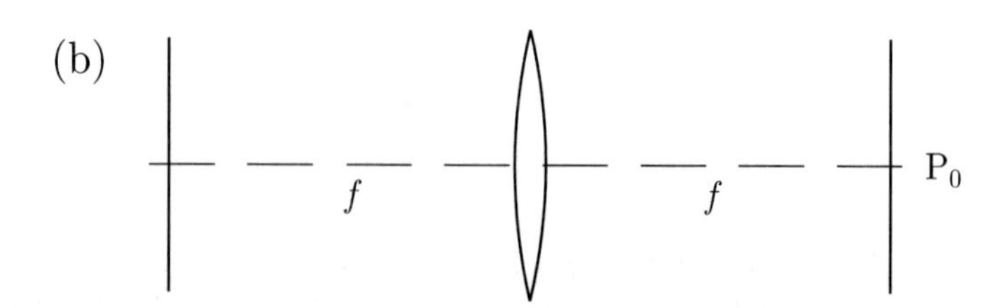

Fig. 3.19 Two pairs of surfaces on which the fields are Fraunhofer–Fourier related. (a) Two curved surfaces that meet the requirement. For reasons you are asked to invent in problem 3.28, each surface has its centre of curvature on the other surface. If these surfaces are made reflecting the result is a 'symmetrical confocal cavity', described in Chapter 8. Either surface may be the S_1 of the text. (b) Two plane surfaces on which the fields are related by a Fourier transform.

3.18 Kirchhoff boundary conditions: a harder look

Return to §3.4 and to an optical system containing an aperture. For evaluating a Kirchhoff diffraction integral, we usually assume that in the transparent part of the aperture $U_{\text{trans}}(x, y)$ is the same as the amplitude of the incident wave. This assumption is very dubious:

(a) because it's not plausible that an opaque obstacle could have so little effect on $\boldsymbol{E}$ or $\boldsymbol{H}$—nor that different kinds of obstacle (say razor blades or black cardboard) could have the *same* effect;
(b) because the assumption requires $\boldsymbol{E}$ and $\boldsymbol{H}$ to have discontinuities at the aperture that are inconsistent with Maxwell's equations (conflict with the continuity of $\boldsymbol{E}_{\text{tangential}}$ and $\boldsymbol{H}_{\text{tangential}}$);
(c) because it's fairly easy to invent paradoxical examples where the diffracted wave has impossible polarization properties.[35]

Nevertheless, the standard theory predicts diffraction patterns with high accuracy. We have a problem to understand how a dubious theory can be so successful.

We must return to firm ground: Maxwell's equations and the correct electromagnetic boundary conditions. Figure 3.20 shows E-fields calculated (properly, from the full Maxwell equations) for light transmitted by a single slit.[36] The material of the slit jaws has been given zero thickness and infinite conductivity. The slit is long in the y-direction and has width a running from $x = -\frac{1}{2}a$ to $x = \frac{1}{2}a$. It is illuminated normally. For the particular cases computed, the slit has width $a = 5\lambda$.

Figure 3.20(a) shows the electric field in the plane of the slit: the plane that we would have to integrate over if evaluating a Kirchhoff

[35] Problem 3.29 asks you to think about this.

[36] For aficionados: A slit in field E_y has E_y as the only Cartesian component of the $\boldsymbol{E}$-field. For calculation, the slit is first replaced by a Babinet-complementary strip in field H_y (Born and Wolf 1999, §11.3). The edges of the strip are used as the foci of elliptical–hyperbolic coordinates μ, θ, y. The wave equation for H_y separates in these coordinates, giving Mathieu functions of μ and of θ. Eigenvalues and eigenfunctions are found by requiring the θ functions to be periodic with period 2π. Field H_y is expressed as a sum over these eigenfunctions; the coefficients are found by requiring H_y at the strip to obey the boundary conditions. The solution (transformed back to the original slit and E_y) may be evaluated where needed, such as in the plane of the slit ($\mu = 0$) or in the far field ($\mu \to \infty$) (Sieger 1908). The E_x polarization is treated similarly. Although the mathematics is a bit untidy, its structure is entirely 'as expected'. Algebraic properties of the Mathieu functions are given by Abramowitz and Stegun (1965). Numerical evaluation was once an obstacle, but is so no longer, given software such as Mathematica.

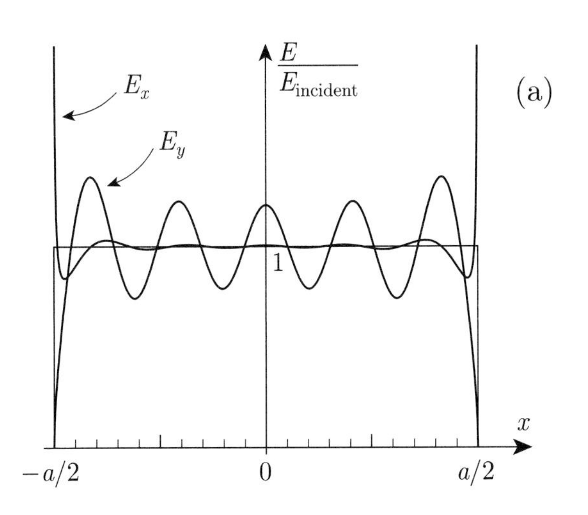

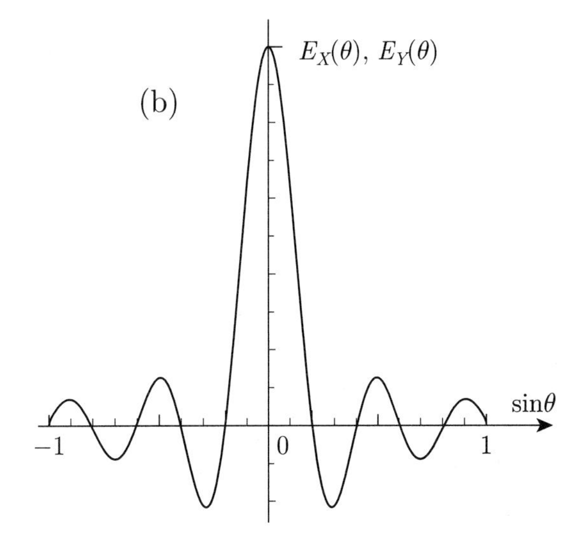

Fig. 3.20 (a) The electric-field amplitude in the plane of an infinitely conducting slit; the slit length lies in the y-direction and its width in the x-direction. The slit is illuminated normally with light whose $\boldsymbol{E}$-field lies either along (E_y) or across (E_x) the length of the slit. For the particular case shown the slit width $a = 5\lambda$. The top-hat curve shows what the field would be according to the usual Kirchhoff assumptions. (b) The amplitude diffracted to the far field, plotted as a function of $\sin\theta$ where θ is the angle to the 'straight-through' direction. Within the thickness of the printed line, it is the same for both choices of polarization direction, and also coincides with the Kirchhoff expression $\sin(\frac{1}{2}ka\sin\theta)/(\frac{1}{2}ka\sin\theta)$. All curves (except the top-hat) have been calculated 'properly' from Maxwell's equations.

integral. The fields drawn[37] are E_x and E_y, for the separate cases where the incident wave has its $\boldsymbol{E}$-field oriented across (E_x) or along (E_y) the slit. If Kirchhoff assumptions applied, both curves (they are normalized to the value of the incident field) would have the top-hat shape that is also drawn.

The field E_y is zero at the edges of the slit, where it is tangential to an ideal conductor. Within the open area of the slit, E_y is 'trying' to follow the Kirchhoff-assumption top-hat curve, but is not succeeding very well.

By contrast, field E_x goes to infinity at the edges of the slit because it is encountering an infinitely curved conducting surface; this infinity is an artefact of modelling the slit jaws as having zero thickness.[38] Apart from these sharp rises near the edges of the slit, this field component agrees relatively well with the Kirchhoff assumption.

Figure 3.20(b) shows the field[39] Fraunhofer-diffracted (i.e. in the far field), again calculated in full from Maxwell's equations, and plotted for the E_y case. Within hardly more than the thickness of the printed line, this curve is the same for both polarization choices, in spite of the differences between the fields at the slit. To similar accuracy, it is also the same as $\sin(\frac{1}{2}ka\sin\theta)/(\frac{1}{2}ka\sin\theta)$. Our customary assumptions work very well indeed for calculating the diffraction pattern.

For neither polarization is the agreement improved if we compare instead with $\frac{1}{2}(1+\cos\theta)\times\sin(\frac{1}{2}ka\sin\theta)/(\frac{1}{2}ka\sin\theta)$, containing an obliquity factor, even though the angle of diffraction θ ranges as large as $90°$. This experience reinforces the attitude we took in §3.5: that the obliquity

[37] Strictly, the field in the plane of the slit is a complex quantity, with a small and oscillating imaginary part. For convenience we have plotted the real part of E for both polarization cases; graphs of $|E|$ are hardly different.

[38] We wish the jaws to be thin in order to avoid complications from the precise shape and thickness of the jaws at the opening. The infinite conductivity is necessary to make the very thin jaws opaque. In turn, the high conductivity makes the jaws ideally reflecting, so all energy that does not pass through the slit is reflected. There is no respectable way to model jaws that are both thin and 'black'.

[39] With a suitable phase choice, this far-field amplitude is almost wholly real, and it is the real part that is plotted. A graph of $|E|^2$ is hardly different from one of $(\mathrm{Re}\,E)^2$.

factor is best ignored for all purposes except disposing of a 'backward' wave.

The graphs in Fig. 3.20 confirm the itemized statements at the beginning of this section: Kirchhoff assumptions give a poor representation of the field in the plane of the slit, yet give a remarkably good approximation to the diffraction pattern. But—again we ask—why?

Enlightenment can come from (in particular) the curve for $E_y(x)$ in Fig. 3.20(a). Let the difference between the actual field and the top-hat be $E_{\text{corr}}(x)$. Our difficulty will be solved if we can find a reason why E_{corr} might make a negligible contribution to the diffracted field. Function $E_{\text{corr}}(x)$ is an oscillating function of x, a standing wave in the direction across the slit, with (in this case) five periods within the slit width $a = 5\lambda$: the wavelength of the oscillation is close to λ. We are reminded of the sinusoidal field in example 3.1: E_{corr} behaves as though it had passed through a grating with a grating period d of the order of a wavelength. For such a grating the grating condition $d \sin\theta = \lambda$ makes θ approach 90°. Little, if anything, can be diffracted to such angles.

More formally: if we were to Fourier-analyse E_{corr} into sinewaves dependent upon x, almost all its components would have wavelengths of λ or less; such components make no contributions to the far-field diffraction.[40] A similar argument can be given for the other polarization.

It seems, therefore, that there are good reasons why we can get away with using Kirchhoff boundary conditions at a diffracting aperture. Nature has been unusually kind to us.

[40] One way to see this is from the grating condition which gives $\sin\theta > 1$, meaning that there is no angle to which this component can be diffracted. An alternative explanation runs as follows. If we insert a field with such a rapid variation with x into the wave equation, we find that it must have exponential decay in the z-direction, with a decay distance of the order of a wavelength. Therefore, these components give rise to fields that are localized in the vicinity of the slit and do not extend out into the far field—or even into the Fresnel-diffraction 'near field'.

Problems

Problem 3.1 (b) The Kirchhoff integral: a paradox
Equation (3.3) tells us that the amplitude $\psi(\boldsymbol{R})$ depends not only on the amplitude $\psi(\boldsymbol{r})$ at location $\boldsymbol{r}$ in the aperture, but also on the gradient of ψ in a direction normal to the surface. Why should the physics require knowledge of both?

Something odd seems to have happened between eqns (3.3) and (3.4). In eqn (3.3) we needed to specify both ψ and $\nabla\psi$ on the boundary surface (which might be the open part of the aperture drawn in Fig. 3.1); but in eqn (3.4) it seems that we need to specify only ψ. What happened to $\nabla\psi$, was it necessary in eqn (3.3), and if so why is it apparently not necessary in eqn (3.4)?[41]

[41] The Kirchhoff integral can be expressed in alternative forms; see Born and Wolf (1999), §8.11.2. If it is known that nothing arrives onto a diffracting aperture at $z = 0$ from infinity on the $+z$ side, the integrand can be shaped to contain either ψ only (Dirichlet boundary condition), or $\nabla\psi$ only (Neumann boundary condition). However, neither of these rearrangements is available in the general case where radiation might arrive from any direction.

The question being raised here can be answered without making use of the result just quoted, though the answer should, of course, be compatible with it.

Problem 3.2 (b) The formal solution of the scalar wave equation

(1) Derive eqn (3.4) from eqn (3.3). You will need to discard quantities of order λ/r_{s} and λ/r_{P}.

(2) [Harder] If you're expert at mathematics, have a go at deriving eqn (3.3) from the scalar wave equation, without help from books.[42]

[42] A derivation is given by Klein and Furtak (1986), appendices A and B.

Problem 3.3 (c) Fresnel diffraction
In comment 2 on p. 48, it is stated that the obliquity factor disposes of the exactly-backward wave because, in a direction pointing straight back towards the source, $\cos\theta_{\text{dif}} = -\cos\theta_{\text{inc}}$, which makes the obliquity factor zero. But this can't be the whole story. For other nearby directions, the obliquity factor is not quite zero, so there could apparently be a little light diffracted backwards. Adapt the discussion of Fresnel zones (problem 3.25 assists with gaining familiarity) to deal with the addition of amplitudes in the near-backward direction, and sort out why the diffracted amplitude really should be zero—or at least so small that it is within the uncertainty of our approximations.

Problem 3.4 (b) Complementary screens
(1) Consider the diffraction from the complementary screens drawn in Fig. 3.3, and consider the Fraunhofer diffraction patterns drawn in parts (c)–(e). Suppose that the unobstructed beam of light is 50 mm wide in the region where the aperture is to be placed. Suppose that the slit and strip both have width 0.1 mm. The lens has focal length 500 mm and the light has wavelength 500 nm. Give a quantitative description of the three diffraction patterns, and confirm the impression given in the text: that the 'spike' is narrow in relation to the rest of the pattern.
(2) [Harder] What about the intensity in the 'wings' of the spike?

Problem 3.5 (a) What 'one dimension' means
Refer to §3.9. Set up the calculation for Fraunhofer diffraction at a long slit (width a in the x-direction, length b in the y-direction), looking at the distribution of the diffracted amplitude in the XZ plane ($\phi = 0$) only. Check the derivation of eqns (3.8) and (3.10) in §3.8, making sure that you agree with the simplifying steps made there.

Problem 3.6 (a) The non-calculus method for diffraction zeros
Fraunhofer diffraction from a slit of width a gives a zero at $\sin\theta = \lambda/a$. To get an exact zero, we must have waves interfering destructively, and doing so exactly. We need to 'see why'.

In Fig. 3.4, take a Huygens source of width dx just above the bottom of the slit, and another just above the middle; they are separated by $\frac{1}{2}a$. Adjust θ so that these elements make cancelling contributions to the amplitude reaching P: $\frac{1}{2}a\sin\theta = \frac{1}{2}\lambda$. Add up the effects of similar paired elements until all elements within the slit width have been used once and once only. They all cancel. So there is a diffraction zero at $\sin\theta = (\frac{1}{2}\lambda)/(\frac{1}{2}a) = \lambda/a$.

Extend this argument to the other zeros, at $\sin\theta = p\lambda/a$.

Compare with problem 2.14(7).

Problem 3.7 (a) Revision on elementary diffraction
(1) Use the diagram of Fig. 3.2. Suppose that light passes normally through a circular aperture with diameter 1 mm and is observed 50 mm downstream at a point on an axis through the centre of circle. Investigate

the range of values taken by $1/r_P$ and show that a negligible error is incurred if $1/r_P$ is taken outside the integral in eqn (3.6). Investigate the range of values taken by kr_P and show that $\exp(ikr_P)$ can *not* be taken outside the integral. Make really sure you understand why the same distance r_P is treated differently in the two cases.[43]

[43] *Comment*: The dimensions given in this part of the problem have been chosen to be quite unusual: r_P on the small side and the hole diameter on the large side, compared with likely choices in an experiment. The effect is to exaggerate the variation of r_P. So if $1/r_P$ can be treated as constant here, it can be in any other likely case.

(2) For the rest of this problem, use the diagram of Fig. 3.4 to define an apparatus layout for Fraunhofer diffraction. To start with, the diffracting aperture will be a long slit of width a. The slit is shown in cross section in Fig. 3.4, with its long dimension normal to the paper. The incident wave has amplitude $U_0\,\mathrm{e}^{-\mathrm{i}\omega t}$ at the slit. Use problem 3.5 to explain why eqn (3.10) gives a correct expression (proportionality constants apart) for the diffracted amplitude $U_{\mathrm{dif}}(\beta)$.

(3) Evaluate the integral in eqn (3.10) for the case described in part (2). *Answer*: $U_0\,\mathrm{e}^{-\mathrm{i}\omega t}\,a\,\mathrm{sinc}(\frac{1}{2}\beta a)$, where $\mathrm{sinc}(x) \equiv (\sin x)/x$ and $\beta \approx kX/f$.

(4) Resist looking back to Fig. 3.7 until you have given this part a go. Draw a sketch graph of the amplitude, and of the intensity, as a function of the coordinate X defined in Fig. 3.4. Plot for both positive and negative values of X.

Now check: Does your central maximum have the same width as the 'side' maxima (it shouldn't)? How high is the first 'side maximum' relative to the central maximum? Do you agree with the statements made in sidenote 26 on p. 58? What fraction of the energy do you think is under the central maximum, and what under the rest of the pattern?

(5) Suppose that the width of the departing light beam in Fig. 3.4 is limited only by the outline of the lens itself (taken as a square of side a, to keep things simple). Then the lens's f-number is f/a. Take the 'width' of the diffraction pattern as the distance from centre to the first zero on one side, as is usual in optics. Show that the width is $\approx f\lambda/a$. Express this as $\lambda \times (f\text{-number})$. This is in agreement with eqn (3.21).

(6) What happens when the slit width a is doubled: to the amplitude at the centre of the diffraction pattern; to the intensity at the centre of the diffraction pattern; to the width of the pattern? A slit of doubled width should transmit double the energy: is this consistent with what you have found?

(7) Look at some angle θ away from the centre of the pattern at $\theta = 0$. Show that the intensity at this fixed θ is an oscillating (a $\sin^2$, not a sinc^2) function of a.

(8) A double slit consists of two slits of width a, whose centres are separated by distance d. Show that the integral in eqn (3.10) evaluates to[44]

$$(\text{constant}) \times 2a\,\mathrm{sinc}(\tfrac{1}{2}\beta a) \times \cos(\tfrac{1}{2}\beta d). \tag{3.22}$$

Compare this expression with eqn (2.24).

(9) Give an account of what happens to the amplitude diffracted from a double slit when the slit width a is increased to approach the slit separation d. Check that the right things happen to the amplitude in expression (3.22). Compare with problem 2.13(5).

[44] *Comment*: Part (8) explains a piece of optical jargon. People sometimes make a verbal distinction between interference and diffraction in the following way. The sinc factor in eqn (3.22), arising from each building-block slit, is said to represent *diffraction* from each slit; and the cosine factor is said to represent *interference* of light from one slit against light from the other. This usage is quite helpful, given what we now understand about the convolution theorem. (X-ray crystallographers make a similar distinction between the structure factor contributed by each basis, and the reciprocal lattice contributed by repetitions of those bases. This isn't surprising because they too are being assisted by the convolution theorem.)

At the same time, we must be aware that the usage just introduced is merely a convenience associated with Fraunhofer diffraction from repeated units. There is at root no fundamental distinction being made here between interference and diffraction. It remains as true as it always was that diffraction consists of the *interference* of Huygens secondary waves where they are superposed at places downstream from their Huygens sources.

(10) Interpret the amplitude transmission function $T(x)$ for the two slits in terms of a convolution: a single-slit top-hat function convolved with two δ-functions. Interpret the shape of eqn (3.22) as a consequence of the convolution theorem. Compare with problem 2.19(3) and eqn (2.24).

Problem 3.8 (b) The proportionality constant isn't constant
The derivation given in §3.8 of eqn (3.10) included dropping the factor e^{ikd}, where d is the distance measured along an optical path from O to P in Fig. 3.4. The factor e^{ikd} is usually neglected in elementary work, for the reason given: it has no effect on the intensity. However, the factor under discussion is not usually constant, and it will be important if we wish to pursue the diffracted light further downstream from the plane at P_0.

(1) Invent a ray diagram that shows, without calculation or difficult geometry, what is the apparatus arrangement that makes the factor e^{ikd} constant; i.e. it makes distance d from O to P independent of the X-coordinate of P.

(2) [Harder] Show that the light arriving at X in Fig. 3.4 has (to order X^2)

$$d = \text{constant} + X^2/2R, \qquad \text{where} \qquad R = f^2/(f - \text{OL}).$$

Check this expression for consistency with the ingenious arrangement you found in part (1).

Problem 3.9 (a) Fraunhofer geometry
Derive expression (3.11) for r_{P} in the two geometries:[45]

(1) where d is extremely large and no lens is used

(2) where a lens is used and the diffracted amplitude is required in the focal plane of the lens.

[45] *Hint*: Look first at the plane containing the $\boldsymbol{k}$-vectors of the waves arriving and leaving. Show that light diffracted via point $\boldsymbol{r}$ in the aperture acquires a phase lag of $(\boldsymbol{k}_{\text{in}} - \boldsymbol{k}_{\text{out}}) \cdot \boldsymbol{r}$. Finally, use the xyz coordinate system of §3.9 and work out the scalar product in that system. Compare with similar expressions involving $\exp\{i(\boldsymbol{k}_{\text{in}} - \boldsymbol{k}_{\text{out}}) \cdot \boldsymbol{r}\}$ in X-ray diffraction from crystals.

Problem 3.10 (b) The coefficient of the diffraction integral
In §3.9, eqn (3.13) has a coefficient $1/(i\lambda f_2)$ patched in front of it. This needs investigation.

Use the diagram of Fig. 3.21, in which we wish to find the amplitude of light diffracted to points like P_2 in plane 2. Start by making plane 2 *not* the focal plane of the lens. A possible procedure should now be recommending itself: find the image P_1 of P_2, then transfer the diffraction problem to the 'object space' of the lens. All optical paths from the aperture to P_1 are straight and uninterrupted.

(1) Use eqn (3.6) to show that the amplitude diffracted to plane 1 is (or would be if the lens wasn't in the way)

$$U_{\text{dif}}(X_1, Y_1) = \frac{1}{i\lambda} \int U_{\text{trans}}(x, y) \frac{\exp(ikr_{\text{P1}})}{r_{\text{P1}}} \, dx \, dy,$$

where r_{P1} is the optical path from point (x, y) in the plane of the aperture to point P_1.

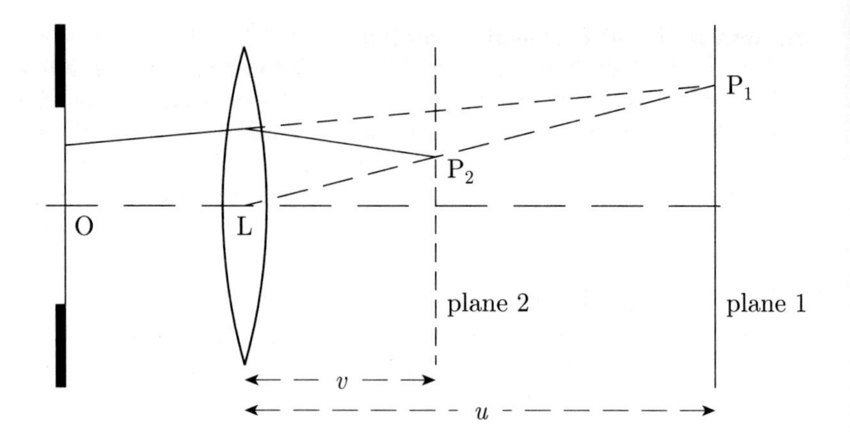

Fig. 3.21 The geometry needed for problem 3.10. It is desired to calculate the amplitude diffracted to plane 2. Plane 1 is the image of plane 2 formed by the lens, so light that will reach P_2 appears (when viewed from the upstream side of the lens) to be travelling towards P_1.

(2) Argue that energy conservation requires the amplitude of the light in plane 2 to be greater in magnitude than that in plane 1 by a factor u/v, where distances u and v are indicated in Fig. 3.21 (both positive as drawn).

(3) Argue that the phase factor for the wave arriving at point P_2 should be modified from $\exp(ikr_{P1})$ to $\exp(ikr_{P2})$, where r_{P2} is the optical path from (x, y) to point P_2. Argue further that the difference between r_{P1} and r_{P2} is independent of x, y *because* P_1 *and* P_2 *are object and image.*

(4) Put parts (2) and (3) together to argue that the amplitude diffracted to plane 2 is

$$U_{\text{dif}}(X_2, Y_2) = \frac{u}{v}\frac{1}{\text{i}\lambda}\int U_{\text{trans}}(x, y)\frac{\exp(ikr_{P2})}{r_{P1}}\,dx\,dy.$$

(5) None of what has been done so far has had reference to the Fraunhofer condition; it would apply 'anyway'. Now let $u \to \infty$. Show that in this limit $u/(v\, r_{P1}) \to 1/f$.

Problem 3.11 (a) Energy conservation
Consider light passing through a general two-dimensional aperture and forming a Fraunhofer diffraction pattern in the focal plane of a lens. Using Kirchhoff assumptions about the fields in the aperture, show that the power passing through the aperture is exactly accounted for by the power arriving at the diffraction pattern.[46]

[46] *Hint*: Parseval's theorem.

Problem 3.12 (a) Careful English
The second viewpoint of §3.10 emphasized the rôle of the aperture function $T(x, y)$ in *limiting the width* of the beam passing through the optical system. The cases being thought about at the time were circular or rectangular apertures, which do just that. But a diffraction grating does not limit the width of the transmitted beam, and yet it spreads light into the grating orders. Devise a more correct statement.[47]

[47] *Hint*: Fourier and example 3.1.

Problem 3.13 (b) Perspective drawings
All the 'perspective' drawings in this chapter have been drawn isometrically, that is, parallel lines in the object are drawn parallel on the paper, rather than drawn converging towards a vanishing point. Check that this rule has been followed. In particular, check that the rather awkward-looking ray shown in Fig. 3.6 has been drawn to conform to the ray-optics rules.

Problem 3.14 (a) The rectangular aperture
(1) Calculate the Fraunhofer diffraction pattern produced when collimated light is passed through a rectangular aperture with sides a, b, and then focused by a lens with focal length f. Show that[48]

$$U_{\text{dif}} = \frac{U_0\,\mathrm{e}^{\mathrm{i}kd-\mathrm{i}\omega t}}{\mathrm{i}\lambda f} \times ab \times \frac{\sin(\frac{1}{2}ka\sin\theta)}{(\frac{1}{2}ka\sin\theta)} \times \frac{\sin(\frac{1}{2}kb\sin\phi)}{(\frac{1}{2}kb\sin\phi)}.$$

[48] Angles θ and ϕ are defined in Fig. 3.6.

Sketch the pattern, including the 'nodal' lines (i.e. the lines where the amplitude is zero). Pay particular attention to the proportions around the centre of the pattern.

(2) Sketch the pattern for the case $b \gg a$, making sure you get the long and short dimensions 'the right way round'.

(3) Describe carefully the conditions under which the diffracted amplitude factorizes, as it does here, into (function of θ) $\times$ (function of ϕ).

Problem 3.15 (b) The 'width' of a diffraction pattern
Use Fig. 3.7(b). The 'width' of the peak is conventionally taken as $\Delta\rho = \pi$. Calculate the range of ρ that gives the FWHM of the peak in $\{(\sin\rho)/\rho\}^2$, and assess the reasonableness of the convention.

Do the same for the circular-aperture pattern drawn in Fig. 3.11(b).

Problem 3.16 (b) The circular aperture
The apparatus layout is shown in Figs 3.5 and 3.22. A collimated beam of light is incident normally from 'behind' the circular aperture, and we are to calculate the amplitude diffracted to the focal plane of the lens. The incident light has complex amplitude $U_0\,\mathrm{e}^{-\mathrm{i}\omega t}$, uniform over the open area of the aperture.

(1) Argue that it suffices to calculate the diffracted amplitude at points on the Y-axis because the optical system has rotational symmetry about the z-axis.

(2) In the diagram, the area of the circular aperture has been divided into strips of width $\mathrm{d}y$ parallel to the x-axis. Use the geometry of problem 3.9(2) to show that the optical path to P from all points on the strip is the same: $d - y\sin\phi$, where d is the optical distance (via the lens) from the centre of the aperture to P. (Yes, this holds for all points along the length of the strip; think about it.)

(3) Use the properties just obtained, together with eqn (3.13), to show

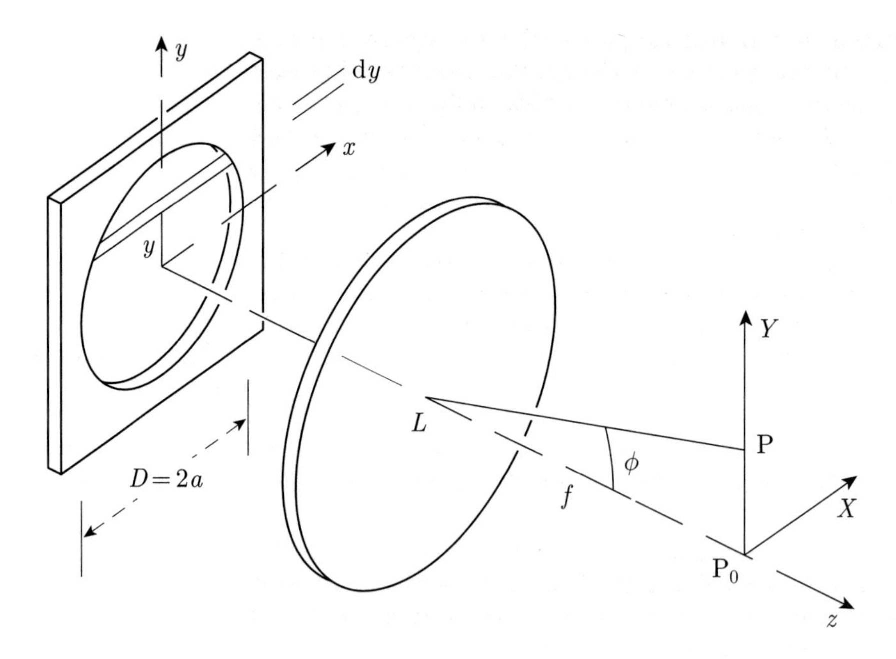

Fig. 3.22 Fraunhofer diffraction caused by passage of light through a circular aperture. The diagram defines the layout used in problem 3.16. All optical paths from the strip of width dy to point P are equal to each other, and are (constant) $- y\sin\phi$.

that the amplitude diffracted to P is

$$U_{\mathrm{P}} = \frac{1}{\mathrm{i}\lambda f}\int_{-a}^{a} U_0\,\mathrm{e}^{-\mathrm{i}\omega t}\,\mathrm{e}^{\mathrm{i}k(d-y\sin\phi)}\,2\sqrt{a^2-y^2}\,\mathrm{d}y.$$

Use the substitution $\xi = y/a$ to tidy this integral into

$$U_{\mathrm{P}} = \frac{2U_0\,a^2\,\mathrm{e}^{\mathrm{i}(kd-\omega t)}}{\mathrm{i}\lambda f}\int_{-1}^{1}\mathrm{e}^{-\mathrm{i}(ka\sin\phi)\xi}\sqrt{1-\xi^2}\,\mathrm{d}\xi. \tag{3.23}$$

(4) The integral here is a function of $\rho = ka\sin\phi$. Look up the properties of Bessel functions in some work of reference such as Abramowitz and Stegun (1965), and identify the integral with $\pi\mathrm{J}_1(\rho)/\rho$.

(5) Tidy the expression for the diffracted amplitude so that it contains the area of the aperture and a function of ϕ that is 1 at the centre of the pattern:

$$U_{\mathrm{P}} = \frac{U_0\,\mathrm{e}^{\mathrm{i}(kd-\omega t)}}{\mathrm{i}\lambda f}\times\pi(D/2)^2\times\frac{2\mathrm{J}_1(\frac{1}{2}kD\sin\phi)}{\frac{1}{2}kD\sin\phi}. \tag{3.24}$$

Here D is the *diameter* of the circular aperture, and is used in place of a to make the result more directly comparable to that for a square of side D.

(6) The diffracted amplitude has zeros where $\sin\phi = 1.22\lambda/D$, $2.23\lambda/D$, $3.24\lambda/D$, aiming for (integer $+\frac{1}{4}$)λ/D when the integer is large. The central maximum is 22% wider than the central maximum for a rectangular aperture of the same width D. Look at Fig. 3.22 and imagine

a square aperture of side D replacing the circle: the square transmits a little more light 'in the corners'. So the circular aperture is in places 'narrower' than the rectangle. Argue that a wider diffraction pattern is to be expected for the circle, and that a width greater by 20% or so is rather easy to live with.

Problem 3.17 (a) Two pinholes
Describe, with as much quantitative detail as you can but without doing any lengthy algebra, the Fraunhofer diffraction pattern caused by passage of light through two circular pinholes. Concentrate particularly on listing those places where the amplitude is zero.[49]

[49] *Hint*: Remember the convolution theorem.

Problem 3.18 (a) The eye ring
The **eye ring** of a telescope or microscope is a real image of the objective lens as formed by the eyepiece. It is the best place for the observer to place her eye pupil.

(1) Assume that the eye ring and the eye pupil of a telescope coincide in location and have the same diameter. Show that the diffraction limit on resolution is the same whether we regard the diffracting aperture as the outline of the objective or that of the eye pupil.

(2) Identify the penalties that are exacted if the eye ring is made either much larger than, or much smaller than, the eye pupil.

Problem 3.19 (a+) Plane wave incident at an angle[50]

[50] This problem is classified as a+ to indicate that it is of particular importance.

(1) Consider the situation as drawn in Fig. 3.4, with a plane wave of amplitude $U_0\,\mathrm{e}^{-i\omega t}$ incident normally on a long slit of width a. Evaluate Kirchhoff integral (3.10) and show that $U_{\text{dif}}(\beta) \propto U_0\, a \,\mathrm{sinc}(\frac{1}{2}a\beta)$ with $\mathrm{sinc}(x) \equiv (\sin x)/x$. Compare your result with eqn (2.23).

Henceforth, the slit is illuminated by a plane wave travelling 'uphill' at angle α to the z-axis: $\boldsymbol{k} = (k\sin\alpha, 0, k\cos\alpha)$.

(2) Show that the amplitude $U_{\text{trans}}(x, y)$ of the light in the plane of the aperture is $U_0\,\mathrm{e}^{-i\omega t}\exp(i\beta_0 x)$, where $\beta_0 = k\sin\alpha$. Because it is independent of y, $U_{\text{trans}}(x, y)$ will in future be abbreviated to $U_{\text{trans}}(x)$.

(3) Calculate the amplitude diffracted to the focal plane of the lens.

(4) Use the variable $\beta = k\sin\theta$ as in eqn (3.10), and show that the diffracted amplitude $U_{\text{dif}}(\beta)$ is a 'shifted' version of that calculated in part (1), displaced along the β-axis by $\beta_0 = k\sin\alpha$.

(5) Derive the result of part (4) by using the 'shifting theorem' of problem 2.11, or perhaps more suitably that of problem 2.16.

(6) Draw a ray diagram,[51] and show that ray optics would bring all the light to a point focus in the lens's focal plane and at an angle α away from the axis. Check this against parts (3) and (4): the light 'tries' to come to a point focus at the ray-optics place, but spreads a bit because of diffraction.

[51] A model may be seen in Fig. 12.4 on page 277.

(7) Compare the diffracted amplitude $U_{\text{dif}}(\beta)$ with the $\delta_N(x)$ of problem 2.7. Show that as we enlarge the slit width, $a \to \infty$, $U_{\text{dif}}(\beta)$ tends to a δ-function at $\beta = \beta_0$. Check for consistency[52] by showing that the

[52] *Hint*: Look back at eqn (2.28), and at the comment on signs in §3.8.

Fourier transform of $e^{i\beta_0 x}$ (having this form for *all* x) is $2\pi\,\delta(\beta-\beta_0)$.

Comment 1 These results are full of important insights into Fourier transforms.

Comment 2 Our $U_{\text{trans}}(x)$ is almost exactly $e^{i\beta x}$ with a single β, and its diffracted field $U_{\text{dif}}(\beta)$ has a peak around this β (namely β_0). The peak indicates that there is a lot of this β present in $U_{\text{trans}}(x)$ and not much of anything else. The peak in the diffracted field is telling us very clearly about the original $e^{i\beta_0 x}$.

Comment 3 The peak of $U_{\text{dif}}(\beta)$ just fails to be a δ-function because $U_{\text{trans}}(x)$ starts and stops at $x=\pm a/2$. But if we make a very large, $U_{\text{trans}}(x)$ becomes more like an uninterrupted $e^{i\beta_0 x}$, and the Fourier transform becomes more concentrated around $\beta=\beta_0$. Again, the diffracted field puts on display the presence of the original $e^{i\beta_0 x}$; by increasing a we make this even clearer by removing a distraction.

Comment 4 Any general amplitude $U_{\text{trans}}(x)$ is built up from many sinewaves like $e^{i\beta x}$, and what is displayed in the lens's focal plane as $U_{\text{dif}}(\beta)$ is the spectrum of these sinewave components: each β going to its own place.[53] We can learn to look at $U_{\text{dif}}(\beta)$ and understand its relation to the $e^{i\beta x}$ components in the original $U_{\text{trans}}(x)$. That's what it means that the Fourier transform of $U_{\text{trans}}(x)$ is on display in the lens's focal plane.

[53] This idea is developed in a different way in problem 3.30.

Comment 5 The converse of comment 2. The spread of diffracted light around $\beta=\beta_0$ tells us that $U_{\text{trans}}(x)$ starts and stops at $x=\pm a/2$. The range of βs must be just what we would need, if we *reconstructed* $U_{\text{trans}}(x)$, to synthesize the original $U_{\text{trans}}(x)$ with its start and stop at $x=\pm a/2$.

Comment 6 We could quite easily use the amplitude $U_{\text{dif}}(\beta)$ to perform a reconstruction ('synthesis') of $U_{\text{trans}}(x)$ by introducing a second (optical) Fourier transform. All it needs is a second lens .... This idea is explored in Chapter 12.

Problem 3.20 (a) A grating that gives only three orders

Consider the physical situation described in example 3.1 on p. 61.

(1) Confirm the algebra outlined in example 3.1.

(2) Show that each grating order consists of a single-slit diffraction pattern, where the 'slit' is the boundary of the grating. Show that the angular width $\delta\theta$ of a first-order peak corresponds to diffraction at a slit of width $D\cos\theta$ (where D is the width of the grating), and interpret this result in terms of the width of the light beam leaving the grating.

(3) How would you specify a new $T(x)$ in order to make a grating that gives orders $0, \pm1, \pm3, \pm4$ only?

Problem 3.21 (b) Fresnel zones and Fresnel number

(1) A plane wave is incident normally on a circular aperture of diameter a, with diffraction observed in a plane distant Z downstream from the aperture. Use Fig. 3.14 to write down an expression for the distance

r_P from a secondary source at $(x, y, 0)$ to a point of observation P at (X, Y, Z). Make a binomial expansion of this expression, treating quantities like x/Z, X/Z as small. Retain terms up to those 'quadratic' in x, y and in X, Y. Identify the terms that are 'linear' in x and y and so do familiar things in a Kirchhoff integral. Identify similarly the terms that are 'quadratic' in x and y and so make the Kirchhoff integrand have a new behaviour. Show that those quadratic terms are unaffected (to the order of approximation being recommended) by the location of X, Y (for given Z). Therefore, let us agree to concentrate attention on point P_0, with $X = Y = 0$. This removes the 'linear' terms and thereby highlights the quadratic terms whose nuisance value we are to investigate.

(2) The Fresnel half-period zones are defined in §3.13. Show that, in our small-angle approximation, all such zones have the same area.

(3) Show that the Fresnel number is

$$N = \frac{1}{4}\frac{d_R}{Z},$$

in agreement with eqn (3.17), where d_R is the Rayleigh distance.

(4) Show that, when observation is made at distance $Z = d_R$, the Fresnel number $N = \frac{1}{4}$, and the greatest deviation from constancy in path from the aperture to P_0 is $\lambda/8$.

Comment: The Fresnel number needs handling with some intelligence. Think again about 'one-dimensional' Fraunhofer diffraction at a single slit. If P is placed at the first zero of the pattern, an observer looking back at the aperture from P sees optical paths differing by one wavelength; he may say that he sees two exposed Fresnel zones. Yet the quadratic contributions to optical paths are zero for Fraunhofer conditions. A Fresnel number (defined as the number of zones exposed when looking back from a general point P) is not then an ideal indicator of the *quadratic* contributions to the optical paths. That explains why this problem concentrates on the centre of the pattern, at P_0, where the Fresnel number does measure the quadratic contribution, uncomplicated by any linear part.

(5) Repeat part (1) for the geometry of Fig. 3.15, and show that the quadratic contributions to the path $S \to P$ are the same as those in path $S \to P_0$.

(6) Fill in the missing steps in the calculation of extreme optical paths $S \to P_0$ in §3.14.

(7) Adapt the idea of a Fresnel zone to the case of Fig. 3.15, and show that the Fresnel number is

$$N = \frac{1}{4}\left(\frac{d_R}{d_s} + \frac{d_R}{d_P}\right). \tag{3.25}$$

Problem 3.22 (a) Diffraction is Fraunhofer at the image of the source

(1) Argue that in Fig. 3.16 the diffraction pattern should be unaltered if the aperture is translated along the optical system between object and image—though the aperture width needs to be changed in such a way as to keep the light beam within the same ray-optics outline. Figure 3.17 may help.

(2) Look back at all the diagrams and descriptions of Fraunhofer diffraction in this chapter, and check that in most cases there was an object–image relation between the source and the plane where the diffraction was displayed. This was not, however, the case with Fig. 3.13. Make a drawing showing that even this arrangement can be considered as a limiting case of the arrangement of Fig. 3.16.

(3) Use Fig. 3.16. Investigate the depth of focus within which the Fraunhofer condition holds, according to the criterion in eqn (3.19).

(4) Use Fig. 3.16 again. Use ray optics to understand how a focus error degrades image quality. Find the depth of focus within which focus error degrades the image by the same amount as does diffraction. Show that the result is the same as that of part (3): we have Fraunhofer conditions if focus error widens the image by less than its diffraction width.

(5) Make a careful statement of a recipe by which you can achieve Fraunhofer conditions in the laboratory.[54]

[54] *Hint*: Part (5) is not quite as trivial as it sounds: the best route is not a single-step process.

Problem 3.23 (b) A brute-force check on §3.14

In Fig. 3.16, work out the incident amplitude $U_{\text{trans}}(x, y)$ directly on a *plane* surface $z = 0$ spanning the aperture, and show that

$$U(x,y) = \text{constant} \times \mathrm{e}^{-\mathrm{i}\omega t} \times \exp\left\{ -\mathrm{i}k \left(\frac{x^2 + y^2}{2d_{\mathrm{P}}} \right) + \text{quantities of 4th order in } x, y \right\}. \quad (3.26)$$

Insert this into a Kirchhoff integral aimed at finding the amplitude at $\mathrm{P} = (X, Y, Z)$, where Z is a general distance; don't attempt to evaluate the nasty integral! Show[55] that the quadratic terms disappear from the phase factor in the integrand if we choose $Z = d_{\mathrm{P}}$. You may assume that quantities like x/d_{P}, X/d_{P}, x/Z, X/Z are all small quantities of the same order.

[55] *Comment*: The integration could have been taken over a spherical surface centred on P_0, and then the quadratic terms would have disappeared in a more obvious way. However, it will be a useful insight for the future (Chapter 7) to know that a wave whose phase contains $k(x^2 + y^2)/2R$ is a spherical wave of radius R (converging as here if the coefficient of $x^2 + y^2$ is negative, diverging if the coefficient is positive).

Problem 3.24 (a) Fresnel conditions inside a laser cavity

Calculate the Rayleigh distance for a laser beam of wavelength 633 nm that is initially plane and of diameter 0.5 mm. What is the Fresnel number at a distance of 0.5 m? You should find that the near-field condition (meaning Fresnel, not Fraunhofer) applies to the radiation field within most of the length of a laser's own cavity.

Problem 3.25 (c) Fresnel diffraction from a circular aperture

A circular aperture of radius a is illuminated normally by a monochromatic plane wave of amplitude $U_0 \exp(-\mathrm{i}\omega t)$. The diffracted amplitude is required at a point P_0 distant Z downstream and on the axis (the mathematics is ugly for points off the axis).

(1) Find the distance r_P to P_0 from a point in the aperture and distant r from the axis. Expand r_P in powers to r/Z keeping terms up to order r^2. Use the result in a Kirchhoff integral to show that the amplitude diffracted to P_0 is

$$U_0 \exp(ikZ - i\omega t)\{1 - \exp(ika^2/2Z)\}. \qquad (3.27)$$

(2) Show that the amplitude at P_0 is zero whenever $ka^2/2Z$ is an integer multiple of 2π. Show that this occurs when an even number of half-period zones is exposed.[56]

(3) Show that, when just one half-period zone is exposed, the amplitude at P_0 is $2\,U_0 \exp(ikZ - i\omega t)$, which is double the amplitude that the original wave would have had there if it had been allowed to continue unobstructed.

(4) Re-express the last result in the terms in which it is usually quoted: when the contributions from all zones of an unobstructed wave are totalled, the result is half the amplitude contributed by the first zone. All other contributions cancel against each other. This explains the cancellation referred to in sidenote 6 on p. 48.

(5) A paradox is contained in eqn (3.27). As we increase the radius a of the aperture, the amplitude at P_0 oscillates without ever tending towards a limit. Talk yourself out of this.

[56] Therefore, alternate Fresnel zones give contributions that cancel against each other, and the cancellation is (almost) total.

Problem 3.26 (a) Diffraction in non-optical physics

(1) A certain radio telescope has a dish aerial 10 m across. Calculate its angular resolution when receiving radio waves at 10 GHz. What is the objective diameter of an optical telescope (assumed ideal) possessing the same angular resolution?

(2) The designer of a loudspeaker makes the speaker cone stiff so that it moves like a rigid circular piston of diameter 300 mm. He tries long and hard to make the speaker have a level frequency response from 20 Hz to 10 kHz when the sound intensity is measured on the axis of the speaker. He finds, however, that the response measured off-axis is deficient in high frequencies. Estimate the highest frequency for which the speaker can be used if the frequency response is to be much the same on the axis and at 45° away from the axis. (The speed of sound in air is 330 m s^{-1}.)

(3) A nuclear-structure physicist throws 30 MeV neutrons at nuclei in a solid target. Neutrons are absorbed and scattered by the target nuclei, and the physicist analyses his results using the textbook relation

$$\alpha = N\sigma,$$

where α is the absorption coefficient, N is the number density of targets and σ is the absorption cross section. Is he right to trust this relation? To identify the point that might worry: do things go wrong if some nuclei are 'hiding behind' others and so manage to absorb less than their due?[57]

[57] *Hint*: Rayleigh distance.

(4) A semiconductor fabricator manufactures field-effect transistors by **ion implantation.** That is, she introduces dopant atoms by hurling ions (of course ions, neutral atoms couldn't be accelerated by an $\boldsymbol{E}$-field) into silicon from a particle accelerator. She needs to dope with arsenic, and gets the right doping profile if the arsenic ions are accelerated to 100 keV. Investigate whether a statement like $\alpha = N\sigma$ (for collisions that bring arsenic ions to rest) can be trusted for this case.[58]

[58] *Answer*: No. Silicon atoms can hide behind each other, and some dopant atoms can travel down 'holes' in the crystal in a phenomenon known as **channelling**. The semiconductor manufacturer has to orient the crystal structure of the target silicon so as to minimize this nuisance effect.

Problem 3.27 (b) Diffraction and technology: the CD
A compact disc (CD) bears information encoded in the form of 'pits' arranged along a spiral track on a plastic disc; see Chapter 16. The disc is read by focusing a laser onto the disc. The spot of light so formed covers only one turn of the spiral track and can resolve the presence or absence of a pit along the track. The wavelength is 0.78 μm (from a semiconductor laser) and the light is focused by a lens into a cone of semi-angle 27° (a 'numerical aperture' of 0.45—see Chapter 12). The track runs from radius 25 mm to 58 mm. Estimate the number of bits that can be held on a CD and be read by the laser. Check your estimate by finding the playing time of a music CD, given that information is read off it at 4.32 Mbit s^{-1}.

Problem 3.28 (b) Surfaces on which fields are related by a true Fourier transform
Section 3.17 describes pairs of surfaces on which the field amplitudes are linked by a Fourier transform. The defining condition is a modification of statement (2) in §3.10:

- The optical path from a point on the first surface to a point on the second must be a bilinear function of whatever coordinates describe those points.

A 'bilinear' function? Look again at problem 3.21(1) and problem 3.23. We were happy with a path of the form $d - xX/Z - yY/Z$, because the added constant d did no harm, and the bilinear xX pieces were just what we needed to make a Fourier-type exponential in a Kirchhoff integral. But we had to exclude 'quadratic' terms involving x^2 or y^2 or xy. If we had been really careful, we'd have excluded quadratic terms involving X^2, Y^2 and XY too, because they have no place in a proper Fourier transform, even if they don't mess up the integration. The pairs of surfaces we require are identified by ridding ourselves of the unwanted 'quadratic' terms.

(1) Problem 3.21 tells us how to proceed. We need a zero Fresnel number when looking back at the first surface from a point P_0 at the middle of the second surface. Argue that this makes the first surface a sphere (of any radius) centred on P_0. This is shown in Fig. 3.19(a). Turn the diagram left-to-right, and argue that terms quadratic in the coordinates of the second surface are likewise removed by making that surface a sphere whose centre lies on the first surface.[59] (An algebraic check will be made when we look at this again in problem 8.12.)

[59] *Comment*: There are still fourth-order and higher terms in the coordinates of the two surfaces, so our result is 'exact' up to quadratic terms only.

(2) Return to Fig. 3.4 and problem 3.8. Invent an argument (adapting the ideas just given) to show that a true Fourier-transform[60] relationship exists between the fields on the left- and right-hand focal planes of the lens shown in Fig. 3.19(b). Show that (in a small-angle approximation) the complex amplitude on the right-hand plane, coordinates $(x_{\rm F}, y_{\rm F})$, is related to that on the left-hand plane, coordinates $(x_{\rm o}, y_{\rm o})$, by

$$U_{\rm F}(x_{\rm F}, y_{\rm F}) = \frac{{\rm e}^{ikd}}{{\rm i}\lambda f} \int U_{\rm o}(x_{\rm o}, y_{\rm o}) \exp\{-ikx_{\rm o}x_{\rm F}/f - iky_{\rm o}y_{\rm F}/f\}\, {\rm d}x_{\rm o}\, {\rm d}y_{\rm o}, \tag{3.28}$$

where d is the optical path along the axis from one plane to the other.

[60] *Comment*: The form of eqn (3.28) shows that the Fourier-transform relationship is to be understood as involving scaled variables, with $x_{\rm o}$ linked to $kx_{\rm F}/f$ rather than to $x_{\rm F}$ itself, and with a multiplying factor of ${\rm e}^{ikd}/({\rm i}\lambda f)$. This is in conformity with the introduction of the scaled variable β in §3.8 and with the comment at the end of §3.9.

Problem 3.29 (c) 'Impossible' polarization in diffraction
This problem picks up the challenge implicitly issued in §3.18 item (c).

Consider a single slit illuminated by a plane wave, as shown in Fig. 3.4. Let the $\boldsymbol{E}$-field of the incident radiation lie in the x-direction. On Kirchhoff assumptions the $\boldsymbol{E}$-field in the plane of the slit is likewise in the x-direction. Some light is diffracted at angle θ to the forward direction. Discuss the polarization of that diffracted light.[61]

[61] *Hint*: E_z.

Problem 3.30 (e) An alternative to Kirchhoff integrals
Problem 3.19 points us to a way of bypassing the Kirchhoff integral altogether. Here is an alternative procedure.[62]

[62] The ideas in this problem are taken from Harvey (1979).

(a) Assume, with Kirchhoff, that the amplitude $U_{\rm trans}(x, y)$ in the transparent part of an aperture is equal to that of the incident wave.
(b) Perform a Fourier analysis of $U_{\rm trans}(x, y)$ into its sinewave components ${\rm e}^{{\rm i}(\beta x + \gamma y)}$.
(c) Each ${\rm e}^{{\rm i}(\beta x + \gamma y)}$ component is just what we would have had if there was a plane wave incident obliquely (infinite in width, not curtailed by the aperture).
(d) Therefore, each ${\rm e}^{{\rm i}(\beta x + \gamma y)}$ component 'continues' as a plane wave,[63] in a direction determined by β and γ, and with the amplitude calculated in step (b).
(e) The amplitude anywhere downstream from the aperture can be found by adding up these plane waves, or what they become after passing through a lens.

[63] This 'continuation' has been investigated in problem 3.19.

(1) Show that a Fraunhofer case of diffraction, dealt with in this way, gives the usual answers.

(2) A plane wave is incident normally on a slit and diffracts. Use the new procedure to set up a calculation of the diffracted amplitude at a *finite* distance, and with no lens. The nasty Fresnel mathematics reappears, as it must, though it happens in a way different from that in problem 3.23.

(3) Think about statement (d) above. Given that there is a field whose amplitude in the xy plane is proportional to ${\rm e}^{{\rm i}(\beta x + \gamma y)}$, and no more information than this, what can be said about the direction of travel of the incident radiation? Compare with problem 3.1.

4 Diffraction gratings

4.1 Introduction

A 'conventional' spectrometer[1] receives light through an entrance slit, separates out (*disperses*) the frequencies in the light by some means, and focuses each frequency to its own place at the output. When the light contains discrete frequencies, a set of bright lines is seen at the output, each of which is an image of the entrance slit.[2] The present chapter is concerned with spectroscopic instruments that incorporate a diffraction grating as the dispersing element.

The optical system of a spectrometer forms an image of the entrance slit, with diffraction affecting the formation of that image. The diffraction, therefore, takes place under the Fraunhofer condition, according to the formulation in §3.14.[3]

[1]'Conventional' meaning a prism or grating spectrometer (or spectrograph or monochromator), not something like a Fabry–Perot or a Fourier transform spectrometer.

[2]It is this appearance that explains the jargon by which a discrete frequency is referred to as a spectral 'line'.

[3]Our topic could, therefore, have been appropriately dealt with within Chapter 3, but there is enough to say about gratings that they are best given a chapter to themselves.

4.2 A basic transmission grating

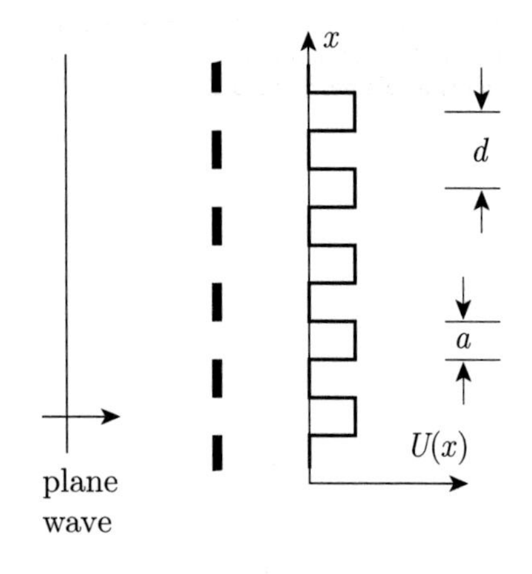

Fig. 4.1 The amplitude $U(x)$ of light just downstream from a transmission grating illuminated normally. The diagram defines the width a of a transparent ruling and the separation d of such rulings. The rulings are long and uniform in a direction normal to the paper.

Figure 4.1 shows the electric-field amplitude $U(x)$ transmitted by a part-transparent grating that is illuminated normally 'from the rear'. The Fraunhofer diffraction pattern can be calculated by elementary methods (explicitly summing the amplitudes diffracted from the N slits). Or we can fabricate $U(x)$ as the convolution of a single-slit transmission function (width a) with an array of δ-functions (spacing d), and then apply the convolution theorem. Either way we find

$$\frac{U(\beta)}{U(0)} = \underbrace{\left(\frac{\sin(\frac{1}{2}\beta a)}{(\frac{1}{2}\beta a)}\right)}_{\text{single-slit envelope}} \times \underbrace{\left(\frac{\sin(\frac{1}{2}N\beta d)}{\sin(\frac{1}{2}\beta d)}\right)}_{\text{multiple-slit pattern}}, \tag{4.1}$$

where $\beta = k\sin\theta$. The intensity is, of course, proportional to $|U(\beta)|^2$. Look back to §2.11 for a reminder of the similarity between this and X-ray scattering from crystals. See problem 4.1 for the derivation of eqn (4.1).

4.3 The multiple-element pattern

In future we'll refer to each repeated unit of the grating as an 'element', because (§4.4 onwards) those elements are not always—or even usually—slits.

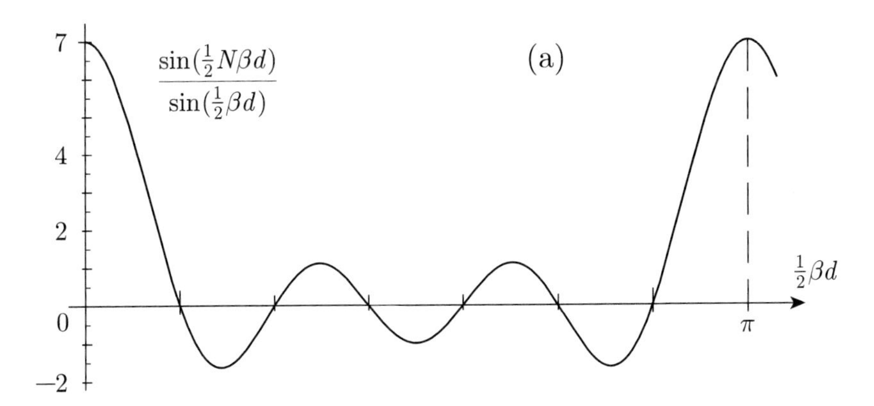

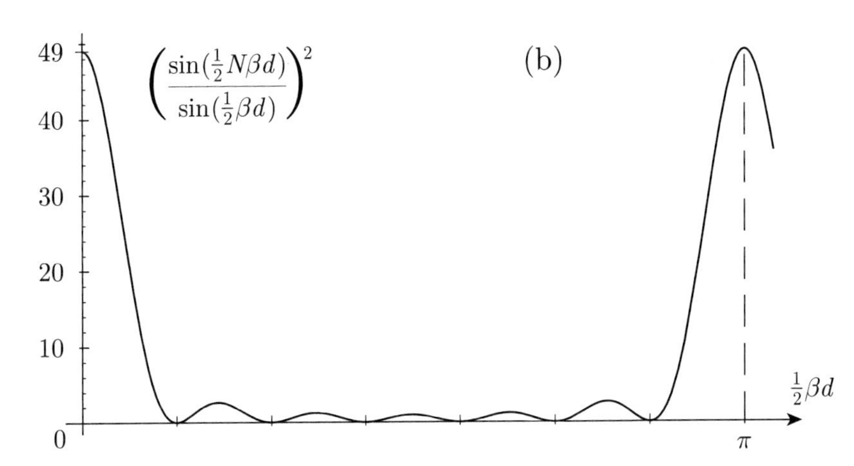

Fig. 4.2 The amplitude (a) and intensity (b) of the multiple-element pattern Fraunhofer-diffracted from a grating with $N = 7$ transparent slits. A real grating has far more than seven elements, but even seven are enough to exhibit the general behaviour of the multiple-element pattern.

Figures 4.2 (a) and (b) show the amplitude and intensity (respectively) of a multiple-element pattern, drawn for the case $N = 7$. (This is, of course, a ridiculously small number of elements for an ordinary grating—problem 4.1(7), but the aim is to show that even with so few elements the general characteristics are showing up.)

For the case of general N,

$$\frac{\sin(\frac{1}{2}N\beta d)}{\sin(\frac{1}{2}\beta d)} = \begin{cases} 0 \text{ when the numerator is zero} \\ \textit{except} \text{ when the denominator is } \textit{also} \text{ zero} \end{cases}$$
$$= \begin{cases} 0 \text{ when } (\frac{1}{2}N\beta d) = m\pi \\ \textit{except} \text{ when } (\frac{1}{2}\beta d) = p\pi \text{ so that } m = Np, \end{cases}$$

where m and p are both integers. Investigation shows that there is a strong maximum,[4] said to be of order p, when $\frac{1}{2}\beta d = p\pi$, or equivalently

$$\left.\begin{array}{ll} kd\sin\theta = 2\pi p & \text{phase delay across one element} = 2\pi p; \\ d\sin\theta = p\lambda & \text{path difference across one element} = p\lambda. \end{array}\right\} \quad (4.2)$$

For all orders p other than the zeroth, different wavelengths are diffracted

[4] A strong maximum of intensity; a strong maximum or minimum or amplitude. See problem 4.2(4).

to different angles θ, which is, of course, why gratings are useful for spectroscopy.

Note that λ is the wavelength *in the medium* surrounding the grating, so $\lambda = \lambda_{\text{vacuum}}/n$, where n is the refractive index of that medium. And likewise $k = 2\pi/\lambda_{\text{medium}}$.

As is illustrated in Fig. 4.2, the strong maxima of intensity constitute the grating orders. In between neighbouring orders there are

- $(N-1)$ zero values of amplitude and intensity
- $(N-2)$ weak maxima of intensity.

These and other properties of the multiple-element pattern are explored in problem 4.2.

4.4 Reflection grating

Real diffraction gratings rarely resemble the idealization of §4.2. The transmission grating at normal incidence is everybody's starting point though, just because the diagram is easy to draw. An opaque grating with a reflecting surface is, however, far more practical.

Figure 4.3(a) shows in cross section a possible reflection grating (not an efficient shape as we'll see). The rays drawn indicate the directions of travel for (parallel beams of) light incident and diffracted. The angles α and θ are to be taken as positive if the ray directions lie away from the grating normal in the senses indicated (and negative if not—a sign convention is being set up).

Waves diffracted from corresponding points on neighbouring grating elements have path difference $d(\sin\alpha - \sin\theta)$. Thus, the condition for there to be a grating maximum of order p is

$$d(\sin\alpha - \sin\theta) = p\lambda_{\text{medium}}. \tag{4.3}$$

Here p is of course an integer and, within the sign convention established, p can be positive or negative.

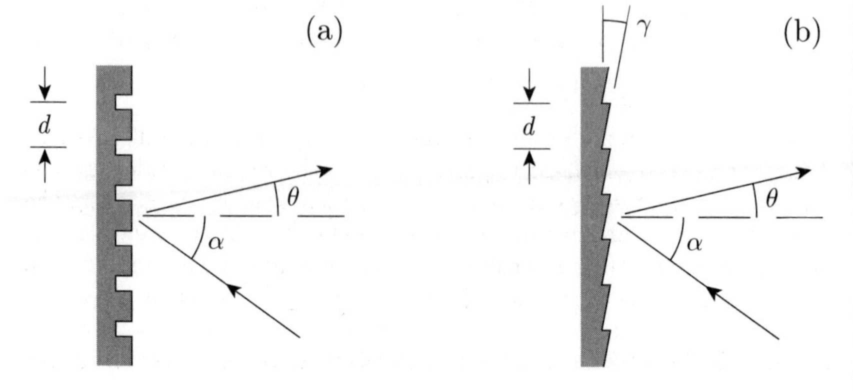

Fig. 4.3 (a) A possible reflection grating, and the angles used in eqn (4.3). (b) A blazed grating shown in cross section. Each element is a tiny mirror (think of one of the slats of a venetian blind), rotated through angle γ. Light diffracted from each element is now thrown more nearly into the direction θ at which a grating order is being collected.

4.5 Blazing

Each element of a reflection grating acts as a tiny mirror. Light reflected from that mirror goes into directions centred around the mirror-reflection direction, though with a quite wide angular spread. In Fig. 4.3(a), the mirror-reflection direction lies at $\theta = \alpha$, the direction of order $p = 0$. The most intense grating order is thus the zeroth, which carries no useful spectroscopic information.

The fact, exhibited in eqn (4.1), that

$$U(\beta) = \begin{pmatrix} \text{envelope; characteristics of the individ-} \\ \text{ual elements control this term only} \end{pmatrix} \times \left(\frac{\sin(\frac{1}{2}N\beta d)}{\sin(\frac{1}{2}\beta d)} \right)$$

means that we have freedom to control the envelope of the diffraction pattern, by shaping the individual elements, without disrupting the locations of the grating orders.

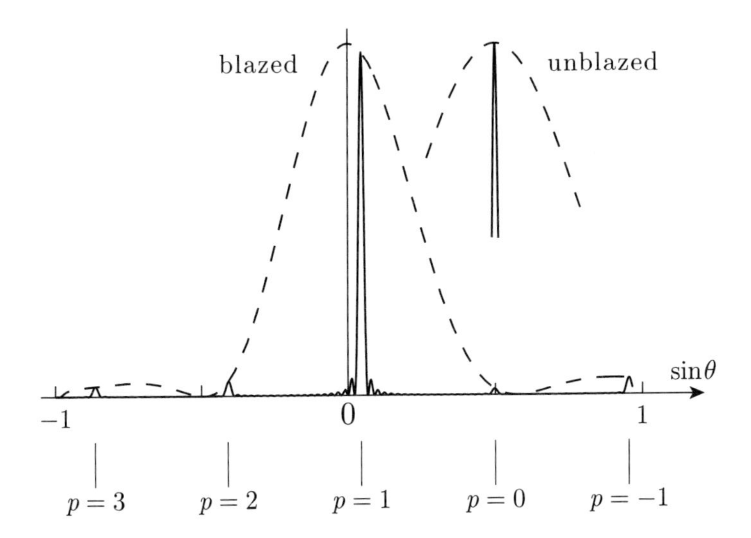

Fig. 4.4 The intensity diffracted to angle θ when a grating is illuminated by light arriving at angle α to the grating normal. The 'blazed' case is drawn with blaze angle $\gamma = \frac{1}{2}\alpha$ so that maximum intensity is available at directions close to $\theta = 0$. However, the wavelength has been chosen to be a little less than that which gives $\theta = 0$ so as not to show too special a case.[5]

Figure 4.3(b) shows a way in which this idea can be exploited. Now all the little mirrors are tilted over, rather like the slats of a venetian blind, through an angle γ. The single-element envelope is rotated, downwards as the diagram is drawn, through 2γ, giving enhanced intensity to any grating order leaving around angle $\theta = \alpha - 2\gamma$. At the same time, the *multiple*-element pattern, whose peaks represent the grating orders, lies in completely unaltered directions.

We choose to shift the single-element envelope until the diffracted intensity is peaked—at least roughly—around the required order; in particular, we direct energy away from the useless zeroth order. Since the grating elements are narrow, the single-element diffraction envelope is broad, problem 4.1(6), and can cover a whole grating order comfortably.

Figure 4.4 shows the effect of blazing a grating for optimum display of the first-order spectrum, $p = 1$. The envelope, the factor arising from the diffraction of light from a single grating element, has been shifted to the

[5] The graphs are drawn for incidence angle $\alpha = 30°$ and (for the blazed case) blaze angle $\gamma = 15°$. Allowance has been made for the effect of the blazed rulings in partly shadowing each other. However, given experience in connection with Fig. 3.20, the obliquity factor has been omitted; it is in any event too small a detail to be of concern (§4.10). Grating-order peaks are drawn for a number N of elements equal to 20, which is of course unrealistically small. The 'unblazed' curve is drawn for transparent elements of width $a = (2/3)d$ where d is the element spacing. The 'unblazed' intensity ought really to be drawn smaller by a factor $(2/3)^2$ in order to allow for energy absorbed in the spaces between the transparent slits.

left so as to have its maximum (intensity) around $\theta = 0$. Notice that the 'diffraction' envelope encompasses little more than one grating order.[6] Orders other than $p = 1$ are, therefore, quite strongly suppressed, which must mean that order $p = 1$ gets nearly all of the available energy.

An ideal blazed grating imposes phase variations onto the reflected light without imposing changes of intensity. If we have any difficulty seeing why this results in diffraction (at all), work problem 4.3.

[6] This has happened because each grating element has width d, equal to the spacing between elements. By contrast, a grating such as that in Fig. 4.1 must have slits that are narrower than d, with the consequence that the envelope function is wider, and the available energy (which is less anyway because some has been absorbed) is spread wastefully across several orders.

4.6 Grating spectrometric instruments

Perhaps the commonest way of mounting a grating into a complete spectroscopic instrument is the Czerny–Turner arrangement sketched in Fig. 4.5. Light entering the entrance slit is collimated by a concave mirror and directed towards the plane grating. Light diffracted from the grating forms a plane wave that is focused by another concave mirror onto the exit slit. Only one wavelength[7] leaves the grating in the right direction to reach the exit slit, so the instrument is, in the first instance, a **monochromator**, a variety of tunable filter. Tuning is achieved by rotating the grating.

[7] This statement isn't quite right, because the grating has more than one order. If λ is transmitted in order p, then so is $\lambda/2$ in order $2p$, for example.

A rather different spectrometric instrument is shown in Fig. 4.6. In that case, glass lenses collimate the light and focus it in the camera. This arrangement is optimized for the display, and photographic recording, of a spectrum, meaning a substantial range of wavelengths perhaps over the whole visible range. We have a **spectrograph**.

It would be tedious to catalogue a number of different grating spectrometers, so we confine discussion here to just the two arrangements

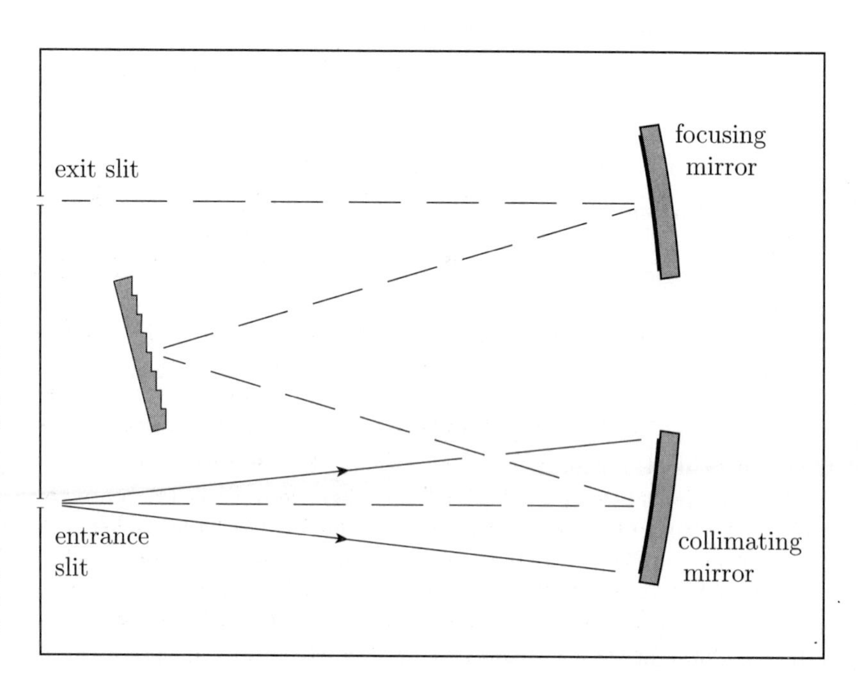

Fig. 4.5 A Czerny–Turner monochromator incorporating a plane reflection grating. Light entering through the entrance slit is collimated by a concave mirror to form a parallel beam incident on the grating. Light diffracted from the grating departs as a plane wave, so a second concave mirror is needed to focus it onto the exit slit. Rays are sketched from the entrance slit to the collimating mirror, but are otherwise omitted to minimize clutter.

This optical layout is highly suitable for a monochromator because the concave mirrors can be made (problem 4.4) to give near-ideal focusing—for one ideal-focus point each.

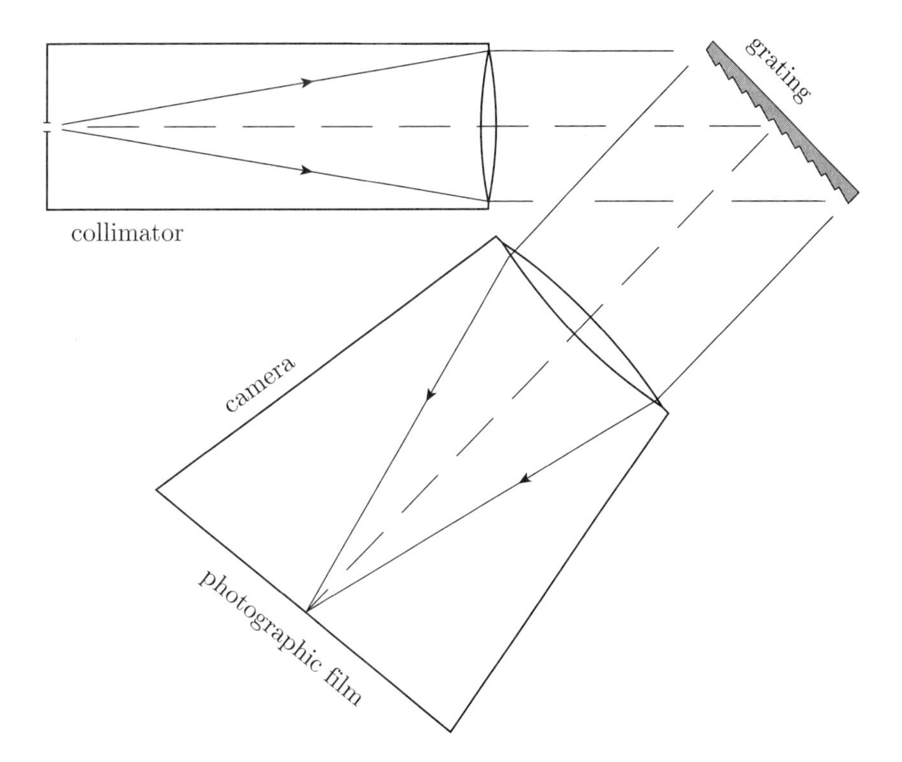

Fig. 4.6 An example of a grating spectrograph. The glass lenses are really composite (two or more glass elements) because they must be designed to image correctly over a range of wavelengths. The photographic film often lies at an angle to the camera-lens axis, as suggested, because the camera lens is designed with knowledge of where each wavelength will go, and it is designed to focus the spectrum onto the flattest possible surface even if that surface is tilted.

already mentioned.

Why would we choose one spectrometer layout rather than the other? The considerations are mostly fairly obvious.

- The use of mirrors, rather than lenses, in the Czerny–Turner gives the instrument a wide wavelength range. Background knowledge: glass transmits over a limited wavelength range, and even a carefully designed composite lens can be 'corrected' to give a good-quality spectrum over a specific range only.
- A price is exacted by the use of mirrors: the image quality is good only close to the nominal focus, so the Czerny–Turner arrangement cannot display a broad spectrum for photographing. (It is, however, often used with a CCD[8] detector to achieve a worthwhile degree of parallel processing of nearby wavelengths.)
- The converse of the last point. A lens, made from two or three pieces of glass, has enough surfaces (with choosable radii) and materials (with choosable refractive indices) to permit its designer to achieve good-quality imaging of a length of spectrum. Lenses must, therefore, be used for most spectrographs where a substantial length of spectrum is to be photographed.

[8]Charge-coupled device. An array of photodetectors made by integrated-circuit methods on a piece of semiconductor. 'Charge-coupled' refers to the way in which electrical signals are read out from the array of detectors.

A pretty trick, used with the Czerny–Turner arrangement, is the sine bar. Rotation of the grating can be achieved in such a way that a dial attached to the driving mechanism reads 'real nanometres'. See problem 4.5.[9]

[9]As always, there are exceptions. For honesty then: A Czerny–Turner-type instrument can give a photographable spectrum provided its f-number is quite high, say above 10. In such a case, the two concave mirrors are often combined into a single mirror, the Ebert layout. But a price is exacted for this versatility: the high f-number means a sacrifice in the ability of the instrument to collect energy.

The more serious exception is the case of a concave grating, which combines in one optical element the functions of the grating and both focusing mirrors. Photographing the spectrum is the common way of using a concave grating, and a wide range of wavelengths can be recorded at once. But you can buy several plane gratings for a Czerny–Turner at similar cost to a good concave grating, so the concave grating tends to have rather a specialist niche; it's used when the number of reflecting surfaces must be minimized, e.g. in awkward regions of the ultraviolet.

4.7 Spectroscopic resolution

A spectroscopic instrument (grating spectrometer, Fabry–Perot, tuned circuit, proportional counter, ...) has the task of separating the frequencies presented to it. 'Resolution' specifies how well it does so.

Several methods are available for quantifying the resolution. The most complete is to obtain the **instrumental line profile**: the apparent frequency distribution that results from sending light of a single frequency through the instrument. The instrumental line profile of a grating spectrometer is investigated in problems 4.7 and 4.8.

A more elementary way of specifying resolution is to give $\delta\nu$, the separation of two just-resolved frequencies. In optical work, it is usual to work with $\bar{\nu} \equiv 1/\lambda_{\text{vacuum}} = \nu/c$, often called the **wavenumber**, rather than with frequency ν in hertz; $\bar{\nu}$ measures frequency in periods per metre, while ν is in periods per second. Within this language, there are two common measures of resolution:[10]

$$\text{resolvable wavenumber interval} = \textit{instrumental width} \equiv \delta\bar{\nu}, \qquad (4.4)$$

$$\textit{chromatic resolving power} \equiv \lambda/\delta\lambda. \qquad (4.5)$$

[10]The name 'instrumental width' reminds us that $\delta\bar{\nu}$ is the width of the instrumental line profile (in wavenumber units). A sensible definition makes this width the same as the separation of two frequencies just resolved.

There is an obvious resemblance between the definition of chromatic resolving power and that for the quality factor Q used for describing resonance in electric circuits.

An uncertainty-principle property of $\delta\bar{\nu}$ makes it the more useful quantity.[11] If we make a measurement over a time interval δt, we can hope to measure a frequency to within $\delta\nu \approx 1/\delta t$. Equivalently, we can measure $\bar{\nu}$ to within $\delta\bar{\nu} = \delta\nu/c \approx 1/(c\,\delta t) = 1/(n\,\delta x)$, where δx is the *distance* over which the measurement is made. Most optical measurements of frequency are in fact measurements of wavelength, and are done by interference (more or less evidently); then $n\,\delta x$ is the optical path difference over which interference is made to happen. A general property then should be:[12]

[11]The preference for working with instrumental width will be even stronger when we consider the Fabry–Perot in Chapter 5 and the Fourier-transform spectrometer in Chapter 10.

The author sometimes says to students that the chromatic resolving power is 'used only by the writers of textbooks'. This is somewhat unfair, as a number of insights into the physics are best expressed in terms of the chromatic resolving power. We have already mentioned the resemblance to the electrical quality factor Q. And chromatic resolving power links neatly to the solid angle within which a Fabry–Perot or a Fourier-transform spectrometer can collect light; see eqn (11.43) and problem 5.15.

$$\text{instrumental width} = \delta\bar{\nu} \approx \frac{1}{\text{longest optical path difference}}. \qquad (4.6)$$

[12]The optical path difference is, of course, being imposed in some kind of interference arrangement. Given the accuracy needed when specifying resolution, the distinction between path and optical path is often ignored.

A brute-force justification of eqn (4.6), in the case of a grating instrument, is obtained by working problem 4.6.

Not only is eqn (4.6) the most insightful way of looking at resolution, it also reduces the process of calculating resolution to little more than mental arithmetic; see the working in example 4.1.

While dealing with general statements, we derive a general theorem on spectroscopic instruments. Such an instrument will be called a 'spectrometer', it being left open whether it might actually be a spectrograph or a monochromator. Light forming the final image of the entrance slit passes through a camera lens focused for infinity, and that lens constitutes a telescope to which the discussion of §3.16 applies. Two spectral frequencies will be just resolved as separate if they have an angular separation $\geq \lambda_{\text{med}}/D$, where D is the width the light beam passing through the lens.[13] The separation in (vacuum) wavelength is given by $\delta\lambda_{\text{vac}} = (\lambda_{\text{med}}/D)\text{d}\lambda_{\text{vac}}/\text{d}\theta$, where $\text{d}\theta/\text{d}\lambda_{\text{vac}}$ is the *angular dispersion* provided by the instrument. Expressing things in terms of wavenumber $\bar{\nu}$ yields

$$\delta\bar{\nu} = \frac{1}{nD\lambda_{\text{vac}}}\frac{\text{d}\lambda_{\text{vac}}}{\text{d}\theta}. \tag{4.7}$$

This is our general expression for the resolution of any spectrometric device.

For the special case of a grating we can obtain $\text{d}\lambda/\text{d}\theta$ by differentiating eqn (4.3):[14]

$$\text{instrumental width } \delta\bar{\nu} = \frac{1}{n\,Nd\,|\sin\alpha - \sin\theta|}; \tag{4.8}$$

$$\begin{aligned}\text{chromatic resolving power} &\equiv \lambda/\delta\lambda\\ &= (\text{number } N \text{ of lines on grating})\\ &\quad \times (\text{order } p \text{ of spectrum}).\end{aligned} \tag{4.9}$$

To check the working here, do problem 4.6. Additional insights into resolution are to be found in problems 4.9–4.11.

[13] Any factor 1.22 is ignored here because the exiting beam may not be circular and what's important is the order of magnitude.

[14] Although the point is of no particular importance, we may note that the refractive index n appears in eqn (4.8), but not in eqn (4.9).

Example 4.1 Spectrometer design

Equations (3.21) and (4.8) reduce spectrometer design to little more than mental arithmetic. To show this, let's outline how we could achieve a given specification. Suppose[15] we need a resolution of $\delta\bar{\nu} = 1\,\text{cm}^{-1}$ at a wavelength $\lambda = 0.5\,\mu\text{m}$.

(1) To achieve the given resolution from an ideal grating, we need an interference path of $1/\delta\bar{\nu} = 1\,\text{cm}$.

(2) Allow that the entrance and exit slits must have finite width. If they are too large, they will seriously degrade the resolution; if they are too small, they will admit little energy. A rough engineering compromise is to choose slit widths so that the resolution is degraded about equally by the slits and by diffraction. With an **error budget** adding up to $\delta\bar{\nu} = 1\,\text{cm}^{-1}$, we must make the grating and slits contribute half each to this. So double the interference path to 2 cm.

(3) Guess $|\sin\alpha - \sin\theta|$ at around 0.5; the grating width (Nd) needs to be $(2\,\text{cm})/0.5 = 4\,\text{cm}$.

(4) Suppose the grating is designed to give a 'normal spectrum', meaning that the spectrum is formed at angles θ around zero. This means $\sin\theta \approx 0$ and $\sin\alpha \approx 0.5$, giving α the value 30°. Looks reasonable.

[15] *Comment*: It is likely to be a long time before spectroscopists can be persuaded to abandon the cm^{-1} (inverse centimetre) for the SI inverse metre, as the unit of preference when giving the photon energies of spectral transitions. It is, therefore, appropriate to quantify the resolution limit in the same customary units. The arithmetic is simplest if we use centimetres throughout—however much that may offend SI principles.

(5) With a normal spectrum, the light leaving the grating forms a beam of width $Nd = 4\,\text{cm}$. To focus this, we'd choose the camera lens to have a focal length of perhaps 40 cm, aiming to have an exit-side f-number of $(40\,\text{cm})/(4\,\text{cm}) = 10$. An acceptable exit-slit width is estimated at

$$\begin{aligned}(\text{exit-slit width}) &= (\text{width of diffraction pattern at exit slit}) \\ &= \lambda \times f\text{-number}, \end{aligned} \tag{4.10}$$

by eqn (3.21). Then the slit width is $\lambda \times 10 \approx 5\,\mu\text{m}$.

(6) With incidence angle $\alpha = 30°$, the beam entering the grating has width $(4\,\text{cm}) \times \cos\alpha = 3.5\,\text{cm}$.

(7) The collimator lens is likely to have the same focal length as the camera lens, so its focal length is again 40 cm. Then the entrance-side f-number is $(40\,\text{cm})/(3.5\,\text{cm}) = 12$, and the consistent width[16] for the entrance slit is $12\lambda \approx 6\,\mu\text{m}$.

(8) The grating equation $d(\sin\alpha - \sin\theta) = p\lambda$ rearranges straightforwardly for $\theta \approx 0$ to $d/p = \lambda/\sin\alpha = \lambda/0.5 = 1\,\mu\text{m}$. The best choice, if it's feasible, of grating order p is 1, because that value minimizes the risk of overlapping orders. If this choice can be made, we end up with $d = 1\,\mu\text{m}$; equivalently, the grating has 1000 lines per millimetre, and 40 000 lines overall. Nothing here looks too difficult, so we settle for these values. Everything is now chosen.[17]

[16] For another way of understanding the relation between the two slit widths, see problem 4.6(7).

[17] A round of refining the design would, of course, take place before we spent any serious money. We might decide to settle for 500 lines per millimetre, working in second order, in order to have a grating with greater versatility. We must include an allowance for imperfections in the grating and in the focusing optics, whether lenses or mirrors .... And caution suggests building in safety factors as well.

4.8 Making gratings

Many books describe a traditional ruling engine, as pioneered by Rowland. Briefly, a diamond cutter is dragged repeatedly across a metal blank while the blank is slowly advanced by a micrometer screw. The machine needs a *very* stable environment, and even then the gratings it makes are likely to have periodic errors, which lead to 'ghosts' in the spectrum. See problem 4.12 for an explanation of ghosts.

What's of more interest is the way in which modern techniques can improve things or use entirely different approaches. We mention two.

- A mechanical ruling engine is still used, but servo control reduces susceptibility to environmental disturbances and mechanical imperfections. For example, you can polish one end of the grating blank (or its support) and use an interferometer to measure where it is. An electrical signal proportional to the blank's positioning error is used in an error-correcting mechanism to put things right. Only when the blank is correctly positioned is the diamond cutter allowed to scribe the next grating element. See Stroke (1963) (oldish now but nice pictures); and Palmer (1975).
- Holographic gratings. We split light from a blue or ultraviolet laser into two broad plane waves, and intersect those waves so as to form an interference pattern on the grating blank. The blank has been

coated with **photoresist**.[18] After exposure, chemical etching of the photoresist dissolves away preferentially the part that was exposed (positive photoresist) or not exposed (negative photoresist). So the surface of the photoresist acquires ripples where the fringes were. Evaporate aluminium on top to give high reflectivity and you have a grating; see Schmahl and Rudoloph (1976). Fine rulings are as easy to make as coarse: all we do is intersect the plane waves at a large angle. Blazing is not normally attempted, but is not so advantageous when the grating is fine. If the rulings are close, there may be (say) only orders 0 and 1 geometrically possible, so there aren't many places for light energy to go. You can also try to adjust the groove depth so that destructive interference removes the zeroth order ....

In the case of holographically made gratings there is no need to restrict attention to conventional plane gratings. It is almost as easy to photorecord fringes on a concave blank so as to make a concave (focusing) grating. An interesting idea is instead to distort one of the fringe-forming waves, and record fringes of non-uniform spacing on a plane blank; the variable fringe spacing gives focusing properties to the finished grating. Plane-but-focusing gratings can now be made with smaller aberrations than the traditional concave grating. You can, of course, combine the best of all of these ideas. Schmahl and Rudolph give a picture of a grating for use in the far ultraviolet, made holographically on a surface that is curved in two different ways (i.e. like a piece cut from the surface of a toroid).

[18] Photoresist is a plastic whose organic molecules become cross-linked to a greater or to a lesser degree—depending on the variety of photoresist—when damaged by energetic photons. It is the material commonly used in the preparation of printed circuits and integrated circuits, where it forms a protective layer over selected areas of copper or semiconductor.

Common-or-garden gratings (such as the reader may see in a teaching laboratory) are usually 'replicas', made by casting plastic on a ruled-metal master grating, then peeling off the replica and sticking it onto a glass backing. The plastic acquires a grooved surface mimicking that of the master, and is coated with aluminium to make it reflect. In this way, a metal master can generate many copies with only a small loss of quality. By contrast, each holographically made grating is ordinarily an original.

4.9 Tricks of the trade

4.9.1 Normal spectrum

When a grating spectrograph is used to photograph a spectrum, it is customary to arrange the grating so that the spectrum is displayed in directions not too far from the grating normal. Such an arrangement is called a 'normal spectrum' (as opposed to 'normal incidence'). The reason may be seen in eqn (4.3). When θ is small, $\sin\theta$ approximates to θ, and there is an approximately linear relation between wavelength and position on the photographic film. This near-linear relationship assists with obtaining a good calibration.[19]

[19] The spectroscopist is more interested in frequency ν, or wavenumber $\bar{\nu}$, than in wavelength, because frequency is proportional to photon energy. So, in an ideal world, one might wish for a linear relation between $\bar{\nu}$ and θ. However, this is not available. It remains that a near-linear relation between *something* and θ assists calibration, and it is not burdensome to find $\bar{\nu} = 1/\lambda$ at a later stage in the analysis.

4.9.2 Correct illumination

We need to collect the maximum amount of optical energy into the spectrometer from the light source. Look ahead to Chapter 11 and eqn (11.3). The power we can collect is

$$(\text{power}) = (\text{radiance of source}) \times (\text{area}) \times (\text{solid angle}),$$

where the area–solid angle product (known as **étendue** or **throughput**) must be evaluated at a location along the optical system where all points on the chosen area send light into all directions within the solid angle.[20] The entrance slit is such a place if the source is correctly focused on it.

For the spectrometer the power collected[21] is

$$\begin{aligned}&(\text{source radiance}) \times (\text{area of entrance slit})\\ &\times (\text{solid angle subtended by collimator lens at entrance slit}). \qquad (4.11)\end{aligned}$$

[20]Étendue is defined and explained in Chapter 11. Here we ignore the n^2 factor in eqn (11.1), as we take it that the surrounding medium is air with $n = 1$. Likewise, we ignore a factor $\cos\theta$ by making a small-angle approximation.

[21]Strictly, eqn (4.11) should contain the radiance, not of the source, but of the image of the source formed at the entrance slit. However, eqn (11.8) shows that the two are equal, apart from 'insertion loss' at the condenser lens.

The force of eqn (4.11) is this: Any arrangement that 'fills the slit with light' and 'fills the grating with light' achieves optimum energy collection, for a given instrument looking at a given source—and any other arrangement meeting these two conditions achieves the same energy collection. (This is a good time to work problem 4.13. See also problem 11.10.)

Figure 4.7(a) shows the commonest arrangement: a **condenser lens** places a real image of the source at the entrance slit. The magnification of the lens is chosen so that the image of the source is larger than the slit; then the whole area of the slit is illuminated and there is an allowance

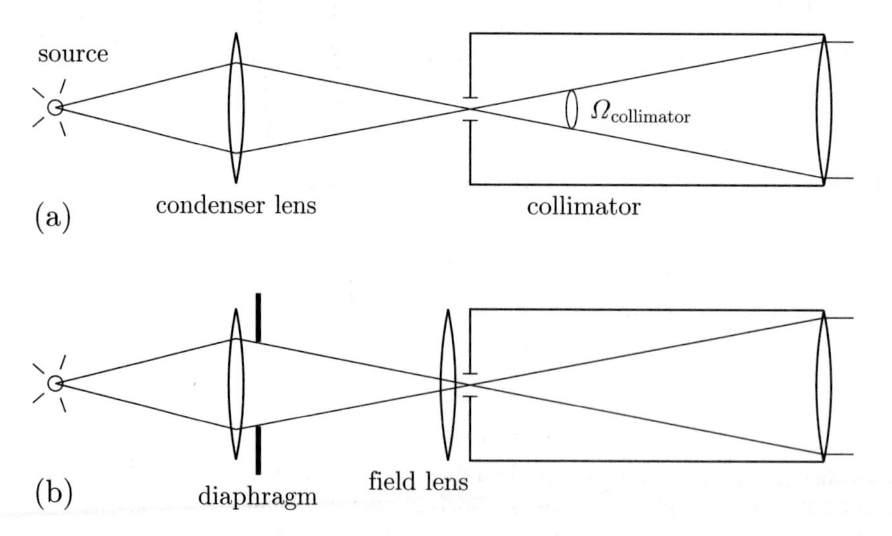

Fig. 4.7 Use of a condenser lens to gather light efficiently into a spectrometer. Imaging the source onto the entrance slit is done in such a way as to 'fill the slit with light' and 'fill the grating with light'. (a) A simple arrangement, usually adequate but not quite ideal. Although it looks satisfactory in this cross section, light spills above and below the collimator lens, hits the inner wall of the collimator, and may end up where it shouldn't. (b) The field lens images the outline of the condenser lens (or of a diaphragm if the condenser is overlarge) onto the outline of the collimator lens, thereby preventing any overspill.

for small movements. The diameter and placing of the lens are chosen so that light is sent towards the slit in the whole range of directions that the collimator can accept. (Draw rays 'backwards' from the periphery of the collimator, through the slit and make sure that they all meet the condenser lens within its outline.)[22]

The arrangement of Fig. 4.7(a) is drawn in cross section, with the length of the slit normal to the paper. It's assumed that the source is long in the same direction, so that it sends light into all points along the length of the slit. This, however, means that some light spills above and below the outline of the collimator lens; that light should be absorbed by black paint on the inside wall of the collimator, but there is a risk that it gets where it isn't wanted. In any critical application, it's better to use additionally the **field lens** shown in Fig. 4.7(b). The field lens forms a real image of the condenser lens on the collimator lens, outline matching outline, so that the collimator lens is only just filled.[23]

[22] If the source is very broad, or can be placed very close to the entrance slit, rays projected back from the collimator through the slit to the source may all meet the source, and then no condenser lens is needed. However, it is rather unusual for this to be possible, and the arrangement of Fig. 4.7(a) is a creative response to the usual situation. Putting a real image at the entrance slit 'puts the source right at the slit without its having to be there'.

[23] *Aside*: Figure 4.7(b) illustrates a rather useful general trick. Suppose we have to get light through a number of holes, separated along the length of an optical system. Here those holes are: the outline of the source; the outline of the condenser; the outline of the slit; and the outline of the collimator lens. Light passing close to the axis of the system gets through as we hope, but off-axis rays have a reduced area to get through. (Problem 4.14 asks you to check this.) The cure is to put a lens in each hole, and design things so that lens L_p images the outline of lens L_{p-1} onto the outline of lens L_{p+1}. Then all light that gets through the first two holes is guided to get through all the others.

4.9.3 Shortening exposure times with a spectrograph

In §4.9.2 we stated a rule for optimizing the total amount of energy (strictly power) collected by a spectrometer. When we photograph a spectrum, what matters is the power per unit area of film, not the total power. So we could gain if we could reduce the area onto which the light is sent. Place a cylindrical lens in front of the photographic film, so as to form a reduced-height image of the entrance slit, yet without spoiling the location or image-width of the spectral lines. See problem 4.15.

4.9.4 Vacuum instruments

Air absorbs strongly in the vacuum ultraviolet: $0.4\,\text{nm} \lesssim \lambda \lesssim 200\,\text{nm}$. You can operate a spectrometer in that region only if the air is removed. Also, of course, you have to put the source and detector up against the spectrometer so that there is no external air path. Commercially available spectrographs and monochromators are often fitted with seals so that they can be evacuated at need. Difficulty: minimize the effect of pressure change on the mechanical adjustment of the instrument .... Take seriously the idea of replacing the air with a non-absorbing gas, rather than working in a vacuum. Why might helium be a good choice? At what wavelength does helium start to absorb?

4.9.5 Double monochromator

Some experiments involve looking for weak spectral lines close (in wavelength) to intense light. An example is Raman scattering, where a sample is illuminated by a laser and a small fraction of the laser light is shifted in frequency by inelastic scattering within the sample. Even a well-constructed monochromator is likely to scatter 10^{-4} of unwanted incident light into its output, and this is enough to swamp a weak Raman-scattered line. The cure is to use two monochromators in series,[24]

[24] Sometimes even three.

thereby improving the rejection (on the figure above from 10^{-4} to 10^{-8}). It is, of course, essential for the two monochromators to be tuned to the same frequency, and for them to track each other's tuning precisely. So the two monochromators are mounted together in a single box, with identical grating drives worked from one shaft: precision mechanics.

4.9.6 An inventor's paradise

Even a high-specification monochromator has quite serious limitations.

- One of them is geometrical: all light that can't be crammed into the tiny entrance slit is wasted. Problem 11.14 gives a detailed study of the energy gathering of a basic monochromator, and shows that it falls below that of a 'whole-fringe instrument' by a factor of order 200.
- Wavelengths are examined one after the other, and all light not at the wavelength currently selected is wasted. In other words, we are using sequential processing when it would be better to use parallel processing. Of course, a photographic film achieves parallel processing by recording all wavelengths simultaneously, but its efficiency for detecting photons is poor[25] and its response to intensity is non-linear (and requires calibration as it varies from batch to batch of film).

[25] A 'quantum efficiency' of 1% is often quoted, meaning that 1% of incident photons result in a developed grain of silver bromide. However, this figure seems usually to be rather optimistic.

Getting round these limitations is truly an inventor's paradise.

The reason for using a monochromator—at all, rather than a spectrograph—is usually to use photoelectric recording of the intensity as a function of wavelength. Photodetectors (between them, not any single detector) cover a wider frequency range than photographic film; they have a response that is linear with intensity; they have a much higher quantum efficiency (up to $\sim 50\%$ for semiconductors at the most favourable wavelength); and above all their electrical output is ideal for transmission to a computer for storage and analysis. However, a simple photodiode or photomultiplier can record only one intensity at a time, which means combining it with a monochromator that transmits only one wavelength at a time. Sequential processing. One problem to be addressed, therefore, is how to combine photoelectric recording with parallel processing.

Here are some ideas.[26]

[26] I make no attempt at giving a complete catalogue here. That would be impossible as any such catalogue would rapidly go out of date. Moreover, some techniques are very sophisticated and would require lengthy explanation. The aim is rather to show by a few examples how good ideas can be generated.

(1) One of the simplest ideas is to use an image intensifier[27] followed by photographic film. An image intensifier has a photocathode from which electrons are released into vacuum by the photoelectric effect. These electrons undergo secondary-emission amplification as in a photomultiplier, but in a way that preserves spatial information. At the end of the amplification process, electrons are accelerated onto a phosphor screen, rather than being collected by an anode. The phosphor screen glows to give a greatly intensified image of the original light distribution over the cathode. And, as

[27] An image intensifier, followed by a television camera, is the basic component in infrared surveillance.

mentioned, we photograph the output from this screen. The effect of the whole scheme is to achieve the parallel processing of photographic film, combined with the quantum efficiency of the photocathode. It is not possible to cover a really broad spectral range because of the limited size and shape of the intensifier, but the effective number of 'information channels' being examined simultaneously can easily be several hundred.

(2) A related idea is to use what is, in effect, a surveillance camera. An image intensifier gathers light and passes an amplified version of it onto a CCD or a vidicon television camera. The image intensifier is chosen for the desired wavelength sensitivity, which may well not be that of the CCD alone, so the combination does more than just the CCD. (If the CCD's wavelength range happens to be suitable, the intensifier may be dispensed with.) The pixels on the CCD are read out sequentially and the charge stored in each (proportional to the light energy received) is recorded in computer memory. The readout process is repeated regularly with time, and can be used to investigate a time-dependent light source; or successive readouts can be averaged to improve signal/noise performance. The number of 'parallel channels' is determined by the number of pixels across the CCD array, and can be up to about a thousand.[28]

[28] A digital camera for 'entertainment' use has an array of pixels (picture elements) amounting to a little more than 1000 each way, giving a few million over a roughly square area.

(3) It is also possible to attack the geometrical limitation set by the smallness of the spectrometer's entrance slit. To start with, imagine that we were to supply a spectrometer with two entrance slits side-by-side and two exit slits side-by-side. For a given wavelength, the energy transmission would be doubled. Of course, there would be cross-talk with light of some unwanted wavelength passing through the 'wrong' exit slit. But what if we could invent some kind of orthogonal function set, so that cross-talk was suppressed? This is the idea behind the 'Hadamard grille'. An array of slits replaces both the entrance slit and the exit slit of a monochromator, and the instrument is scanned with a single photodetector behind the exit array. A typical array has 255 slits.

(4) The Hadamard grille is just one of a number of devices that address either the parallel-processing problem or optimize the signal/noise performance of the detection equipment. For further information, a good introduction may be found in Harwit and Decker (1974).

4.10 Beyond the simple theory

It is well known to professional spectroscopists that grating spectrometers behave in ways that lie outside the theoretical descriptions given so far in this chapter. The grating orders appear in the right places (otherwise measurements of wavelength would really be problematic), but the intensity delivered (coming from the 'envelope' factor) varies: the diffracted intensity from a grating exhibits 'anomalies'. Anomalies can be demonstrated simply by scanning a monochromator illuminated by white light and observing unexpected changes in the signal from the photodetector as a function of wavelength. Another simple demonstration is to show that the power diffracted by a grating depends upon whether the $\boldsymbol{E}$-field of the illuminating light lies along or across the rulings—a phenomenon that is not predicted by Kirchhoff boundary conditions, but should perhaps 'feel' plausible when we think that electric currents are being forced in the one case to flow along the grooves and in the other across them.

The anomalies that have been most studied are those that can be accounted for by assuming the reflecting surface of the grating to be an ideal electrical conductor and then doing a careful solution of Maxwell's equations in the presence of this corrugated conducting boundary. In effect, such a treatment identifies inadequacies in the Kirchhoff boundary conditions postulated in §3.4, and attempts to calculate corrections. Accounts may be found in Maystre (1984) and Popov (1993). It is inevitable that these theoretical studies are somewhat complicated and lie outside the terms of reference of this book. However, the reader embarking on spectroscopic studies should be warned to expect anomalies and be prepared to undertake a careful calibration before trusting intensity measurements.

Another category of anomaly arises from the solid-state physics of reflection from a metal surface. A metal is a less-than-ideal electrical conductor at frequencies of order 10^{15} Hz. It is hardly surprising then that the reflectivity of the metal surface is frequency dependent, especially when the surface has a complicated shape. Some discussion of this, in relation to plasmon excitations, is given by Popov (1993).

Problems

Problem 4.1 (a) Grating basics

(1) Discuss carefully, making reference to the conditions given in Chapter 3, why the diffraction produced by a grating within a spectrometer is always observed under the Fraunhofer condition.

(2) Derive the diffracted amplitude of eqn (4.1) using both of the methods suggested in §4.2.

(3) Show that the factorization on display in eqn (4.1) is a direct consequence of the convolution theorem of eqn (2.30).[29]

[29] *Comment*: Part (3) shows that the factorization of $U(\beta)$ in eqn (4.1) has an inevitability about it. Look back also at eqn (3.22) and problem 2.19 for the case of a 'grating' with only two elements.

(4) Draw a ray diagram that explains the interference physics expressed in statements (4.2).

(5) Think of at least three reasons why a reflection grating is a more practical device than a transmission grating.

(6) A reflection grating, perhaps of the type shown in Fig. 4.3(a), is designed so that light of wavelength 500 nm, incident at $\alpha = 30°$, is diffracted to $\theta = 0$ in first order. Calculate the required spacing d between grating elements. Use this value of d to estimate the angle through which light reflected from *one element* spreads by diffraction. See if you agree with 'quite wide' near the beginning of §4.5.

(7) Use the value of d obtained in part (6) to make an order-of-magnitude estimate of the number N of elements on a typical grating such as is used in a spectrometer. (It's an unspoken assumption in much of this chapter that N is 'large', and the large-N limit is often taken without comment.)

Problem 4.2 (a) The multiple-element expression

(1) Show that, when $\frac{1}{2}\beta d = p\pi$, both numerator and denominator of the multiple-element function $\sin(\frac{1}{2}N\beta d)/\sin(\frac{1}{2}\beta d)$ are zero, and that their ratio, evaluated as a limit, is $\pm N$.

(2) Show that the sign of the ratio in part (1) is $(-1)^{(N-1)p}$.

(3) Confirm the statements made in §4.3 that between adjacent grating-order peaks there are $(N-1)$ zeros and (therefore) $(N-2)$ maxima or minima (of amplitude; maxima of intensity).

(4) Sketch your own graph, resembling Fig. 4.2 but for the case $N = 8$. (This sounds trivial, but there is a difference to be got right.)

(5) Argue that the values of $\sin(\frac{1}{2}N\beta d)/\sin(\frac{1}{2}\beta d)$ at the maxima and minima well away from the grating orders are of order unity only.

(6) [Harder] Consider the (realistic) case of large N. Investigate the behaviour of $\sin(\frac{1}{2}N\beta d)/\sin(\frac{1}{2}\beta d)$ in the vicinity of one of the grating orders. Allow $(\frac{1}{2}N\beta d)$ to range over a few π on either side of $Np\pi$. Show that the expression under investigation resembles the amplitude diffracted from a single slit.[30]

(7) Show that the width of the 'single slit' referred to in part (6) is $Nd\cos\theta$, and give a physical interpretation of this value. (Something similar happened in problem 3.20(2).)

(8) Think about light diffracted into one of the orders of a transmission grating, say the first order. The wavefront that leaves the grating is chopped up into little pieces. Yet when we assemble the (secondary waves from) those pieces in the Fraunhofer diffraction pattern the light behaves merely as if it had had its width restricted by a single slit almost as wide as the grating. Something here seems to be 'too good': without the chopping-up, light wouldn't be going in this direction at all; yet once that's dealt with the chopping-up seems to do nothing. Can you be comfortable with this?[31]

[30] *Hint*: Think about the range occupied by $(\frac{1}{2}\beta d)$ and suggest an approximation that is valid *for the denominator only*.

[31] *Hint*: Problem 3.26 part (3) involves somewhat similar physics.

Problem 4.3 (b) A sinewave phase grating

Most of the examples of diffraction given in Chapter 3 involved apertures that were either transparent or opaque. Given this experience, it can sometimes be hard to see why something like a blazed grating should diffract at all: the phase of the reflected light is being changed, but not the amplitude $|U|$; why does that result in diffraction?[32]

This problem offers a model that may help. Consider then a 'phase grating' whose transmission function is

$$T(x, y) = \exp\{iA\cos(\beta_0 x)\}, \tag{4.12}$$

where A is a real constant. If this aperture is illuminated normally, the transmitted amplitude $U(x, y)$ has constant magnitude $|U|$, but bears a

[32] The Fourier-transform property of Fraunhofer diffraction offers some help here. We know from other areas of physics (perhaps electronics) that any modulation of a sinewave generates 'carrier wave plus sidebands'. This is as much the case for phase modulation as for amplitude modulation. Yet another application of this idea is given in problem 4.12.

periodic phase variation with period $d = 2\pi/\beta_0$; it really does deserve the name 'grating'.

(1) Take $A \ll 1$ and show that eqn (4.12) approximates to

$$T(x, y) \approx 1 + \mathrm{i}A\cos(\beta_0 x). \tag{4.13}$$

(2) Calculate the Fraunhofer diffraction pattern that results from using this grating at normal incidence. Except for the factor i, there should be a strong similarity to everything that happened in example 3.1.

(3) Describe qualitatively what would happen if we had a grating in which the transmission function had the form

$$T(x, y) = \exp\{\mathrm{i}f(x)\},$$

where $f(x)$ is some general periodic function of x with the same period d as above. One possibility for the choice of $f(x)$ is, of course, the blazed-grating profile.[33]

[33]This problem will be referred to in Chapter 12 when we think about how a microscope can be designed to render visible phase variations such as those imposed by this grating.

Problem 4.4 (b) Mirror geometry for Czerny–Turner
Use Fig. 4.5 to explain the optical layout.

(1) Consider the case where the mirrors are spherical, and be prepared to use small-angle approximations. Show that the mirrors have a common centre. It follows that the Ebert arrangement mentioned in sidenote 9 (p. 92) is possible.

(2) Show that the mirrors of a Czerny–Turner monochromator give perfect collimation (at least for rays in the plane of the diagram) if they are 'off-axis paraboloids'. Identify the focus and symmetry axis of each paraboloid. In practice we need to use paraboloids (rather than spheres which are easier to make and cheaper) only when the f-number $\lesssim 6$.

Problem 4.5 (b) Sine-bar drive for a Czerny–Turner monochromator
Figure 4.5 shows that the angle between incident and diffracted light is fixed by the design of a Czerny–Turner instrument. Let that angle be 2δ. Let the grating normal make an angle ϕ to the bisector of that angle as shown in Fig. 4.8(a). Then angles α and θ (defined in Fig. 4.3) are related to these angles by $\alpha = \delta + \phi$ and $\theta = \delta - \phi$.

(1) Show from eqn (4.3) that the grating condition can be expressed as

$$p\lambda = (2d\cos\delta)\sin\phi.$$

There is, therefore, a linear relation between wavelength λ and $\sin\phi$.

(2) Show that the device drawn in Fig. 4.8(b) can achieve what we need: $\sin\phi$ is proportional to the linear displacement of the nut when the screw is rotated, so there is a linear relationship between wavelength and the rotation of the screw. A counter attached to the screw shaft can be made to display 'real nanometres' if the sine bar is given the correct length. (Not shown on the diagram: the length of the sine bar is in fact made adjustable.)

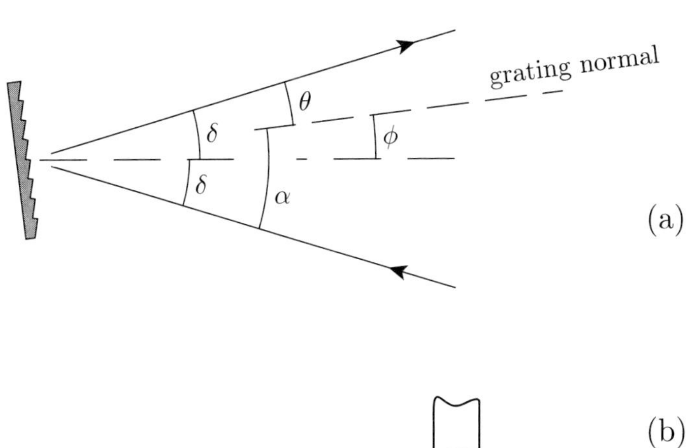

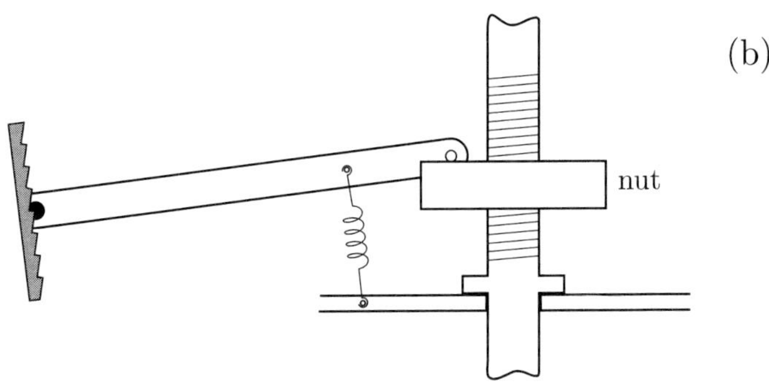

Fig. 4.8 A sine-bar grating drive as discussed in problem 4.5. Angles δ are fixed by the design of the spectrometer. Angle ϕ is the angle between the grating normal and the fixed direction bisecting the ray directions. (a) This defines the angles. The grating rulings (blazed) are hinted at, though they are of course drawn far too large. (b) The sine-bar layout. The pivot is partly concealed by the grating. The grating has its reflecting surface located above the pivot so that the grating does not move sideways as it rotates. The (non-rotating) nut linking the sine bar to the micrometer screw is hinted at only, and the reader will easily identify impracticalities in what is drawn, which has been kept simple.

(3) Show that the non-negligible radius of the peg at the 'driven' end of the sine bar does not compromise the geometrical relationships.

Problem 4.6 (a) Resolution of a grating spectrometer
For parts (1)–(3), take the spectrometer's entrance slit and exit slit (if any) to have negligible width.

(1) Find the *angular dispersion* $\equiv \mathrm{d}\theta/\mathrm{d}\lambda_{\mathrm{vac}}$ for a grating used with light incident at angle α to the grating normal.

(2) Derive eqn (4.7).

(3) Use the Rayleigh criterion (help in eqns 3.20) to find the condition for two spectral frequencies to be just resolved. Hence, derive eqns (4.8) and (4.9).

(4) Let collimator and camera lenses both have the same focal length f. Show that the entrance-side and exit-side f-numbers are $f/(Nd\cos\alpha)$ and $f/(Nd\cos\theta)$, respectively, if light fills the width Nd of the grating.

(5) Follow the reasoning of example 4.1(5) to show that the exit slit degrades resolution by the same amount as does diffraction if its width is $\lambda_{\mathrm{med}}f/(Nd\cos\theta)$.

(6) Imagine that light is sent 'backwards' through the spectrometer from a thin exit slit to form a diffraction pattern at the entrance slit. Match the entrance-slit width to the diffraction pattern after the fashion of part (5). Show that the resulting entrance-slit width is $\lambda_{\mathrm{med}}f/(Nd\cos\alpha)$.

(7) The slit widths calculated in parts (5) and (6) are unequal; a grating instrument is not symmetrical between input and output. Now find the

ratio of slit widths by another route: arrange[34] that the entrance slit and exit slit have widths that make them ray-optics images of each other.[35]

[34] *Hint*: Find $(\partial\theta/\partial\alpha)_\lambda$.

[35] The agreement between part (7) and the two preceding parts gives us the rather unsurprising result that the monochromator may be used 'either way round'; the resolution and light-gathering are the same.

Problem 4.7 (b) Instrumental line profile and instrumental width
Take the physical setup of problem 4.6(1)–(3), and express the result another way. Imagine a wavelength (or frequency) scale to be drawn across the exit plane, so that the user can read off the wavelength of radiation directed there by the grating. When monochromatic light is input, it produces a diffraction pattern at the exit plane. It *looks* as though light with a frequency distribution has been passed through a spectrometer with perfect resolution. This apparent frequency distribution is the instrumental line profile. Write down the apparent frequency distribution, and show that its width (centre to first zero) is the same as the $\delta\bar{\nu}$ of eqns (4.6) and (4.8). Fill out the argument given in §4.7 for the relation:

$$\text{least resolvable wavenumber interval } \delta\bar{\nu} = \text{instrumental width.} \tag{4.14}$$

Problem 4.8 (b) Monochromator with wide slits
Monochromatic light illuminates a grating monochromator whose entrance and exit slits are both much *wider* than the diffraction pattern formed within the instrument. This is the exact opposite of the limit taken in problem 4.6(1)–(3), but it was implicit in problem 4.6(7). The slit widths are arranged so that the image of the entrance slit in the exit plane has the same width as the exit slit. Show that the intensity recorded as the instrument is scanned—the instrumental line profile—is triangular.[36]

[36] *Hint*: Remember problem 2.20.

Problem 4.9 (e) Any filter stores energy
A filter receives an input signal $V_{\text{in}}(t)$ containing a range of frequencies $\Delta\nu_1$ and it outputs signal $V_{\text{out}}(t)$ containing frequencies in the (smaller) range $\Delta\nu_2$. The physical nature of the filter does not matter; it is anything compatible with the description just given.

(1) Argue that the filter must store energy within itself.[37]
(2) Discuss how the following store energy when functioning as filters:[38]
 (a) a grating spectrometer
 (b) a Fabry–Perot
 (c) an electrical *RLC* filter
 (d) [harder] a prism spectrometer.

[37] *Hint*: Imagine an input signal lasting for a time $\tau_1 = 1/\Delta\nu_1 < 1/\Delta\nu_2$. What can be said about the duration of the output?

[38] *Comment*: This problem contains a useful insight. If you invent something new that you think might be a filter, it's a good idea to check whether it can conceivably store energy. If it doesn't, it can't be a filter after all.

Problem 4.10 (e) A very short pulse input to a diffraction grating
Imagine we have a grating monochromator, with very narrow entrance and exit slits. (If it helps you may imagine an unblazed transmission grating used at normal incidence, though this isn't necessary.) Apply to its input a *very* short pulse of electromagnetic radiation (say 10^{-18} s long). Think about the radiation arriving at the exit slit, on a timescale of 10^{-15} s (ask Mr Tompkins[39] to slow down the speed of light to less than a micrometre per second).

[39] The allusion is to Gamow (1965).

(1) Describe what you would see, looking at the grating through the exit slit, with Mr Tompkins' assistance.[40]

(2) Draw a sketch graph of the E-field at the exit slit as a function of time, making quantitative

(a) how many square-wave periods it has,
(b) its total duration, and
(c) the ratio of the on and off times.

(3) Now think about the (time) frequencies in the wavetrain leaving the monochromator. Without heavy algebra obtain:

(a) the fundamental frequency contained in the wavetrain,
(b) an estimate of the frequency width that the wavetrain must have because it is of finite duration.

(4) Describe in words how you could work out the strength of the harmonics (multiples of the fundamental frequency) present in the wavetrain.

(5) How do the harmonics of part (4) relate to things you have seen before? Would the harmonics be absent if we used the grating of example 3.1 (p. 61)?

(6) Suppose the grating has rulings that consist of transparent and opaque stripes of equal width.[41] Then the wavetrain has ups and downs with a width ratio of 1. What does this imply about the harmonics?

(7) Obtain the result of part (6) from eqn (4.1).

[40] *Answer*: A thin bright line, moving across the grating, sometimes visible through a slit, sometimes hidden between slits.

[41] *Hint*: Have you met a symmetrical square wave before, perhaps in electronics? It has a curious property.

Problem 4.11 (b) The 'finesse' of a grating

(1) Show that, provided distances are measured along a graph with $\sin\theta$ for its horizontal axis,[42]

$$\begin{aligned}\text{finesse} &\equiv \frac{\text{separation of grating orders}}{\text{width of one grating-order peak}}\\ &= (\text{number } N \text{ of lines on the grating}).\end{aligned}$$

(2) The result of part (1) comes close to telling us how many spectral lines could be fitted side-by-side into one grating order. But that isn't quite right: when you change the wavelength you change the width of each diffraction pattern as well. Continue plotting against $\sin\theta$ and show that the number of lines that fit into the first order (just touching according to a Rayleigh criterion,[43] and not overlapping any of the same lines in second order) is $N \ln 2$.

[42] Compare the 'finesse' introduced here with the finesse of a Fabry–Perot étalon, in the form given in eqn (5.7). It has been defined in an exactly similar way.

[43] A spectrum doesn't normally consist of such a regular array of lines. And if it did, we should wish to apply a safety factor, rather than having lines just touching. The figure $N \ln 2$ is, however, the correct expression to which such a safety factor should be applied.

Problem 4.12 (c) Grating ghosts

A diffraction grating is ruled on metal by a ruling engine with an imperfect lead screw.[44] The rulings are in places spread out, in places bunched up. In consequence, the rth ruling falls at position $x_r = rd + a\sin(br)$. As an idea of orders of magnitude, we might have a of order $\lambda/50$ and b of order 0.01.

(1) Show that the spacing of adjacent rulings takes all values between $(d-ab)$ and $(d+ab)$, depending on r. It's assumed of course that $ab \ll d$.

[44] The 'lead screw' is the micrometer screw that is turned to advance the metal blank between the inscribing of one element and of the next. 'Lead' is pronounced 'leed', not 'ledd', the idea being that the metal blank is being encouraged to move forward.

(2) Show that the diffraction pattern (take normal incidence for simplicity) contains not only the usual main maxima at $d \sin\theta = p\lambda$, but also weaker maxima at

$$\frac{d\sin\theta}{\lambda} = p \pm (b/2\pi), \quad p \pm 2(b/2\pi), \quad \ldots .$$

These maxima are the ghosts.

Hint: It's easy to get bogged down in Bessel functions if you don't approximate. Write

$$\exp\{ika\sin\theta\sin(br)\} = 1 + ika\sin\theta\sin(br) + \cdots .$$

(3) Show that, if the ghosts are misinterpreted as additional spectral lines, they form an array with uniformly spaced apparent wavelength.[45]

Comment: Grating ghosts are, of course, bad news for the spectroscopist: ghosts may be misinterpreted as additional spectral lines. Nevertheless, we can feel that things are not as bad as they might have been. Had we not worked this problem, we might have imagined that an imperfect grating would generate a smudgy mess, degrading the spectrometer's resolution. But it has turned out that the resolution is hardly affected: the 'carrier' is not broadened.[46] Moreover, we have seen in part (3) that ghosts have a characteristic spacing on the recorded spectrum which helps in identifying them for what they are.

[45] *Comment*: Shifting pieces back and forth slightly in a waveform would in other contexts be called **phase modulation**. It's well known in electronics that a phase modulated wave can be analysed into an unshifted frequency (the **carrier wave**) and discretely shifted frequencies either side of the carrier (**sidebands**). So the ghosts are an optical manifestation of these sidebands.

[46] *Aside*: There is a passing resemblance to the Mössbauer effect here. The Mössbauer effect can be described (semiclassically) as the presence of an unbroadened carrier wave in the (phase-modulated) γ-ray emission from a nucleus that is being vibrated by phonons within a solid.

Problem 4.13 (a) Getting light into a spectrograph
Use Fig. 4.7(a) on p. 96.

(1) Use eqn (11.3) to show that, once the entrance slit is 'filled with light' and the collimator lens is 'filled with light', then no improvement is possible in the gathering of power from a source of given radiance. The étendue of the instrument is being fully utilized.

(2) A student setting up a spectrograph places the condenser lens so as to form a reduced-size image of the source on the entrance slit. He does this (probably) because light scattered from the slit jaws is made bright, so he has the impression that he is squeezing a lot of light into the instrument. Draw a ray diagram to demonstrate that light is gathered from a large area of the source, but light gathered from any point on the source occupies only a small solid angle. There is a trade-off, and this one isn't looking good any more.[47]

[47] A lot of light probably *is* being squeezed into the spectrograph, but it occupies too wide a cone of angles, and much of it fails to get through the collimator lens. Or, worse, it ends up where it shouldn't.

(3) Having seen what was wrong with part (2) the student tries the other extreme, placing the condenser lens to form a magnified image of the source on the entrance slit. Draw a ray diagram to show that now the source delivers light into a greater solid angle but only a small area of the source is being used. The trade-off is again looking unfavourable.

(4) A spectrograph is used to *photograph* a line spectrum. The entrance slit is wide enough that a distinct ray-optics image of its shape is formed at the output of the spectrograph. The spectral lines are narrow enough not to blur this image appreciably. Show that narrowing the entrance slit reduces the amount of light energy entering the spectrograph, but the exposure time is not lengthened.

(5) Surely you can't narrow the entrance slit for ever and pay no penalty? At what width of entrance slit does the result of part (4) cease to hold? Does this width resemble one you've met before, perhaps in problem 4.6(6)?

(6) Return to part (4) and consider whether the photographic exposure time is independent of slit width when recording a *continuous* spectrum.

(7) Parts (4) and (6) have given us very different results for two extreme cases. What happens 'in the middle', perhaps when we look at a rather fuzzy spectral line?

Problem 4.14 (b) Vignetting

(1) Draw Fig. 4.7(a) again but 'in the other section' so that the length of the entrance slit lies in the plane of the paper. The source may be thought of as a long thin gas discharge tube that can supply light to all points along the entrance slit. Draw the rays that pass through the top end of the slit, from upper and lower extremities of the condenser lens, and see where they go inside the collimator. Confirm the statement in §4.9.2 that some of this light misses the collimator lens.

(2) Address the same situation in another way. One aim of 'correct illumination' is to fill the collimator lens with light. Draw rays 'backwards' from the top of the collimator lens, through top and bottom of the slit, back towards the condenser lens. Do some of these 'rays' pass outside the outline of the condenser lens? If so, the collimator lens is receiving less light than it should. If not, you've probably drawn the condenser lens too big .... Investigate.

(3) Draw Fig. 4.7(b) again but 'in the other section'. Draw rays on your diagram to show that the difficulty addressed in parts (1) and (2) is now solved.

(4) Draw the arrangement envisaged in sidenote 23 on p. 97.

Problem 4.15 (b) Shortening exposure times

Invent a layout for the cylindrical lens described in §4.9.3. Can you arrange for the shortened slit image to have 'sharp ends', meaning a sudden transition from light to dark?

Problem 4.16 (b) Fraunhofer diffraction from a random array

A plane monochromatic light wave is incident normally on a screen that is pierced with a large number N of small identical pinholes forming a random array within a square area of side D. For parts (1)–(5) the diameter of a pinhole may be taken to be $\lesssim\lambda$, where λ is the wavelength of the light. Light transmitted by the array forms a diffraction pattern that is observed under the Fraunhofer condition.

(1) Show that in the forward direction ($\theta = 0$),

$$\text{the diffracted amplitude} = N \times (\text{the amplitude from one pinhole})$$
$$\text{and so the intensity} = N^2 \times (\text{the intensity due to one pinhole}).$$

(2) Show that the result of part (1) is valid over angles θ up to about $\theta_{\rm m}$ where $\sin\theta_{\rm m} \approx \lambda/D$.

(3) Show that for $\sin\theta \gtrsim \lambda/D$, the diffracted amplitude is 'expected to be' zero.

Comment: The phrase 'expected to be' is designed to deal with the fact that we are given no detailed information about this particular random array. You can overcome this lack by imagining that you have a large number—an ensemble—of random arrays, all different, but satisfying the statistical tests for randomness. It is possible to find the average—the *ensemble average*—amplitude by averaging the amplitudes produced at angle θ by all the arrays in the ensemble. We'll write $\langle U(\theta)\rangle$ for the ensemble average of the diffracted amplitude $U(\theta)$. Then the result asked for in part (3) is $\langle U(\theta)\rangle = 0$.

Compare the formation of $\langle U(\theta)\rangle$ with making an 'expectation value' in quantum mechanics, hence the choice of words 'expected to be'.

Henceforth, we'll make the further assumption that the amplitude diffracted by our pinhole array is not very different from the ensemble average. Compare this with the similar assumption, made in statistical mechanics (especially in the microcanonical ensemble), that the macrostate you have is the most probable one.

(4) Show that, for $\sin\theta \gtrsim \lambda/D$, the average $\langle |U(\theta)|^2\rangle \neq 0$, but instead[48]

$$\langle |U(\theta)|^2\rangle = N \times \begin{pmatrix} |\text{amplitude}|^2 \text{ contributed by} \\ \text{one pinhole at the same angle } \theta \end{pmatrix}.$$

[48] *Comment*: This is an example of a general rule (problem 9.3): when the contributing waves have random phases, add their intensities. Equation (2.27) and problem 2.15 quoted the result in another form. In problem 2.15, at the end, did you find a restriction on ω, doing the same job as $\sin\theta \gtrsim \lambda/D$ does here?

(5) Why are the results of parts (3) and (4) compatible?

(6) Consider now a random array of identical (and identically oriented) apertures, where the apertures have some arbitrary shape (not necessarily negligibly small compared to a wavelength). How does the diffraction pattern differ from that found previously?

(7) What happens if the apertures of part (6) are rearranged onto a set of lines parallel to Oy? The lines are regularly spaced by a in the x-direction ($\lambda \ll a \ll D$), but the apertures are randomly positioned along the lines in the y-direction.

(8) What happens if the apertures are further rearranged until they are regularly spaced in both x- and y-directions, i.e. they lie on a regular lattice?

(9) Does the result of part (1) conflict with the conservation of energy?

The Fabry–Perot

5

5.1 Introduction

This chapter examines a particular case of multiple-beam interference: where light is bounced repeatedly between two highly reflecting mirrors. It is assumed that the reader has already encountered interference 'by division of amplitude' in its two-beam case, as happens when light is reflected from opposite faces of a slab or traces the two paths through a Michelson interferometer. It is further assumed that the following are familiar:[1]

- The optical path difference is $n\,2d\cos\theta$ when a light beam divides at the surfaces (separation d) of a slab of refractive index n, and travels inside the slab at angle θ to the normal.
- In the case where the reflecting surfaces are parallel, interference fringes 'of equal inclination' are localized at infinity.

[1]An unusual derivation of the optical path difference is outlined in problem 5.1, and fringe localization is explained in §11.11.

We have already encountered a case where multiple beams of light are interfered: those beams that pass through the many elements of a diffraction grating and are combined 'downstream'.[2] In both cases, the advantage of 'multiple' is a sharpening of the fringes, with a consequent enhancement of resolution, according to the value of a quantity called the finesse.

[2]Interference at a grating can be made to resemble that with a Fabry–Perot, as is shown in problem 5.11.

The Fabry–Perot is worth understanding in its own right as a spectroscopic instrument. It is also a useful model to have in mind when understanding thin films (Chapter 6) and laser cavities (Chapter 8).

5.2 Elementary theory

Figure 5.1 shows a light ray (meaning a wave whose wavefronts are normal to the given ray) arriving from bottom left and incident on the first of two identical partly reflecting mirrors. Some light passes through into the space between the mirrors. There it is repeatedly reflected, with a small fraction transmitted at each encounter with a mirror.

At the first surface, the incident wave of amplitude U_0 has a fraction t_1 of its amplitude transmitted. At the second surface, a fraction t_2 is transmitted and a fraction r_2 is reflected. After each round trip, the light that remains between the mirrors has undergone two reflections (amplitude reflection coefficient r_2) and a phase lag of $k\,2d\cos\theta$ resulting from travel through distance $2d\cos\theta$. The outcome is a set of transmitted wave amplitudes, the first few of which are marked on Fig. 5.1.

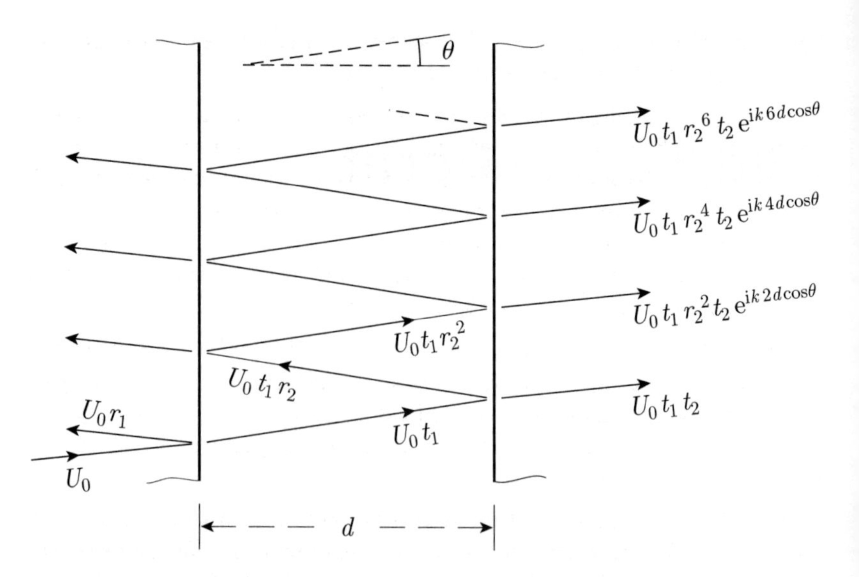

Fig. 5.1 The path of a representative wave through a Fabry–Perot. The wave is partially reflected and partially transmitted at each of the two reflecting surfaces. Amplitude transmission coefficients t_1, t_2 are defined on the diagram, as are the 'outside' amplitude reflection coefficient r_1 and the 'inside' reflection coefficient r_2. The commonest configuration has an air gap between two spaced glass blocks, and such a structure is hinted at in the drawing.

The coefficients t_1, t_2, r_1, r_2 may be complex if there are phase shifts at the mirrors, as there may well be. Let the 'internal' intensity reflection coefficient be $R = |r_2|^2$ and let $r_2^2 = R\,e^{-i\alpha}$. The total phase lag between successive transmitted beams is now $\delta = k\,2d\cos\theta - \alpha$, where this allows for the round-trip travel and the effect of two reflections. (The reflection phase shift α is included for honesty only and will sometimes be ignored to avoid cluttering the discussion.)

The total amplitude transmitted through the Fabry–Perot is now

$$U_{\text{out}} = U_0\,t_1 t_2 \{1 + R\,e^{i\delta} + R^2\,e^{2i\delta} + \cdots\} = \frac{U_0 t_1 t_2}{1 - R\,e^{i\delta}}, \tag{5.1}$$

where we use the sum of a geometric progression.

The transmitted intensity $I_{\text{transmitted}}$ evaluates to

$$\frac{I_{\text{transmitted}}}{I_{\text{incident}}} = \frac{|U_{\text{out}}|^2}{|U_0|^2} = \left(\frac{|t_1 t_2|}{1-R}\right)^2 \frac{1}{1 + \dfrac{4R}{(1-R)^2}\sin^2(\delta/2)}. \tag{5.2}$$

The prefactor $|t_1 t_2|/(1-R)$ is of little interest: it varies only slowly with θ (depending on the physics of the reflecting surfaces) and not at all with d, so it will henceforth be absorbed into I_{max}, the maximum transmitted intensity. Summarizing and tidying a little, we now have:

$$\frac{I_{\text{trans}}}{I_{\text{max}}} = \frac{1}{1 + (4\mathcal{F}^2/\pi^2)\sin^2(\delta/2)} \qquad \text{the Airy function} \tag{5.3}$$

$$\mathcal{F} = \pi\sqrt{R}/(1-R) \qquad \text{the finesse} \tag{5.4}$$

$$\delta = k\,2d\cos\theta - \alpha = (\omega/c)n\,2d\cos\theta - \alpha \qquad \text{the phase lag.} \tag{5.5}$$

Here n is the refractive index of the medium between the reflecting surfaces, and θ is the angle between the rays and the normal *between the reflectors.* Some straightforward mathematical steps have been omitted in the above, and are rehearsed in problem 5.2.

Comment 1 The single incident ray represents, as usual, a broad wavefront arriving from some source, details of which are not shown. Since only one ray is drawn, we are not told whether the incident wavefront is plane or spherical or of some other shape; we shall see that it does not matter. Because the incident wave is broad, the transmitted waves are broad also, and can be taken to overlap each other even though their rays are drawn with a lateral displacement.[3]

Comment 2 Fabry–Perot devices[4] are most commonly made from two glass blocks, with the reflecting surfaces facing 'inwards' to an air gap. The amplitude transmission coefficients t_1 and t_2 are necessarily unequal, because the light in one case travels from glass to air and in the other from air to glass.

Comment 3 The finesse $\mathcal{F}$ has been constructed from the group of factors $4R/(1-R)^2$ in a way that may not seem obvious. We shall see that the finesse $\mathcal{F}$, as defined, is the most useful quantity for specifying the quality of the interference fringes.[5]

Comment 4 No attempt has been made here at simplifying or evaluating the prefactor $|t_1 t_2|/(1-R)$. Since the reflecting surfaces might be as different as evaporated aluminium or several layers of dielectric, no general simpler expression could exist.[6]

Comment 5 Although the calculation leading to eqns (5.3)–(5.5) is in every textbook, we should not accept it uncritically as the only possible way forward. Problem 5.3 offers a rather different analysis. Problem 5.4 asks whether eqn (5.3) meets tests for reasonableness.

5.3 Basic apparatus

The addition of amplitudes that was performed in obtaining eqn (5.1) represents, of course, the interference of the successively reflected waves, each differing in phase from the one before. Equation (5.5) shows that the phase difference varies with θ only, if d and n are held fixed. Whether we get light or dark then depends upon θ.

In order to display the interference, we need to gather all light of a given θ and put it into its own place, separate from all other θs. Figure 1.3 explains how this may be done, because a lens focused for infinity selects angles in just the way we need. A minimal arrangement of apparatus is, therefore, that shown in Fig. 5.2. In the focal plane of the lens there are fringes in the form of bright circular rings,[7] narrow if the finesse $\mathcal{F}$ is large.

Since the optics has its correct geometry only when we look in the focal plane of the fringe-forming lens, we say that the fringes are *localized at infinity*. (For more on fringe localization, see §11.11.)

By contrast, restrictions on the 'upstream' side of the Fabry–Perot are conspicuously absent. The single input ray drawn in Fig. 5.1 is

[3] Under the conditions envisaged in this comment, interference takes place everywhere that there is light on the downstream side of the Fabry–Perot. If the interference is constructive, light is transmitted; if not, it is reflected. The lens introduced in §5.3 is not required to make the interference happen, but to unscramble what would otherwise be a confusing mess.

[4] A pair of reflecting plates mounted either side of a fixed-width spacer constitutes a Fabry–Perot **étalon**. An arrangement in which one plate may be moved on rails to vary d is called a Fabry–Perot **interferometer**. In this book we write 'Fabry–Perot' when (as is usually the case) the discussion applies to devices of either construction; we refer to an 'étalon' only when concerned specifically with a fixed-d device.

The French word 'étalon' has two meanings: a standard, and a stallion. It is usually clear which is to be found in a laboratory.

[5] Some books write $F = 4R/(1-R)^2$ and work with this F as well as (or instead of) $\mathcal{F}$. Here we avoid F in an effort at minimizing confusion.

[6] There is a theorem, investigated in problem 6.5(4), that $|t_1 t_2|/(1-R) = 1$ for the special case of a loss-free reflecting surface. This property applies even if the 'surface' is composite so long as it is loss-free. Our grounds for discarding the prefactor as uninteresting were not, however, based on any assumption that the reflectors might be loss-free.

[7] The fringe pattern must be circular because the optical system has axial symmetry about an axis normal to the étalon plates and passing through the centre of the lens.

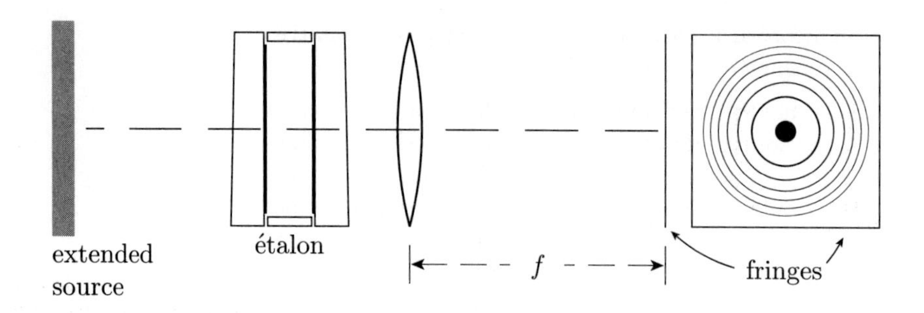

Fig. 5.2 A 'minimal' apparatus for exhibiting Fabry–Perot fringes. The étalon plates and spacer are shown, but the mount for holding them together is omitted to avoid clutter. The 'fringe-forming' lens selects light travelling in each direction θ and puts it in its own place in the focal plane; it maps directions onto positions.

consistent with any wavefront geometry in the input light. And Fig. 5.2 shows a broad illuminating source with no 'preparation' of any kind of the light from that source. There are in fact usually good reasons for elaborating the apparatus of Fig. 5.2, but those reasons are secondary, such as optimizing the brightness of the fringes, rather than getting fringes to happen at all. Such matters are discussed in problem 5.5. See also sidenote 24 on p. 122.

5.4 The meaning of finesse

The intensity transmitted by a Fabry–Perot, given in eqn (5.3), has peaks that occur where $\sin(\delta/2) = 0$, which means that $\delta = p\,2\pi$, where p is an integer. Equivalently, and ignoring α for simplicity, there are

$$\text{maxima of transmitted intensity where} \quad n\,2d\cos\theta = p\lambda_{\text{vacuum}}. \qquad (5.6)$$

Figure 5.3 shows $I_{\text{trans}}/I_{\text{max}}$ plotted versus the phase difference δ, covering a range of δ that includes the maxima of orders p to $p+1$.

We shall find the FWHM, the full width at half maximum height, of the pth peak by finding where $I_{\text{trans}}/I_{\text{max}} = \frac{1}{2}$. From eqn (5.3), this happens at $\delta = p\,2\pi + \varepsilon$ where

$$\pm\pi/(2\mathcal{F}) = \sin(\delta/2) = (-1)^p \sin(\varepsilon/2),$$

so that

$$\varepsilon = 2\arcsin(\pm\pi/2\mathcal{F}).$$

In the cases of interest, $\mathcal{F}$ is made to be large (perhaps 30 or more, achieved by using high reflectivity R), so ε is small[8] (though δ is not), and ε can be evaluated by using the power series for arcsin. The result is: $\varepsilon = \pm\pi/\mathcal{F} + \text{order } \mathcal{F}^{-3}$. The FWHM of the peak, measured along the δ axis, is thus $2\pi/\mathcal{F}$ and

$$\frac{\text{separation of peaks}}{\text{FHWM of one peak}} \approx \frac{2\pi}{2\pi/\mathcal{F}} = \mathcal{F}. \qquad (5.7)$$

This result explains why the finesse $\mathcal{F}$ is the useful measure of Fabry–Perot fringe quality.

[8] Phase shift δ is *not* small because the integer p may well be of the order of thousands.

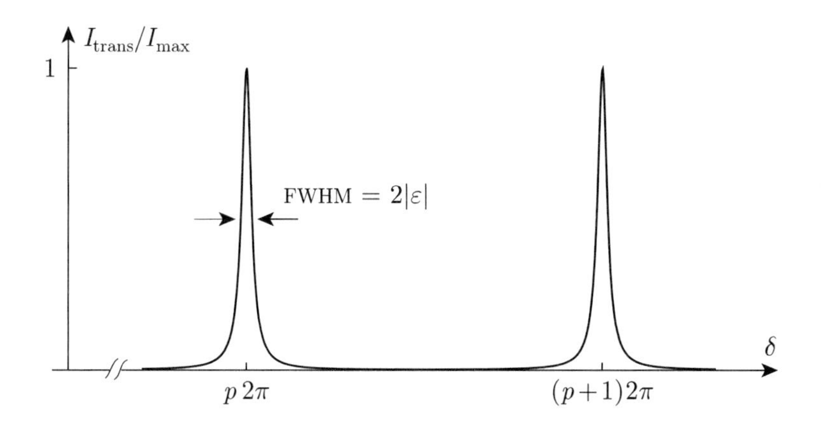

Fig. 5.3 A graph of $I_{\text{trans}}/I_{\text{max}}$, the normalized intensity transmitted, plotted against phase shift δ. The graph is drawn for a finesse of 20, a modest value, but one that helps with displaying the details of the curve.

5.5 Free spectral range and resolution

The main use for a Fabry–Perot is as a spectroscopic device for investigating a group of closely spaced spectral lines.

Figure 5.4 shows a graph of the transmitted intensity I_{trans} plotted against $n2d\cos\theta$ (not δ this time, note).[9] We apply two slightly different frequencies ν, ν' (vacuum wavelengths λ, λ', wavenumbers $\bar{\nu}$, $\bar{\nu}'$). Frequency ν forms a system of rings; ν' forms a second system of rings, interlacing those for ν. Our aim is to understand how measurements can reveal that two distinct frequencies are present and yield a value for the frequency difference $(\nu' - \nu)$.

[9]The graph of Fig. 5.4 can be understood as describing families of rings if we regard $\cos\theta$ as varying while (nd) is held fixed. But the abscissa has been chosen so that the graph is equally applicable to cases where one of the other variables is being scanned. We shall see that in practice it more usual to scan n while holding $d\cos\theta$ fixed.

Figure 5.7 shows a plot that is the result of scanning the refractive index n.

5.5.1 Free spectral range

Imagine that ν' starts equal to ν, and that we are able gradually to tune ν' while watching the two ring families as they separate. To keep free from confusion, we want the peak for $p\lambda'$ to stay between $p\lambda$ (where it started) and $(p+1)\lambda$. The largest permitted change brings $p\lambda'$ just up

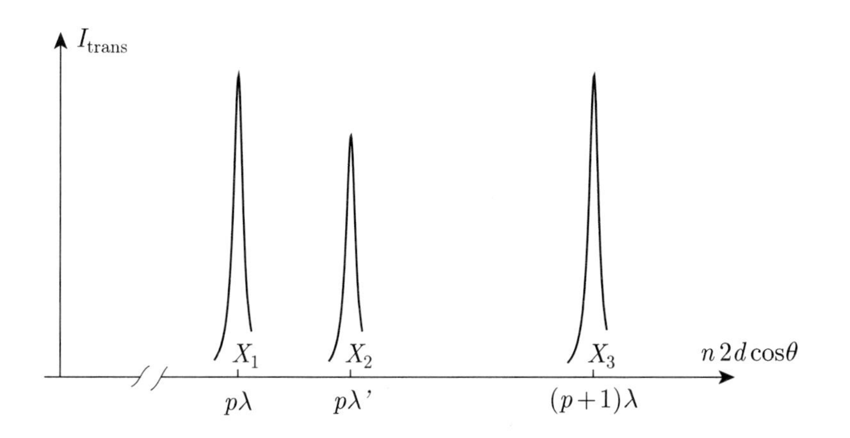

Fig. 5.4 The intensity I_{trans} transmitted by a Fabry–Perot, drawn as a function of $(n\,2d\cos\theta)$, when two wavelengths λ, λ' are input. Wavelength λ' is imagined to be adjustable so that point X_2 can be moved along the graph. To avoid confusion, X_2 should not normally be allowed to reach or pass X_3, a condition that fixes the free spectral range.

to $(p+1)\lambda$, at which point $\lambda' = \lambda'_F$, $\bar{\nu}' = \bar{\nu}'_F$, with

$$p\lambda'_F = (p+1)\lambda, \qquad p/\bar{\nu}'_F = (p+1)/\bar{\nu}, \qquad p(\bar{\nu} - \bar{\nu}'_F) = \bar{\nu}'_F = 1/\lambda'_F.$$

The **free spectral range** is this largest permitted change of frequency, quantified in terms of wavenumber $\bar{\nu} = 1/\lambda_{vac} = \nu/c$:

$$\text{free spectral range} = \bar{\nu} - \bar{\nu}'_F = \frac{1}{p\lambda'_F} = \frac{1}{n\,2d\cos\theta} \approx \frac{1}{2d}. \tag{5.8}$$

The approximation in the final step is permitted only now, after nearly equal quantities have been subtracted.

5.5.2 Resolution

Return to Fig. 5.4. Imagine tuning ν' starting from ν, but this time moving the $p\lambda'$ peak only just enough to separate it from $p\lambda$ by the width (FWHM) of one peak. We take this to be the setting at which the two frequencies are just resolved.[10] Equation (5.7) tells us that the FWHM is $1/\mathcal{F}$ of the separation between peaks (adjacent, belonging to the same wavelength). So $p\lambda'$ has been tuned by $1/\mathcal{F}$ of the distance from $p\lambda$ to $(p+1)\lambda$, that is, by $1/\mathcal{F}$ of free spectral range:

$$\text{least resolvable wavenumber difference } \delta\bar{\nu} = \frac{\text{free spectral range}}{\text{finesse}}. \tag{5.9}$$

Problem 5.6 draws attention to some subtleties that entered into the correct construction of the reasoning above.

The 'least resolvable wavenumber difference' is more commonly referred to as the *instrumental width*, meaning the apparent width in frequency of monochromatic light, as signalled by the FWHM of the peaks in the output from the Fabry–Perot.[11] Equation (5.9) can be recast as

$$\text{instrumental width} = \delta\bar{\nu} = \frac{1}{n\,(2d) \times (\text{finesse})}. \tag{5.10}$$

Comparing this with eqn (4.6), we see that[12]

$$n\,(2d) \times (\text{finesse}) = \begin{pmatrix}\text{effective longest optical path} \\ \text{difference for interference}\end{pmatrix}. \tag{5.11}$$

For this reason, the finesse is often thought of as the 'effective' number of round trips undertaken by the light as it bounces between the two reflecting surfaces.

Finally, we return to eqn (5.7) and Fig. 5.4. Two adjacent peaks belonging to a single frequency are separated (with our choice of abscissa) by λ. The width (FWHM) of a peak is $1/\mathcal{F}$ of this. So the number of additional peaks (belonging to other frequencies) that could just be fitted edge-to-edge[13] into one order is $\mathcal{F}$. The finesse $\mathcal{F}$, therefore, gives us a measure of the complexity of a spectrum which the Fabry–Perot could be expected to display, given an ideal choice of free spectral range.

Problems 5.8–5.10 investigate the resolution and the instrumental width from several points of view that complement that given here.

[10] This choice is sometimes called the Taylor criterion, in contrast to the Rayleigh criterion used in connection with a diffraction grating.

[11] Instrumental width has already been met in connection with a grating in eqn (4.8) and problem 4.7, and the idea is further developed in problem 5.8.

[12] As with eqn (4.6), the accuracy we need when specifying resolution does not usually justify our bothering with the refractive index in eqn (5.11).

[13] We can't, of course, expect to separate and analyse peaks if they really are fitted in so tightly. A safety factor of perhaps 3–5 is needed. But the finesse correctly gives us the estimate to which such safety factors should be applied.

5.6 Analysis of an étalon fringe pattern

Use Fig. 5.4 again. If frequency ν' has its pth fringe as shown, between $p\lambda$ and $(p+1)\lambda$ and not at some more remote place, we can obtain ν' by simple proportion:

$$\frac{X_2 - X_1}{X_3 - X_1} = \frac{p\lambda' - p\lambda}{(p+1)\lambda - p\lambda} = p\lambda'\frac{\lambda' - \lambda}{\lambda\lambda'} \approx 2d(\bar{\nu} - \bar{\nu}') = \frac{\bar{\nu} - \bar{\nu}'}{1/(2d)}$$

and so

$$\bar{\nu} - \bar{\nu}' = \left(\text{free spectral range } \frac{1}{2d}\right) \times \left(\text{the ratio } \frac{X_2 - X_1}{X_3 - X_1}\right). \quad (5.12)$$

The quantities on the right of eqn (5.12) can all be measured, so the frequency difference $(\nu - \nu')$ can be obtained from experiment.

We may well not know that the fringe labelled $p\lambda'$ is in fact of order p. Then eqn (5.12) is replaced by[14]

$$\bar{\nu} - \bar{\nu}' = \frac{1}{2d} \times \left(\frac{X_2 - X_1}{X_3 - X_1} + \text{integer}\right). \quad (5.13)$$

[14]This expression results from making a slight approximation, see problem 5.12.

The added integer (which might be positive, negative or zero) must be identified by obtaining additional experimental information, as is explained in the solution to example 5.1. In the remainder of this section we assume, for simplicity, that the additional integer is known to be zero.

One possible measurement procedure is to photograph the ring-fringe pattern, in which case n and d are held constant and the angles θ are obtained by measuring the ring radii r. Then $X = n\,2d\cos\theta \propto 1 - \frac{1}{2}\theta^2$ in a small-angle approximation and

$$\frac{X_2 - X_1}{X_3 - X_1} = \frac{\theta_1^2 - \theta_2^2}{\theta_1^2 - \theta_3^2} = \frac{r_1^2 - r_2^2}{r_1^2 - r_3^2}. \quad (5.14)$$

Data are used in this way in the solution of example 5.1.

The method of choice (reasons in §5.9), however, is usually to place the étalon inside a box where the pressure P of the air can be scanned, after the fashion of Fig. 5.8. A pinhole at the focus of the fringe-forming lens allows light to reach a photodetector only if it has $\cos\theta$ very close to 1. Increasing the pressure causes each 'ring' in turn to appear at the centre of the pattern and grow and, while it is emerging through the centre, a peak of intensity is recorded by the photodetector. To a good approximation the refractive index n of air is such that $(n-1) \propto P$ (Clausius–Mossotti relation), so for this method

$$\frac{X_2 - X_1}{X_3 - X_1} = \frac{n_2 - n_1}{n_3 - n_1} = \frac{P_2 - P_1}{P_3 - P_1}.$$

The pressures P_1, P_2 and P_3 are the pressures at which the photodetector records a maximum of intensity. Once a tracing on a chart recorder (or a computer screen) has been obtained, it is not hard to identify which pressure should be called P_1, P_2 or P_3 (apart from the difficulty of the added integer). Measurements of the pressures can be used to evaluate the ratio $(P_2 - P_1)/(P_3 - P_1)$, which can then be substituted into eqn (5.12) to give the frequency difference.

Example 5.1 Analysis of a Fabry–Perot pattern

A certain spectral line is known to consist of two equally intense components with a wavenumber separation Δ less than $20\,\mathrm{m}^{-1}$. The components are examined using a Fabry–Perot etalon in vacuum whose plate separation is 25 mm. The fringes are photographed and the diameters of the smallest rings are found to be in millimetres:

$$1.82, \quad 3.30, \quad 4.84, \quad 5.57, \quad 6.60, \quad 7.15.$$

What are the possible values of Δ indicated by this experiment? Suggest a further experiment which could be carried out to resolve the ambiguity.

[Part of a question: Oxford Physics Finals, paper A2, 1995]

Solution: Figure 5.5(a) displays the raw values of ring-fringe radius r along a 'number line'. It can be seen that there are two interlaced families of rings, but analysis is awkward because the radii crowd together as r is increased. This crowding is removed by plotting against r^2 as recommended in eqn (5.14); Fig. 5.5(b) shows the result. Each family of rings now has a set of regularly spaced values of r^2, and it's easier to assign the rings correctly to their two families. What remains is to see how best to put these values into the ratio in eqn (5.14).

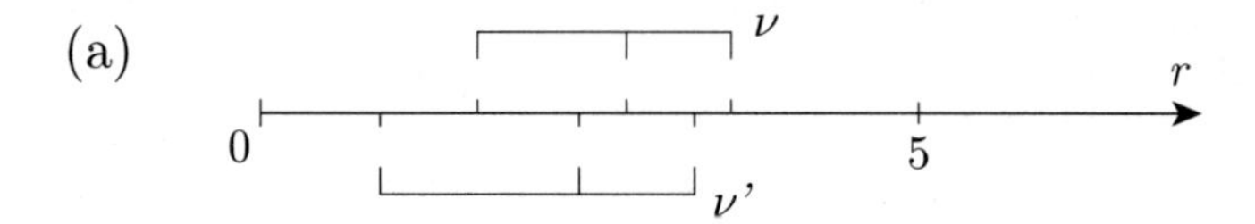

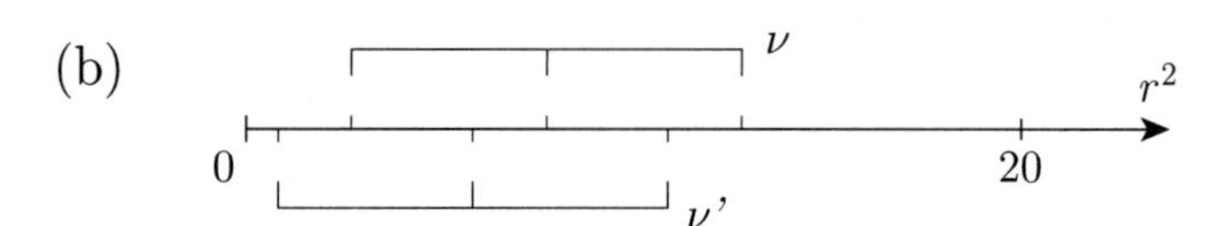

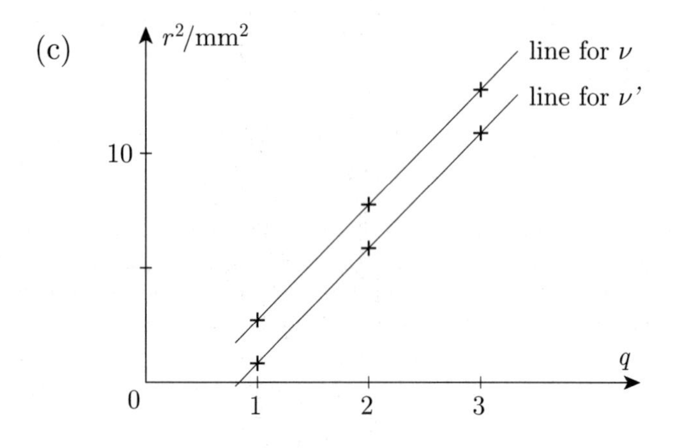

Fig. 5.5 Diagrams used in the solution of example 5.1. (a) Plotting the ring-fringe radii along a 'number line' gives values that crowd together for large radius r. (b) Plotting r^2 instead gives values that are uniformly spaced for any one frequency; a reflection of the fact that the optical path between the étalon plates varies as $\cos\theta \approx 1 - \frac{1}{2}\theta^2$. The ring fringes now divide cleanly into two families corresponding to the two frequencies being input. (c) A plot of the r^2 values for the two ring families against a 'counting number' q.

Figure 5.5(c) shows the values of r^2 for the two families of fringes (separated now and labelled ν, ν') plotted against a 'counting number' q which counts rings from 1 starting at the middle of the pattern.[15]

The notation of eqn (5.14) suggests that we ought to obtain the ratio $(r_1^2 - r_2^2)/(r_1^2 - r_3^2)$ by taking specifically chosen fringe radii. However, the story of Fig. 5.5(c) is somewhat different. The numerator is the difference between the heights of the two straight lines, and the denominator is the gradient of either line. Standard methods[16] for the analysis of straight-line graphs will now be recommending themselves. We find: difference of heights $= 1.895\,\text{mm}^2$; gradient $= 5.030\,\text{mm}^2$; $(\bar{\nu} - \bar{\nu}') = 7.535\,\text{m}^{-1}$.

There remains the ambiguity referred to in the question. We do not know yet how the orders of interference for the ν' set of rings relate to those of the ν set. Figure 5.5(c) has been drawn on the simplest assumption: that the two bracketed sets of radii of Fig. 5.5(b) use the same set of p-values. That is, the added integer in eqn (5.13) is assumed to be zero—and it may not be.

The étalon's free spectral range is $\delta\bar{\nu}_{\text{fsr}} = 1/(2 \times 25\,\text{mm}) = 20\,\text{m}^{-1}$. The possible values of $(\bar{\nu} - \bar{\nu}')$ permitted by eqn (5.13) are

$$(\bar{\nu} - \bar{\nu}') = 7.535\,\text{m}^{-1} + (\text{integer}) \times (\text{free spectral range of } 20\,\text{m}^{-1}).$$

In this, all choices of integer except 0 and -1 make $|\bar{\nu} - \bar{\nu}'|$ greater than the limit of $20\,\text{m}^{-1}$ conveniently specified in the question. So only two assignments are in contention, yielding $(\bar{\nu} - \bar{\nu}') = +7.535\,\text{m}^{-1}$ and $-12.465\,\text{m}^{-1}$.

To resolve the ambiguity: do a second experiment, using another étalon with a different thickness of spacer. The second étalon will yield a new set of possible values of $(\bar{\nu} - \bar{\nu}')$. Do some trial-and-error calculations before choosing the new spacer to make sure that whichever interpretation of the original experiment turns out to be right, the other will be definitely excluded by the second experiment.

[15] Note that q is *not* the order of interference p for two reasons: first, p is a number of order 10^4 while q is of order 1; second, as q increases, θ increases, $\cos\theta$ decreases, so p decreases—the integers count in opposite directions.

[16] Theory requires the two straight lines to be parallel, except for a slight correction, investigated in problem 5.12, that is here well within the experimental uncertainties. The 'standard methods' should, therefore, be modified to force the two lines to have a common gradient.

5.7 Flatness and parallelism of Fabry–Perot plates

Figure 5.6 shows a Fabry–Perot with a (highly exaggerated) lump on one of the plates. The effective number of round trips made by the light between the plates, by eqn (5.11), is $\mathcal{F}$. So the lump is visited about $\mathcal{F}$ times, and it introduces an error in the optical path of $\mathcal{F} \times (2h)$, where h is the height of the lump (a factor 2 for both coming and going). If the error in the optical path is to disrupt the interference by a negligible amount, it must be less than a wavelength, so we require $\mathcal{F}\,2h \lesssim \lambda$. Thus, the permitted variation in plate separation d (which might be caused by departures from flatness or from a failure in setting the plates parallel) is

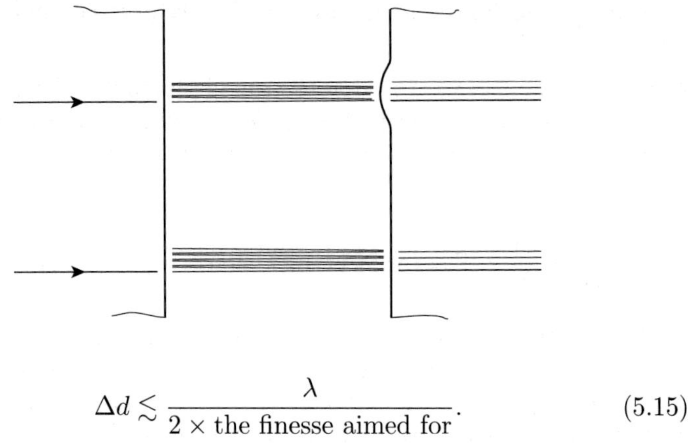

Fig. 5.6 The effect of a non-uniform spacing d between a pair of Fabry–Perot plates, illustrated by showing a shallow lump on one plate. The lump is visited about $\mathcal{F}$ times and contributes a path error of about $2\mathcal{F}\times$ its height.

$$\Delta d \lesssim \frac{\lambda}{2 \times \text{the finesse aimed for}}. \tag{5.15}$$

Problem 5.13 asks you to give a more careful derivation of eqn (5.15).

In practice, a Fabry–Perot has its finesse limited both by departures from flatness and by the limited reflectivity of the plates. To achieve an overall finesse of (say) 30, we should construct an 'error budget' by which the combined imperfections do not bring the finesse below 30. As a start, we might make the two imperfections equally damaging, in which case we should require a 'reflectivity finesse' of 60 accompanied by a 'flatness finesse' of 60. Equation (5.15) shows that we are asking for surfaces flat (and adjusted parallel) to around $\lambda/120$, which is quite hard—and expensive—to achieve. Do not demand a higher finesse than you really need ....

5.8 Designing a Fabry–Perot to do a job

The quantities to be chosen are: the plate spacing d; the reflectivity R; the tolerance Δd on plate flatness. Here is a recommended procedure:

(1) Choose d so that the free spectral range $(= 1/2d)$ is greater, but only a little greater, than the frequency range expected. If $1/2d$ is too big, the fringes of one order will be crowded together with a lot of empty space before the next order, and the finesse will have to be made unnecessarily large to separate them.

(2) Choose the finesse to resolve the structure expected. Since $\mathcal{F}$ is the greatest number of peaks that will fit (very snugly!) into an order, we shall be able to resolve N equally spaced frequencies if $\mathcal{F} = N \times$ (a safety factor of say 3 to 5).

(3) Check that this $\mathcal{F}$ is not so large that it demands an unattainable flatness.

(4) Aim[17] for a reflectivity $R \approx 1 - \pi/(2\mathcal{F})$ and flatness $\Delta d \approx \lambda/(4\mathcal{F})$, in which I've inserted safety factors of 2 in the spirit of §5.7.

The above procedure fixes d and $\mathcal{F}$ (whence R and the surface flatness) one at a time in a rational sequence. It makes a nice model for the clean design of any piece of apparatus. Things worked so well because I isolated two independent physical quantities: free spectral range (related

[17] *Comment*: Don't aim for a reflectivity much closer to 1 than is calculated in item (4). You can achieve a higher reflectivity by depositing additional quarter-wave layers on the plates. But each added layer contributes slight variations of thickness that degrade the flatness.

There is an additional and important reason for not pursuing an over-high reflectivity finesse: The smaller we make the instrumental linewidth, the less light energy is transmitted through our apparatus to its detector. In the extreme case, a very small instrumental width causes a small fraction of the energy to be selected from within the incoming linewidth, while doing nothing to improve resolution. In different words: the overall line profile is the convolution of the incoming line profile with the instrumental profile; the resolution is controlled by the wider; the energy gathering is controlled by the narrower. If energy gathering is of concern—and in real cases it always seems to be—then the partitioning of the error budget suggested in §5.7 will usually be about right.

to d only), and finesse $\mathcal{F}$ (related to R only), which could be decided upon separately.

Comment: The above uses the physics determining the Fabry–Perot's resolution (of course), but makes no *direct* use of either the instrumental width $\delta\bar{\nu}$ given by eqn (5.9), or the chromatic resolving power[18] given by

$$\frac{\lambda}{\delta\lambda} = \frac{2d}{\lambda} \times \mathcal{F} = (\text{order of interference } p) \times (\text{finesse}). \tag{5.16}$$

[18]This expression for the chromatic resolving power is derived in problem 5.7.

Both resolution quantities $\delta\bar{\nu}$ and $\lambda/\delta\lambda$ contain the product $(d\mathcal{F})$, and if we set a requirement on either of these we still have to find d (and thence $\mathcal{F}$) separately by deciding upon a free spectral range. Such a route is less straightforward than that recommended.

Example 5.2 Design of an étalon for recording a hyperfine structure

Figure 5.7 shows a curve such as might result from pressure-scanning an étalon to investigate a spectral line split by hyperfine structure.[19] The two tallest peaks, 6 and 6′, belong to the same frequency, and have $n\,2d\cos\theta = p\lambda$ and $(p+1)\lambda$ as in Fig. 5.4. The remaining five peaks belong to other frequencies, and all have the same order p, so peaks 1–6 constitute a spectrum (with wavelength increasing towards the right as in Fig. 5.4).

Note particularly the empty space that has been left between peaks 1 and 6′. This empty space is large enough that we can be sure there is no peak to the right of 1 that is in danger of being missed. But given that, the spectrum near-fills the free spectral range. (More correctly, the free spectral range has been made only a little larger than the range of frequencies to be measured.)

The curve displayed in Fig. 5.7 is calculated, but it has been modelled on an actual case described by Walther (1962).[20]

[19]The numbers attached to the peaks are the values of the total angular-momentum quantum number F for the upper levels of the transition (the hyperfine-structure splitting of the lower levels is tiny).

The curve has been calculated assuming that the spectral components have Lorentzian line profiles so that the peak widths arising from the source's line broadening and from the étalon's limitations can be handled together by assigning an effective value of finesse (see sidenote 30 on p. 127); that finesse has been given the value 32.

Although the curve presented is a calculated one, it has been proportioned to resemble one of the beautiful recordings reported by Walther (1962) in measurements on the hyperfine structure of ^{55}Mn. There are differences of detail because in Walther's experiment the line broadening was not wholly Lorentzian.

[20]For those who, like the author, have difficulty with German, a description of the essentials of Walther's experiment can be found in Woodgate (1980).

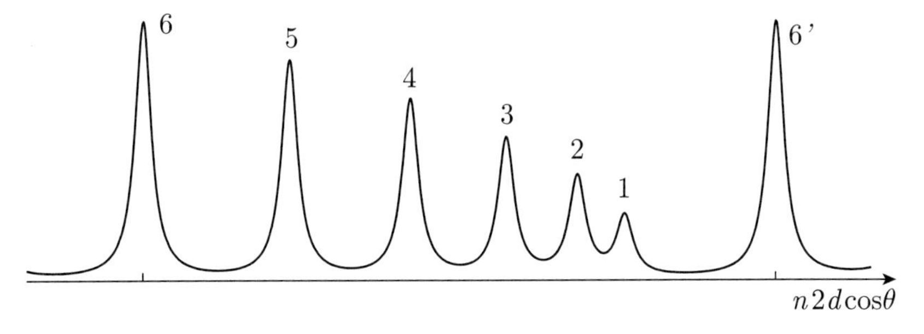

Fig. 5.7 A recording, after the fashion of Fig. 5.4, such as might be obtained by pressure-scanning an étalon illuminated with a spectral line split by hyperfine structure. This means that $n\,2d\cos\theta$ is scanned by varying n. The choice of free spectral range, in relation to the frequency range of the spectrum, is exemplary.

5.9 Practicalities of spectroscopy using a Fabry–Perot

(1) When do you use it? No one would claim that a Fabry–Perot is as convenient as, say, a grating spectrograph. So the answer is: when you need more resolution than a grating can give.[21] How do you tell? $\delta\bar{\nu} = 1/(\text{longest interference path})$. With a large grating, say 20 cm across with light incident at 30° to the normal, the longest interference path is ~10 cm and $\delta\bar{\nu} \sim 0.1\,\text{cm}^{-1}$. This is a theoretical ideal in which the grating is the only limiting factor. Example 4.1 on p. 93 has shown that one or two factors of 2 have to be introduced to allow for practicalities and imperfections, so $\delta\bar{\nu} \sim 0.5\,\text{cm}^{-1}$ is a rough estimate for the best we can do with a grating of ordinary size and shape. You can, of course, do better with a bigger grating. But this numerical estimate shows how inconveniently you'd have to enlarge things to win much.

By contrast, a modest Fabry–Perot might have $d \sim 5\,\text{mm}$ and $\mathcal{F} \sim 20$, giving (longest interference path) $= 20\,\text{cm}$, a figure that would be much more costly to achieve with a grating. The price we pay, of course, is a free spectral range of only $1/2d$; for $d = 5\,\text{mm}$ the free spectral range is only $1\,\text{cm}^{-1}$. So the Fabry–Perot is best suited for examining a group of spectral lines close together: hyperfine structure; a Zeeman splitting.

(2) Do we use an étalon or an interferometer? Given the strict requirement on the parallelism of the plates, the mechanism for translating a plate in a Fabry–Perot interferometer has to be rigorously designed to move its plate without tilt. This isn't out of the question, but it does require special attention. Section 5.6 and example 5.1 show that the fixed-but-choosable spacing of an étalon is not usually an obstacle to measurement and analysis. The étalon is, therefore, likely to be the configuration of choice.

(3) The spacer. The spacer between étalon plates must be made from a material with unusually low thermal expansion: fused quartz or a specially concocted low-expansion glass (or possibly a low-expansion alloy such as invar[22]). There are two reasons why a near-zero expansion coefficient is required: we do not want the plate separation to change if the temperature of the apparatus changes; and we do not want the plates to move out of parallel if one side of the spacer is warmer than the other. For a reason why such temperature changes might happen, see problem 5.14(4); for the reason why temperature differences may be serious, see problem 5.14(5). The same considerations require the étalon plates themselves to be made from fused quartz or a low-expansion glass.

(4) Auxiliary equipment: a monochromator in tandem. For light with $\lambda = 500\,\text{nm}$, $\bar{\nu} = 20\,000\,\text{cm}^{-1}$, so $1\,\text{cm}^{-1}$ is a small slice out of a spectrum. We are likely to need a monochromator (grating or prism) to separate our bunch of spectral lines from all the others.

[21] The Fabry–Perot has become a less common laboratory apparatus with the advent of sophisticated modern spectroscopic techniques that give high resolution combined with removal of Doppler broadening of the spectral lines. Such methods include saturated absorption spectroscopy and two-photon spectroscopy. Nevertheless, the Fabry–Perot remains worthy of understanding for the reasons listed in §5.10.

[22] Invar is an alloy consisting of 64% iron, 36% nickel, which happens to have an expansion coefficient close to zero. A variant alloy contains 4% of cobalt (Kaye and Laby 1995).

(5) Pressure scanning. Étalon fringes can be photographed—and if the source flickers there's little else you can do—but the method of choice is to record photoelectrically the intensity at the centre of the fringe pattern ($\theta = 0$), as (nd) is scanned in some way. The commonest method is to put the étalon in a box and change the pressure, hence also the refractive index n, of the enclosed gas. The pressure can be raised linearly with time by using a constant-flow source (the aerodynamic equivalent of a constant-current generator).

An apparently attractive alternative is to separate the plates with a piezoelectric spacer and scan the plate spacing by changing a voltage applied to the spacer. This idea sounds attractive, but a piezoelectric has a relatively large thermal expansion coefficient, and is likely to tilt the plates as well as translating them (remember how small is the permitted variation of d). (A thin piezoelectric, forming only part of the spacer, could, however, be used to impose a small modulation on d while the main scan is done by gas pressure ....)

(6) Pinhole size. A pinhole can't select $\theta = 0$ exactly because it would have zero area. Problem 5.15 investigates the permissible size of the pinhole.

(7) Unwanted reflections. There are unavoidable weak multiple-reflection effects within the thickness of each plate, with light bouncing

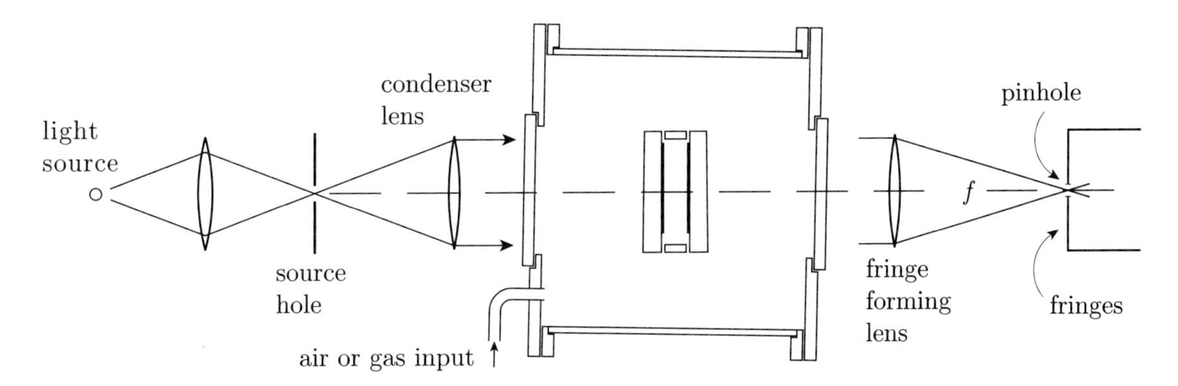

Fig. 5.8 For reasons itemized in §5.9 an étalon apparatus is usually elaborated from that of Fig. 5.2 to something like that shown here. Light from the source is gathered by the condenser lens, so arranged that with the fringe-forming lens it places a real image of the source in the same plane as the fringes. The fringes are formed at the collimator 'slit' (actually replaced by a pinhole) of a monochromator, details of which are not shown. At the output of the monochromator there is a photodetector whose electrical output is used to record the fringe intensity as a function of time. Also not shown are: all details of the étalon's mount and its support inside the pressure box; and the air-tight seals.

The windows are shown attached to the 'outsides' of the box's end plates. This arrangement is appropriate to a case where the pressure in the box is always at atmospheric or lower; the windows are pressed against their sealant. If pressures above atmospheric are to be used, the windows should be on the inside.

In this figure, and in Fig. 5.2, the étalon plates are shown slightly 'wedged', with the outer surfaces a little out of parallel with the inner surfaces. As explained in §5.9, this has the effect of throwing unwanted reflections off to the side.

The 'source hole' is made a little larger than the image, formed 'backwards' by the two lenses, of the pinhole at the output. (A little larger so that slight movements do not result in zero transmission!) It is the smallness of this source hole that permits unwanted reflections to fall wholly outside the area of the pinhole and thereby be discarded.

between the highly reflecting surface and the outer face. In effect each plate is itself a low-quality étalon and generates its own set of ring fringes. If the outer surface is parallel to the inner, these rings coincide with the 'wanted' ring pattern and degrade it. Anti-reflection coatings help, but not enough on their own. Therefore, the outer faces are deliberately cut at a small angle to the inner ones, to throw unwanted reflections away at an angle where they can be identified and allowed for.[23] Likewise, if the étalon is being pressure-scanned, it is mounted at a small angle to the container's windows (Fig. 5.8).

(8) Auxiliary equipment: a condenser lens. This is a lens which, in combination with the fringe-forming lens,[24] produces an image of the source at the same location as the fringes. Reasons why such a lens is often desirable are rehearsed in problem 5.5.

[23] When fringes are photographed these additional rings are not easily rejected, though they may be recognized for what they are. The arrangement shown in Fig. 5.8 does reject stray reflections, and it constitutes an additional, though minor, reason why pressure scanning is the preferred method of measurement.

[24] The wording is careful here. Given that the fringe-forming lens must be 'focused for infinity', the condenser will send 'parallel' light towards the Fabry–Perot. However it would be wrong to conclude that the Fabry–Perot 'needs' parallel light to be input if it is to work.

When all the above considerations have been taken into account, a spectroscopic apparatus involving a Fabry–Perot is likely to look somewhat like Fig. 5.8. Problem 5.14 asks you to think about some of the reasons why the design has ended up this way.

5.10 The Fabry–Perot as a source of ideas

As mentioned at the beginning of this chapter, the Fabry–Perot is worth understanding because it helps to give a foundation for future work. Topics that build on the Fabry–Perot include:

(1) Interference filter. This can be thought of as a thin Fabry–Perot étalon, though the mathematical methods for analysing it are somewhat different from those developed here. See §6.8.

(2) Laser cavities. The concept of a laser cavity grew out of the Fabry–Perot, though a different geometry and yet another line of analytical attack are now usual. See Chapter 8.

(3) In §8.10 we shall meet a hybrid between the Fabry–Perot and a laser cavity, the 'confocal' Fabry–Perot.[25] This instrument has a niche as an 'optical spectrum analyser', and is often used for displaying the mode structure of laser light.

[25] The existence of the confocal Fabry–Perot illustrates a general point. Apparatus design is not set in stone once it has been described in a textbook. It is always open to us to rethink and to invent. The attitude here is similar to that of §4.9.6.

Problems

Problem 5.1 (b) The path difference $2d\cos\theta$

Elementary derivations of the path difference $2d\cos\theta$ proceed by drawing a ray diagram and cancelling pieces of optical path. Let's show that there's another way.

Let a plane wave with wave vector $\boldsymbol{k} = (k\sin\theta, 0, k\cos\theta)$ travel between plates whose surfaces are normal to the z-axis. Show that, for fixed x, y, the phase change of the wave in travelling through $\Delta z = d$ is $kd\cos\theta$. Complete the reasoning from here.

Problem 5.2 (a) Fabry–Perot basics

(1) Fill in the omitted steps in the derivation of eqn (5.3).

(2) Explain why in comment 2 of §5.2 the amplitude transmission coefficients t_1 and t_2 had to be unequal.

(3) Textbooks on electromagnetism derive 'Fresnel's equations', which give the amplitudes of light transmitted and reflected when a plane wave is incident on a boundary between two media. Why would it be quite wrong to try to prove $|t_1 t_2|/(1-R) = 1$ from Fresnel's equations?

(4) Refer to the condition for maximum transmitted intensity, eqn (5.6). Get a rough idea of orders of magnitude by assuming that a Fabry–Perot has a plate spacing of 5 mm and the wavelength of light is 500 nm. What order of magnitude does this give to the integer p? Why is it necessary, in §5.4 to avoid assuming that δ is a small angle, and carefully to make ε small instead?

(5) Rework the calculation of §5.2 using different reflection coefficients r_2, r_2' for the two reflecting surfaces. Show that eqns (5.3) and (5.4) continue to hold with $R\,\mathrm{e}^{-\mathrm{i}\alpha} = r_2 r_2'$ replacing the previous definitions of R and α. It is, therefore, not critical for the two reflectors to be accurately matched in their reflectivity.

(6) Suppose we want a Fabry–Perot to have a finesse of 30. Use eqn (5.4) to find what reflectivity R is required.[26]

[26] *Hint*: You could solve a quadratic for $\sqrt{R}$, but all the accuracy you need can be achieved using a simpler route. Think about it.

Problem 5.3 (a) Alternative calculation!

Thin films are described in Chapter 6. In a thin-film context we think about just two travelling waves in any layer, one going to the right and the other going to the left. This isn't an approximation in which only the first two rays of Fig. 5.1 are considered (or it shouldn't be). The amplitudes of the two waves are worked out from the electromagnetic boundary conditions and they come out 'exactly'. No set of rays repeatedly reflected, no geometric progression.

In §5.2, each successive ray/wave reflected between the mirrors is there because we didn't get the reflection physics right previously: the boundary conditions were not satisfied. Each new ray partially corrects the error, and you get the correct physics only when the rays (and their geometric progression) are summed to infinity. If we had learned about thin films first we should find the treatment of §5.2 quite extraordinary!

Let's treat the Fabry–Perot in a manner that looks more like what we would do with a thin film. Refer to Fig. 5.9, in which five waves are drawn. The wave labelled a is the only wave there is travelling to the right between the mirrors: the sum of all the right-going waves in Fig. 5.1. Similarly for the wave going to the left and the wave transmitted. Let the waves have complex amplitudes

$$\left.\begin{aligned} \text{amplitude inside, to right} &= a\,\exp(\mathrm{i}k_z z), \\ \text{amplitude inside, to left} &= b\,\exp(-\mathrm{i}k_z z), \\ \text{amplitude transmitted} &= c\,\exp\{\mathrm{i}\kappa_z (z-d)\}. \end{aligned}\right\} \qquad (5.17)$$

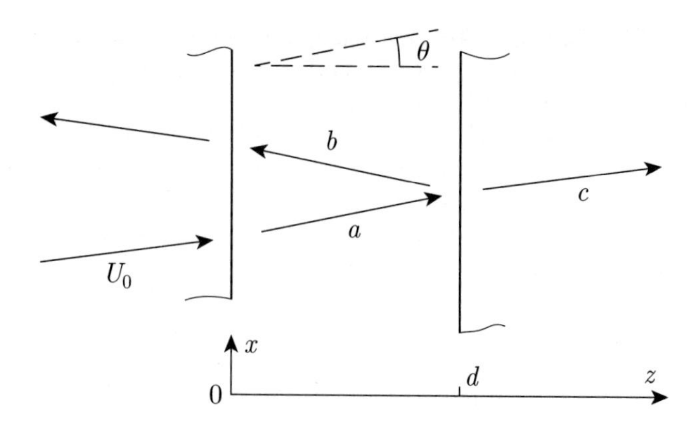

Fig. 5.9 The wave amplitudes for the 'alternative' calculation of problem 5.3. The amplitude coefficients are marked against each wave; more complete expressions are given in eqn (5.17).

Here $k_z = k\cos\theta$, and a factor $\exp(-\mathrm{i}\omega t + \mathrm{i}k_x x)$ is common to all amplitudes and has been dropped.

The physics of transmission and reflection at the left-hand mirror is expressed now by

$$a = U_0 t_1 + r_2 b.$$

Write down equations that deal with the physics at the right-hand mirror. Eliminate a and b between the three equations and show that

$$c = U_0 \frac{t_1 t_2}{1 - r_2^2 \exp(\mathrm{i}2k_z d)} \exp(\mathrm{i}k_z d).$$

Show that this is the same as eqn (5.1) except for the inclusion of the final exponential factor, which should more correctly have been included in eqn (5.1) anyway.

Comment: This problem is no more difficult to set up than the original calculation of §5.2. And it fits much more comfortably into the scheme of things that is used everywhere else in this book. The conventional treatment of the Fabry–Perot is looking even more outlandish.

Nevertheless, I don't really recommend the present calculation as a substitute for the conventional one—for the Fabry–Perot. The conventional treatment emphasizes that the physics involves multiple reflections, adding up to an 'effective total optical path' of about $(2d) \times \mathcal{F}$. And we have the insight of problem 5.11 that both links the physics to multiple reflections and explains the fringe localization in terms of an array of equally spaced sources. (With a little reluctance) I have to accept that the treatment of §5.2 is the most insightful way of understanding the Fabry–Perot. Thin films are another story ....

Problem 5.4 (c) A paradox: energy conservation

Problem 5.2(6) has shown that the (power) reflectivity R of each Fabry–Perot plate is likely to be in excess of 90%. Consider then the following argument.

In Fig. 5.1 the incident ray arriving from bottom left must have over 90% of its energy reflected towards top left when it encounters the left-hand reflector for the first time. With so much energy lost so soon, there is no possibility that the other rays emerging after reflection towards top left could be strong enough, even in combination with each other, to give complete destructive interference against the first ray. So, no choice of angle θ or plate spacing d could give zero power reflected towards top left. But it is claimed in connection with eqn (5.3) and sidenote 6 on p. 111 that 100% of the incident power is transmitted in the bright fringes. If all this is true, we have a violation of energy conservation.

Where is the above argument in error?

Problem 5.5 (b) The optical system for displaying Fabry–Perot fringes

(1) The lamp used for a particular spectroscopic investigation is a gas discharge tube with radius about 1 mm. Draw a ray diagram to show that if this lamp is used in the arrangement, unmodified, of Fig. 5.2 the fringe pattern will be dim. (It may help to refer to eqn (11.3).)

(2) Add to the apparatus of Fig. 5.2 a condenser lens between the lamp and the étalon. The lens is to be adjusted so that, together with the fringe-forming lens, it forms a real image of the lamp at the same location as the étalon fringes. Show that this optimizes the optical power per unit area falling on the bright fringes.

(3) The lamp used for a particular spectroscopic investigation is a gas discharge similar to that of part (1), but placed between the poles of a magnet for study of the Zeeman effect in the emitted spectrum. Explain why a condenser lens, similar to that of part (2) and focused in the same way, is desirable for this setup (for a reason separate from maximizing the power collected).[27]

[27] *Hint*: The region over which the $\boldsymbol{B}$-field is reasonably uniform and, therefore, readily predictable is likely to be only a few millimetres across.

Problem 5.6 (b) Free spectral range and resolution

(1) Figure 5.4 was drawn with a different abscissa from that of Fig. 5.3: $n\,2d\cos\theta$ instead of δ. What would go wrong if Fig. 5.4 were drawn with δ as its horizontal axis?

(2) Figure 5.4 was drawn with a different ordinate from that of Fig. 5.3: I_{trans} instead of $(I_{\text{trans}}/I_{\text{max}})$. What would go wrong if Fig. 5.4 were drawn with the same vertical axis as Fig. 5.3?

(3) Rework the discussions of free spectral range, of resolution and of fringe analysis, this time including the additional phase angle α coming from the reflectors. Show that the discussions become more cluttered, but that no points of principle are affected—so long as α is constant with frequency and with angle θ over the ranges considered.

Problem 5.7 (e) Chromatic resolving power and Q

(1) Use eqn (5.9) to show that the chromatic resolving power for a Fabry–Perot is given by eqn (5.16).

(2) In electric circuit theory, a resonant circuit (or a resonant cavity if we are working in the microwave region) is assigned a quality factor Q that quantifies the sharpness of the resonance. In the first instance, the

definition of Q is

$$Q = \frac{\text{frequency of resonance}}{\text{FWHM of resonance}},$$

which shows that Q is defined in exactly the same way as the chromatic resolving power $\lambda/\delta\lambda = \nu/\delta\nu$.

At radio frequencies, Q is usually thought of as related to the R, L and C of a resonant circuit. But at microwave frequencies such 'lumped' circuit elements don't exist, and an alternative expression for Q is used:

$$Q = \omega_0 \times \frac{\text{energy stored}}{\text{rate of energy loss}}.$$

Apply this definition of Q to the light energy rattling between the mirrors of a Fabry–Perot, and show that it is the same as (order of interference)×(finesse) provided that the finesse is large.[28]

[28] *Suggestion*: Think of a short pulse of optical energy, occupying time τ such that its length $c\tau$ in space is much less than the separation of the plates. (Similarity to the δ-function pulse of problem 5.9.) This pulse carries energy E. Each time the pulse encounters a mirror it loses a fraction $(1 - R)$ of its energy. Take $(1 - R) \ll 1$ so that the energy leaks out over a large number of round trips. You should be able to obtain expressions to feed into the second definition of Q. You won't get the factor $\sqrt{R}$ in the finesse, because your calculation is taken to only the first order in the small quantity $(1 - R)$.

Problem 5.8 (a) The instrumental line profile
Give a new interpretation to the Fabry–Perot fringe pattern in the following way. Monochromatic light of angular frequency $\omega_{\rm in}$ is input. The Fabry–Perot is to be thought of as giving infinitely sharp fringes. Equation (5.3) continues to hold, but with the dependence of intensity on δ interpreted as representing an apparent spectrum of frequencies present in the light.[29] To make things simple, we restrict attention to a range of δ given by $\delta = p\,2\pi + \varepsilon$, where ε is small compared to π. Show from eqn (5.5) that, close to the peak of order p,

$$(\omega_{\rm in}/c)n\,2d\cos\theta - \alpha = p\,2\pi + \varepsilon,$$
$$(\omega/c)n\,2d\cos\theta - \alpha = p\,2\pi;$$
$$\text{hence} \qquad (1/c)(\omega - \omega_{\rm in})n\,2d\cos\theta = -\varepsilon,$$

[29] This is precisely the definition of 'instrumental line profile' given in §4.7.

where $\omega_{\rm in}$ is the (angular) frequency that is being input and ω is the frequency that is attached by our interpretation to the non-zero intensity on the side of the peak. Substitute the above expression for ε into eqn (5.3) to obtain

$$\frac{I_{\rm t}}{I_{\rm max}} = \frac{1}{1 + \dfrac{4\mathcal{F}^2}{\pi^2}\sin^2(\frac{1}{2}\varepsilon)} \propto \frac{1}{(\omega - \omega_{\rm in})^2 + \left(\dfrac{\pi c}{\mathcal{F}n\,2d\cos\theta}\right)^2}.$$

This expression represents the instrumental line profile. Recognize it as having the shape of a Lorentzian spectrum, and show that the FWHM $\delta\bar{\nu} = 1/(n\,2d\mathcal{F})$. This expression for $\delta\bar{\nu}$ agrees with eqn (5.10), and gives the reason for assigning the name 'instrumental width' to $\delta\bar{\nu}$.

This problem can be compared with the treatment of a grating in problem 4.7.

Problem 5.9 (b) Resolution revisited
Consider a Fabry–Perot to which is applied a short pulse of light whose electric field is $U_0\,\delta(t)$. For simplicity, take all refractive indices and reflectivities to be real and frequency independent, and take the light to be incident normally.

(1) Show that the output (the light transmitted) consists of pulses separated in time by $T = n\,2d/c$. Show that each round trip decreases the pulse amplitude by $r_2^2 = R$.
(2) Show that the transmitted amplitude can be written as

$$U_{\text{out}}(t) = U_0\, t_1\, t_2 \sum_{p=0}^{\infty} \delta(t - T/2 - pT)\, \mathrm{e}^{-\gamma(t-T/2)} \tag{5.18}$$

(for $t > 0$), where γ is defined by $\mathrm{e}^{-\gamma T} = R$.
(3) Show that the output pulses have fallen in amplitude by a factor e ($\approx 2.7\ldots$) after $N = 1/(-\ln R)$ round trips. Show that when $R \approx 1$, $N \approx 1/(1-R) \approx \mathcal{F}/\pi$.
(4) Hence show that the output pulse train has a length (optical path), from beginning to the 1/e point, of $(n\,2d\,\mathcal{F})/\pi$, not too far from the $n\,2d\,\mathcal{F}$ claimed in eqn (5.11). We, therefore, confirm yet again the correctness for the Fabry–Perot of the physical statement in eqns (4.6) and (5.10) that the instrumental width is related to the longest optical path difference being imposed within the spectrometric apparatus.

Problem 5.10 (c) The Airy function is a sum of Lorentzians
Problem 5.8 has exhibited a resemblance between each peak of an Airy pattern and a Lorentzian spectrum. In fact, that resemblance is much closer than has so far been apparent. Let's see why.

Use the same physical picture as in problem 5.9: a short pulse is input. Equation (5.18) can be recast as

$$U_{\text{out}}(t) = U_0 \frac{t_1\, t_2}{\sqrt{R}} \left(\sum_{p=-\infty}^{\infty} \delta(t - T/2 - pT) \right) \times \begin{cases} \mathrm{e}^{-\gamma t} & \text{when } t > 0 \\ 0 & \text{when } t < 0. \end{cases}$$

Draw sketch graphs of the two factors here. The output amplitude has the form

(infinite comb of δ-functions) $\times$ (one-sided exponential decay).

The convolution theorem, used backwards, suggests that $\widetilde{U_{\text{out}}}(\omega)$ should be a convolution of: an infinite comb of frequencies; and a building block that is the transform of the exponential. It is well known (problem 10.6) that a Lorentzian spectrum results from a radiation process in which the emitted wave has an exponentially decaying amplitude, so it is no surprise if a Lorentzian is present. However, we still need to calculate the intensity, in order to put the Lorentzian properly on display.

Do what you can to tidy things up here. There is the outline of an elegant proof in Koppelmann (1969) eqn (1.13).[30]

[30] *Comment*: The property just derived, that the Airy function is a sum of side-by-side Lorentzians, is a useful insight when Fabry–Perot spectra are analysed. If a spectral line has two contributions to its profile, both of Lorentzian shape, the result is known to be another Lorentzian whose width is the sum of the contributing widths (problems 10.7 and 10.8). It follows that a Lorentzian line, input to a Fabry–Perot, will give an Airy pattern whose width is the sum of the instrumental width and the line's own width. A carefully measured instrumental width can be subtracted off to give the linewidth of the spectral line under investigation.

Problem 5.11 (b) Ways of looking at 'multiple-beam interference'
(1) Consider light radiating from a point source on the axis of a Fabry–Perot and passing through the plates with no other optical components on the 'source side'. Use ray optics to show that reflections inside the

Fabry–Perot produce a set of virtual images of the source, located at multiples of $2d$ behind the source, and getting weaker in amplitude by a factor R each time. This gives us a new way of thinking about the multiple-beam interference: the beams being interfered can be thought of as originating from equally spaced point sources in a line. In principle, there is an infinite number of these sources, but they get weaker as they get further from the real source. We have seen that the effective number of sources is about $\mathcal{F}$.

(2) Argue that if the beams of light from the many sources are to interfere in the same way, that is to have the same path difference from one to the next, all light collected must travel at the same angle θ. We require a fringe-forming lens focused for infinity.

(3) Contrast the result of part (2) with the case of two-beam interference, as at a parallel-sided slab or a Michelson interferometer set up with 'parallel' mirrors. In the 'Michelson' case, we need a fringe-forming lens if light originates from an extended source but not if it originates from a point source (§11.11). Yet, in part (2) we need a lens even for a point source. Think through the two cases until you can be comfortable.

(4) The way we are describing multiple images of a point source makes the Fabry–Perot optics resemble the optics of a diffraction grating—which indeed is another device that interferes multiple beams of light. Consider then a diffraction grating that extends from $x = 0$ to $x = L$ and has very narrow rulings running in the y-direction. Light is incident normally, in the z-direction, but the grating is 'shaded' so that the transmitted light has amplitude $U_0\,\mathrm{e}^{-\eta x}$ ($\eta L \gg 1$), as well as being broken up by the grating slits. Show that the Fraunhofer diffraction pattern formed has its intensity in the form of an Airy function. Check that the FWHM of each grating-order peak relates to η in a way that is consistent with what is expected by analogy with a Fabry–Perot.

(5) Just to show that the last result isn't an accident, repeat the calculation for the case of a 'sinewave grating' similar to that of example 3.1 on p. 61: $U(x) = U_0\,\frac{1}{2}(1 + \cos\beta_0 x)\,\mathrm{e}^{-\eta x}$. Show that the diffraction pattern is the sum of three Lorentzians.

(6) The inconvenient feature of a Fabry–Perot is the possibility of getting confused if the frequency range of the light supplied exceeds the Fabry–Perot's free spectral range. Gratings have not bothered us in a similar way .... Yet, we have just invented a problem in which a grating and a Fabry–Perot are mathematically identical. Think again about a grating (it can be 'unshaded' now), and assign to it quantities equivalent to (a) free spectral range and (b) finesse. Does this agree with what you found in problem 4.11?

Problem 5.12 (b) The unknown integer in eqn (5.13)
In Fig. 5.4, assume that the peak labelled $p\lambda'$ in fact belongs to order $(p - q)$. Show that eqn (5.12) is replaced by

$$\bar{\nu} - \bar{\nu}' = \frac{1}{2d} \times \left(\frac{X_2 - X_1}{X_3 - X_1} + q\frac{\lambda'}{\lambda} \right). \tag{5.19}$$

Argue that the conditions in which an étalon is used[31] make it satisfactory to replace λ'/λ by 1.

Show that the two straight lines drawn in Fig. 5.5 are not quite parallel, but their gradients differ by about 4 parts per million (depending on the mean wavelength which isn't given).

[31] *Suggestion*: Use the orders of magnitude in §5.9 item (4).

Problem 5.13 (a) The 'flatness finesse'
Section 5.7 gives a somewhat informal derivation of eqn (5.15). Give a more careful calculation of the acceptable Δd, meaning a variation of plate spacing caused by surface imperfections or failure of alignment. Use the graph of Fig. 5.4 and find what Δd shifts one of the peaks by its FHWM.

Problem 5.14 (b) Earthy practicalities

(1) The pressure-changing arrangement shown in Fig. 5.8 looks unduly complicated. Why do we not make an air-tight seal at the edges of the étalon plates and just pressurize the étalon itself? Given that the étalon is not in fact sealed but is inside a box, as in Fig. 5.8, does any precaution have to be taken to ensure that the air between the plates is at the same pressure as that outside?

(2) A 'solid étalon' is a single slab of glass coated with high-reflectance mirrors on its two surfaces. Such things exist, but are not met very often. Of course, a solid étalon can't be pressure-scanned, but it should be as good as a two-plate étalon if the fringes are to be photographed .... Why is a solid étalon nonetheless a rarity?

(3) The refractive index of air is 1.000292. An étalon is to be pressure-scanned through at least three orders, while the pressure of the air is increased by 1 atm. What is the smallest usable plate spacing d?

If it is necessary to pressure-scan a thinner étalon than this, people use carbon dioxide instead of air[32] in the étalon box. What is now the smallest usable plate spacing? [At STP, $n(CO_2) = 1.000451$.]

[32] Before carbon dioxide can be introduced, the air in the box must be removed by a vacuum pump. It is then simplest to admit CO_2 to give pressures between vacuum and atmospheric (rather than higher pressures). It is largely for this reason that Fig. 5.8 shows an apparatus designed for pressures below atmospheric.

(4) Discuss the thermodynamics of admitting air into the étalon box, raising the pressure by one atmosphere

(a) starting from vacuum,
(b) starting from atmospheric pressure.

Do you now see why the étalon must be made from materials with low thermal expansion?

(5) An étalon with plate spacing 5 mm is to have a 'parallelism finesse' of 60. The spacer is made from fused quartz with linear expansion coefficient $4 \times 10^{-7}\,K^{-1}$. What is the largest permitted temperature variation across the spacer?

(6) We've discovered that the reflecting surfaces of a Fabry–Perot need to be polished to a much higher quality than 'ordinary' optical flatness. What kind of tolerance (i.e. nearer to $\lambda/100$ or $\lambda/4$) applies to

(a) the outer surface of the plate on the side nearer the source,
(b) the outer surface of the plate on the exit side,
(c) the surfaces of the fringe-forming lens?

Problem 5.15 (b) The pinhole size for pressure scanning
In §5.6 it was mentioned that in pressure scanning we use a pinhole to pick out just the centre of the ring-fringe pattern. Now the pinhole must have a non-zero diameter. We wish to make the hole as large as possible in order to gather as much energy as we can onto the photodetector. And we wish to make the hole as small as is necessary to avoid loss of resolution. An engineering compromise is called for, rather similar to that already seen in example 4.1(2) on p. 93. This problem investigates.

(1) Consider the entire ring-fringe pattern, not just the centre. Show that all bright ring fringes have the same area and, therefore, carry the same optical power from a uniformly bright source. This shows that in a scanning instrument the best place to put the pinhole is in the centre of the ring pattern: away from the centre, it could intercept only a small fraction of a ring.

(2) A pinhole of finite size transmits light covering a range of $\cos\theta$, and, therefore, a range of optical-interference paths $2d\cos\theta$ and a range of phase shifts $\delta = (\omega/c)n\,2d\cos\theta\;[-\alpha]$. Use the ideas that led up to eqn (5.9) to assign a greatest acceptable range to δ for a given value of the finesse.

(3) Arrange the result of part (2) into the form

$$\frac{\text{pinhole diameter}}{\text{focal length of fringe-forming lens}} = \left(\frac{8}{p\mathcal{F}}\right)^{1/2},$$

where p is the order of interference, equal to $n\,2d/\lambda_{\text{vac}}$ when the instrument is set to transmit wavelength λ_{vac}. So far, no attempt has been made at introducing any safety factor.

(4) Suppose that we introduce an 'error budget' after the fashion of example 4.1(2). A first shot would be to make the three sources of imperfection (reflectivity, flatness, pinhole) contribute equally to degrading the resolution. Show that, with this choice,

$$\frac{\text{pinhole diameter}}{\text{focal length of fringe-forming lens}} = \left(\frac{8}{3p\mathcal{F}}\right)^{1/2}, \tag{5.20}$$

where $\mathcal{F}$ is the (overall) finesse we intend to end up with.

(5) An important quantity for a spectroscopic instrument is the product

$$(\text{solid angle accepted}) \times (\text{chromatic resolving power}).$$

We shall see in Chapter 11, eqn (11.16), that this takes the value of 2π for a 'whole-fringe instrument'.[33] Show that for the Fabry–Perot,

$$(\text{solid angle accepted}) = 2\pi/(p\mathcal{F}_{\text{pinhole}}),$$
$$(\text{chromatic resolving power}) = p\mathcal{F}_{\text{overall}}.$$

(a) Show that if the finesse of the instrument is limited only by the pinhole, $\mathcal{F}_{\text{pinhole}} = \mathcal{F}_{\text{overall}}$, and then the product of solid angle and resolving power takes the theoretical value of 2π.

[33] Used, as we show below, under conditions where the only limitation on the resolution comes from the pinhole itself.

(b) Show that if the 'engineering compromise' of part (4) is adopted then the product of solid angle and resolving power takes the value $2\pi/3$.

(6) Show from eqn (5.20) that the instrument's performance is independent of the focal length chosen for the fringe-forming lens. If the focal length is doubled, the allowable pinhole diameter is doubled too, and there is no change in either the resolution or the solid angle within which light is collected.

(7) Part (4) causes us to reconsider the way that safety factors are put into the design equations given in §5.8 item (4). Make the appropriate changes.

Problem 5.16 (e) Quality control
An interesting problem with an étalon is: how do you evaluate how good it is? If you want a very high resolution, the étalon you are making may well be the best instrument you have, so there's a problem finding a better instrument against which to check it. Here is a pretty answer. Use four plates of similar quality. Build two into an étalon of the desired spacing d. Build the other two plates into an étalon of length $5d$, say. Put the two étalons in series and illuminate with a fairly narrow spectral line. The long étalon filters the light down to a frequency range that is narrow enough to be used for stringently testing the short étalon.

Check the validity of this claim. Devise a procedure for executing the test. Reference: Stacey *et al.* (1974).

Problem 5.17 (c) A paradox: frequency dependences
Each plate of a Fabry–Perot is coated with dielectric thin films (as in Chapter 6) to be highly reflecting for red light. If one plate is examined in white light, it looks red by reflected light. It also looks blue by transmitted light; why?

Two such plates are assembled into a Fabry–Perot, and again examined in white light. The same colours are seen. But the theory says that, in the bright fringes seen in transmission, 100% of the incident light is transmitted, independent of the reflectivity of the plates, and so independent of the frequency. Whence then the blue colour of the transmitted light?

6 Thin films

6.1 Introduction

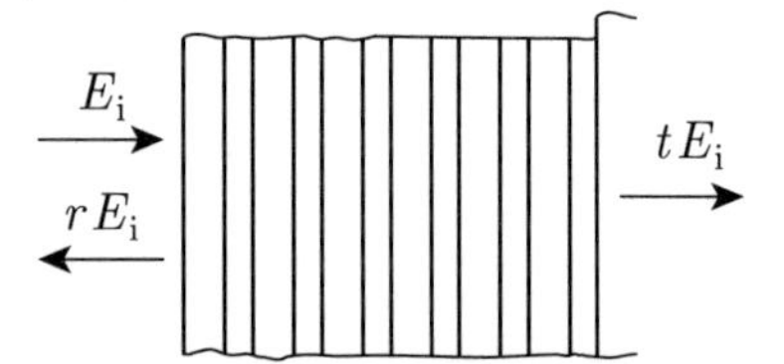

Fig. 6.1 An example of a thin-film structure, such as is made to have high reflectivity. Mirrors with this kind of construction are commonly needed for building lasers.

Thin-film structures provide an important technology for making optical components with a desired surface reflectivity, either highly reflecting for such applications as laser mirrors, or minimally reflecting for the surfaces of a lens. The physics of operation and design is, of course, that of interference. There are also very insightful links with the transmission lines of §1.8.

Consider Fig. 6.1, which is a miniature version of Fig. 6.4 on p. 137. A stack consisting of several thin layers of transparent dielectric has been deposited on a glass substrate. Light is incident from the left; some is reflected, some is transmitted. Within each of the thin layers there will be two waves, one travelling to the right, one travelling to the left. The amplitudes of all these waves are determined physically by Maxwell's electromagnetic equations, and in particular by the boundary conditions: $\boldsymbol{E}_{\text{tangential}}$ and $\boldsymbol{H}_{\text{tangential}}$ are continuous.

In Fig. 6.1 there are 12 thin layers with 13 boundaries, so the whole system has 27 electric-field amplitudes and 27 magnetic-field amplitudes, linked by 27 electromagnetic relations (1.14) and 26 boundary conditions. Yet, among these amplitudes there are only two that we wish to calculate: the reflected amplitude $E_r = rE_i$ and the transmitted amplitude $E_t = tE_i$. All the amplitudes inside the thin layers are of no interest and are to be eliminated; we need a systematic way of performing that elimination. Two different calculation methods are explored in §§6.2–6.5.

6.2 Basic calculation for one layer

Figure 6.2 shows a representative layer situated somewhere inside a stack such as that of Fig. 6.1. As forecast, there is one electromagnetic wave travelling to the right, and one to the left.[1] The physical reason why there is a wave travelling to the left is, of course, that there are reflecting boundaries at $z = 0$ and further to the right. The treatment given here will be restricted[2] to the case where all waves travel along the z-axis, positive or negative, and the wave amplitudes are independent of x and y.

Fields $E_x(-l)$, $H_y(-l)$ are present on the plane $z = -l$. Similarly, there are fields $E_x(0)$, $H_y(0)$ present on the plane $z = 0$. These fields are continuous at the boundaries, so we shall, in due course, link them to fields in the neighbouring layers, but that is for later; at the moment we

[1] *Comment*: It's important that the two waves drawn on Fig. 6.2 have 'exactly' the amplitudes written, because their coefficients will be worked out exactly from the electromagnetic boundary conditions. The picture we are using is very different from that of Fig. 5.1, where an infinite set of reflected waves was added up to give a physically correct total. In those terms, each wave considered here is that total, not a mere one of the contributing waves. Look back to problem 5.3; the waves we deal with in the present section are like those appearing in eqn (5.17).

[2] Oblique incidence is treated in problem 6.7.

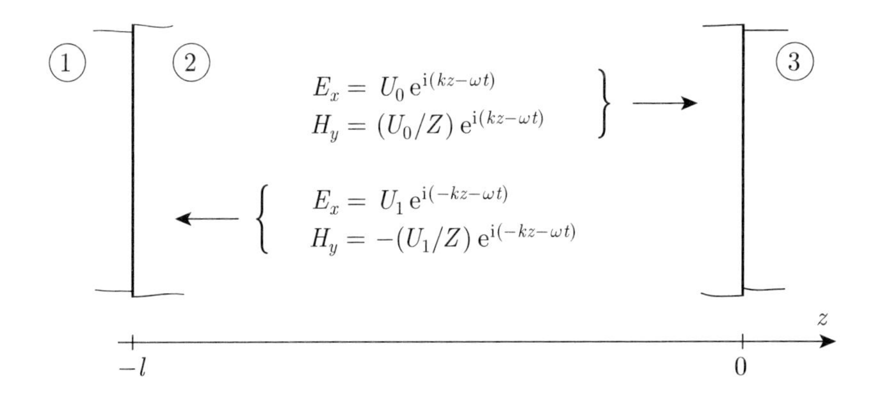

Fig. 6.2 A single layer somewhere in the middle of a stack of layers. The field amplitudes indicated constitute two travelling waves, travelling towards $\pm z$.

consider the physics inside the given layer only. The notation indicates that we are making strict use of a Cartesian coordinate system, and that the electromagnetic field is linearly polarized with all $\boldsymbol{E}$-fields in the x-direction (give or take a sign). The $\boldsymbol{E}$- and $\boldsymbol{H}$-field components have already been related by making use of eqn (1.14). The origin of z has been placed at the right-hand side of the layer because this choice yields slightly less cluttered algebra.

The fields on the bounding surfaces of the layer can be written in terms of the electric-field amplitudes U_0 and U_1 of the right-going and left-going waves:

$$\left.\begin{aligned} E_x(-l) &= U_0\,e^{-ikl} + U_1\,e^{+ikl}, & E_x(0) &= U_0 + U_1, \\ H_y(-l) &= \frac{U_0}{Z}\,e^{-ikl} - \frac{U_1}{Z}\,e^{+ikl}, & H_y(0) &= \frac{U_0}{Z} - \frac{U_1}{Z}. \end{aligned}\right\} \tag{6.1}$$

Here factors $e^{-i\omega t}$ have been dropped throughout.[3] Also dropped for now are subscripts attached to k and Z that identify which of several layers these quantities refer to. Eliminating U_0 and U_1 yields

$$\left.\begin{aligned} E_x(-l) &= \cos(kl)\,E_x(0) - i\frac{Z}{Z_0}\sin(kl)\,Z_0H_y(0), \\ Z_0H_y(-l) &= -i\frac{Z_0}{Z}\sin(kl)\,E_x(0) + \cos(kl)\,Z_0H_y(0). \end{aligned}\right\} \tag{6.2}$$

In these equations[4] no assumptions have been made yet about the physical properties of the layer under consideration,[5] even whether it absorbs. Therefore, k and Z might be complex, and the sine and cosine should be understood as functions of a possibly-complex argument.

At optical frequencies $\mu_r = 1$ and by eqn (1.16) $Z_0/Z = n$ so that

$$\left.\begin{aligned} E_x(-l) &= \cos(kl)\,E_x(0) - (i/n_{\text{layer}})\sin(kl)\,Z_0H_y(0), \\ Z_0H_y(-l) &= -in_{\text{layer}}\sin(kl)\,E_x(0) + \cos(kl)\,Z_0H_y(0). \end{aligned}\right\} \tag{6.3}$$

Equations (6.3) are our basic statements that tell us how the physics of the layer relates the fields on one of its bounding surfaces to the fields on the other. We shall almost always think of eqns (6.3) in the context of a non-absorbing layer, but the equations are formally correct for an absorber if we let k and n be complex.

[3] This omission is for eqns (6.1) only.

[4] *Comment*: There is a choice as to whether the equations are set up to contain H_y or Z_0H_y. The present author prefers corresponding terms in the equations to have the same dimensions, and has placed factors Z_0 to make that so. Impedance Z (that of the layer) cannot be used instead of Z_0 as it would make impossible the eliminations of §6.3.

[5] Well, it is assumed to be LIH.

6.3 Matrix elimination of 'middle' amplitudes

Equations (6.3) are asking to be put into a matrix form, so we write

$$\begin{pmatrix} E_x(-l) \\ Z_0 H_y(-l) \end{pmatrix} = \begin{pmatrix} \cos(kl) & -(\mathrm{i}/n)\sin(kl) \\ -(\mathrm{i}n)\sin(kl) & \cos(kl) \end{pmatrix} \begin{pmatrix} E_x(0) \\ Z_0 H_y(0) \end{pmatrix}. \qquad (6.4)$$

To see the usefulness of the matrix presentation, consider the two adjacent layers in Fig. 6.3(a). We have statements of the form

$$\begin{pmatrix} E(-l_2) \\ Z_0 H(-l_2) \end{pmatrix} = \begin{pmatrix} \alpha & \beta \\ \gamma & \delta \end{pmatrix} \begin{pmatrix} E(-l_1) \\ Z_0 H(-l_1) \end{pmatrix},$$

$$\begin{pmatrix} E(-l_1) \\ Z_0 H(-l_1) \end{pmatrix} = \begin{pmatrix} a & b \\ c & d \end{pmatrix} \begin{pmatrix} E(0) \\ Z_0 H(0) \end{pmatrix}.$$

Here each square matrix has the form of that in eqn (6.4), with insertion of the appropriate k, refractive index and layer thickness. The 'middle' field amplitudes can now be eliminated (because $\boldsymbol{E}_{\text{tangential}}$ and $\boldsymbol{H}_{\text{tangential}}$ are continuous):

$$\begin{pmatrix} E(-l_2) \\ Z_0 H(-l_2) \end{pmatrix} = \begin{pmatrix} \alpha & \beta \\ \gamma & \delta \end{pmatrix} \begin{pmatrix} a & b \\ c & d \end{pmatrix} \begin{pmatrix} E(0) \\ Z_0 H(0) \end{pmatrix}; \qquad (6.5)$$

two layers in succession are handled by multiplying their matrices. By extension, the square matrix for any number of layers can be found by multiplying the matrices for the individual layers. Thus, for a stack of thickness L consisting of any number of intermediate layers, we have[6]

$$\begin{pmatrix} E(-L) \\ Z_0 H(-L) \end{pmatrix} = \begin{pmatrix} A & B \\ C & D \end{pmatrix} \begin{pmatrix} E(0) \\ Z_0 H(0) \end{pmatrix}, \qquad (6.6)$$

[6] Several general properties of the $ABCD$ matrices are obtained in problem 6.6. The case of a strongly absorbing layer, such as a Polaroid for the polarization it blocks, is investigated in problem 6.8.

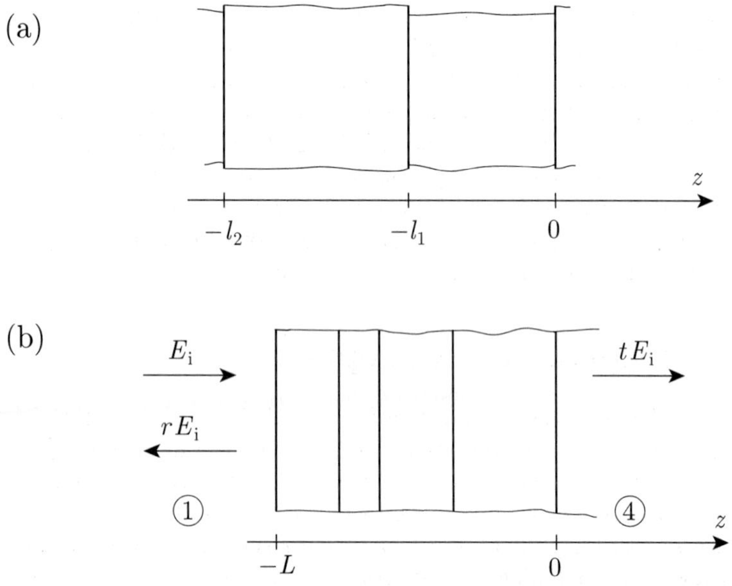

Fig. 6.3 (a) Two adjacent layers are each described by a square matrix of the form given in eqn (6.4). The behaviour of the pair of layers is obtained by multiplying their individual matrices. (b) Knowledge of the matrix representing a composite structure enables us to work out the amplitudes rE_{i} (reflected) and tE_{i} (transmitted).

where *the square matrix is the product of the matrices for the individual layers*—taken in the correct order of course.[7,8]

6.4 Reflected and transmitted waves

The matrix elimination of §6.3 is useful because it enables us to obtain the reflected and transmitted amplitudes. We spell out now how this may be done.

Refer to Fig. 6.3(b), which shows the reflected and transmitted waves arising when a wave is incident from the left (medium 1) onto a stack on whose far side is medium 4. The stack is described by a matrix such as that in eqn (6.6), and we need not enquire what is its internal structure. In detail the amplitudes of the three waves[9] are:

$$\begin{aligned} E_{x\,\text{incident}} &= E_{\text{i}} \exp\{\text{i}k_1(z+L) - \text{i}\omega t\}, \\ H_{y\,\text{incident}} &= (E_{\text{i}}/Z_1) \exp\{\text{i}k_1(z+L) - \text{i}\omega t\}, \\ E_{x\,\text{reflected}} &= rE_{\text{i}} \exp\{-\text{i}k_1(z+L) - \text{i}\omega t\}, \\ H_{y\,\text{reflected}} &= -r(E_{\text{i}}/Z_1) \exp\{-\text{i}k_1(z+L) - \text{i}\omega t\}, \\ E_{x\,\text{transmitted}} &= tE_{\text{i}} \exp\{\text{i}k_4 z - \text{i}\omega t\}, \\ H_{y\,\text{transmitted}} &= t(E_{\text{i}}/Z_4) \exp\{\text{i}k_4 z - \text{i}\omega t\}. \end{aligned}$$

We may substitute the listed fields into eqn (6.6) and cancel through $E_{\text{i}}\,\text{e}^{-\text{i}\omega t}$ to obtain[10]

$$\begin{pmatrix} 1+r \\ (Z_0/Z_1)(1-r) \end{pmatrix} = \begin{pmatrix} A & B \\ C & D \end{pmatrix} \begin{pmatrix} t \\ (Z_0/Z_4)t \end{pmatrix}. \tag{6.7}$$

(The remaining exponentials have been set up so that they are all 1 on the boundaries.) The quantities (Z_0/Z_1) and (Z_0/Z_4) can be tidied into refractive indices n_1 and n_4, but the important feature of eqn (6.7) is already on display: the matrix equation can be decomposed into two simultaneous equations for the two unknowns r and t. The reflection and transmission coefficients (for amplitude) are, therefore, available to us, provided only that we know the $ABCD$ matrix describing the stack.[11]

The fact that the entire problem is solved (in principle) once the $ABCD$ matrix is evaluated confirms the promise implicit in §§6.2 and 6.3: the matrices give a reasonably painless route for eliminating all the 'middle' amplitudes within a stack of layers.

6.5 Impedance concepts

The calculation methods developed in §§6.2–6.4 are very powerful and elegant, but they need to be supplemented by the physical insights that come from thinking about impedances. The concepts of input impedance, impedance matching and impedance mismatching originate from electrical transmission lines, and we shall freely apply to thin films the transmission-line analogy set up in §1.8 and problem 1.10.

[7] *Comment*: It is best to remember the way that eqn (6.6) was obtained, rather than to remember a mnemonic for the matrix order. It is tempting to notice that the matrices being multiplied in eqn (6.5) are in the same order on the paper as the optical layers. However, there may be occasions when we wish to find $E(0)$ in terms of $E(-L)$ using the inverse matrix, and then the order would be reversed.

[8] *Comment*: This matrix formulation gives a structured way of handling any 'layer' problem, however complicated the layers. It is equally applicable to (for example) waves in a sequence of joined transmission lines, or a beam of quantum particles encountering a sequence of potential steps.

[9] It is assumed that medium 4 contains a wave travelling towards $+z$ only. This makes the fields particularly simple on the plane $z = 0$. The same simplifying assumption is made in §6.5, where it is used to obtain the input impedance of the substrate.

[10] You are asked to check the algebra here, and in the earlier parts of this chapter, in problem 6.1.

[11] *Comment*: The development given in §§6.2–6.4 has been restricted to the case of light at normal incidence. This was for simplicity only, to display the principles in the most uncluttered way. It is not hard to apply the same ideas to non-normal incidence, but the physics then divides into two cases, according to the polarization of the incident light; see problem 6.7.

An important property can be deduced quickly by reworking the final step of §6.4 in a slightly different way. At $z = -L$, we have

$$E_x(-L) = (1 + r)E_{\mathrm{i}},$$
$$H_y(-L) = (1 - r)E_{\mathrm{i}}/Z_1,$$

and so

$$Z_1(1 + r)/(1 - r) = E_x(-L)/H_y(-L). \tag{6.8}$$

Now $E_x(-L)/H_y(-L)$ is the input impedance Z_{in} which the stack presents to a wave arriving from the left, and so we have

$$(1 + r)/(1 - r) = Z_{\mathrm{in}}/Z_1,$$

whence

$$r = \frac{Z_{\mathrm{in}} - Z_1}{Z_{\mathrm{in}} + Z_1}, \tag{6.9}$$

which exactly resembles eqn (1.30). The reflection coefficient r is determined by the value of $Z_{\mathrm{in}} = E_x(-L)/H_y(-L)$ on the *surface* at $z = -L$. Whatever[12] may be happening to the right of $z = -L$ is important only in so far as it controls the value of Z_{in}.

[12]Compare this statement with the way that the reasoning was set up in §6.2. There we studiously avoided saying anything about the physics taking place in the region to the right of $z = 0$, being content that fields $E_x(z = 0)$ and $H_y(z = 0)$ existed on that boundary.

Having gained this insight, we return to eqns (6.2) and divide them to obtain the input impedance of a representative layer:

$$Z_{\mathrm{in}} = \frac{E_x(-l)}{H_y(-l)} = \frac{\cos(kl)\,E_x(0) - \mathrm{i}Z_{\mathrm{layer}}\sin(kl)H_y(0)}{-\mathrm{i}\sin(kl)E_x(0)/Z_{\mathrm{layer}} + \cos(kl)\,H_y(0)}$$
$$= \frac{Z_{\mathrm{load}} - \mathrm{i}\,Z_{\mathrm{layer}}\tan(kl)}{1 - \mathrm{i}(Z_{\mathrm{load}}/Z_{\mathrm{layer}})\tan(kl)}. \tag{6.10}$$

In eqn (6.10), Z_{load} is the value of the ratio $E_x(0)/H_y(0)$ existing on the plane $z = 0$. It is the input impedance of whatever lies to the right of $z = 0$. If $Z_{\mathrm{load}} = E_x(0)/H_y(0)$ at $z = 0$ is known, then the input impedance Z_{in} at $z = -l$ can be calculated using eqn (6.10).[13]

[13]Note that $k = n_{\mathrm{layer}}(\omega/c)$, so its value depends upon which layer we are considering.

At the right-hand end of a thin-film structure will be a medium, such as the glass substrate shown in Fig. 6.4, in which an electromagnetic wave travels away to the right only. For that wave, E_x/H_y is the characteristic impedance of the substrate, so it is known. This provides us with a known load impedance for the final thin-film layer, so eqn (6.10) can be used to calculate the input impedance of that layer. The load impedance presented to the *next* layer is now known, and the process may be continued 'leftwards' through the entire stack.[14] In this way, repeated application of eqn (6.10) can achieve the same results as the repeated matrix multiplications of eqn (6.5). However, it will be obvious that the algebra is less well structured and can get very untidy, except for specially simple cases. Fortunately, almost all cases of interest *are* specially simple.

[14]The fact that we work from right to left when thinking about impedances is the reason why the author set up the matrix of eqn (6.4) so that it too works from right to left.

Two special-case properties can be derived from eqn (6.10), or by going back to eqns (6.2), or by processing eqns (6.1) directly:

$$\text{quarter-wave}^{15}\text{ layer:} \quad \tan(kl) = \infty, \quad Z_{\mathrm{in}} = Z_{\mathrm{layer}}^2/Z_{\mathrm{load}}, \tag{6.11}$$
$$\text{half-wave layer:} \quad \tan(kl) = 0, \quad Z_{\mathrm{in}} = Z_{\mathrm{load}}. \tag{6.12}$$

[15]Note the terminological difference between a quarter-wave *layer* and a quarter-wave *plate*. A quarter-wave plate is a device used for introducing a relative phase shift between two polarization states of light.

These two statements are so simple that they permit us to analyse many thin-film[16] structures (consisting only of quarter-wave or half-wave layers) without undue mathematical complication. More importantly, they can give a direction to the designing of a structure to achieve a required behaviour. This is illustrated in the next section.

[16]The methods of this chapter are general, valid for layers of any thickness. So matrices and impedances are equally applicable to a Fabry–Perot, for example. Indeed, impedance ideas are helpful in problem 5.4 for resolving an apparent paradox involving energy conservation. However, for reasons explored in problem 5.3, the 'traditional' approach to obtaining the Fabry–Perot equations (5.2)–(5.5) is more insightful.

6.6 High-reflectivity mirrors

We wish to make a structure that reflects nearly 100% of radiation incident normally from air. Equation (6.9) implies that

$$R \equiv \frac{\text{intensity reflected}}{\text{intensity incident}} = \left| \frac{Z_{\text{in}} - Z_{\text{air}}}{Z_{\text{in}} + Z_{\text{air}}} \right|^2 . \tag{6.13}$$

The way to achieve high reflectivity is now clear: we need an 'extreme' impedance mismatch, that is, we need $Z_{\text{in}}/Z_{\text{air}}$ to be far from 1, either very large or very small. We shall see that repeated quarter-wave layers, having alternately high and low refractive indices, achieve just such an 'extreme' input impedance.

A suitable multilayer structure is shown in Fig. 6.4. The thin-film layers are deposited on a glass substrate; twelve are shown but the diagram is intended to be representative for any number. Each layer is given a thickness equal to a quarter of the optical wavelength in that layer. This means that the thicknesses in metres are unequal, a fact emphasized on the diagram.

To analyse the behaviour of the structure of Fig. 6.4, we work from right to left.

(1) Medium 4 carries a wave to the right only, so at any plane in that medium, $E_x/H_y = Z_4$. In particular, this ratio holds at plane A, so Z_4 is the 'load impedance' presented to layer BA.
(2) Quarter-wave layer BA has characteristic impedance Z_3 and is loaded by Z_4. Then, from eqn (6.11),

$$(E_x/H_y)_{\text{B}} = \text{input impedance of layer BA} = Z_3^2/Z_4.$$

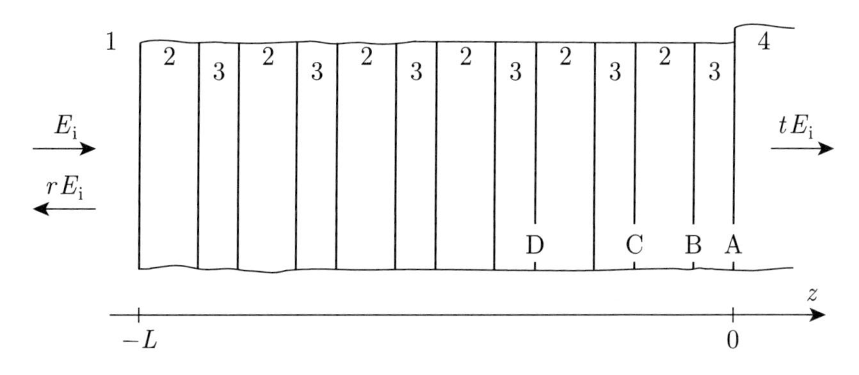

Fig. 6.4 A possible arrangement of thin-film layers, deposited on a glass substrate to form a high-reflectance mirror. Such a structure may be used to give the high-reflectance coating for a Fabry–Perot or for one of the mirrors defining a laser cavity.

Each layer has thickness equal to a quarter of the optical wavelength in the medium, so the actual thickness is $\lambda/4n_2$ or $\lambda/4n_3$. To emphasize this, the layers are drawn with unequal thicknesses.

The structure shown here is unrealistic in that the number of thin-film layers in practice is usually odd.

(3) Layer CB is a quarter-wave layer loaded by impedance Z_3^2/Z_4, so

$$\left(\frac{E_x}{H_y}\right)_{\text{plane C}} = \frac{Z_2^2}{Z_3^2/Z_4} = \left(\frac{Z_2}{Z_3}\right)^2 Z_4. \tag{6.14}$$

(4) The two layers from C to A can now be looked at in a different way. They can be thought of as a composite 'transformer': they produce an input impedance $(Z_2/Z_3)^2 Z_4$ —and this is true whatever is Z_4 because we have given Z_4 no specific value. The two layers from D to C are identical, so they do the same thing:

$$(E_x/H_y)_{\text{D}} = (Z_2/Z_3)^2 \times (Z_2/Z_3)^2 Z_4 = (Z_2/Z_3)^4 Z_4.$$

(5) The pattern is now clear:

$$2p \text{ layers give} \qquad Z_{\text{in}} = (Z_2/Z_3)^{2p} Z_4. \tag{6.15}$$

The reflectivity R for a stack of $2p$ layers is now obtained by substituting this expression into eqn (6.13). Since the thin-film layers have been assumed not to absorb, what is not reflected is transmitted.

(6) Should we wish to use an odd number of layers (and an odd number is usual for reasons explored in problem 6.2), we can use eqn (6.15) on all layers except the last (or the first). Then we can find the overall input impedance by applying eqn (6.11) to deal with the final layer.

(7) The equations now obtained can be used to find the number of layers needed to achieve a specified reflectivity, or can be used some other way round.

Equation (6.15) shows that, to get a high reflectivity without using too many layers, the layers should have Z_2 and Z_3 as different as possible, so we require two substances with widely different refractive indices. The layers will have to be deposited by evaporation in vacuum, and they must be reasonably hard when deposited so that the coatings are not too fragile. These considerations often lead to the use of:[17]

low index	magnesium fluoride MgF_2	$n = 1.38$
	or cryolite[18] Na_3AlF_6	$n = 1.35$
	or silicon dioxide SiO_2	$n = 1.46$

combined with

high index	titanium dioxide TiO_2	$n \approx 2.34$
	or zinc sulphide ZnS	$n = 2.35$.

The required reflectivity is (usually) achieved with the least number of deposited layers if the stack starts and ends with a high-refractive-index layer. Using a crown-glass substrate ($n = 1.52$) coated with TiO_2 and SiO_2, we can get a (power) reflectivity of 0.999 by using a total of 17 layers. An indication of how good this is can be obtained by comparing it with evaporated aluminium whose reflectivity is only about 91%.[19] (Of course, the very high reflectivity of the stack is achieved over a limited range of wavelengths only, but this limitation is not usually a difficulty, given the likely applications.)

[17] Quoted values for the refractive indices of coatings vary a little, quite apart from their dependence upon frequency, because the index can in some cases be adjusted, in its second decimal place, by controlling the rate of deposition.

[18] Cryolite and zinc sulphide are 'traditional' materials because they are quite easy to evaporate. However, they are not very hard and are slightly hygroscopic so they tend to deteriorate. Current practice is to prefer silicon dioxide with titanium dioxide (for most work in the visible frequency range), the evaporation technology being adapted, as described below, to cope with the more challenging substances.

[19] When fresh; falling to about 88% after a few days.

Example 6.1 Design of a mirror for a laser

We wish to make a mirror with reflectivity at least 0.995 by depositing quarter-wave layers of TiO_2 and SiO_2 onto glass (refractive index 1.52), starting and finishing with TiO_2.

Solution: Given that all impedances are real, eqn (6.13) can be rearranged to read[20]

$$0.005 = 1 - R = \frac{4x}{(1+x)^2}, \tag{6.16}$$

where $x = Z_{\text{in}}/Z_{\text{air}}$. If there are $(2p+1)$ layers with n_3 first and last,

$$x = \frac{Z_{\text{in}}}{Z_{\text{air}}} = \frac{Z_3^2}{Z_0 Z_4}\left(\frac{Z_3}{Z_2}\right)^{2p} = \frac{n_4}{n_3^2}\left(\frac{n_2}{n_3}\right)^{2p}. \tag{6.17}$$

We are to solve eqn (6.16) for x, then use eqn (6.17) to find p.

With high-index layers deposited first and last, the input impedance of the stack is very much less than Z_0, so $x \ll 1$. We do not need to solve a quadratic to find x from eqn (6.16) because

$$4x = 0.005(1+x)^2 \approx 0.005, \qquad \text{giving} \qquad x = 1.25 \times 10^{-3}$$

to sufficient accuracy. Equation (6.17) now gives

$$\frac{1.52}{2.34^2}\left(\frac{1.46}{2.34}\right)^{2p} = 1.25 \times 10^{-3}.$$

Taking logarithms of both sides now yields $p = 5.73$, which must be rounded up to $p = 6$, and the total number of layers[21] to be used is $(2p+1) = 13$.

The reasons why it was best to choose an odd number of layers, and to start and finish with TiO_2, are explored in problem 6.2.

[20] The number of layers required is determined physically by how far R departs from 1, so it makes good sense to obtain x from an expression for $(1-R)$, rather than R.

Equation (6.16) is correct with x set equal to either $Z_{\text{in}}/Z_{\text{air}}$ or $Z_{\text{air}}/Z_{\text{in}}$. We have made the choice, looking ahead, that makes x small for the problem in hand.

[21] Using $(2p+1) = 13$ we find that the reflectivity is 99.6%, a little higher than intended. If it is necessary to achieve precisely a required reflectivity then adjustments can be made. As an example, we might consider whether an even number of layers would just happen to give the value required, even though it makes less efficient use of layers. This recourse does not help here, as the reader may check.

6.7 Anti-reflection coatings

To make an anti-reflection coating, we must do the exact opposite of what we did in §6.6: we must prepare a coating which, when deposited on glass, gives an input impedance Z_{in} as close as possible to that of air.

Equation (6.11) shows that a quarter-wave layer of characteristic impedance Z_2 deposited on glass of impedance Z_{glass} gives an input impedance of $Z_{\text{in}} = Z_2^2/Z_{\text{glass}}$. If the input impedance is to be the same as the impedance Z_0 of air, then we need $Z_2^2/Z_{\text{glass}} = Z_{\text{air}}$. Equivalently, since the materials considered are non-magnetic at optical frequencies, $n_2 = \sqrt{n_{\text{air}}\, n_{\text{glass}}}$.

Problem 6.3 explores the practicalities of using a single quarter-wave layer to achieve near-zero reflectivity. Suitable materials with sufficiently low refractive index are not available; and if we use what is available the reflectivity remains above 1%. For critical purposes (a zoom lens may have up to 20 pieces of glass with many surfaces), we need to do better. For such applications, we need to use multiple layers: often six or more.

A way forward (possible, but there are better) can be seen by referring back to eqn (6.14). With two quarter-wave layers, whose characteristic impedances are Z_2 and Z_3, the input impedance is $Z_{\text{in}} = (Z_2/Z_3)^2 Z_{\text{glass}}$. There is plenty of choice now for the values of Z_2 and Z_3 that can be used in combination to achieve $Z_{\text{in}} = Z_0$. As example: suppose we deposit onto crown glass a quarter-wave layer of Al_2O_3 followed by a quarter-wave layer of MgF_2 (indices 1.67 and 1.38). At the design frequency, the reflectivity is 0.03%.

A more difficult problem is the variation of performance with frequency. A layer's thickness can be equal to a quarter wavelength at only one frequency, so calculations based on eqn (6.11) apply to the 'design wavelength' only.[22] To see how bad things get away from the design wavelength, and how they might be improved, requires detailed calculation; a computer is helpful. It is here that the matrix methods of §§6.3 and 6.4 come into their own. The details of design and optimization are, however, beyond the scope of this book.

[22]The two-layer coating described in the last paragraph is called in the trade a V-coating, because a graph of reflectivity versus wavelength has a rather sharp minimum. If a half-wave layer of refractive index 2.15 is interposed between the Al_2O_3 and the MgF_2, the coating (Stetter *et al.* 1976) is made broad-band, with a low reflectivity extending over most of the visible region; the reflectivity has a W-shaped graph with two minima and the structure is known as a W-coating.

With all but the most elaborate anti-reflection coatings, there is a noticeable residual reflection that varies with the frequency of the light. It is a matter of common experience that camera lenses, or spectacles, look coloured in reflected light. In the case of spectacles, the residual colour is usually chosen according to the appearance that the wearer prefers. In other cases, the colour may be purple (the 'traditional' colouring where a single quarter-wave layer gives near-zero reflection in the middle of the visible spectrum but reflects more at red and blue) or yellowish (three or more layers giving near-zero reflection at both ends of the spectrum but more in between—though mostly still less than for the single layer).

6.8 Interference filters

The principle of an interference filter is shown in Fig. 6.5. The central layer has thickness equal to a half wavelength, or a small integer mul-

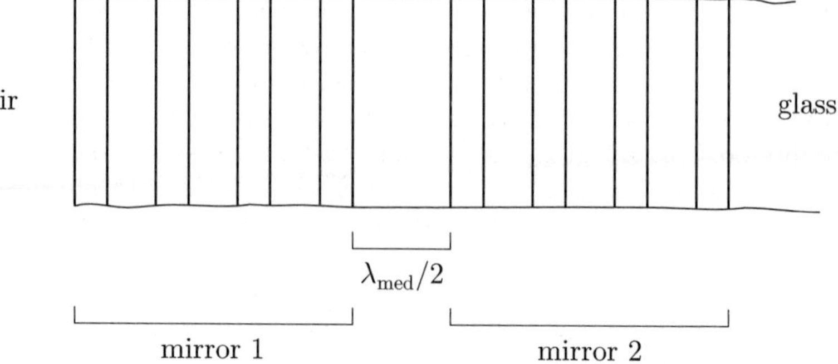

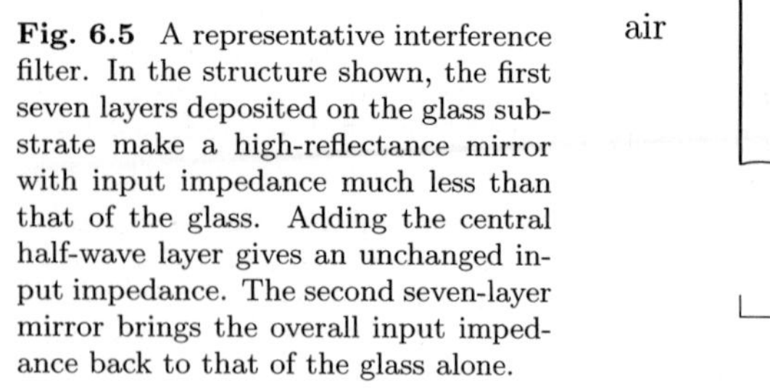

Fig. 6.5 A representative interference filter. In the structure shown, the first seven layers deposited on the glass substrate make a high-reflectance mirror with input impedance much less than that of the glass. Adding the central half-wave layer gives an unchanged input impedance. The second seven-layer mirror brings the overall input impedance back to that of the glass alone.

tiple of a half wavelength; it acts rather like a very thin Fabry–Perot. On either side of this is a stack of quarter-wave layers forming a high-reflectance mirror. The whole structure is made by depositing it on a glass substrate. If all the layers are ideally made for frequency ν_0, the structure transmits at ν_0 like uncoated glass; see problem 6.4(1). The multilayer mirrors give the thin Fabry–Perot a high finesse, so it transmits over a narrow range of frequencies[23] around ν_0.

If there were no dispersion in the dielectrics, doubling the frequency would make all layers into half-wave or one-wave, and the 'filter' would transmit again. Dispersion (meaning the frequency dependence of refractive index) prevents this from being accurately the case, but significant transmission does occur. The cure is to deposit the layers on coloured glass, whose pigment absorbs the unwanted frequencies.

6.9 Practicalities of thin-film deposition

Thin films are made by evaporation of the required materials, onto a substrate (in laboratories usually glass or fused quartz, but also, in commercial practice, plastics) in vacuum. A rather old-fashioned apparatus (described first for simplicity) is sketched in Fig. 6.6(a). Quantities of zinc sulphide and cryolite are held in small containers that are known, from their shape, as boats. The boats are heated alternately to deposit the desired alternating layers of the two substances.[24] The glass must be coated from below, otherwise the stuff falls out of the boat!

How do we know when the correct thickness has been deposited? Figure 6.6(b) indicates how the thickness may be monitored. A light beam can be transmitted through the structure under preparation, but it is usually more convenient to pass it through an extra piece of glass placed nearby. Each deposition is stopped when the transmission through the monitor reaches a minimum or maximum.[25]

It may well happen that the distance r_2 from boat to monitor is not the same as the distance r_1 from boat to sample. In such a case the thicknesses deposited are determined by an inverse square law.[26] Using a monitoring wavelength that is scaled from the design wavelength by a factor $(r_1/r_2)^2$ corrects for this. Alternatively, the distances r_1 and r_2 can be adjusted to suit a chosen monitoring wavelength.

How do we get the whole area of substrate coated to the same thickness? Not all parts of it are at the same distance from the boat, and the inverse square law does harm. The solution is to mount the substrate in a wheel, somewhat as shown in Fig. 6.7, and rotate the wheel during deposition so that any given area is sometimes nearer, sometimes farther from the source. If a uniform coating thickness is really critical, as it is for a Fabry–Perot, the substrate is rotated in a planetary arrangement.[27]

For a laser mirror one is usually happy to deposit a large number of layers, even if that number is 30 or more, without devoting any effort to minimizing the number. By contrast, the flatness requirement with a Fabry–Perot demands that no more layers are used than are really

[23] I think this is probably right. However, you may well share my unease at attributing all the frequency dependence to one layer among so many. Another explanation might lie along the following lines. At the centre frequency, reflector 2 has an input impedance that is arranged to be very far from that of the glass substrate (much less, as the diagram is drawn). The half-wave layer leaves the impedance unchanged. Reflector 1 transforms the impedance back up to that of glass. In effect, the entire structure engineers an elaborate cancellation of large factors. Away from the centre frequency, the layers are not exactly quarter- or half-wavelength, and the cancellation is no longer perfect. It's reasonable to believe that the input impedance is brought well away from that of glass by a smallish detuning.

[24] The thermal capacity of the boats and their contents is such that evaporation does not stop when the heating is stopped. Therefore, the deposition is controlled by opening and closing a shutter between each boat and the substrate.

[25] It is also common to monitor by allowing the vapour to fall onto a crystal of quartz that is the frequency-determining element in an electronic oscillator. The oscillator frequency (typically about 6 MHz) falls as mass is added to the crystal. Previous experience calibrates the relationship between frequency change and film thickness for each material. The quartz crystal is particularly useful for monitoring the rate of deposition, which can be kept within required limits by the operator or by a feedback-control system.

[26] There is also a dependence upon $\cos^2\theta$, where θ is the angle between the evaporation direction and the normals to both substrate and source material.

[27] 'Planetary' means that the glass substrate is rotated rather as the Earth rotates, not only following an orbit round the Sun but also rotating about its own axis.

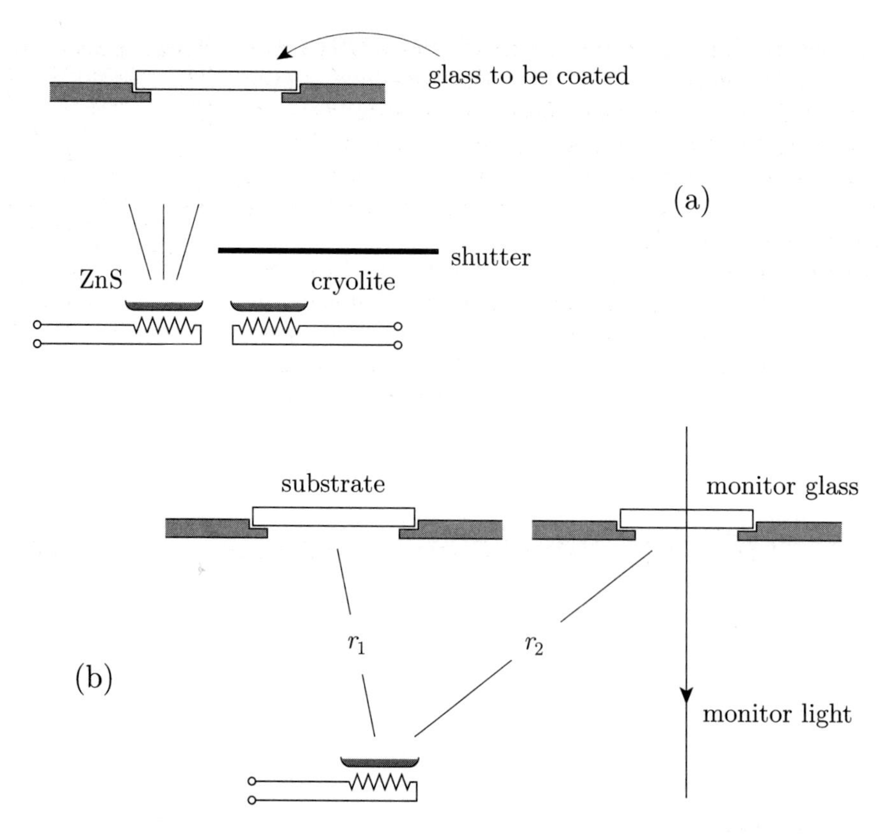

Fig. 6.6 The essentials of a vacuum deposition arrangement. All vacuum equipment is omitted for simplicity. (a) Materials such as zinc sulphide and cryolite are heated so as to evaporate them alternately onto the glass substrate. (b) The inverse square law (combined with cosine-squared factors) is used to work out the film thickness on the substrate from that on the monitor.

necessary (nine is a likely number—see problem 6.2(8)), and the two refractive indices must be chosen accordingly.

What if the final layer is not very hard? An elegant solution is to put on top a half-wave layer (which leaves the reflectivity unaffected) of some convenient hard material such as silica (SiO_2).

Hard materials like TiO_2 or silica evaporate at high temperatures. It's not practical to heat such a material in a boat as described above. The solution is to use an electron beam to heat and evaporate a small portion in the middle of the material to be evaporated, thereby using the rest of the substance as part of its own crucible. A possible arrangement (based on that in the Oxford Physics Department) is shown in Fig. 6.7. The electron gun is located below the crucible, away from contamination, and the electron beam is guided onto the evaporant by a magnetic field.

As a concrete example, we describe the deposition of TiO_2. Bulk TiO_2 (in one of its crystal structures) has a refractive index of 2.70. But if TiO_2 is deposited onto cold glass it forms a spongy structure with a refractive index of only 1.9. Measures must be taken to encourage atoms to migrate over the surface of the substrate and consolidate the layer. Heating the substrate is one way to do this,[28] but a better way is to irradiate the coating during deposition with a beam of oxygen ions. The ions have only a small heating effect (plastic substrates are not damaged), and their function is to knock atoms away from loosely bound positions so that they migrate until they settle into more secure locations. A TiO_2 coating that has been made with ion irradiation has a refractive index in the region of 2.34.

[28]Heating consolidates TiO_2 to a refractive index of 2.2. Interstices still remain in the evaporated film, and when the coating is exposed to air it absorbs water vapour; the refractive index rises to 2.3. All this makes it a complex matter to achieve a layer with a desired optical thickness. Ion-beam irradiation (ion-beam-assisted deposition, IBAD) produces a denser coating whose refractive index does not change significantly when exposed to air.

When TiO_2 is being deposited, oxygen is used for the ion beam because it encourages the TiO_2 to form a coating with the right chemical composition. Beams or argon or nitrogen may be used when depositing other substances (more cooperative materials than TiO_2 still are hardened by ion-beam irradiation). The ion beam has a further use, because

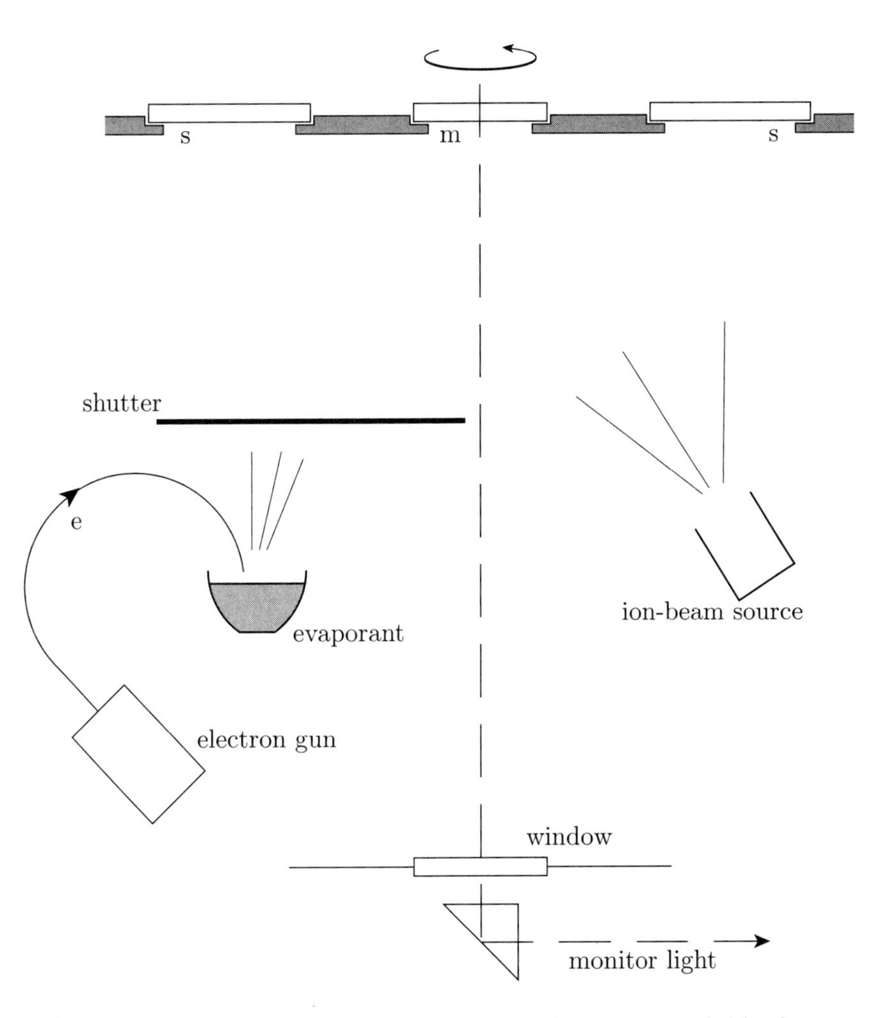

Fig. 6.7 A more detailed representation of a coating apparatus, incorporating features needed for the evaporation of refractory materials such as TiO_2. Section 6.9 explains the reasons for rotating the wheel (shaded) bearing several substrates, for electron-beam heating of the evaporant, and for irradiating the substrates and coatings with an ion beam. Only one crucible is shown, but it is common for there to be several, perhaps four, arranged in a turret so that each in turn can be brought under the electron beam. s: substrates to be coated; m: glass on which the thickness of the evaporated layer is monitored.

before deposition it is applied to bombard and clean the substrate.

Hard coatings are on for good. There is no way to remove them other than abrasive polishing, and it is usually more economical to start again with a fresh substrate. This is generally the case with laser mirrors. For Fabry–Perot plates, where the specially flat substrate is expensive, the best recourse[29] is usually to use 'soft' coatings (meaning, probably, ZnS and cryolite, or, in the near ultraviolet, Sb_2O_3 and cryolite). These coatings are less durable, but once damaged by scratches or moisture they can be easily removed chemically[30] and the substrate can be reused.

The coating materials mentioned so far are those suitable for use at visible frequencies. Each material has a somewhat limited range of frequencies within which it is usefully transparent, and outside that range we must choose something else. For example, hafnium oxide (paired with silica) may be used in place of titanium oxide in the ultraviolet for wavelengths above 240 nm; and antinomy oxide, Sb_2O_3 (paired with cryolite) replaces ZnS for wavelengths above 300 nm. We do not catalogue materials here as the reader seeking such technical information is best served by specialist treatises.[31]

[29] Another, rather surprising, reason for preferring soft coatings is that they subject the substrate to less mechanical stress.

[30] They float off in warm water.

[31] The standard work on thin films is Macleod (2001).

Problems

Problem 6.1 (a) Thin film basics

(1) Work and check all the calculations in §§6.2–6.5.

(2) Perform the manipulation referred to after eqn (6.7): write out two simultaneous equations for r and t, and solve them to find the reflection and transmission coefficients for amplitude. Why is it *not* in general the case that $|t^2| + |r^2| = 1$ even when the layer structure is non-absorbing?

(3) Show that the matrices that describe quarter-wave and half-wave layers are

$$\text{quarter-wave:} \quad \begin{pmatrix} 0 & -\mathrm{i}/n \\ -\mathrm{i}n & 0 \end{pmatrix} \qquad \text{half-wave:} \quad \begin{pmatrix} -1 & 0 \\ 0 & -1 \end{pmatrix}.$$

Are the negative signs in the matrix for a half-wave layer any cause for anxiety?

(4) Suppose we start again and derive eqns (6.3), not for a general layer, but for the final thin layer of Fig. 6.3 or Fig. 6.4, which we'll now call layer 3. On the right-hand surface of that layer the fields $E_x(0)$ and $H_y(0)$ are related via the characteristic impedance $Z_4 = Z_0/n_4$ of the glass substrate. Argue that the first of eqns (6.3) now can be written as

$$E_x(-l) = \cos(k_3 l)\, E_x(0) - (\mathrm{i}/n_3)\sin(k_3 l)\, Z_0 H_y(0)$$

added to any multiple of

$$0 = E_x(0) - (1/n_4) Z_0 H_y(0),$$

with a similar range of choices for the second equation. There is no unique way to assemble a matrix statement like eqn (6.4) when $E_x(0)$ and $H_y(0)$ are coupled.[32]

[32] *Comment*: Yet only the form given in eqn (6.4) is capable of delivering the elimination of eqn (6.5). In §6.2, $E_x(0)$ and $H_y(0)$ had to be treated as—to be—independent quantities. The clearest way of guaranteeing this was to exclude from §6.2 all mention of the physics in 'other' layers and in particular that in layer 4.

Problem 6.2 (a) Design of a high-reflectance mirror

(1) Show that, for light incident from a medium of characteristic impedance Z_1, the reflectivity at a boundary (input impedance Z_{in}) is the same whether $Z_{\text{in}} = XZ_1$ or $Z_{\text{in}} = Z_1/X$. We get high reflectivity by making Z_{in} 'far from' Z_1 by a large numerical factor X, but whether larger or smaller does not matter.

(2) Why, in these impedance terms, is a metal a fairly good reflector at optical frequencies?

(3) Find how many layers of TiO_2 and SiO_2 you need to deposit on glass ($n = 1.52$) to get a (power) reflectivity $R = 0.995$. Try the effect of using
 (a) an even number of layers with SiO_2 deposited first,
 (b) an even number of layers with TiO_2 deposited first,
 (c) an odd number of layers with SiO_2 deposited first,
 (d) an odd number of layers with TiO_2 deposited first.

(4) Part (3) shows that there are differences in the effectiveness of the quarter-wave layers depending on the order of deposition.[33] Show that depositing a single layer of SiO_2 on glass gives $X = Z_1/Z_{\text{in}} = 1.40$, while depositing a single layer of TiO_2 on glass gives $X = Z_1/Z_{\text{in}} = 3.60$. This should make sense: SiO_2 has a refractive index rather close to that of glass, so next to the glass it does little good—in the extreme case, it would be useless to deposit a material with $n = 1.52$ directly onto glass.[34]

[33] The reason for choosing high-index material first and last in example 6.1 (p. 139) will now also be apparent.

[34] *Comment*: We are trying to make a structure whose input impedance is as far from that of air as possible; either larger or smaller will do. The characteristic impedance of glass is less than that of air, so it gives us a head start towards low impedances. The economical thing to do is to go further in the same direction, and that is what a first layer of TiO_2 does. Subsequent pairs of layers take the input impedance to even smaller values. By contrast, a first layer of SiO_2 raises the impedance to value close to that of air, and subsequent pairs of layers raise the impedance further. It is this 'crossed over' behaviour of the impedances that explains the poor results in part (6) when the low-index layer is deposited first.

(5) Invent an argument similar to that in part (4) to show that a final layer of SiO_2 does little good because its refractive index is too close to that of air.

(6) Check these understandings directly by calculating $X =$ (the larger of Z_1/Z_{in}, Z_{in}/Z_1)
 (a) for 12, 13, 14, 15 layers with SiO_2 deposited first,
 (b) for 12, 13, 14, 15 layers with TiO_2 deposited first.
You should end up seeing clearly that it is the high-index material that does most good—just because its refractive index is further from both n_{air} and n_{glass}. The low-index material is (necessary) stuffing between the high-index layers.

For the reasons you have discovered in parts (3)–(6), multilayer mirrors almost always have an odd number of thin-film layers, and with the high-index material deposited first and last.

(7) Consider whether the conclusion of part (6) is altered if the substrate has a higher refractive index than either coating material.

(8) The plates for a Fabry–Perot are to be made from a substrate of fused quartz ($n = 1.46$) coated with ZnS and cryolite. What 'reflection finesse' can be obtained using
 (a) 7 layers,
 (b) 9 layers?

Problem 6.3 (a) Anti-reflection coatings

(1) Show from eqn (6.10) that the input impedance of a thin-film layer (non-absorbing layer on a non-absorbing substrate) is real only when the layer's thickness is a quarter wavelength, or an integer multiple of a quarter wavelength. Argue that if the reflectivity is to be zero when a single layer is coated on glass, that layer must have thickness of a quarter wavelength—or an odd multiple of a quarter wavelength. (Why not a half wavelength?)

(2) A layer intended to behave like a quarter-wave layer may in principle have a thickness that is any odd-integer multiple of a quarter wavelength. Show that for minimum sensitivity to frequency the most suitable choice of integer is 1.

(3) Suppose we wish to make an anti-reflection coating on crown glass ($n = 1.52$). The ideal coating material would have refractive index $\sqrt{1.52} = 1.23$. It is hard to find a material with so low a refractive index, so we wonder if we can get away with using MgF_2 ($n = 1.38$). What reflectivity is achieved by depositing a quarter-wave layer of MgF_2?

Problem 6.4 (a) Interference filter

(1) Prove the property quoted in the first paragraph of §6.8: at the design frequency an interference filter transmits the same amount as does the uncoated substrate.[35]

(2) The final paragraph of §6.8 says that you need a filter to make a filter. Why bother with the thin-film structure at all; why not use coloured glass to do the whole job?

[35] *Hint*: Start at the central half-wave layer. It has no effect on any impedances. Then a replacement structure in which the half-wave layer was omitted would have the same input impedance—at this one frequency. See if this same trick can be used again ....

Problem 6.5 (e) General property of a loss-free reflector

Figure 6.8 shows a reflector that may well be composite but is symmetric under time reversal—which implies in particular that it is loss-free.[36] Shown on the diagram are the electric-field amplitudes of waves incident, transmitted and reflected, given so as to define amplitude reflection and transmission coefficients. (The notation is similar to that in Fig. 5.1.) The incident wave must have its $\boldsymbol{E}$-field either in the plane of the paper, or normal to it; we need not specify which, though different values of r_1, t_1, r_2, t_2 apply to the two cases.

[36] A number of statements in this problem will say 'loss-free' in quotes to mean 'symmetric under time reversal', which is a bit of a mouthful and differs only in rather unusual cases.

In Fig. 6.8(b) a wave is incident from the right at the same angle to the normal, and with the same polarization and amplitude, as the transmitted wave $t_1 E_\mathrm{i}$ of Fig. 6.8(a); it will be a reverse-going replica of that transmitted wave. A reversed version of the reflected wave $r_1 E_\mathrm{i}$ (not shown on the diagram) will likewise be applied and we shall require that the outcome is a reverse-going version of E_i.

(1) Let the incident-wave amplitude be $E_\text{incident} = E_\mathrm{i} \exp(\mathrm{i}\boldsymbol{k}_\mathrm{i} \cdot \boldsymbol{r} - \mathrm{i}\omega t)$. To 'run the video backwards' we start by reversing the sign of t in the exponent.[37] Argue that the same physical wave is described if we additionally complex-conjugate the amplitude. Argue that this conjugation is necessary if the whole process is to yield the expression we require for the reverse-going incident wave, having exponent $-\mathrm{i}\boldsymbol{k}_\mathrm{i} \cdot \boldsymbol{r} - \mathrm{i}\omega t$.

[37] Reversing the sign of t in Maxwell's equations has additional consequences, as the magnetic fields in the waves reverse direction. This complication does not disrupt the argument being developed.

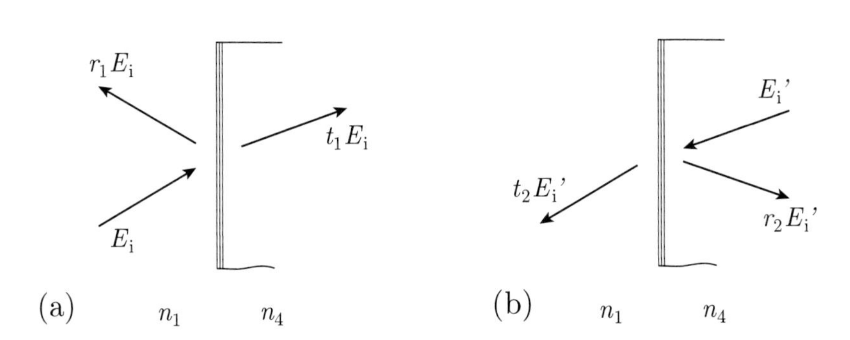

Fig. 6.8 (a) A wave with electric-field amplitude E_i is incident at a possibly-oblique angle onto a 'loss-free' reflector which may be composite. Some of the wave is reflected, some transmitted. (b) The situation of (a) is time-reversed. A wave of amplitude E_i' is the time reversal of that labelled $t_1 E_\mathrm{i}$. Another wave, not shown, is the time reversal of $r_1 E_\mathrm{i}$. Given that the reflector is symmetric under time reversal, the two time-reversed waves must yield the time reversal of E_i, and no wave in the direction of $r_2 E_\mathrm{i}'$.

(2) Build on the ideas of part (1) to show that

$$r_1^* t_1 + r_2 t_1^* = 0, \tag{6.18}$$
$$r_1^* r_1 + t_1^* t_2 = 1. \tag{6.19}$$

Because they have been built very explicitly on the fact that the reflector is symmetrical under time reversal[38] (implying loss-free), eqns (6.18) and (6.19) must be restricted to reflectors with that property. Equations (6.18) and (6.19) are often known as Stokes relations.

(3) Show that both eqn (6.18) and eqn (6.19) can be used to derive the property

$$|r_1|^2 = |r_2|^2, \tag{6.20}$$

with the obvious interpretation: a 'loss-free' reflector has the same reflectivity 'from both sides'.

(4) Use eqn (6.19) to show that the 'prefactor' $|t_1 t_2|/(1-R)$ of eqn (5.2) is 1 if the Fabry–Perot coatings are 'loss-free'. Notice that this holds for all angles of incidence, so it is independent of the angle θ of §5.2.

(5) Use eqn (1.23) to show that for a 'loss-free' reflector at normal incidence

$$n_1(1 - r_1^* r_1) = n_4\, t_1^* t_1$$

and that this taken with eqn (6.19) leads to[39]

$$n_1\, t_2 = n_4\, t_1. \tag{6.21}$$

[38] The reflector is not symmetrical under time reversal if it is subject to a $\boldsymbol{B}$-field—other than that in the light itself—as magnetic fields are reversed in sign by a reversal of t. Specifically, a Faraday effect (or other magneto-optical effect) in the material of the reflector would necessitate re-examination of the argument being set up in this problem. We do not consider such complications here.

[39] Problem 6.6(8) shows that at normal incidence eqn (6.21) remains true even if the reflector absorbs. A generalization to non-normal incidence is explored in problem 6.7(3)

Problem 6.6 (c) Properties of thin-film matrices

The matrix equation (6.4) for a single thin layer can be written in the form

$$\begin{pmatrix} E_x(-l) \\ Z_0 H_y(-l) \end{pmatrix} = \begin{pmatrix} a & b \\ c & d \end{pmatrix} \begin{pmatrix} E_x(0) \\ Z_0 H_y(0) \end{pmatrix};$$

the matrix elements a, b, c, d can be read off from eqn (6.4).

(1) Show that the *abcd* matrix is unitary. Make sure you do not sneak in any assumption that k or Z is real. The unitary property applies to absorbing layers just as much as to non-absorbing.[40]

[40] A unitary property usually means that the size of something is staying the same. The author therefore first guessed that unitarity might be something to do with energy conservation, but this isn't the case. It seems that thin-film matrices simply *are* unitary. The unitary property (with 4×4 matrices now) even applies to a material that rotates the plane of polarized light by optical activity or the Faraday effect.

(2) Show that a non-absorbing layer has a, ib, ic and d all real.

(3) Let

$$\mathsf{M} = \begin{pmatrix} A & B \\ C & D \end{pmatrix}$$

be the matrix representing the overall effect of a number of layers. Show that this matrix is unitary whether the layers absorb or not. Show that if all layers are free from absorption A, $\mathrm{i}B$, $\mathrm{i}C$ and D are all real.[41]

[41] *Hint*: Multiply two representative constituent matrices, then generalize.

(4) Show that if

$$\begin{pmatrix} E_x \\ Z_0 H_y \end{pmatrix}_{\text{plane 1}} = \begin{pmatrix} A & B \\ C & D \end{pmatrix} \begin{pmatrix} E_x \\ Z_0 H_y \end{pmatrix}_{\text{plane 2}}$$

the inverse transformation between the same two planes is

$$\begin{pmatrix} E_x \\ Z_0 H_y \end{pmatrix}_{\text{plane 2}} = \begin{pmatrix} D & -B \\ -C & A \end{pmatrix} \begin{pmatrix} E_x \\ Z_0 H_y \end{pmatrix}_{\text{plane 1}}.$$

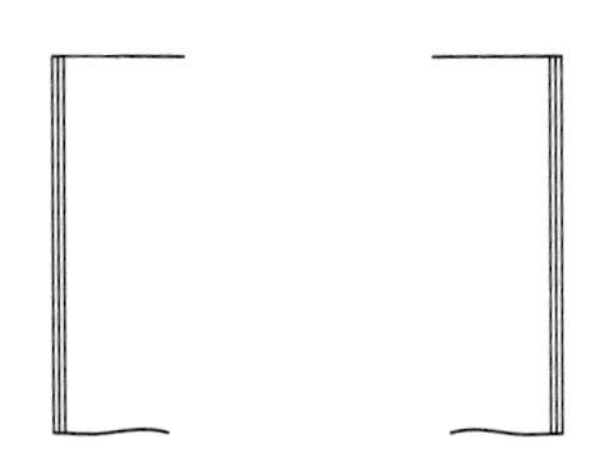

Fig. 6.9 The two structures shown are identical except that the one on the right has been rotated to face the opposite way. Problem 6.6(5) relates the $ABCD$ matrices for the two arrangements.

(5) Let the left-hand structure of Fig. 6.9 be described by

$$\begin{pmatrix} E_x \\ Z_0 H_y \end{pmatrix}_{\text{left side}} = \begin{pmatrix} A & B \\ C & D \end{pmatrix} \begin{pmatrix} E_x \\ Z_0 H_y \end{pmatrix}_{\text{right side}}. \tag{6.22}$$

Show that for the structure, rotated by otherwise identical, on the right[42]

$$\begin{pmatrix} E_x \\ Z_0 H_y \end{pmatrix}_{\text{left side}} = \begin{pmatrix} D & B \\ C & A \end{pmatrix} \begin{pmatrix} E_x \\ Z_0 H_y \end{pmatrix}_{\text{right side}}. \tag{6.23}$$

[42] *Hint*: The obvious way to do this is to use part (4) to obtain the inverse matrix; then rotate everything (axes and all) through 180°; then rotate the axes alone back again. If you adopt this route, do it formally, as it's easy to get muddled.

(6) Use eqns (6.22) and (6.23) to obtain expressions, in terms of A, B, C and D, for the reflection and transmission coefficients r_1, r_2, t_1, t_2 (notation as in problem 6.5), for a general reflector at normal incidence.

(7) Use the expressions of part (6) to verify that statements (6.18) and (6.19) hold—for normal incidence on a non-absorbing structure.

(8) Use the expressions of part (6) to show that eqn (6.21) holds generally at normal incidence, even if the structure is not loss-free.

Problem 6.7 (c) Thin films at oblique incidence

(1) Light travels in a thin layer as in Fig. 6.2, but at angle θ to the normal. Show that, if the $\boldsymbol{E}$-field is normal to the plane of incidence, eqn (6.4) holds[43] but the square matrix has n replaced by $n\cos\theta$.

[43] As with the derivation of eqn (6.4), it should not be assumed that n or k is real. The replacements given here apply whether or not the layer absorbs.

(2) Show that, if the $\boldsymbol{E}$-field lies in the plane of incidence, matrix equation (6.4) holds but the square matrixz[44] has n replaced by $n/\cos\theta$.

[44] The square matrices for oblique incidence are different for the two polarization directions. It is this fact that makes possible the construction of a polarization-dependent beam splitter. An application for such a device is mentioned in §16.4.

(3) Light is incident at angle θ_i to the normal on the structure of Fig. 6.8, and is part-transmitted at angle θ_t. Show that, for both polarizations, eqn (6.21) is replaced, whether or not the structure absorbs, by

$$n_1 t_2 \cos\theta_\mathrm{i} = n_4 t_1 \cos\theta_\mathrm{t}. \tag{6.24}$$

(4) Show that eqn (6.24) implies a reciprocity relation between the structure's power transmission coefficients.

Problem 6.8 (e) Polaroid and other absorbers
Consider a layer that absorbs almost totally. It might be a Polaroid absorbing its 'unwanted' polarization. Show that the square matrix of eqn (6.4) continues to have the same form, and remains unitary, though e^{-ikl} tends to infinity.

Suppose we have a Faraday rotator that rotates the plane of polarization of light through 45°. With two Polaroids it can be made into an 'optical isolator': a window that transmits light in only one direction. Discuss whether this device has a unitary 'thin-film' matrix.

Problem 6.9 (c) A thin-film mirror as a periodic structure
The stack of thin-film layers shown in Fig. 6.4 is a periodic structure with period equal to the combined thicknesses of two adjacent layers. Couldn't we treat the stack in the way that periodic structures are handled elsewhere: in solid-state physics, for example?

For a monatomic linear chain of atoms held together by springs, we can express any motion of the atoms in terms of waves, one travelling in each direction for each possible frequency. Something similar happens for electrons moving in the periodic potential of a lattice (Bloch waves). And something similar happens with an electrical filter made from a periodic arrangement of inductors and capacitors.

We ought, therefore, to be able to take a thin-film periodic structure and obtain eigenfunctions (for a chosen frequency) representing two waves, one travelling in each direction. Any electromagnetic disturbance (of this frequency) is a sum over the two eigenfunctions, with coefficients determined by boundary conditions at the ends of the stack. Here is another route—surely an orthodox one—for handling the algebra.

In the notation of Fig. 6.4, we find the travelling-wave eigenfunctions by requiring that

$$\begin{pmatrix} E \\ Z_0 H \end{pmatrix}_{\mathrm{A}} = a \begin{pmatrix} E \\ Z_0 H \end{pmatrix}_{\mathrm{C}}.$$

There are two solutions for a, corresponding to the two possible directions of travel, and for each we can find an **iterative impedance**,[45] meaning the value of E/H at a plane like C or A.

In language borrowed from electrical filters, there are certain ranges of ω for which the stack has a **pass band** (in which case a is a complex number of modulus 1), while for other ranges the stack has a **stop band** (a real and $|a| < 1$ for a wave travelling towards $+z$).

Develop an analysis given these clues. Show that when the layers approximate to quarter-wave layers the frequency applied lies in a 'stop band'. Show that when the layers are exactly quarter-wave, the iterative impedance is either infinite or zero, depending on which layer has the higher refractive index and on which way the wave is going. Obtain eqn (6.15) by using the new formalism.

Comment: Unfortunately, the algebra is messy, so we have not discovered an exciting new way of streamlining the algebra of this chapter. It remains, however, worth pointing out that the analogy with other periodic structures exists.

[45] The term *iterative impedance* is taken over from electrical filters. The impedance takes different values according as we evaluate it at plane A or plane B. This should not be disconcerting, as the iterative impedance of an LC filter depends on whether the filter starts with a T-section or a Π-section. See, e.g., Bleaney and Bleaney (1976), Chapter 9.

7 Ray matrices and Gaussian beams

7.1 Introduction

A Gaussian[1] beam is a beam of light whose profile varies in a Gaussian way with radial distance from its central axis. Such beams are of great importance in modern optics because the output from a laser is very often of this form.

[1]We should avoid confusing a Gaussian beam with 'Gaussian optics' which is a name sometimes given to ray optics within the 'paraxial' approximations used in Chapter 1.

This chapter starts with an introduction to matrix methods. Although the connection is not at first obvious, it turns out that these matrices are just what we need to describe the way in which a Gaussian beam propagates.

7.2 Matrix methods in ray optics

Figure 7.1 shows a light ray propagating in the yz plane.[2] The ray's properties at plane $z = z_1$ are fully specified by giving the values of (y_1, θ_1): its distance y_1 from the axis and the angle θ_1 that it makes with the axis. Tracing such a ray through an optical system to plane z_2 consists in finding values for (y_2, θ_2) in plane z_2 starting from (y_1, θ_1) in plane z_1. This is asking for a matrix formulation.

[2]Any plane through the z-axis can be the yz plane for present purposes. At first sight, it might seem better to give a ray's distance from the z-axis as r, where $r^2 = x^2 + y^2$, since all directions at right-angles to the z-axis are equivalent. But the algebra will require y to change sign if the ray crosses the axis, which r does not.

There are a number of advantages, as we shall see later, in making one refinement. It will be convenient[3] to work with y and $(n\theta)$ rather than y and θ. The matrices are shapelier if they link $(y_1, n_1\theta_1)$ to $(y_2, n_2\theta_2)$ and they also have the useful property of being unitary.[4]

[3]The treatment is restricted to small angles and to non-skew rays. These would be serious limitations if we were pursuing ray optics itself, perhaps trying to analyse aberrations in some lens system. But they do no harm here.

[4]The ideas in this chapter, in particular the sign conventions and the preference for $(n\theta)$ over θ, owe a great deal to Gerrard and Burch (1975).

When following a ray through an optical system from one z to another, we shall need to attach algebraic signs to distances, angles and radii

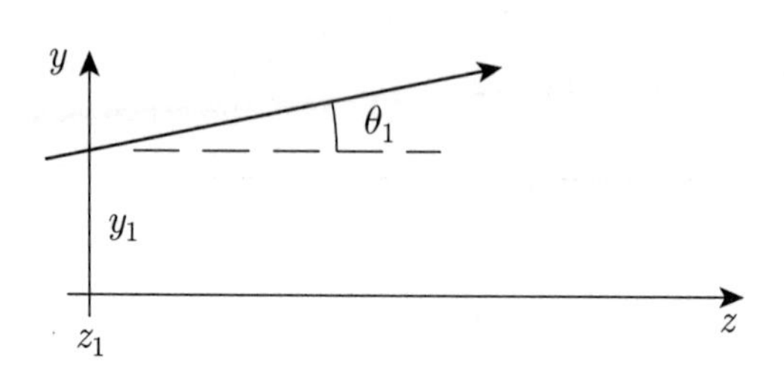

Fig. 7.1 A ray passing through plane $z = z_1$ of an optical system whose axis is the z-axis. The ray lies at distance y_1 from the z-axis and makes (small) angle θ_1 with the axis. Discussion is confined to non-skew rays: those which lie in the plane of the diagram, or any rotation of it about the z-axis. Rays meeting these conditions are fully specified by giving the two quantities (y_1, θ_1).

of curvature,[5] and the sign convention will be the Cartesian version introduced in §1.9.2.

[5] Radii of curvature are awkward, as discussed in §7.8.

7.3 Matrices for translation and refraction

Figure 7.2(a) shows a ray travelling through distance $z_2 - z_1$. The values of y and θ are related by

$$y_2 = y_1 + (z_2 - z_1)\theta_1 \qquad \text{and} \qquad \theta_2 = \theta_1.$$

These statements can be put into matrix form as

$$\begin{pmatrix} y_2 \\ n\theta_2 \end{pmatrix} = \begin{pmatrix} 1 & (z_2 - z_1)/n \\ 0 & 1 \end{pmatrix} \begin{pmatrix} y_1 \\ n\theta_1 \end{pmatrix}, \tag{7.1}$$

where $\begin{pmatrix} 1 & (z_2-z_1)/n \\ 0 & 1 \end{pmatrix}$ is the **ray transfer matrix** for translation[6] through a distance $z_2 - z_1$.

[6] Notice that $(z_2 - z_1)/n$ is not the optical path from z_1 to z_2; irritatingly the refractive index is underneath, not on top.

Figure 7.2(b) shows a ray refracted at a spherical surface[7] separating media with refractive indices n_1 and n_2. Using small-angle approximations throughout we have

$$\begin{aligned} i_1 &= \theta_1 + y_1/R_s, \\ i_2 &= \theta_2 + y_1/R_s, \\ \text{Snell's law:} \quad n_2\, i_2 &= n_1\, i_1, \end{aligned}$$

[7] The radius of curvature R_s of the refracting surface is treated as positive when the surface is convex to the left. This is in conformity with the convention introduced in connection with Fig. 1.4.

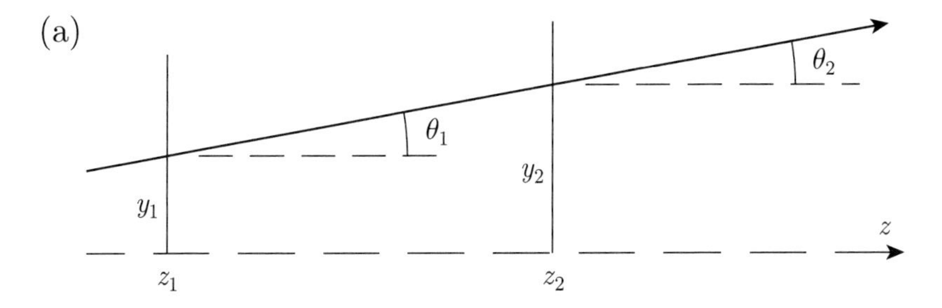

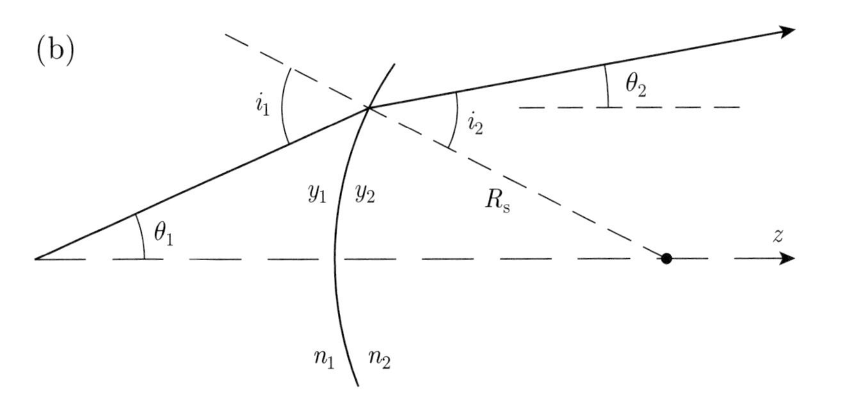

Fig. 7.2 All transformations encountered by a ray can be decomposed into translations, refractions and reflections. (a) A ray undergoing translation from z_1 to z_2. (b) A ray undergoing refraction at a spherically curved surface of radius R_s separating media with refractive indices n_1, n_2.

whence

$$y_2 = y_1 \qquad \text{and} \qquad (n_2\theta_2) = (n_1\theta_1) - (n_2 - n_1)y_1/R_s,$$

which rearranges into the matrix form

$$\begin{pmatrix} y_2 \\ n_2\theta_2 \end{pmatrix}_{\text{just right of surface}} = \begin{pmatrix} 1 & 0 \\ -P & 1 \end{pmatrix} \begin{pmatrix} y_1 \\ n_1\theta_1 \end{pmatrix}_{\text{just left of surface}} . \tag{7.2}$$

Here $P = (n_2 - n_1)/R_s$ is the *power* of the surface, as defined in eqn (1.34).

Two successive transformations, say a translation through $h = z_2 - z_1$ followed by a refraction, are handled by multiplying matrices:

$$\begin{pmatrix} y \\ n\theta \end{pmatrix}_{z_2+} = \begin{pmatrix} 1 & 0 \\ -P & 1 \end{pmatrix} \begin{pmatrix} y \\ n\theta \end{pmatrix}_{z_2-} = \begin{pmatrix} 1 & 0 \\ -P & 1 \end{pmatrix} \begin{pmatrix} 1 & h/n \\ 0 & 1 \end{pmatrix} \begin{pmatrix} y \\ n\theta \end{pmatrix}_{z_1} . \tag{7.3}$$

It is obvious that this generalizes: *a succession of transformations is dealt with by multiplying their matrices*—in the proper order of course.[8]

Problems 7.1–7.3 provide practice with these matrices.[9]

[8] As with the thin-film matrices of Chapter 6, it is best to remember what happened in obtaining eqn (7.3) rather than asking for a mnemonic. But for those who prefer to remember a rule, the matrices are written in the *reverse* order to that of the optical components on the page.

[9] Matrix methods are not restricted to media that are uniform or piecewise uniform. Problem 14.2 deals with the propagation of rays in a medium, such as an optical fibre, whose refractive index is 'graded'.

7.4 Reflections

Consider reflection from a plane mirror as shown in Fig. 7.3. We aim to relate the $(y, n\theta)$ for the reflected ray to that of the incident ray, both evaluated very close to the plane $z = z_1$ of the mirror. The physics of reflection gives simply $y_2 = y_1$ and $\theta_2 = -\theta_1$, so the obvious way of relating the (y, θ) matrices would be via a square matrix $\left(\begin{smallmatrix} 1 & 0 \\ 0 & -1 \end{smallmatrix}\right)$. Such a matrix is not unitary, and it is best to do things a different way.

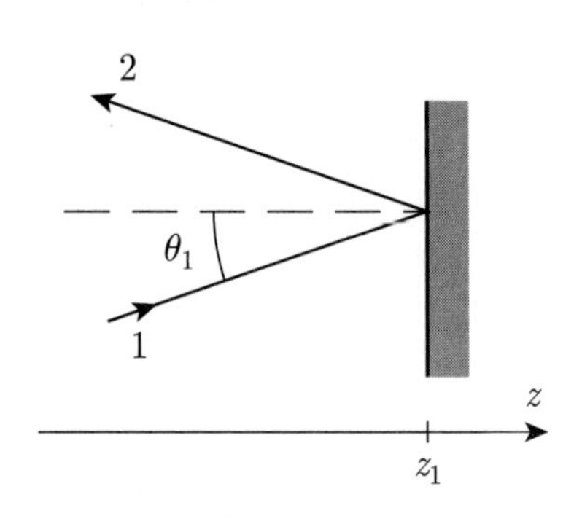

Fig. 7.3 A ray reflected from a plane mirror. The reflected ray has $\theta_2 = -\theta_1$.

We give the left-going wave a negative value of 'refractive index' n_2 (reversed in sign but unchanged in magnitude). Thus, we describe the reflection by $n_2 = -n_1$, $\theta_2 = -\theta_1$ and $(n_2\theta_2) = (n_1\theta_1)$. With this understanding,

$$\begin{pmatrix} y_2 \\ n_2\theta_2 \end{pmatrix} = \begin{pmatrix} 1 & 0 \\ 0 & 1 \end{pmatrix} \begin{pmatrix} y_1 \\ n_1\theta_1 \end{pmatrix}, \tag{7.4}$$

in which the square matrix is once more unitary.

The convention[10] just enunciated looks very peculiar. We can make ourselves comfortable with it in either of the following ways.

(1) We shall often 'unfold' an optical system containing a mirror, so that ray 2 of Fig. 7.3 continues towards top right, following the mirror image of the actual path towards top left. In the unfolded diagram, there is no change in the sign of θ and there is no risk of writing a non-unitary matrix. It is useful to set up the real reflected ray so that it is described in the same way as its image. A negative n_2, cancelling the negative sign in θ_2, achieves this.[11]
(2) If we were interested in 'rays' only as lines drawn on a piece of paper, we should have no wish to attach different signs to the same line. However, we are going to use our matrices on waves (in

[10] The convention was anticipated in problem 1.14(9).

[11] When waves travel towards $-z$, the refractive index n in the $(z_2 - z_1)/n$ element of the translation matrix of eqn (7.1) must be given a negative sign likewise. See problem 7.1(7). The same applies to the refractive indices n_1, n_2 in the power P of a refracting surface; see problem 7.1(8).

particular on Gaussian beams) which have amplitude proportional to $e^{ikz-i\omega t} = e^{in(\omega/c)z-i\omega t}$. It is not silly to put $k = n\omega/c$ always; then a wave going towards $-z$ has negative k which must now go with negative n. Equation (1.12) was written in just such a way as to anticipate what we are now doing.

As with refraction at a plane surface, detailed in problem 7.1 parts (3) and (4), the unit matrix in eqn (7.4) seems to signal 'no change'; we do have to insert the changed n 'by hand', knowing what is the physical system under consideration.

The convention of reversing the sign of n for left-going waves is equally applicable to reflection from a spherical mirror. We do not spell out a description here; the reader can easily supply the omission with the aid of problem 7.4. The more important case of a Gaussian beam reflected from a spherical mirror is dealt with in problem 7.12 parts (4) and (5).[12]

[12] *Aside*: Although the emphasis in this chapter has been on applying ray matrices to light, we should mention that similar matrices are used by particle physicists when designing quadrupole magnets for focusing beams of particles. Such magnets are used within an accelerator and also between the accelerator and a target.

7.5 Spherical waves

A spherical wave makes a half-way step between the rays we have dealt with up to now and the Gaussian beams to be considered below.

We consider only those spherical waves whose centres lie on the z-axis; this means that all rays composing the wave are non-skew. Figure 7.4 shows one representative ray that has come from a source at z_1. Suppose we somehow come to know $(y, n\theta)$ for this ray at some plane z; then we know where the centre lies: it is at z_1, where

$$z - z_1 = y/\tan\theta \approx n \times (\text{upper element})/(\text{lower element}). \qquad (7.5)$$

Putting this another way, the radius q of the wavefront drawn in Fig. 7.4 is given by

$$\frac{q}{n} \approx \frac{z - z_1}{n} \approx \frac{y}{n\theta} = \frac{\text{upper element}}{\text{lower element}} \text{ of matrix } \begin{pmatrix} y \\ n\theta \end{pmatrix}_z . \qquad (7.6)$$

The matrix methods we use to chase the values of $(y, n\theta)$ through an optical system can be adapted in a straightforward way to chase through the value of q. Let

$$\begin{pmatrix} y \\ n\theta \end{pmatrix}_{z=z_2} = \begin{pmatrix} A & B \\ C & D \end{pmatrix} \begin{pmatrix} y \\ n\theta \end{pmatrix}_{z=z_1}$$

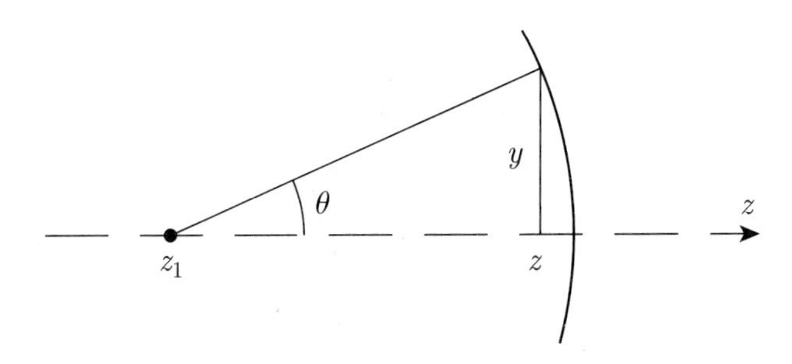

Fig. 7.4 A spherical wave radiating from a point on the z-axis at $z = z_1$.

represent an optical system's transformation from z_1 to z_2. Separating the matrix equation into its constituents and dividing upper and lower elements gives

$$(q_2/n_2) = \frac{A(q_1/n_1) + B}{C(q_1/n_1) + D}. \tag{7.7}$$

This result is called the $ABCD$ rule.[13]

There is a presentational problem with the $ABCD$ rule: matrices have disappeared; it has ceased to be obvious—though it's still true—that two successive applications of the rule are equivalent to one application using the matrix product. We need to re-present things so that matrices are again on show. Formalize what we did in obtaining eqn (7.7) by removing separate mention of y and $(n\theta)$:

$$\begin{pmatrix} q_2/n_2 \\ 1 \end{pmatrix} = (\text{cancel me}) \times \begin{pmatrix} A & B \\ C & D \end{pmatrix} \begin{pmatrix} q_1/n_1 \\ 1 \end{pmatrix}. \tag{7.8}$$

This presentation is somewhat unusual mathematics but it preserves the matrix shape. At the same time, when we expand it to

$$\begin{aligned} q_2/n_2 &= (\text{cm}) \left\{ A(q_1/n_1) + B \right\} \\ 1 &= (\text{cm}) \left\{ C(q_1/n_1) + D \right\} \end{aligned}$$

the 'cancel me' (cm) factor (equal to $\{C(q_1/n_1) + D\}^{-1}$ of course) is screaming to be divided away; the $ABCD$ rule is still clearly there.

We shall not manipulate spherical waves in this book, using either matrices or the $ABCD$ rule. But the idea of obtaining q_2 from a matrix statement like eqn (7.8) is just what we need in the remainder of this chapter.

[13] It would simplify things a little if we were to invent a symbol for a 'scaled radius' $\rho = q/n$, and indeed this is sometimes done. However, we prefer to leave the refractive index n explicitly on show, in anticipation of a similar choice with Gaussian beams. Although we make no use of it here, the refractive index n is to be made negative for a wave travelling towards $-z$, and this determines the sign of q in ways consistent with the sign convention built into the $(y, n\theta)$ for rays.

7.6 Gaussian beams

Gaussian beams are important in modern optics because the output from a laser often has the form of a Gaussian beam, for reasons we shall see in Chapter 8.

All our Gaussian beams will travel in the $\pm z$-direction and be centred on the z-axis. The canonical form for such a beam is

$$U = U_0 \, e^{i(kz-\omega t)} \, \frac{\exp(ikr^2/2q)}{q}, \qquad \text{where} \qquad q = z - z_{\text{waist}} - ib, \tag{7.9}$$

and $r = \sqrt{x^2 + y^2}$ is the radial distance from the z-axis. Here U is the amplitude of the wave (a scalar amplitude representing the electric field), q is called the **complex radius**, and z_{waist} and b are real constants.[14]

Equation (7.9) does not at first seem to have an 'obvious' shape. However, the Gaussian beam has a very close mathematical relationship to the spherical wave of Fig. 7.4 (which can be put into a similar mathematical form), however much the two waves differ in their physics. This connection is established in problem 7.5.[15]

[14] The quantity b is a length and is known as the **confocal parameter** because it is half the length of a symmetrical confocal mirror cavity into which the beam fits, as will be seen in Chapter 8.

[15] *Comment*: The family resemblance between a Gaussian beam and a spherical wave is a very worthwhile insight. In particular, the matrix algebra for dealing with rays and waves, developed earlier in this chapter, carries over to Gaussian beams with complex q replacing the real radius q of §7.5. This similarity is neither mysterious nor a coincidence.

7.7 Properties of a Gaussian beam

We look again at eqn (7.9) and remove z_{waist} for now by shifting the origin of z to z_{waist}. Then we have simply $q = z - \mathrm{i}b$. Look first at the quantity in the exponent:

$$\frac{\mathrm{i}kr^2}{2q} = \frac{\mathrm{i}kr^2}{2(z-\mathrm{i}b)} = \frac{\mathrm{i}kr^2(z+\mathrm{i}b)}{2(z^2+b^2)} = \frac{\mathrm{i}kr^2}{2}\left(\frac{z}{b^2+z^2}\right) - \frac{kbr^2}{2(b^2+z^2)}$$
$$\equiv \frac{\mathrm{i}kr^2}{2R(z)} - \frac{r^2}{w(z)^2}. \tag{7.10}$$

When we take the exponential of this we get two factors. Comparison of $\exp(\mathrm{i}kr^2/2R)$ with eqn (7.20) in problem 7.5 shows that R is the radius of curvature[16] of the wavefront. Factor $\exp(-r^2/w^2)$ gives a Gaussian profile; we call w the spot size. Equating coefficients in eqn (7.10) gives[17]

$$R(z) = \frac{b^2+z^2}{z}, \tag{7.11}$$

$$w(z)^2 = w_0^2\left(1+\frac{z^2}{b^2}\right), \qquad \text{with} \qquad w_0^2 = \frac{2b}{k} = \frac{|b|\lambda}{\pi}. \tag{7.12}$$

[16] Note the sign attached to R, as discussed in §7.8.

[17] *Recommendation*: Don't try to memorize eqns (7.11) and/or (7.12). The one equation that should be learned is eqn (7.9). Once you've seen the derivation of its properties you can quickly repeat the steps given here.

Next look at the factor $1/q$:

$$\frac{1}{q} = \frac{1}{z-\mathrm{i}b} = \frac{\mathrm{i}\,\mathrm{e}^{-\mathrm{i}\alpha}}{\sqrt{b^2+z^2}} = \mathrm{i}\frac{\mathrm{e}^{-\mathrm{i}\alpha}}{b}\frac{w_0}{w(z)}, \qquad \text{where} \qquad \tan\alpha = \frac{z}{b}. \tag{7.13}$$

Combining all the pieces, we now have

$$\begin{aligned}
U = {} & (U_0\mathrm{i}/b) && \text{constants; amplitude on axis at waist}\\
& \times w_0/w(z) && \text{amplitude gets smaller as beam spreads}\\
& \times \mathrm{e}^{\mathrm{i}(kz-\omega t)} && \text{wave travels towards } z\times\text{sign}(k)\\
& \times \mathrm{e}^{-\mathrm{i}\alpha(z)} && \text{additional phase change with } z\\
& \times \exp\{-r^2/w(z)^2\} && \text{Gaussian profile with spot size } w(z)\\
& \times \exp\{\mathrm{i}kr^2/2R(z)\} && \text{wavefronts have radius of curvature } R(z).
\end{aligned} \tag{7.14}$$

The meanings of the various factors appearing in eqn (7.14) are illustrated in Fig. 7.5. The beam is narrowest at $z = 0$, where it is said to

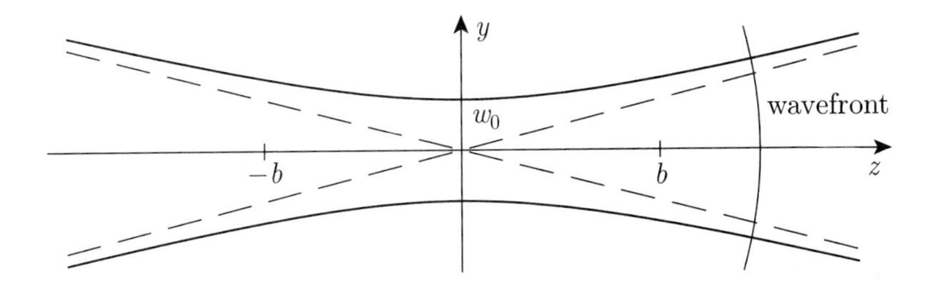

Fig. 7.5 A Gaussian beam shown in yz cross section. The outline of the beam is indicated by plotting $y = \pm w(z)$ versus z. The spot size takes its least value w_0 at the beam's waist, which here is at $z = 0$. The spot size increases from w_0 to $\sqrt{2}w_0$ by $z = \pm b$. The beam outline is a hyperbola, whose asymptotes are shown by broken lines.

have a **waist**. More generally, the waist lies at z_{waist}, which explains the notation used in anticipation in eqn (7.9). On either side of the waist the beam spreads, with a hyperbolic outline whose asymptotes are also shown in the figure. Away from the waist, where the beam diverges (or converges), the wavefronts are of course curved; one representative wavefront is drawn. The properties quoted here are established in the problems, especially problem 7.6.

The behaviour shown in Fig. 7.5 is a manifestation of diffraction: a beam with restricted width is compelled to spread.[18]

[18]It is easy to lose sight of this fact, perhaps because the mathematics presented here does not exhibit a Kirchhoff diffraction integral. However, the amplitude of eqn (7.14) is a solution of the scalar wave equation (1.24) and, therefore, must conform to all the characteristics of a wave. A brute-force check that we have a solution of the scalar wave equation is given in problem 7.7. A check using a Kirchhoff integral is given in problem 7.9.

Further confirmation that we're looking at diffraction, if it is needed, can be obtained from the presence of the wavelength λ in the w_0 of eqn (7.12).

7.8 Sign conventions

The Gaussian-beam expression of eqn (7.9) caters for waves travelling towards $+z$ or towards $-z$. The understandings are the same as those in item (2) of §7.4: k may have either sign; $k = n\omega/c$ always. Then a wave travelling towards $-z$ has negative k, accompanied by negative refractive index n and (we now see from $\omega_0^2 = 2b/k$) negative confocal[19] parameter b.

[19]Note the use of $|b|$, rather than b, in the final expression of eqn (7.12).

The real part of q is $(z - z_{\text{waist}})$; this is independent of the wave's direction of travel. The equations following from eqn (7.9), such as eqns (7.11) and (7.12), can all be used 'mechanically' for both wave directions provided signs for $(z - z_{\text{waist}})$ and b are assigned according to the understandings here.

As a Gaussian beam propagates, its q changes according to eqn (7.16) below. The use of negative n for a wave travelling towards $-z$ ensures that matrix transformation of $(q/n, 1)$ obeys the same rules as those transforming a ray's $(y, n\theta)$ or a spherical wave's $(q/n, 1)$. As with rays, we have to put in the sign of n 'by hand'.

Unfortunately, an inconsistency has crept in, because radius $R(z)$ has been given a sign here that is the exact opposite of that allocated to spherical refracting and reflecting surfaces in §1.9.2. Unfortunately, both of these sign choices are entrenched in their own fields, and are unlikely to be displaced. As with the complex conventions discussed in §2.12, it seems best to be resigned to the inconsistency, rather than to tidy it up and impose an incompatibility—somewhere—with what is customary. The reader is warned to be particularly alert to signs; opposite conventions can collide when a Gaussian beam is refracted as in problem 7.10.

7.9 Propagation of a Gaussian beam

Practical applications of lasers involve propagating a Gaussian beam from place to place, focusing it with a lens or a mirror, and so on. The beam is often narrow enough that diffraction dominates, to the point where the ray-optics approximations given in earlier chapters of this book are no longer adequate.

An example of Gaussian-beam behaviour that contrasts with previous experience is explored in problem 7.8: when a beam is focused by a lens, it reaches a smallest diameter at the beam waist, but this waist lies 'early' of the ray-optics focus.[20]

The focus shift is fairly easy to work out; it can be calculated using eqns (7.11) and (7.12) only. For other cases we need more powerful methods: a systematic procedure for handling Gaussian-beam transformations. The key lies in finding the changes of complex radius q as the beam develops.

To see why q is so important, refer back to eqn (7.9) and the definition of q as $q = z - z_{\text{waist}} - ib$. We can express this in words as

$$q = \begin{pmatrix} \text{how far we are to the} \\ \text{right of the beam's waist} \end{pmatrix} - \mathrm{i}(\text{confocal parameter}). \qquad (7.15)$$

If we have somehow come to know q for some value of z, the real part tells us where the beam's waist lies, and the imaginary part tells us the confocal parameter b. Knowing b, we can obtain the beam's waist spot size w_0 from eqn (7.12). Knowing the distance from the beam waist we can find the spot size $w(z)$ at z from eqn (7.12) and the wavefront radius from eqn (7.11). We even know which way the beam is travelling from the sign of b. *All* the information we need about the beam's shape can be obtained from knowledge[21] of the one quantity q.

The rule for obtaining the q in one plane z_2 from that in another z_1 exactly resembles eqn (7.8):

$$\begin{pmatrix} q/n \\ 1 \end{pmatrix}_{z=z_2} = (\text{cancel me}) \begin{pmatrix} A & B \\ C & D \end{pmatrix} \begin{pmatrix} q/n \\ 1 \end{pmatrix}_{z=z_1}, \qquad (7.16)$$

where $\left(\begin{smallmatrix} A & B \\ C & D \end{smallmatrix}\right)$ is identical to the ray transformation matrix[22] linking z_1 to z_2. The same understanding applies here as applies to eqn (7.8): to find $q(z_2)$, separate the matrix equation into two algebraic equations and divide through to remove the 'cancel me' factor.

The recipe for following a Gaussian beam through an optical system is now clear:

(1) Find the ray-transformation matrix[23] describing each optical element (translation, refraction, reflection, ...) and multiply the matrices together to find the overall matrix $\left(\begin{smallmatrix} A & B \\ C & D \end{smallmatrix}\right)$.

(2) Use this matrix as in eqn (7.16) to find the output q value from the input q value.

(3) Use eqn (7.15) to interpret this q to find $z - z_{\text{waist}}$ and b. Everything else we need can be found from here using eqns (7.11), (7.12) and (possibly) (7.13).

Of course the reasoning can be traced along other routes, in particular to design an optical system that will give a required output.

[20] Such a focus shift must happen with a narrow beam of any shape, not only Gaussian, but it shows up most clearly here. A non-Gaussian beam undergoes changes of shape (Fresnel and Fraunhofer diffraction patterns) as it approaches a focus, making it difficult to decide where it has its narrowest width.

[21] The one thing that q doesn't tell us is the beam's amplitude constant U_0. For that we need additional reasoning, such as that in problem 7.10(7).

[22] The reason why the square matrix here has to be the same as that for spherical waves in eqn (7.8), which in turn is the same as for rays, lies in reasoning rehearsed in problem 7.5(3). Brute-force confirmation can be found by working problem 7.10.

We can, of course, rearrange eqn (7.16) into the form of an $ABCD$ law, just as we did for a spherical wave in eqn (7.7). But the matrix presentation is preferable.

[23] The matrix may of course be found by any valid method, but it is always simplest to find it from the way that a ray is processed. That is why §§7.2–7.4 began this chapter.

Example 7.1 Achieving a required waist size using a lens

A Gaussian beam has its waist (spot size w_1) distant $f + X$ from a converging lens of focal length f, and travels along the lens axis. Refraction at the lens results in a beam with waist (spot size w_2) distant $f + Y$ from the lens. Show that

$$\frac{Y}{X} = \frac{w_2^2}{w_1^2} \qquad \text{and} \qquad f^2 = \frac{k^2}{4} w_1^2 w_2^2 + XY.$$

It is desired to take a laser beam of wavelength 500 nm with a spot size of 0.2 mm and use a lens to make from it a beam of spot size 0.4 mm. Show that the lens's focal length must exceed 0.503 m. A lens of focal length 0.6 m is chosen; find the distance between the two beam waists.

[Part of a question, Oxford Physics Finals, paper B2, 2002.]

Solution: The layout is shown in Fig. 7.6. All waves travel towards $+z$ and are in air (except inside the lens) so the refractive index $n = 1$ throughout and will be dropped. The equation transforming q reads[24]

[24] The working here is explored in problem 7.11, together with alternative ways through the calculation.

$$\begin{pmatrix} q_2 \\ 1 \end{pmatrix} = (\text{cm}) \begin{pmatrix} 1 & f+Y \\ 0 & 1 \end{pmatrix} \begin{pmatrix} 1 & 0 \\ -1/f & 1 \end{pmatrix} \begin{pmatrix} 1 & f+X \\ 0 & 1 \end{pmatrix} \begin{pmatrix} q_1 \\ 1 \end{pmatrix}$$

which evaluates to

$$\begin{pmatrix} q_2 \\ 1 \end{pmatrix} = (\text{cm}) \begin{pmatrix} -Y/f & -(XY - f^2)/f \\ -1/f & -X/f \end{pmatrix} \begin{pmatrix} q_1 \\ 1 \end{pmatrix}.$$

Multiplying out the matrix product and dividing to remove (cm), we find

$$q_2 = \frac{(XY - f^2) + Y q_1}{X + q_1}. \tag{7.17}$$

The beam has waists[25] at the beginning and end of our transformation, so $q_1 = -ib_1$ and $q_2 = -ib_2$. Cross-multiplying and separating real and imaginary parts gives

[25] **Waists.** It is important to remember that eqns (7.10)–(7.14) use a 'local' origin for z at the beam waist. When the beam is refracted, it acquires a new waist location and a new local origin. Equations like $R = (b^2 + z^2)/z$ must not be used 'mechanically' with z measured from the old origin.

$$\frac{Y}{X} = \frac{b_2}{b_1} = \frac{w_2^2}{w_1^2} \qquad \text{and} \qquad f^2 = XY + b_1 b_2 = XY + \frac{k^2}{4} w_1^2 w_2^2, \tag{7.18}$$

where we use the fact that w_1 and w_2 are the waist spot sizes and use $b = k w_{\text{waist}}^2/2$ from eqn (7.12).

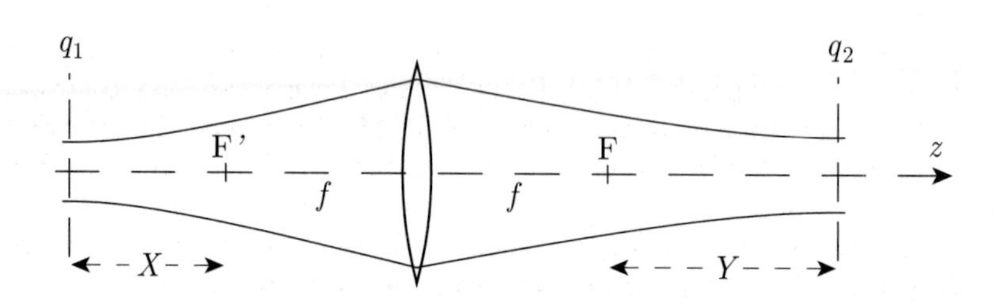

Fig. 7.6 A Gaussian beam has its waist distant $f + X$ to the left of the thin lens shown, and is refracted by the lens to a new waist distant $f + Y$ to the right. The beam propagation is discussed in example 7.1.

The input beam has $b_1 = kw_1^2/2$ which evaluates to $b_1 = 0.251\,\text{m}$. The given spot sizes are in the ratio $w_2/w_1 = 2$ so $b_2/b_1 = w_2^2/w_1^2 = 4$, and we also have $Y/X = 4$. Since Y/X is positive, XY is positive and eqn (7.18) tells us that

$$f^2 \geq b_1 b_2 = b_1(4b_1) = 4b_1^2 \qquad \text{so that} \qquad f \geq 2b_1 = 0.503\,\text{m}.$$

Given a lens with focal length 0.6 m, we have

$$X^2 = XY\frac{X}{Y} = \frac{XY}{4} = \frac{f^2 - b_1 b_2}{4} = \frac{f^2}{4} - b_1^2,$$

giving $X = \pm 0.164\,\text{m}$ and $Y = 4X = \pm 0.655\,\text{m}$. The negative choice for Y makes $f + Y$ negative, so that the output beam has its waist at a 'virtual image' on the wrong side of the lens. Therefore, we discard the negative possibilities for X and Y and accept $X = 0.164\,\text{m}$, $Y = 0.655\,\text{m}$. The distance between the beam waists works out to $X + 2f + Y = 2.02\,\text{m}$.

7.10 Electric and magnetic fields

'Electromagnetic waves are transverse' was proved for a *uniform* plane wave in eqn (1.11). But a Gaussian beam is not uniform, even where it's plane.

We need to go back to first principles: $\operatorname{div} \boldsymbol{D} = 0$, which for an LIH medium implies $\operatorname{div} \boldsymbol{E} = 0$. When we substitute into this an E_x equal to the U of eqn (7.9), we find that E_z is not zero: the $\boldsymbol{E}$-field is not wholly transverse but has a longitudinal component.[26] This is the case even at the beam's waist where the wavefront is plane. Indeed, the gradient $\partial E_x/\partial x$ is largest where the beam is narrowest; similarly for the magnetic field. For a careful confirmation, see problem 7.13.

We can understand the longitudinal field in a pictorial way. Lines of $\boldsymbol{D}$ (and here of $\boldsymbol{E}$) are continuous; in the absence of free charge they do not start or stop but have to form closed loops. As we move away from the z-axis, E_x gets weaker, so the lines of $\boldsymbol{E}$ must have turned into some other direction, which can only be the z-direction if the wave is polarized with $E_y = 0$. An impressionistic sketch of this behaviour is shown in Fig. 7.7.[27]

A Gaussian beam is commonly referred to as a TEM_{00} mode. (In Chapter 8, we shall meet more complicated transverse modes[28] that are similarly called TEM_{lm}.) The nomenclature is derived from microwave electronics where waves are classified as TE (*T*ransverse *E*lectric field but magnetic field partly longitudinal),[29] TM (*T*ransverse *M*agnetic field but electric field partly longitudinal) and TEM (*T*ransverse *E*lectric and *M*agnetic, both fields wholly transverse). Since a Gaussian beam has longitudinal components of both $\boldsymbol{E}$ and $\boldsymbol{B}$, it should not be described as TEM (the '00' part is right). Unfortunately, it seems to be too late to stop this misleading terminology from becoming universal.

[26] $\partial E_z/\partial z = -\partial E_x/\partial x \neq 0$.

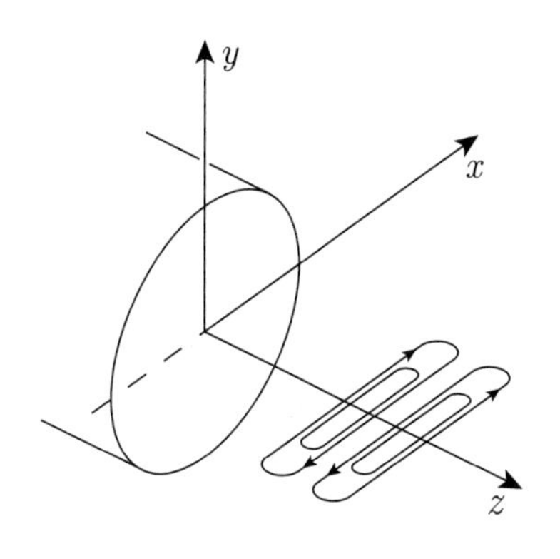

Fig. 7.7 An impression of how the lines of $\boldsymbol{E}$ must form loops in planes parallel to the xz plane, given a Gaussian beam that is primarily polarized with $\boldsymbol{E}$ in the x-direction. Lines of $\boldsymbol{B}$ form 'vertical' loops in a similar way. Since the fields reverse sign every half-wavelength, the loops are about half a wavelength long in the z-direction.

[27] Pictures of field lines like Fig. 7.7 are habitually drawn by people working in the microwave frequency region.

[28] Here 'transverse' refers to the variation of the field with x and y, not to the vector direction of the field.

[29] The higher-order generalizations of the Gaussian beam are described by eqn (8.1). A very few linear combinations of these are either TE or TM; see §14.3. Modes that propagate in optical fibres are likewise not TEM, and only rarely TE or TM. Less exotically, it is TE waves (almost always though TM is possible) that propagate in metal-walled waveguide. Examples of TEM waves are: *uniform* plane waves, and the wave in a loss-free coaxial line.

Problems

Problem 7.1 (a) Matrices for refraction

(1) Fill in all the algebraic steps in §7.3.

(2) Draw a diagram corresponding to that of Fig. 7.2(b) but with a concave refracting surface. Show that eqn (7.2) and the statements leading up to it apply without change, though the value attached to R_s is now negative.[30]

(3) Obtain the matrix for refraction at a plane boundary at $z = z_1$ by taking the limit of a curved surface as $R_s \to \infty$.

(4) Draw the diagram for refraction at a plane surface, and derive its matrix directly.[31]

(5) Show that the matrices in eqns (7.1) and (7.2) are unitary, and that, therefore, all matrices built up from them are also unitary.

(6) We could have set up our matrices to transform (y, θ) instead of $(y, n\theta)$. Show that for refraction we would have instead of eqn (7.2)

$$\begin{pmatrix} y \\ \theta \end{pmatrix}_{z_1+} = \begin{pmatrix} 1 & 0 \\ -P/n_2 & n_1/n_2 \end{pmatrix} \begin{pmatrix} y \\ \theta \end{pmatrix}_{z_1-},$$

which is distinctly less memorable and gives a non-unitary matrix. This explains why we went to the trouble of working with column matrices $(y, n\theta)$ rather than (y, θ).

(7) Consider the ray drawn in Fig. 7.2(a), but let it travel towards $-z$. Consider the signs attached to θ and to $(z_2 - z_1)$. Show that eqn (7.1) continues to hold provided that the refractive index n in $(z_2 - z_1)/n$ has reversed sign, as well as those in the $(n\theta)$ column-matrix elements.

(8) Consider the ray drawn in Fig. 7.2(b), but let it travel towards $-z$. Show that eqn (7.2) continues to hold, but the refractive indices in $P = (n_2 - n_1)/R_s$ must both now be treated as negative.

Parts (7) and (8) show that left-going waves are treated by giving reversed signs to refractive indices wherever they occur, whether in column matrices or in square matrices.

[30] *Comment*: This is in line with the statements made in §1.9.2: equations like (7.2) are made to apply to all circumstances, at the cost of rigorously attaching signs (by a sign convention) to all the quantities in them.

[31] *Comment*: The matrix for this is merely a unit matrix. A unit matrix seems to signal 'no change'; we do need to put $n_2 \neq n_1$ 'by hand' into the column matrices in this case.

Problem 7.2 (b) Ray matrices in special cases

Let $\begin{pmatrix} y_2 \\ n_2\theta_2 \end{pmatrix} = \begin{pmatrix} A & B \\ C & D \end{pmatrix} \begin{pmatrix} y_1 \\ n_1\theta_1 \end{pmatrix}$ describe how a complete optical system propagates rays from plane $z = z_1$ to $z = z_2$.

(1) Show[32] that if parallel light in plane 1 is brought to a focus in plane 2 then $A = 0$.

(2) Show that if an object in plane 1 is imaged in plane 2 then $B = 0$.

(3) Show that if parallel light in plane 1 emerges as parallel light in plane 2 then $C = 0$.

(4) Show that if all light leaving a point in plane 1 emerges parallel in plane 2 then $D = 0$.

(5) Which of the above are necessary as well as sufficient? For example, if $B = 0$ does that guarantee we have object and image?

[32] *Hint for all parts of this problem*: Draw a ray diagram!

Problem 7.3 (a) Matrices useful for work with a lens
A thin lens lies between media of refractive indices n_1 and n_3 as shown in Fig. 7.8.

(1) Obtain the matrix for a thin lens, linking $(y_3, n_3\theta_3)$ just to the right of the lens to $(y_1, n_1\theta_1)$ just to the left, by combining the matrices for two spherical surfaces whose separation is negligible. Show that the power P of the thin lens is the sum of the powers of its two surfaces.[33]

(2) Show that the matrix $\begin{pmatrix} 1 & 0 \\ -P & 1 \end{pmatrix}$ obtained in part (1) applies equally well to a thick lens, provided that $(y_1, n_1\theta_1)$ are taken on the first principal plane and $(y_3, n_3\theta_3)$ on the second.

(3) Obtain the matrix that transforms rays from plane $z = z_1$ just left of the lens to a general plane distant h_3 to the right of the lens.

(4) Find the value of h_3 that makes the matrix's A element zero and (problem 7.2(1)) argue that this locates the lens's right-side focal plane.

(5) Use a similar method to locate the lens's left-side focal plane. Show that the two focal lengths are $f_1 = n_1/P$, $f_3 = n_3/P$. Note[34] that the focal lengths are unequal when $n_3 \neq n_1$.

(6) Obtain the matrix that carries rays from the focal plane through F_1 to the other focal plane through F_3. Check against problem 7.2 parts (1) and (4).

(7) Use a matrix method to find the object–image relations for a lens separating media with refractive indices n_1, n_3. Show that these relations have the form[35] (1.33). Show also that they have the Newtonian form

$$XY = f_1 f_3, \qquad \frac{h_i}{h_o} = \frac{Y}{f_3} = \frac{f_1}{X}. \tag{7.19}$$

(8) Use the result of part (5) to show that a thin lens in air ($n_1 = n_3 = 1$) has focal length $f = 1/P$, the same on both sides. Since it is somewhat unusual for a lens to lie between two different media, we rarely need to distinguish two different focal lengths, and we treat $f = 1/P$ as *the* focal length. Use expressions from earlier in this problem to obtain an expression for f in terms of the radii of the refracting surfaces and the refractive indices, and hence confirm eqn (1.35).

(9) Find the matrix describing a thin lens in air (linking planes just left and just right of the lens) by an elementary method, not building on matrices constructed in this chapter. Figure 1.3 may help.

[33] *Comment*: The result of part (1) can of course be obtained without matrices: eqn (1.35) was obtained by making two applications of eqn (1.33). However, the intention here is to exercise the reader's familiarity with matrix methods. It may even be felt that the matrix route is the easier ....

[34] *Comment*: Notice also that the power of a lens depends on the refractive indices of the media on either side of it. If this isn't a comfortable idea, think what would happen if you placed a lens of refractive index n_2 inside a medium of refractive index n_2.

[35] The *form* is that of eqn (1.33), but the power P here takes the value appropriate to the lens under discussion. The power might be that calculated in part (1) if the lens is thin, but everything in parts (2)–(7) applies equally to a thick lens or a composite lens.

Check for consistency between the results obtained here and the properties that apply generally to any reasonable imaging system, as derived in problem 13.8.

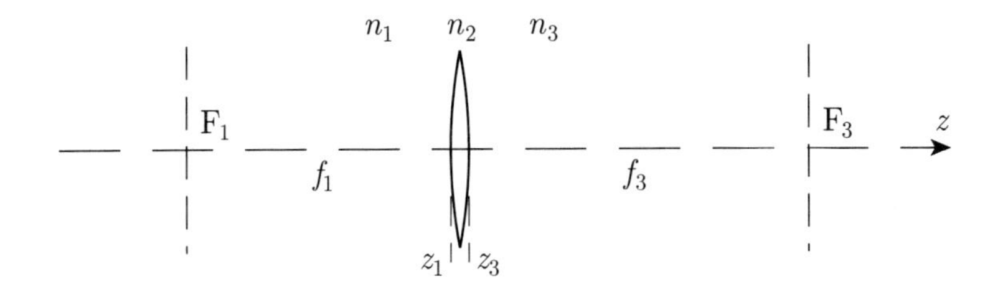

Fig. 7.8 A thin lens of refractive index n_2 lies between media with indices n_1, n_3. Points F_1 and F_3 are the principal foci on the two sides of the lens; the focal lengths f_1, f_3 are unequal as is shown in problem 7.3(5).

Problem 7.4 (b) Reflection of rays at a spherical mirror
Consider a mirror of radius R_s (R_s taken to be positive if the mirror is convex to the left, using the sign convention of §1.9.2). Light rays are reflected from the mirror, and we use the sign convention of §7.4 that makes the refractive index n negative when a wave travels to the left.

(1) Show[36] that the reflection matrix is $\begin{pmatrix} 1 & 0 \\ -P & 1 \end{pmatrix}$ with $P = -2n_1/R_s$.

(2) Check that the statements just made are correct both for a convex mirror ($R_s > 0$) and a concave mirror ($R_s < 0$). Notice that for both cases $P = (n_2 - n_1)/R_s$, which is the same expression as applies to refraction. Thus, refraction and reflection are integrated into a single mathematical description, as forecast in problem 1.14(9).

(3) A parallel beam of light is incident from the left (at angle θ to the axis) onto a concave mirror of radius R ($R > 0$ so $R_s = -R$). Find the matrix transformation that describes reflection followed by translation half way to the mirror's centre of curvature. Show that $y_2 = R\theta_1/2$, and interpret this result by drawing a ray diagram.

[36] *Comment*: Getting the signs right in this problem is quite tricky. That's why I've set it. You understand the ray matrices properly only when you can apply them correctly, even in a muddling case.

Problem 7.5 (a) Spherical wave and Gaussian beam
Refer to Fig. 7.4. The spherical wave has travelled distance R from its centre at z_1, so its amplitude can be written as in eqn (1.25) or its complex form $U = (U_0/R)\exp(\mathrm{i}kR - \mathrm{i}\omega t)$ in which $R = \sqrt{(z-z_1)^2 + r^2}$ and $r^2 = x^2 + y^2$.

(1) Take a 'paraxial' approximation in which $r \ll |z - z_1|$ and write q for $z - z_1$. Show that,[37,38] to second order of small quantities, $R \approx q + r^2/2q$.

(2) Compare the present approximation with that explored in problem 3.7(1), and argue that the full expression for R must be used in the exponent of U though q suffices in the denominator. With these replacements, we have, for a spherical wave,

$$U \approx U_0\,\mathrm{e}^{\mathrm{i}(kz-\omega t)}\frac{\exp(\mathrm{i}kr^2/2q)}{q}, \qquad \text{where} \qquad q = z - z_1, \tag{7.20}$$

and a constant factor $\exp(-\mathrm{i}kz_1)$ has been absorbed into U_0.

(3) Equation (7.20) has a very similar form to eqn (7.9), the only difference being the presence in eqn (7.9) of a constant imaginary part to q. We, therefore, look for a possible reason why it might be acceptable to change the q in eqn (7.20) from real to complex. Here is an outline argument for you to develop if you think it's respectable.

(a) The expression for U in eqn (7.20) is a solution of the scalar wave equation provided $q = (z - \text{a constant})$ and $r^2 \ll q^2$.

(b) Therefore, we could substitute the expression for U in eqn (7.20) back into the wave equation and (with appropriate approximations) verify that it is a solution.

(c) Alternatively, we could substitute U from eqn (7.20) into the wave equation with $q = q(z)$ left as an unspecified function. The result will be a differential equation for q; we know what it must be: $\mathrm{d}q/\mathrm{d}z = 1$. This has solution $q = z - z_1$, where z_1 is a constant of integration.

[37] A similar calculation, and an approximation to the same order, has already been encountered in eqn (3.26) on p. 82.

[38] *Comment on notation*: A sign choice has been sneaked into taking the square root here, so that R is the same as the radius of the spherical wave drawn in Fig. 7.4, including its sign, positive for a wave diverging towards $+z$. Likewise the new q, though not exactly the same as the q of §7.5, agrees with it in sign. However, there has been a change: the spherical wave is now being thought of in proper 'wave' terms and no longer as a bundle of rays. We need no longer mention rays lying in a yz plane—with their y and $n\theta$ taking either sign—nor think of R as related to a y/θ. We have, therefore, looked forward and made a switch to cylindrical polar coordinates with distances from the z-axis measured by the always-positive $r = \sqrt{x^2+y^2}$ rather than y.

(d) A differential equation tells us nothing about its constants of integration except that they are, well, constant. Therefore, a complex or imaginary z_1 could be used in eqn (7.20) and the result would still be formally a solution of the scalar wave equation.

(e) Choose $q = z - z_{\text{waist}} - ib$ and you have eqn (7.9).

(4) The argument just outlined started with an approximation $r^2 \ll q^2$. A corresponding inequality must restrict the Gaussian-beam expression. A reasonable guess might be $r^2 \ll |q^2|$. Argue that if this is right then the Gaussian-beam expression can be a valid approximation for all z if we can restrict attention to values of r satisfying $r^2 \ll \{(z - z_{\text{waist}})^2 + b^2\}$.

(5) Look back through the calculations of this problem to see what is the interpretation of k. Check that, apart possibly from a sign, $k = n\omega/c$. Check that a wave propagating towards $-z$ can be accommodated with no change to eqn (7.9) simply by allowing k to be negative.

(6) Check the statements on signs in §7.8: that we have a consistent set of signs and interpretations if $k = n\omega/c$ always so that negative k is accompanied by negative n and negative b. Check all of the Gaussian-beam equations in this chapter to see that these sign choices lead to correct interpretations of the physics.

Problem 7.6 (a) Interpretation of Gaussian-beam mathematics

(1) Show from eqn (7.12) that a Gaussian beam has its smallest spot size at $z = 0$ (more generally, at $z = z_{\text{waist}}$ if z_{waist} has not been set to zero), and that the least spot size is w_0.

(2) Show from eqn (7.12) that when $z = \pm b$ the spot size is $\sqrt{2}w_0$.

(3) Show that the factor w_0/w in eqn (7.14) ensures that energy is conserved.

(4) Show from eqn (7.11) that the radius of curvature R of the wavefront has its least value (for positive z) when $z = |b|$. Show that this minimum radius is $2|b|$.

(5) Express eqn (7.12), relating w to z, as the equation of a hyperbola. check that the hyperbola has been drawn correctly in Fig. 7.5.

(6) Figure 7.5 shows not only the w–z hyperbola but also a representative wavefront with radius $R(z)$. Show from the equations of §7.7 that the wavefront cuts the hyperbola at right-angles. To what extent is it then correct to describe either branch of the hyperbola as a 'ray'?

(7) Show from eqn (7.11) that the wavefront drawn on Fig. 7.5 must have its centre of curvature to the left of the beam's waist. Check that this agrees with what you would deduce by drawing a tangent to the hyperbola where the wavefront intersects it.

(8) Think of the Gaussian beam as starting from a collimated wave in plane $z = 0$. Interpret eqn (7.12) and Fig. 7.5 as showing that there is a roughly-plane wave of width $2w_0$ in the 'near field' and a diverging wave in the 'far field', the changeover taking place around $z = b$. Compare this with Fig. 3.13. Then b can be understood as something like the Rayleigh distance for a Gaussian beam. Does the algebra agree?

(9) Find the angle between the hyperbola's asymptotes. This is the angular divergence of the beam in the far field ($|z| \gg |b|$). Show that it is $(4/\pi)\lambda/(2w_0)$, and interpret the shape of this expression as indicating a diffraction spread from an initially-plane wave at $z = z_{\text{waist}}$ whose width starts out at about $2w_0$. Does the factor $4/\pi$ worry you when previous diffraction problems have yielded a factor 1 (slit) or 1.22 (circular hole)?

(10) Suppose we think of our Gaussian beam as prepared somewhere to the left of Fig. 7.5 and brought to a 'focus' at its waist. Then, by parts (2) and (8), $|b|$ gives us some measure of the depth of focus. Investigate whether this idea is in line with what was found in problem 3.22(3) and (4).

(11) Everything we have done here has been based on the scalar wave equation (1.24). We have seen in comment 1 on p. 48 a condition for a scalar amplitude to be acceptable: Huygens secondary waves must arrive at a point from a small range of directions so that the $\boldsymbol{E}$-fields being summed are nearly parallel. Now we have not been using Kirchhoff diffraction integrals to describe the propagation of our Gaussian beam because other mathematical methods have been more convenient; but we could and in problem 7.9 we shall. Adding $\boldsymbol{E}$-fields as scalars would fail if we added amplitudes at $z = 0$ arriving from a large range of angles. So the whole of our treatment (most pointedly in part (10)) must be dependent on the beam's far-field divergence angle $w_0/|b|$ being small. Show that this condition is equivalent to $|b| \gg \lambda$. Show that meeting this condition also ensures that the inequality mentioned in problem 7.5(4) is satisfied.

(12) What remains to be understood is the additional phase shift $\alpha(z)$. The wave's phase (lag) angle is $kz - \alpha(z)$ which is not quite linear with z. Over the whole length of the beam, from $z = -\infty$ to $z = +\infty$, α changes by π so its effect is not great, but it's there. Show that the beam's phase velocity is $\omega/(k - \mathrm{d}\alpha/\mathrm{d}z)$. Show that this differs most from ω/k at $z = 0$ where the beam is narrowest. It is the narrowness of the beam that causes its phase velocity to be altered.[39]

[39] To check this statement, it may help to look at other cases where a wave has a width that is not many wavelengths. A good example is a waveguide as used in microwave electronics; another is an optical fibre. In those cases we have a wave whose amplitude behaves like $U = f(x, y)\exp(\mathrm{i}kz - \mathrm{i}\omega t)$ and obeys the scalar wave equation

$$\frac{\partial^2 U}{\partial x^2} + \frac{\partial^2 U}{\partial y^2} - k^2 U = -\frac{\omega^2}{v^2} U.$$

If the partial derivatives are non-zero, as they are here, then $k \neq \omega/v$. In the case of our Gaussian beam, the mathematics has a different shape but the physical outcome is the same. A model optical fibre is investigated in §14.3, and its modes clearly exhibit the property discussed here.

(13) [Harder] Think about the group velocity.

(14) Show that the expression for U in eqn (7.9) has a dependence on x and y that can be expressed as (function of x, z) $\times$ (function of y, z), with each function depending on z through q or w. This rather surprising factorization will reappear in a more significant way in Chapter 8.

Problem 7.7 (c) Validity of our approximations

(1) Check by direct substitution that the U of eqn (7.9) is a solution of the scalar wave equation (1.24) provided that we can neglect quantities of order $(kw_0)^{-2}$ beside 1. So U can be expressed more correctly by

$$U = U_0 \exp(\mathrm{i}kz - \mathrm{i}\omega t)\,\frac{\exp(\mathrm{i}kr^2/2q)}{q}\left(1 + \text{ order } (kw_0)^{-2}\right). \qquad (7.21)$$

(2) Show that the condition $(kw_0)^{-2} \ll 1$ is the same as the condition investigated in problem 7.6(11), needed for a scalar wave equation to

apply, and is equivalent to requiring the beam to have only a small angular divergence in its far field. Show that it also guarantees that the condition guessed at in problem 7.5(4) is satisfied.

Problem 7.8 (b) A consequence of diffraction: focus shift
A collimated Gaussian beam is focused by a lens as shown in Fig. 7.9. If ray-optics rules applied exactly, the light would come to a point focus at the lens's principal focus F. Of course, light never concentrates to a point, as is clear from the discussions of diffraction in Chapter 3. But in Chapter 3, diffraction was always 'patched' onto a ray-optics picture that was treated as 'almost working'. Now we have an opportunity to investigate a case that is mathematically cleaner; and of course we learn more about Gaussian beams in the process.

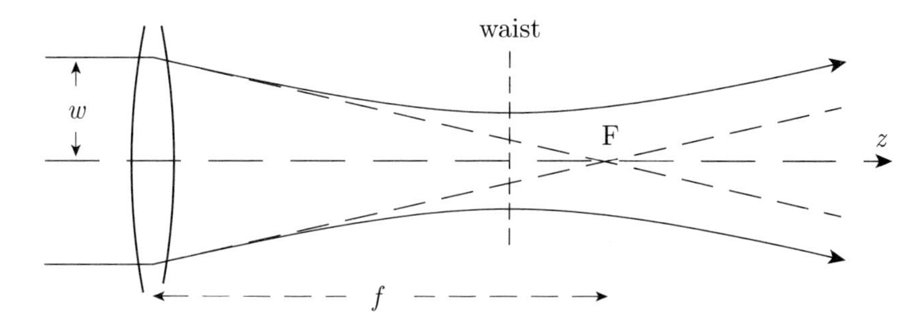

Fig. 7.9 The effect of diffraction on the focusing of a Gaussian beam. The broken straight lines show the geometrical-optics focal point F of the lens, to which the lens is 'trying' to send the light. The beam's hyperbolic outline starts off in that direction but spreads away because of diffraction. The waist occurs to the left of F by a distance calculated in problem 7.8.

(1) Use eqns (7.11) and (7.12) to show that:[40]

$$\text{confocal parameter } b = \frac{kw^2/2}{1+(kw^2/2f)^2} = \frac{\lambda}{\pi}\left(\frac{f}{w}\right)^2 \frac{1}{1+(2f/kw^2)^2}$$

$$\text{distance from waist to F} = \frac{f}{1+(kw^2/2f)^2}$$

$$\text{spot size at waist } w_0 = \frac{w}{\{1+(kw^2/2f)^2\}^{1/2}} = \frac{f\lambda/\pi w}{\{1+(2f/kw^2)^2\}^{1/2}}.$$

[40] *Hint*: It may avoid unnecessary negative signs if you reverse the diagram left-to-right and imagine that the beam travels from waist to lens.

(2) Show that 'conventional expectation' would make the second of these (the focus shift) zero and the third (diffraction-limited focus size) $f\lambda/\pi w$.

(3) To get an idea of how serious the corrections are in practice, find numerical values for the three tabulated quantities in the cases

(a) $w = 1\,\text{mm}$, $\lambda = 500\,\text{nm}$, $f = 50\,\text{mm}$,

(b) $w = 1\,\text{mm}$, $\lambda = 500\,\text{nm}$, $f = 5\,\text{m}$.

Problem 7.9 (c) Fresnel diffraction of a Gaussian beam
The behaviour of a Gaussian beam, given by eqn (7.9), necessarily involves diffraction because the wave is a solution of the scalar wave equation. Nevertheless, it is easy to get the impression that the different

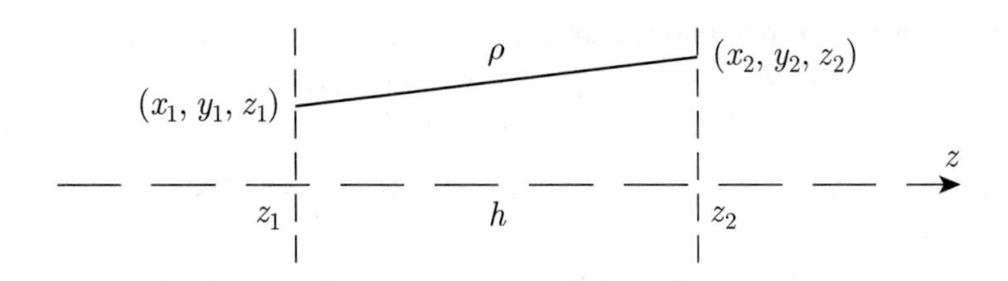

Fig. 7.10 The electric-field amplitude existing on the plane $z = z_1$ is given by a Gaussian-beam expression. The beam is Fresnel-diffracted to the plane $z = z_2 = z_1 + h$, with the outcome calculated in problem 7.9.

mathematical treatment must imply different physics. This problem clinches the connection. The Kirchhoff integral (eqn 3.6) tells us that, in Fig. 7.10,

$$U(x_2, y_2, z_2) = \frac{1}{\mathrm{i}\lambda}\int U(x_1, y_1, z_1)\,\frac{\exp(\mathrm{i}k\rho)}{\rho}\,\mathrm{d}x_1\,\mathrm{d}y_1. \tag{7.22}$$

(1) Express the distance ρ as a power series in x_1/h, y_1/h, x_2/h, y_2/h, where $h = z_2 - z_1$, keeping terms up to second order of these small quantities. Check that you get

$$k\rho = kh\left(1 + \frac{(x_2 - x_1)^2}{2h^2} + \frac{(y_2 - y_1)^2}{2h^2} + \text{fourth order}\right). \tag{7.23}$$

Comment: Notice that the terms in the exponent interpret as follows:

$kh\left(1 + \dfrac{x_2^2}{2h^2} + \dfrac{y_2^2}{2h^2}\right)$	are independent of x_1, y_1 so they do not inconvenience the integration;
$kh\left(-\dfrac{x_1x_2}{h^2} - \dfrac{y_1y_2}{h^2}\right)$	are terms we are comfortable with because they turn up in the Fraunhofer case of diffraction;
$kh\left(\dfrac{x_1^2}{2h^2} + \dfrac{y_1^2}{2h^2}\right)$	terms quadratic in the integration variables.

The terms containing x_1^2 and y_1^2 are what ordinarily make the Fresnel case of diffraction mathematically unpleasant. We met these terms before, in §3.15 and problem 3.23, but then we did little more than find the condition for them to be negligible.

(2) In the case of a Gaussian beam, we can include the $(x_1^2 + y_1^2)$ terms, and the integrations are still manageable. Substitute eqns (7.23) and (7.9) into eqn (7.22), and approximate $1/\rho \approx 1/h$. Show that $U(x_2, y_2, z_2)$ works out to be

$$U(x_2, y_2, z_2) = U_0 \exp(\mathrm{i}kz_2 - \mathrm{i}\omega t)\frac{\exp(\mathrm{i}kr_2^2/2q_2)}{q_2}, \qquad \text{where} \qquad q_2 = q_1 + h.$$

Thus, Fresnel diffraction of a Gaussian beam generates the expected new Gaussian beam in a plane downstream from the start.

$\left[\int_{-\infty}^{\infty} \exp(-\alpha x^2)\,\mathrm{d}x = \sqrt{\dfrac{\pi}{\alpha}}\right.$ even when α is complex provided $\left.\mathrm{Re}\,\alpha > 0.\right]$

Problem 7.10 (a) Direct proof that ray matrices transform q

(1) Let a Gaussian beam propagate through distance $h = z_2 - z_1$. Show from the definition of q in eqn (7.9) that $q(z_2) = q(z_1) + h$, and that this

can be cast into matrix form as

$$\begin{pmatrix} q/n \\ 1 \end{pmatrix}_{z_2} = (1) \times \begin{pmatrix} 1 & h/n \\ 0 & 1 \end{pmatrix} \begin{pmatrix} q/n \\ 1 \end{pmatrix}_{z_1},$$

where the square matrix describing the translation is the same as that in eqn (7.1).

(2) Two media are separated by a spherical boundary that crosses the z-axis at $z = z_0$. The surface has its centre on the z-axis at $z_0 + R_s$, and by the convention of §1.9.2 R_s is a positive quantity if the centre lies to the right of the surface. Show that the boundary surface has equation $z = z_0 + r^2/2R_s$ to the second order in r.

(3) Let an incoming wave have complex amplitude

$$U = U_1 \exp(\mathrm{i}k_1 z - \mathrm{i}\omega t)\, \frac{\exp(\mathrm{i}k_1 r^2/2q_1)}{q_1},$$

where this identifies the meanings of U_1, k_1 and q_1. Show that on the boundary surface of part (2) the incoming wave's complex amplitude has exponent $-\mathrm{i}\omega t + \mathrm{i}k_1(z_0 + r^2/2R_s) + \mathrm{i}k_1 r^2/2q_1$.

(4) The wave of part (3) is refracted at the spherical surface. Write down an expression for the amplitude of the refracted wave using U_2, k_2, q_2 to replace U_1, k_1, q_1 . Require the electric-field amplitude to be continuous on the boundary, so that the (complex) amplitudes match for all values of r. By equating the coefficients of r^2 show that

$$\frac{k_2}{R_s} + \frac{k_2}{q_2} = \frac{k_1}{R_s} + \frac{k_1}{q_1}, \qquad \text{so} \qquad \frac{n_2}{q_2} - \frac{n_1}{q_1} = -\frac{(n_2 - n_1)}{R_s} = -P,$$

where, in the last step, we recognize the power of the refracting surface from the definition of eqn (1.34).

(5) There is a strong similarity between the equations just written and eqn (1.33). However, there is a difference of sign, in that eqn (1.33) has $+P$ on the right. Sort out why.

(6) Rearrange the last equation into the matrix form

$$\begin{pmatrix} q_2/n_2 \\ 1 \end{pmatrix} = (\mathrm{cm}) \times \begin{pmatrix} 1 & 0 \\ -P & 1 \end{pmatrix} \begin{pmatrix} q_1/n_1 \\ 1 \end{pmatrix}, \tag{7.24}$$

and check that the square matrix is the same as that obtained in eqn (7.2).

(7) Show that $(U_2/q_2)\exp(\mathrm{i}k_2 z_0) = (U_1/q_1)\exp(\mathrm{i}k_1 z_0)$ by returning to part (4). It is, therefore, possible to obtain the strength U_2 of the refracted wave as well as finding its shape via q_2.

(8) Show that refraction at a plane surface leaves the value of q/n unchanged so that $q_2/n_2 = q_1/n_1$.

(9) Pursue an argument similar to that of §7.3 to show that two transformations (e.g. a translation followed by a refraction) are described by

$$\begin{pmatrix} q/n \\ 1 \end{pmatrix}_{\text{final}} = (\mathrm{cm}) \times \begin{pmatrix} A & B \\ C & D \end{pmatrix} \begin{pmatrix} q/n \\ 1 \end{pmatrix}_{\text{initial}},$$

where the $ABCD$ matrix is the product of the individual transformation matrices, combined in the same way as in eqn (7.3).

Problem 7.11 (a) Refraction of a Gaussian beam at a lens

(1) Confirm the calculations in example 7.1 as far as eqn (7.18).

(2) Rearrange eqn (7.18) to give

$$XY = \frac{f^2}{1 + b_1^2/X^2}.$$

(3) An alternative form of the thin-lens formula eqn (1.31) is the Newton formula $XY = f^2$ of eqn (1.37). Compare the equation just obtained with the Newton formula. It is different because of the focus shift encountered in problem 7.8: the beam waists do not lie at the same places as the ray-optics object and image. Discuss what happens when $b_1 \ll f$, and discuss the physical meaning of this limit.

(4) Show that for any b_1, $X = 0$ implies $Y = 0$. This *never* agrees with ray optics. Draw a diagram showing what happens for $X = 0$ and $b_1 \ll f$, and show that (in spite of the superficial appearance of the mathematics) the light waves do behave almost as we would expect from a ray-optics picture.

(5) The case discussed in part (4) is that drawn in Fig. 3.19(b). It was pointed out in problem 3.28(2) that there is an exact Fourier-transform relationship linking the fields on the lens's two focal planes. Is there a connection between the Fourier transformation and the waist–waist property?

(6) In part (1), you were encouraged to multiply three matrices to find the transformation from q_1 to q_2. Now use a quicker route. Multiply the first two matrices only and use them to find the value of q on the exit surface of the lens. Statement (7.15) tells us that we can find everything we need from the real and imaginary parts of this q. Use this to obtain the results of part (1) without doing the third matrix multiplication.

(7) Now use an even quicker route. Show that on the left-hand focal plane the beam has $q_1 = X - ib_1$ and on the right-hand focal plane it has $q_2 = -Y - ib_2$. Exploit the matrix worked out in problem 7.3(6) that carries us from focal plane to focal plane.

Problem 7.12 (c) Reflection of a Gaussian beam

(1) A Gaussian beam is reflected from a plane mirror at $z = z_1$ as shown in Fig. 7.11. Use methods similar to those of problem 7.10 to show that the boundary condition $U = 0$ can be satisfied if there is a reflected wave

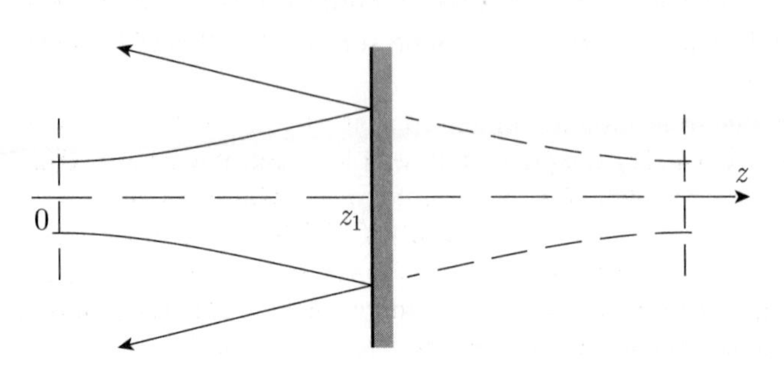

Fig. 7.11 A Gaussian beam reflected from a plane mirror. Problem 7.12 investigates this reflection.

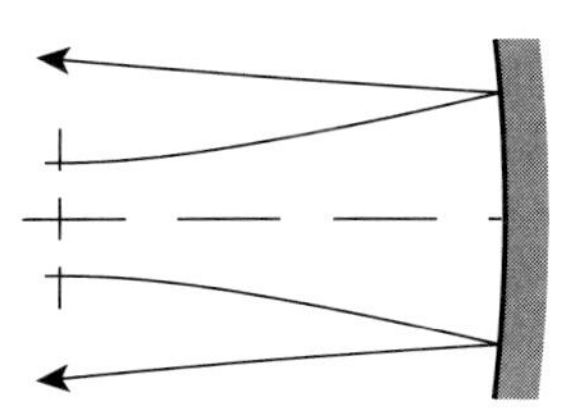
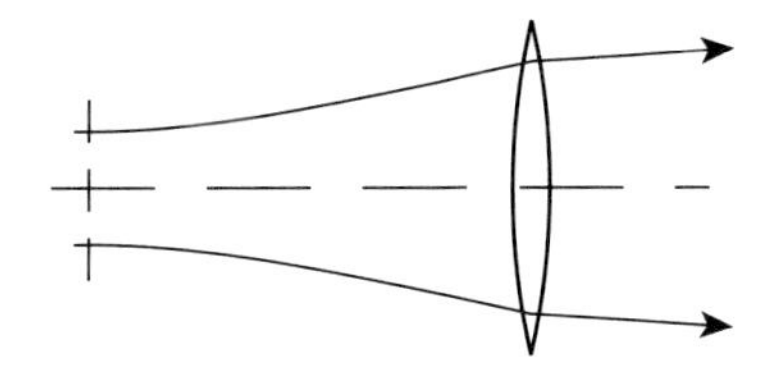

Fig. 7.12 A Gaussian beam reflected from a spherical mirror, and the lens refraction that is equivalent to it. The equivalence is shown in problem 7.12.

with $q_2/k_2 = q_1/k_1$; and since the reflection requires $k_2 = -k_1$ we must have $q_2 = -q_1$.

(2) Interpret the result $q_2 = -q_1$ to show that the reflected beam diverges from a waist at the mirror image of the original waist.[41]

(3) The reflected beam is given 'refractive index' $n_2 = -n_1$ after the fashion set in §7.4. Show that the $ABCD$ matrix describing the reflection is a unit matrix. Refer to sidenote 31 on p. 160, and make sure you can be comfortable with this.

(4) Adapt the discussion of problem 7.10 parts (3) and (4) to the reflection of a Gaussian beam at a spherically curved surface. Show that the $ABCD$ matrix has the form found in eqn (7.24) with $P = -2n_1/R_s$, consistent with $n_2 = -n_1$.

(5) Show[42] that reflection by a spherical mirror of power P is equivalent to refraction by a lens of the same power P, in that both are described by the same square matrix and both yield the same final value of q/n. Figure 7.12 illustrates two configurations that are equivalent by this result.

[41] Incidentally, it was the $q_2 = -q_1$ here that led us to define b as negative for a left-going wave, a step prepared for in eqn (7.12).

[42] *Comment*: Waves travelling towards $-z$ are always awkward. Part (5) establishes a general theorem that you can always 'unfold' a mirror problem into an equivalent lens problem, replacing each mirror of power P by a lens of power P. This opportunity will be exploited in the next chapter when we deal with an optical cavity bounded by a pair of mirrors.

Problem 7.13 (b) Longitudinal fields

(1) Take the U in expression (7.9) as giving the value of E_x, the x-component of the electric field in a Gaussian beam. Confirm the statements in §7.10 that E_z cannot be zero, and the same is true for B_z.

(2) On the same assumptions as those used in part (1), show that $E_z/E_x \sim \lambda/w$, which is small but is not zero. Show that this is the case even at $z = 0$ where the wave is plane; indeed λ/w is largest there because w is smallest.[43]

(3) [Harder] It could be objected that the arguments of this problem are not convincing, because they rely on an approximation: eqn (7.9) is founded on the approximation that λ/w must be small. Investigate this objection using the algebra set up in problem 7.7 and show that the estimate of E_z/E_x in part (2) is trustworthy to the order of accuracy of our expressions.

[43] *Comment*: In Fig. 7.5 we show a Gaussian beam and a representative one of its spherical wavefronts. We might expect for a 'transverse' wave that the $\boldsymbol{E}$-field must lie in the wavefront surface, perpendicular to the hyperbolic rays. A field component E_z is therefore not a surprise on the 'sloping' part of the wavefront: at points $(r > 0)$ away from the beam axis and $(z \neq z_{\text{waist}})$ away from the beam waist. But this is different from what we have found in part (2), where there is a nonzero E_z even at $z = z_{\text{waist}}$ where wavefronts are plane.

Problem 7.14 (c) Longitudinal fields and the Poynting vector

In §7.10 and problem 7.13 we found that a Gaussian beam must have 'longitudinal' components of $\boldsymbol{E}$ and $\boldsymbol{H}$. This should raise an awkward question: Doesn't that make the Poynting vector face in an unexpected direction, and isn't that a reason to reject the claimed longitudinal fields?

Investigate.

8 Optical cavities

8.1 Introduction

An optical cavity is an essential component in almost every laser. A pair of mirrors, such as is shown in Fig. 8.1, can enclose a steady distribution of electromagnetic field: an eigenfunction of Maxwell's equations in the presence of this boundary. The mirrors are said to form a 'cavity', a name taken from microwave practice, where a closed metal box acts as a cavity resonator—the high-frequency equivalent of a tuned circuit. A Fabry–Perot is in principle a cavity resonator, but other shapes are usually better for lasers. In a laser, a cavity mode (eigenfunction) is maintained in oscillation, and is the source of the output beam.

Optical cavities come in a wide variety of shapes, as one might expect given that lasers are a mature technology. Nevertheless, the principles can be exhibited by describing the simplest case: a cavity consisting of two mirrors facing each other. The cavities discussed in this chapter are all of this two-mirror type.

8.2 Gauss–Hermite beams

The electromagnetic waves that fit into an optical cavity are usually of Gauss–Hermite form,[1] so it will be convenient to introduce these wave shapes now.

The Gaussian beam, encountered in Chapter 7, is the simplest member of a whole family of approximate[2] solutions to the scalar wave equation. These Gauss–Hermite wave amplitudes have the following form:

$$
\begin{aligned}
U = U_0 \exp(\mathrm{i}kz - \mathrm{i}\omega t)\,\frac{\exp(\mathrm{i}kr^2/2q)}{q} \quad & \text{Gaussian-beam expression with } q = z - z_{\text{waist}} - \mathrm{i}b \\
\times\, \mathrm{H}_l\left(\sqrt{2}\frac{x}{w}\right) \times \mathrm{H}_m\left(\sqrt{2}\frac{y}{w}\right) \quad & \text{new factors incorporating the eigenvalues } l,\ m \\
\times\, \exp\{-\mathrm{i}(l+m)\alpha\} \quad & \text{increased phase shift}^3 \text{ with } \tan\alpha = z/b.
\end{aligned}
\tag{8.1}
$$

The first line here is identical to the Gaussian-beam expression of Chapter 7. The earlier interpretations remain: b is the confocal parameter; $w = w(z)$ is the spot size given by eqn (7.12); and $\tan\alpha = z/b$, as in eqn (7.13), defines the α appearing in the addition to the phase. The other factors are new. Functions H_l and H_m are the Hermite polynomials of order l and m. The phase term $\mathrm{e}^{-\mathrm{i}(l+m)\alpha}$ accompanies another

[1] *Comment*: It is striking that the functions in eqn (8.1) resemble closely the wave functions for a quantum-mechanical simple harmonic oscillator. It is natural to ask if there is something in the mathematics that makes these two pieces of physics have similar mathematical structure. There is, though it will not show up in the present chapter. The clue will appear in Chapter 14, especially problem 14.3.

[2] *Comment*: The function set of eqn (8.1) is approximate for the reasons and within the limits identified in problem 7.7. A brute-force confirmation that these functions satisfy the scalar wave equation is rehearsed in problem 8.1. A brute-force proof that they are cavity eigenfunctions (much harder) is worked in problem 8.15.

[3] The additional phase shift, and an associated slight departure of the phase velocity from ω/k, arises from the more rapid variation of U in the x- and y-directions imposed by the oscillating Hermite functions. The reason why this variation affects the phase is explained in sidenote 39 on p. 164.

$e^{-i\alpha}$ that is hiding in $1/q$, so the complete phase shift added to $\omega t - kz$ is $(l+m+1)\alpha$.

Each (l, m) wave of the form (8.1) is referred to as a **transverse mode** of the electromagnetic field. The (l, m) mode[4] is often called a 'TEM_{lm}' mode just as a Gaussian beam is called a TEM_{00} mode.

[4] *Comment*: The 'TEM' part of this notation is regrettable for reasons explained in §7.10 and problem 7.13.

Comment: Expression (8.1) was first obtained by solving the (scalar) wave equation for the field fitting into a mirror cavity (Boyd and Gordon 1961). However the Gauss–Hermite modes can of course be set up without any cavity—they are solutions of Maxwell's equations after all.[5] In fact, a wave of the precise form (8.1) is *not* an eigenfunction for a two-mirror cavity, because such an eigenfunction is a standing wave, a superposition of two waves travelling in opposite directions. We present a travelling wave in eqn (8.1) because is the simplest form to start from.

[5] The Gauss–Hermite modes are used in problem 8.2 to give an account of the 'phase anomaly', an additional phase shift undergone by a light wave—any light wave—when it passes through a focus. The Gauss–Hermite modes make possible a simple analytical treatment of a phenomenon that formerly required lengthy numerical investigation. See Born and Wolf (1999), §8.8.4.

A selection of Gauss–Hermite functions, and their squares, is plotted in Fig. 8.2. Photographs of observed laser-output intensity distributions are shown in Fig. 8.3, where the correspondence with the calculated intensities is clearly exhibited.

8.3 Cavity resonator

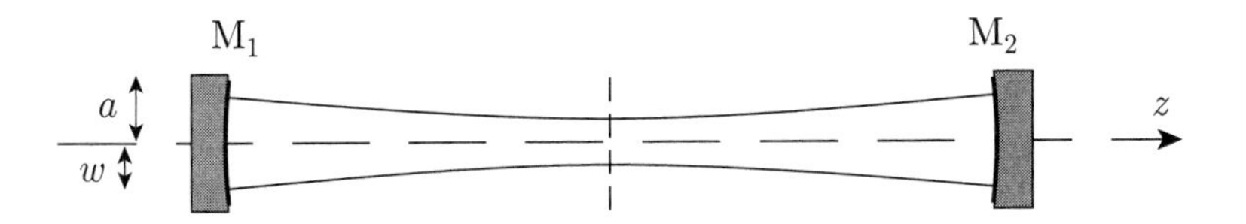

Fig. 8.1 A two-mirror optical cavity. The mirrors M_1, M_2 enclose a standing wave with light reflected back and forth between them. At least one mirror is made concave so as to refocus light that is spreading by diffraction.

An example of an optical cavity is shown in Fig. 8.1. Also shown is the outline of a light beam that fits itself between the mirrors: an eigenfunction of Maxwell's equations for the given boundary. The rays constituting the outline of the beam are curved for the same reason that rays are curved in Fig. 7.5: a beam of limited width spreads by diffraction.

Comment: Someone used to microwaves, or to solving Maxwell's equations in the presence of conducting boundaries, may feel uncomfortable at seeing a 'cavity' with open sides in Fig. 8.1: surely a 'cavity' is, in the first instance, a totally enclosed box? This unease can be countered as follows. Imagine a perfectly conducting oblate spheroid[6] whose axis of rotational symmetry is the z-axis. Such an enclosed volume must have genuine eigenfunction solutions to Maxwell's equations. Now it is believable that, for a cavity large compared to the wavelength, some of the eigenfunctions may have fields occupying only a small region around the z-axis, like that drawn in Fig. 8.1. So far as these eigenfunctions are concerned, their fields do not explore the space remote from the axis, so the conducting boundary there can be removed without affecting things. It will be no surprise to find that the Gaussian beam of Chapter 7 is one of these near-axis eigenfunctions.[7]

[6] An oblate spheroid has the squashed-orange shape (like the Earth, not like a rugby ball), generated mathematically by rotating an ellipse about its minor axis. The reason for requiring the spheroid to be oblate will appear in §8.5.

[7] The spheroidal cavity introduced here can be taken more seriously. The scalar wave equation can be solved inside a spheroidal conducting boundary by using elliptical–hyperboloidal coordinates. The Gauss–Hermite functions of eqn (8.1) can be obtained from these solutions by taking the limit where the cavity is large compared to a wavelength.

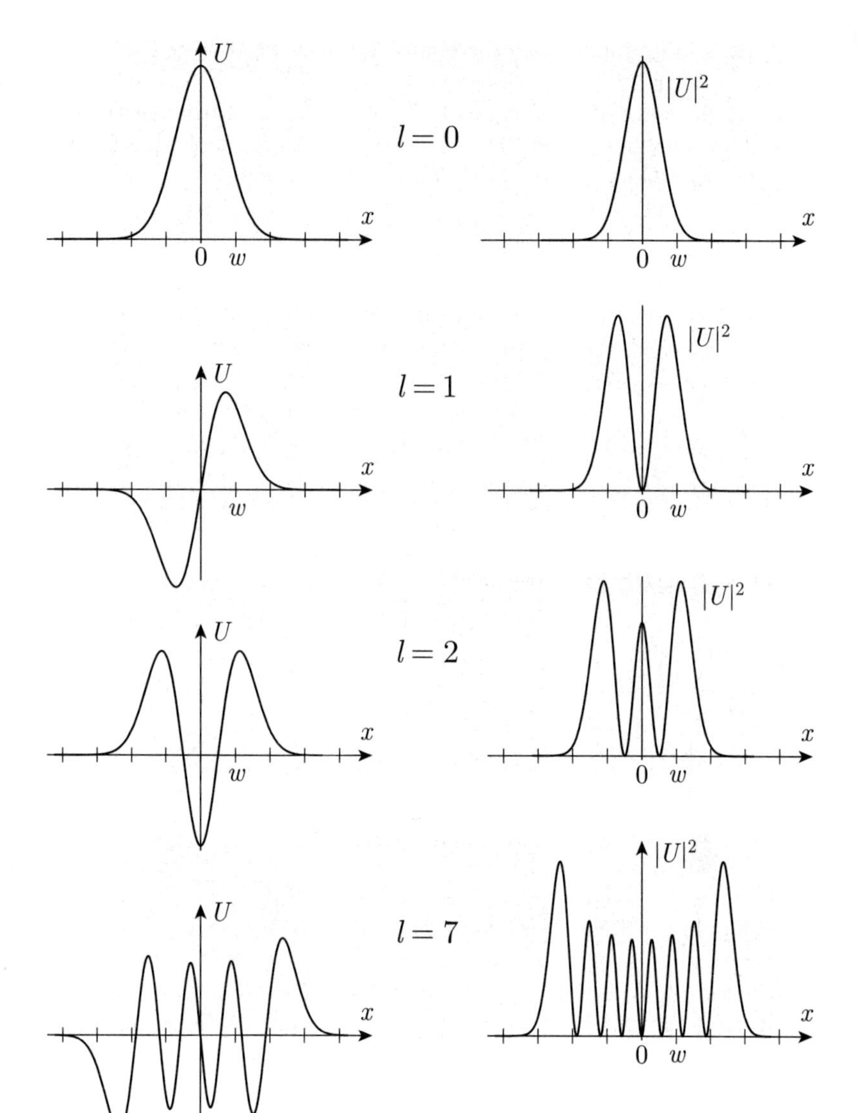

Fig. 8.2 A selection of Gauss–Hermite functions: the real function $\exp(-x^2/w^2)\mathrm{H}_l(\sqrt{2}x/w)$ and its square. The x-dependent part of U in eqn (8.1) contains also the phase factor $\exp(\mathrm{i}kx^2/2R)$; this distraction may be avoided if we think of the amplitudes plotted as being those on a curved surface of radius $R = (b^2 + z^2)/z$.

Gauss–Hermite functions will be familiar from the quantum mechanics of a harmonic oscillator. We draw attention to the behaviour of the amplitude for large l; it has high peaks at its ends and is displayed in books on quantum mechanics to exhibit the approach to a classical limit—the pendulum spends most time at the ends of its travel because it has stopped moving there. The same high peaks are beautifully displayed in Fig. 8.3.

8.4 Cavity modes

The expressions given in eqn (8.1) are just what we need to describe an electromagnetic field enclosed in a mirror cavity. The physical reason can be seen in the factor $\exp(\mathrm{i}kr^2/2R)$ in eqn (7.14), which gives a uniform phase (apart from sign changes in the Hermite functions) on a spherical surface of radius R; the spherical surface is a wavefront. A spherical mirror, with radius chosen to fit the wavefront, returns the beam along its previous path.

Even if we had not already seen Figs. 8.2 and 8.3, experience elsewhere with solving differential equations subject to boundary conditions should lead us to expect eigenfunctions and for them to have certain properties.

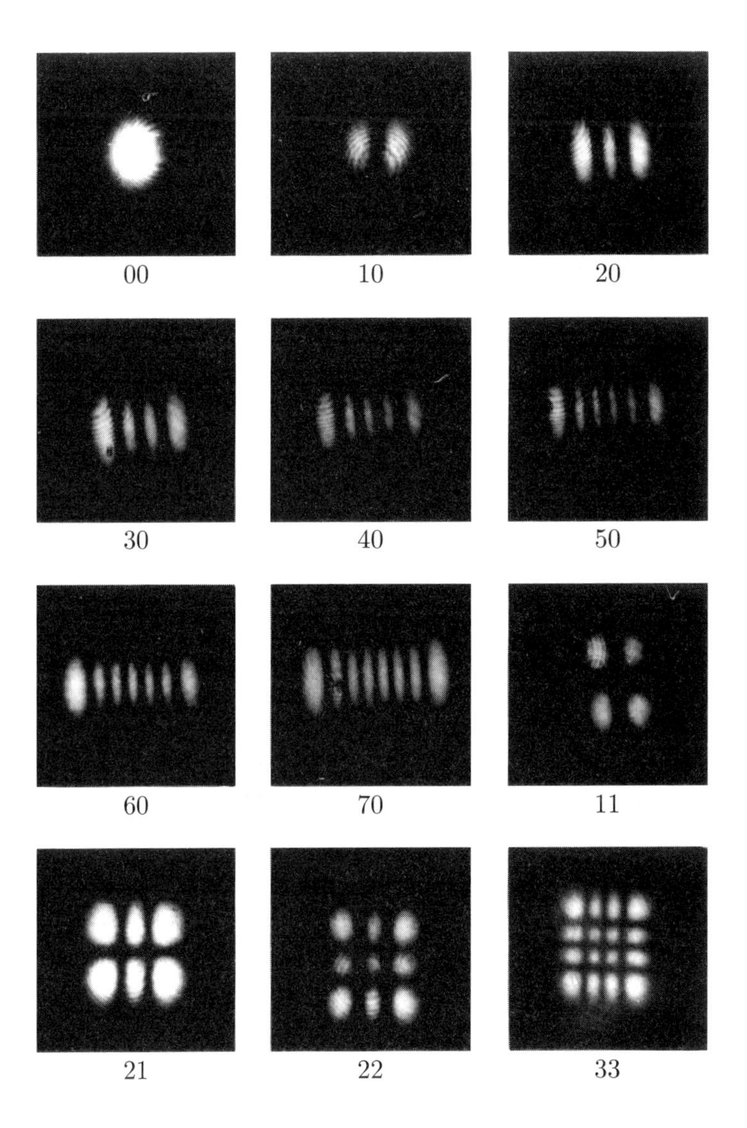

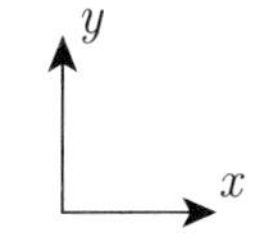

Fig. 8.3 Photographs of the intensity distributions of some of the transverse modes given by eqn (8.1). The reader should check that l is the number of nodal lines parallel to the y-axis while m is the number of nodal lines parallel to x-axis. Axes run in the directions shown: x horizontally and y vertically. Reproduced with permission from Kogelnik and Rigrod (1962), ©1962 IEEE.

In particular, each eigenfunction passes through zero at its **nodes** and the eigenvalue is the number of those nodes (or it can be so defined). It is, therefore, no surprise to learn that $\mathrm{H}_l(\sqrt{2}x/w)$ passes through zero l times. For any z an amplitude U of the form (8.1) has l **nodal lines** parallel to the y axis and m nodal lines parallel to the x axis.[8] This behaviour is clearly displayed in Fig. 8.3.

The mirrors of a cavity impose a boundary condition: $U = 0$ if the mirrors are perfectly conducting.[9] This means that two oppositely travelling waves of the form (8.1) must be superposed to form a standing wave. To achieve $U = 0$ on both mirrors, the value of $|k|$ must take one of a discrete set of values: there are **longitudinal modes** to accompany the transverse modes already encountered. The implications of longitudinal modes for the optical frequency are explored in §8.7.

[8] In three dimensions we have **nodal surfaces**. These surfaces are not plane because w varies with z.

[9] Cavity mirrors are usually made from multi-layer dielectric coatings as described in Chapter 6. High reflectivity is achieved by making the input impedance either very small ('perfectly conducting') or very large compared to Z_0. The discussion here would not be much changed if the mirrors had a very high input impedance, though for reasons given in sidenote 34 on p. 145 this is in any event not likely.

The standing wave between the cavity mirrors implies the existence of a third set of nodal surfaces (curved), p of them, spaced along the z-axis. A cavity mode is, therefore, a mathematical function (of x, y, z) described by three integers (eigenvalues): l, m and p, where l and m are the same integers as in eqn (8.1). There is a similarity to other eigenfunction problems, such as the wave functions for an electron in a hydrogen atom that likewise have three eigenvalues: n, l and m_l.

Laser physicists often talk of transverse and longitudinal modes as if they were different kinds of thing, which seems odd at first. However, in context it makes sense. Often we have a laser whose output beam has definite values of l and m and we are not much interested in whether it is a superposition of modes having more than one value of p. In such a case we simply say we have an (l, m) transverse mode. If attention is then turned to whether the laser's output contains more than one frequency, we say it contains (or not) several longitudinal modes: modes with more than one value of p, probably all having the same common values of l and m.

Finally, we mention that standing-wave functions based on eqn (8.1) are not the only possible eigenfunction set for a mirror cavity. Given axial symmetry about the z-axis, wave solutions must exist that are functions of cylindrical coordinates r, ϕ, z; the existence of an x, y, z set as well is almost an accident (the reason appears in problem 8.15). Both sets are complete: linear combinations of the x, y transverse modes can be built into the form (function of r, z) $\times$ (function of ϕ), and of course conversely.[10] However, the r, ϕ, z functions involve Laguerre functions of r, even less user-friendly than Hermite functions, so we shall follow custom by preferring the eigenfunction set displayed in eqn (8.1). For the record, the Gauss–Laguerre eigenfunctions have the form

$$U = U_0\, \mathrm{e}^{\mathrm{i}(kz-\omega t)}\, \frac{\mathrm{e}^{\mathrm{i}kr^2/2q}}{q}\, \mathrm{e}^{-\mathrm{i}(2n+s)\alpha} \left(\sqrt{2}\frac{r}{w}\right)^s \mathrm{L}_n^{(s)}(2r^2/w^2)\, \mathrm{e}^{\pm \mathrm{i}s\phi}, \quad (8.2)$$

where n and s are integers (though this n is unconnected with refractive index!).[11,12]

[10] A clue to the structure of the linear combinations is to be seen in the exponents $\mathrm{e}^{-\mathrm{i}(l+m)\alpha}$ and $\mathrm{e}^{-\mathrm{i}(2n+s)\alpha}$. If Laguerre solutions are expanded in terms of Hermite solutions (or vice versa), only functions with $(l+m) = (2n+s)$ can appear in the sum.

[11] The notation used for Laguerre polynomials in the literature is somewhat variable. We have followed that of Abramowitz and Stegun (1965). Integers $n \geq 0$ and $s \geq 0$ are independent of each other. In case it helps, we may mention that in the same notation, hydrogen-atom wave functions contain $\mathrm{L}_{n-l-1}^{(2l+1)}(\rho)$ where ρ is a scaled radius, and the n, l quantum numbers are defined in the usual way.

[12] Beautiful photographs of Gauss–Laguerre modes, with a description of their preparation, are given by Padgett *et al.* (1996).

8.5 The condition for a low-loss mode

Only certain mirror geometries permit the electromagnetic field to have the form given in eqn (8.1). This will be no surprise since sidenote 6 on p. 171 mentioned that the continuation of the mirrors into an enclosed surface was to be an oblate spheroid (not just any spheroid).[13] We'll work out the condition for a Gauss–Hermite beam to reproduce itself after a round trip through a mirror cavity. It may make things concrete if we think mainly about a Gaussian beam ($l = m = 0$), though in fact the reasoning applies to any of the Gauss–Hermite modes; see problem 8.4(2).

Although the reason for the terminology will not yet be apparent, we shall say that a beam that reproduces itself constitutes a **low-loss mode**

[13] Problem 8.10 asks you to think about a closed prolate cavity.

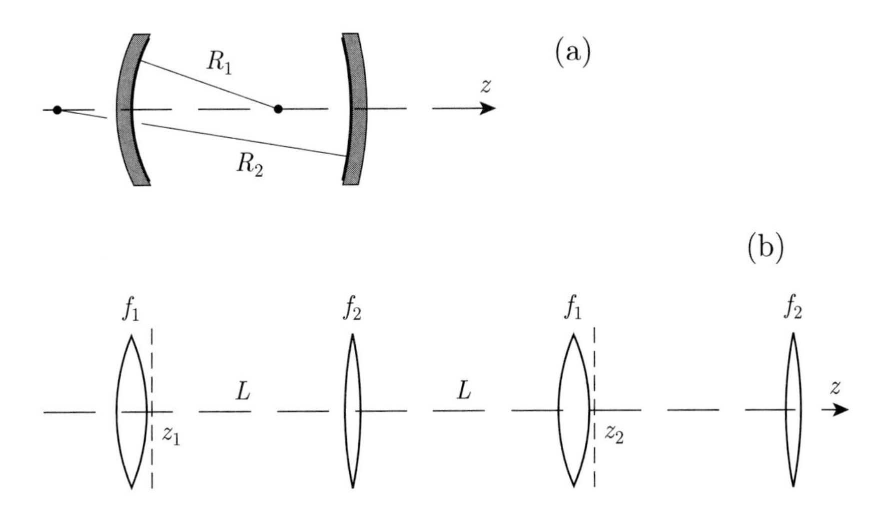

Fig. 8.4 (a) Two mirrors form a cavity that has axial symmetry about the z axis. The mirrors have radii R_1, R_2, which are positive if the mirrors are 'concave towards the middle' as shown. (The cavity drawn is 'unstable'.) (b) For ease of calculation, the mirrors are replaced by a chain of lenses, each having power identical to that of the associated mirror. Then the light travels always towards $+z$ and some of the sign complications of §7.8 are avoided.

for the cavity. In low-gain lasers, such as the familiar He–Ne radiating 633 nm, the laser works only if a low-loss cavity mode is available to receive energy from the population inversion.

The cavity to be discussed is shown in Fig. 8.4(a). Since left-going waves require a lot of irritating attention to signs, we shall replace the mirrors by lenses of equivalent power, and then the optical system is 'opened out' into the lens chain shown in Fig. 8.4(b). The light beam now travels towards $+z$ only. Requiring the beam to reproduce itself after a round trip between the mirrors is equivalent to requiring the field at z_2 to reproduce that at z_1 (or similarly for any other pair of surfaces separated by $2L$).

It should be obvious (if not check problem 1.14 or problem 7.4) that the lenses have focal lengths $f_1 = R_1/2$ and $f_2 = R_2/2$. The radii and focal lengths are positive if the lenses and mirrors are 'converging', which is the case drawn in Fig. 8.4; mirrors curved the other way would have negative radii and negative focal lengths and would be 'diverging'.[14] The optical system will be assumed to be in air with refractive index $n = 1$; and since all waves now travel to the right $n = +1$ always and it can be omitted from matrices like those in eqn (7.16).

For reasons explained in §7.9, everything we need to know about the profile of the beam at z_2 is encapsulated in the value q_2 of the beam's complex radius.[15] Therefore, the condition for the beam to reproduce its original shape after travelling through one period of the lens chain is simply $q(z_2) = q(z_1)$. The matrix methods developed in Chapter 7 are ideal for investigating this. A beam travelling from z_1 to z_2: travels through distance L; is refracted by a lens of power P_2; travels through

[14] The sign assignments for focal lengths and powers are the same as always. However, it should be noted that this is *not* the case for the mirror radii, whose signs are allocated according to a different rule from that of §1.9.2. Moreover, this rule is different again from that which attached a sign to a Gaussian beam's wavefront radius in eqn (7.11)—we are using a *third* convention, effectively 'real is positive'! It would be possible to impose uniformity here, but only at the cost of making too great a break with well established custom. The reader is cautioned to be aware at all times of the sign convention being used.

[15] We are not yet requiring the waves at z_2 and z_1 to match in phase. That would identify a longitudinal mode, and we defer detailed discussion of longitudinal modes until §8.7.

distance L; and finally is refracted by a lens of power P_1. Therefore,

$$\begin{pmatrix} q(z_2) \\ 1 \end{pmatrix} = (\text{cm}) \begin{pmatrix} A & B \\ C & D \end{pmatrix} \begin{pmatrix} q(z_1) \\ 1 \end{pmatrix}$$
$$= (\text{cm}) \begin{pmatrix} 1 & 0 \\ -1/f_1 & 1 \end{pmatrix} \begin{pmatrix} 1 & L \\ 0 & 1 \end{pmatrix} \begin{pmatrix} 1 & 0 \\ -1/f_2 & 1 \end{pmatrix} \begin{pmatrix} 1 & L \\ 0 & 1 \end{pmatrix} \begin{pmatrix} q(z_1) \\ 1 \end{pmatrix},$$

where (cm) stands, as earlier, for the 'cancel me' factor to be divided away as explained in §§7.5 and 7.9.

The beam is required to have $q(z_2)$ identical to $q(z_1)$:

$$q(z_2) = q(z_1) = q = \frac{Aq + B}{Cq + D}.$$

This gives a quadratic equation for q whose solution is

$$2Cq = (A - D) \pm \sqrt{(A + D)^2 - 4(AD - BC)}.$$

Our matrices have all been constructed to be unitary, so we know even without working out the brute-force algebra that $(AD - BC) = 1$. So

$$2Cq = (A - D) \pm \sqrt{(A + D)^2 - 4}. \tag{8.3}$$

A further condition must be imposed: the wave that is propagating along the lens chain must be a Gaussian beam (q complex) and not a spherical wave (q real). We must, therefore, require the square root in eqn (8.3) to be imaginary: $(A + D)^2 < 4$. Working out the matrix product and tidying up yields

$$0 < (1 - L/R_1)(1 - L/R_2) < 1. \tag{8.4}$$

This is the condition for a Gauss–Hermite beam to reproduce itself after travelling through one period of the lens chain; equivalently, it is the condition for the cavity to have Gauss–Hermite beams as its eigenfunctions.[16] A check on the above calculation is made in problems 8.4 and 8.5. Down-to-earth 'reasons why' are revealed in problem 8.3.

The implications of eqn (8.4) are most clearly brought out by plotting the chart of Fig. 8.5. If the values of L/R_1, L/R_2 identify a point in the white area of the chart, condition (8.4) is met, and a low-loss mode exists; and if not not. Problem 8.7 provides practice.

[16] Problem 8.6 looks at the physics in a completely different way, by asking what happens to rays propagating along the lens chain. The same 'stability' condition is obtained, though it is not very obvious that this must happen.

Comment on terminology: A cavity meeting the low-loss condition is commonly said to be 'stable'; one not meeting it is 'unstable'. The condition expressed by eqn (8.4) is then the 'stability condition' for the cavity. This terminology arose in the early days of lasers before the properties of optical cavities were properly understood, and has persisted. It is one of the most unfortunate pieces of jargon in a subject—I mean physics, not just optics—that is already oversupplied with inappropriate terms.[17] The 'stability' of a cavity has nothing in common with the ordinary meaning of 'stable': a pencil balanced on its point is unstable; a pencil lying on its side is stable. If words could be taken at their face

[17] An 'atom', if we believed its name, would not be cuttable into smaller pieces; an 'imaginary' number would be no more than a figment of the imagination; 'oxygen' would be the defining constituent of an acid ....

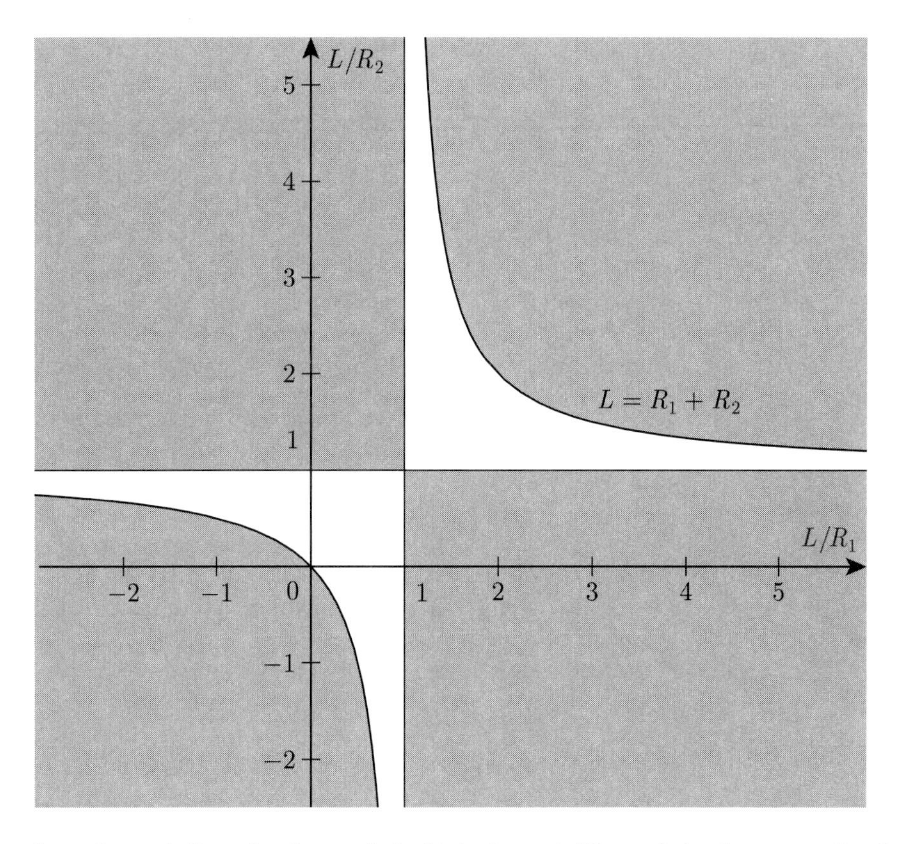

Fig. 8.5 A chart that tells us whether a low-loss mode is obtained or not. The cavity's mirror separation L and mirror radii R_1, R_2 together define a point on the chart. If that point lies in the white area the cavity supports a low-loss mode; if it lies in the shaded area there is no low-loss mode.

value, we might reasonably understand that an 'unstable' cavity is capable of loss-free operation but is excessively sensitive to perturbations. It is neither.

Up to this point, I have avoided 'stable' and its relatives and have talked of a 'low-loss mode' instead. However, this attempt at correctness has led to convoluted sentences, and in future I shall use 'stable' and 'stability' where avoidance would be unduly clumsy.

8.6 Finding the mode shape for a cavity

Equation (8.3) tells us that in plane $z = z_1$,

$$(z - z_{\text{waist}}) - \mathrm{i}b = q = \frac{(A - D)}{2C} - \mathrm{i}\frac{\sqrt{4 - (A + D)^2}}{2|C|}. \tag{8.5}$$

In eqn (8.5) the imaginary part of q has been given a negative sign[18] since we'll choose to discuss a beam travelling[19] towards $+z$. As was emphasized in §7.9, q contains within itself all the information needed for fully describing the shape of the beam. The real part gives $(z_1 - z_{\text{waist}})$, so

[18] Guaranteed by writing $|C|$ in the denominator.

[19] The mode travelling between mirrors is a standing wave: a superposition of two Gauss–Hermite beams travelling in opposite directions. Since the two travelling waves have the same shape—they differ only in the sign of b—we simplify the discussion by talking explicitly about only one of them.

z_{waist} can be found since we know where z_1 is. The imaginary part gives the confocal parameter b. Equations (7.11)–(7.13) can now be used to find the radius of curvature R of the wavefront at z_1, the spot size at z_1, the spot size at the beam's waist, and the additional phase α. The only things not given by eqn (8.5) are: the transverse-mode numbers l, m, which must be known separately; the longitudinal-mode eigenvalue p which is the subject of a separate calculation; and the overall coefficient U_0.

Although eqn (8.5) is available to us for finding full information about a cavity mode, its use is something of a last resort. Often one of the mirrors of a cavity is plane, so the beam's waist must be at that mirror. And there can be other short cuts, one of which can be found in problem 8.8.

8.7 Longitudinal modes

When a beam is enclosed in a mirror cavity of length L, as opposed to travelling along a lens train, it must match itself in phase after a round trip. This means *roughly* that $2L$ = an integer number of wavelengths. However, this statement is not exactly correct, and the correction is surprisingly important in the operation of lasers. The correction comes from the α in eqns (7.13) and (8.1).

Let the mirrors at the two ends of a cavity lie at z_1 and z_2 (both relative to an origin at the beam waist as in Fig. 7.5). Then the condition for the wave to match itself in phase after a round trip is

$$p\pi = kL - (l + m + 1)\{\alpha(z_2) - \alpha(z_1)\}, \tag{8.6}$$

where p is an integer. This condition sets the precise value of k, and from there the precise value[20] of the frequency ν because $\nu = \omega/2\pi = ck/2\pi$. Problem 8.9 asks you to derive eqn (8.6) and investigate some special cases.

[20] For simplicity, the refractive index of the medium between the mirrors has been taken to be 1 here, so the speed of the wave is c.

8.8 High-loss cavities

The insistence on 'low loss' as the justification for the eigenvalue condition (8.4) may seem odd. All the mathematics, and even the pictorial argument of problem 8.3, suggests that we are finding the condition for a mode to exist at all. It is indeed the case that an 'unstable' cavity does not have eigenfunction solutions, but this mathematical statement doesn't quite represent the physics.[21]

[21] Problem 8.10 invites thought about the reason for there to be no low-loss mode when the cavity fails to meet condition (8.4).

First, why should 'loss' be an issue at all? This goes back to the discussion of open-sided cavities in the comment in §8.3. For a 'stable' cavity, energy loss through the open sides is negligible so long as the mirrors are large enough compared to the beam's spot size w. The spot size is determined by two effects in competition: diffraction; and the focusing effect of the mirrors. It is the success of the focusing, in keeping the field away from the edges of the cavity, that we encapsulate

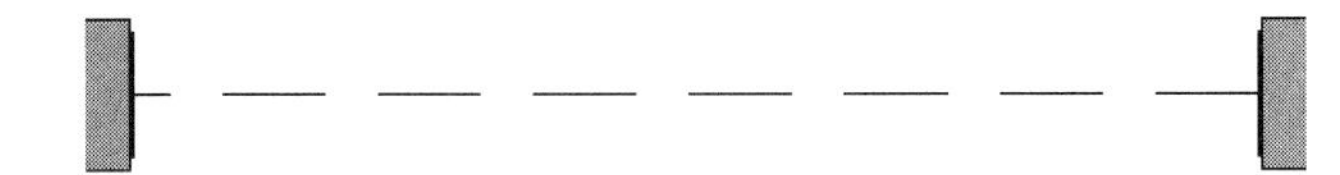

Fig. 8.6 A cavity made from two plane-parallel mirrors.

in 'low-loss'. If now we make the mirrors smaller,[22] only a few w across, some light energy will pass outside the outline of a mirror, an effect known as **diffraction loss**. As long as it is not too severe, diffraction loss does no more than add to other small losses of energy (finite reflectivity of the mirrors, for example). In particular, the outline of the mirror does not exert any control over the beam's spot size.

[22] Or, more conveniently, reduce the effective size of an aperture placed in front of an unchanged mirror.

Rather different physics applies to an 'unstable' cavity such as that drawn in Fig. 8.6. The earliest laser cavities were made like this with plane mirrors, based on the idea of a Fabry–Perot.[23] A remarkable set of computer simulations (Fox and Li 1961) gives a good idea of the physics. Imagine a wave amplitude, arbitrarily chosen, to exist on one mirror. A Kirchhoff diffraction integral is used to calculate the field to which it diffracts on the second mirror. Some part of the diffracted field misses the second mirror and undergoes diffraction loss. What survives is used as the source for a second diffraction back to the first mirror where it is dealt with similarly. The process is repeated many times, until the field amplitude has settled to a steady configuration. It is easy to imagine that just such a process takes place inside a laser when it is first switched on.

[23] 'Unstable' cavities like that drawn in Fig. 8.6 have not become obsolete. They can have advantages with high-gain lasers, where the gain overcomes the loss. The cavity can be designed so that all modes except one are suppressed, and the form of the output beam is thereby controlled.

The outcome of the repeated diffractions is that the light forms itself into a **minimum-loss configuration**, where most but not quite all of

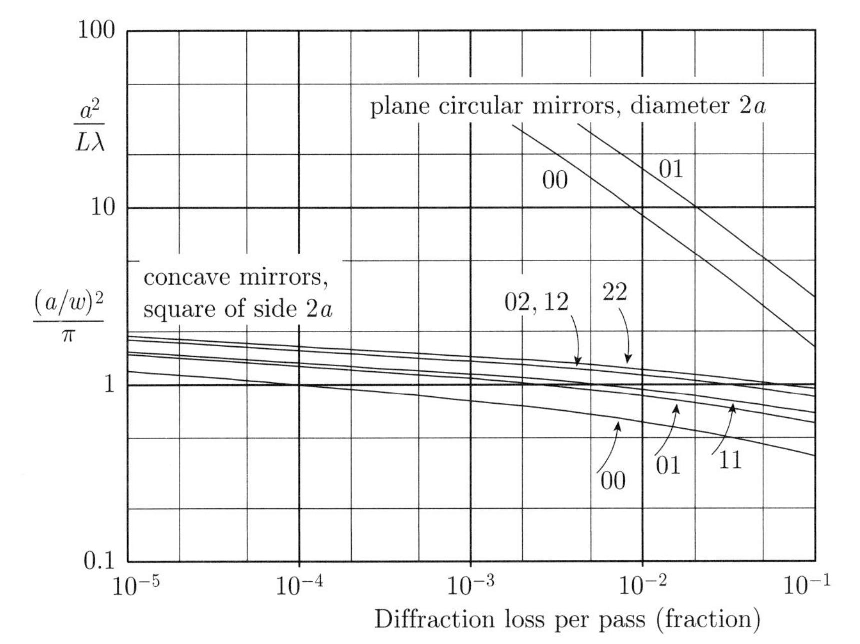

Fig. 8.7 The diffraction loss for two minimum-loss configurations of an 'unstable' cavity with parallel plane circular mirrors (effective diameter $2a$, separation L), and for several cavity eigenfunctions of a 'stable' cavity made from two concave mirrors, effectively square with side $2a$ and separation L. For the square concave mirrors the mode labels are the l, m (either order) of eqn (8.1). For circular plane mirrors, the modes should be classified by integers (n, s) similar to those in eqn (8.2), since the circular boundary has axial symmetry. However, for the two lowest modes there is no difference between $(l, m) = (0, 0)$ and $(n, s) = (0, 0)$, or between $(l, m) = (0, 1)$ and $(n, s) = (0, 1)$. The plane-mirror loss is determined by $a^2/(L\lambda)$. The concave-mirror loss is determined by $(1/\pi)(a/w)^2$ (where w is the spot size at a mirror), which is the same as $a^2/(L\lambda)$ for the special case of confocal mirrors. After Boyd and Gordon (1961).

the energy lies within the reflecting area of each mirror. A minimum-loss configuration is different from one of the eigenfunctions discussed previously in an important respect: its shape is wholly dependent on the shape and size of the mirror's boundary. If we enlarge the mirrors, the field distribution broadens to occupy the increased area, and the diffraction loss is thereby reduced.

It is the existence of minimum-loss configurations that answers the anxiety raised at the beginning of this section. A mode of sorts exists for any cavity, not just a 'stable' cavity. So it was necessary to define the 'stability' condition with reference to the kind of mode sought, rather than the existence of any mode at all.

Figure 8.7 shows the diffraction loss per pass through two kinds of cavity: an 'unstable' one with parallel plane mirrors; and a 'stable' cavity of any shape.[24] The diffraction loss is very much higher for the 'unstable' cavity, because the beam is not being focused by the mirrors to keep it away from the mirror boundaries.[25] We look again at these curves in §8.12.

[24] In the original publications all curves are labelled with ordinate $a^2/(L\lambda)$. This is because the authors were thinking mainly about a symmetrical confocal cavity when considering the stable-cavity case. For stable cavities we have relabelled the ordinate with the expression $(1/\pi)(a/w)^2$ which applies more generally. Problem 8.8(6) shows that the two expressions agree for a symmetrical confocal cavity.

[25] Diffraction loss is always discussed as if light, having spread by diffraction, fell outside the outline of a small mirror. In practice, diffraction loss is controlled (deliberately as shown in §8.12) by means of some aperture elsewhere along the optical path in the cavity. The actual mirror is made of a size that is convenient to handle, say 15–25 mm diameter even though the laser beam may have a spot size of 1 mm or less.

8.9 The symmetrical confocal cavity

'Confocal' means that the two concave mirrors have their principal foci at a common point inside the cavity. When the mirrors are identical, the foci both lie at the cavity centre, and each mirror has its centre at a point on the other.

We have encountered the symmetrical confocal geometry previously: Fig. 3.19(a) on p. 70. Problem 3.28 shows that the fields on the two surfaces are Fourier transforms of each other. A harder look is taken in problem 8.12.

The resonance frequencies of a symmetrical confocal cavity obey

$$\left.\begin{aligned} (l+m)\ \text{even}: &\quad \nu = (c/2L)(\text{integer} + \tfrac{1}{2}), \\ (l+m)\ \text{odd}: &\quad \nu = (c/2L)(\text{integer}). \end{aligned}\right\} \tag{8.7}$$

These equations[26] exhibit the very high degeneracy of the longitudinal modes of a symmetrical confocal cavity; all modes for which $(l+m)$ is even have a common 'comb'[27] of frequencies, and all modes with $(l+m)$ odd have another comb interlacing the first.[28]

A Gaussian beam with $b = L/2$ can fit into a symmetrical confocal cavity of length L, hence the name 'confocal parameter' for b. But problem 8.11 shows that many other configurations are possible, a manifestation of the degeneracy just mentioned.

The symmetrical confocal cavity is usually avoided in laser design. One reason lies in the degeneracy of the longitudinal modes which has undesirable effects: When several modes have the same frequency, so does any linear combination of them. The light adopts a linear combination, either chosen at random or favoured because it can avoid loss or grab more amplification. So the output beam has unpredictable form.

[26] Derived in problem 8.9.

[27] Think of the appearance of a plot against frequency in which each mode is indicated by an upward-facing tick on the horizontal axis.

[28] *Comment*: The added half-integer in eqns (8.7) may look unexpected, somehow 'the wrong way round'. Nevertheless it is right. It is the consequence of the $e^{-i\alpha}$ in eqn (7.14) and the $e^{-i(l+m)\alpha}$ in eqn (8.1). Problem 8.12 shows that these phase shifts are also related to the factor i in the $1/i\lambda$ coefficient of a Kirchhoff integral.

Another reason can be seen from Fig. 8.5. The symmetrical confocal configuration has $L/R_1 = L/R_2 = 1$, right at the 'corner' where stable and unstable regions meet. If the two mirrors do not have exactly equal radii of curvature, the cavity will be 'unstable', and can be made to work correctly only by adjusting L a little away[29] from the confocal setting.

However, everything that makes the symmetrical confocal cavity undesirable for use as a laser cavity gives it fortunate properties for use as a Fabry–Perot, as discussed in the next section.

[29] To below the smaller of R_1, R_2 or above the larger.

8.10 The confocal Fabry–Perot

The confocal Fabry–Perot is a symmetrical confocal cavity, the same layout as is described above in §8.9.

Any symmetrical field distribution $E_s(x, y) = +E_s(-x, -y)$ (for constant z) can be expanded using transverse modes with $(l + m)$ even only, and has resonance condition given by eqn (8.7): $2L = (p + \frac{1}{2})\lambda$. Any antisymmetric field distribution $E_a(x, y) = -E_a(-x, -y)$ consists of modes with $(l+m)$ odd and has resonance condition $2L = p\lambda$. A general input wave is likely to excite both E_s and E_a fields, so the cavity transmits energy, like an ordinary Fabry–Perot, when[30]

$$\nu = (\text{integer}) \times (c/4L). \tag{8.8}$$

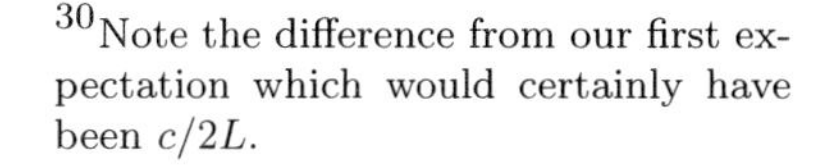

[30] Note the difference from our first expectation which would certainly have been $c/2L$.

These resonance conditions hold for light entering through any part of the left-hand mirror and at any reasonable angle. There is no ring fringe pattern, only light and dark. We don't need to throw away energy by putting a tiny pinhole at a lens focus, nor do we even need a fringe-forming lens. The confocal Fabry–Perot is often a great improvement on the plane-parallel one, both in the amount of light it can transmit and in ease of setting up.[31]

[31] Problem 8.13(2) investigates mirror alignment.

All this, of course, carries some reservations. Look at problem 8.12 and eqn (8.10). All calculations on optical paths between the mirrors have been carried through to order r^2/L^2 only, and they are not reliable in the next order r^4/L^4. Therefore, the resonance condition (8.8) is good only so long as we limit the incident light beam to positions (x, y) and to directions that keep the r^4/L^4 terms $\ll 1$. This is still a big improvement on the plane-parallel Fabry–Perot, where you have to select angles θ (if you're centre-spot scanning) to keep θ^2 terms $\ll 1$.

The confocal Fabry–Perot finds its main application in the 'optical

Fig. 8.8 A confocal Fabry–Perot. A blob drawn on each mirror indicates the location of the other mirror's centre of curvature. Light is shown entering in a nothing-special way, as a reminder that light entering and leaving does not need to be prepared or selected.

spectrum analyser'. One mirror of the cavity is scanned through a wavelength or so (little enough that the confocal condition $L = R$ is negligibly disturbed) with the aid of a piezoelectric mount. We record, versus scanning voltage, the optical power transmitted to a photodetector. See problem 8.13 for the advantages and disadvantages of the device.

8.11 Choice of cavity geometry for a laser

One to avoid: the symmetrical confocal cavity. This cavity is tempting because it is widely discussed in textbooks[32] and seems to permit simple analysis.[33]

[32]The symmetrical confocal cavity has a weight of history behind it, because it was the first configuration to be analysed, both analytically (Boyd and Gordon 1961) and numerically (Fox and Li 1961).

[33]Problems 8.11 and 8.15 show that this simplicity is illusory; to permit tractable analysis a cavity must *not* be symmetrical confocal.

One to think of first: the near-hemispherical cavity. One mirror is plane and one spherical-concave. The spherical mirror has its centre outside the cavity, just beyond the plane mirror, as shown in Fig. 8.9. The beam fitting into the cavity has its waist at the plane mirror, so it is widest at the spherical mirror. The spot size at the spherical mirror can be adjusted to fit the amplifying medium by making small adjustments to $(R - L)$, as is discussed in problem 8.14.

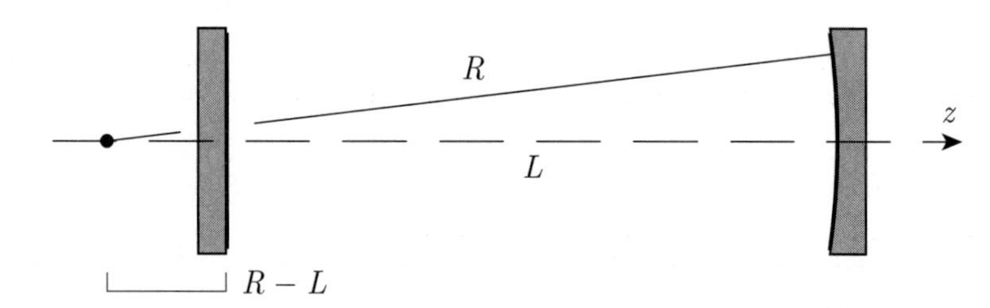

Fig. 8.9 A near-hemispherical optical cavity. One mirror is plane, and the other is spherical. The spherical mirror has its centre of curvature just beyond the plane mirror.

8.12 Selection of a desired transverse mode

Usually a laser is intended to oscillate, and give an output, in a transverse mode with $(l, m) = (0, 0)$, so we wish to suppress other modes. How? The clue is in the horizontal scale of the graphs in Fig. 8.2 and the photographs of Fig. 8.3; the proportions are realistic: modes with high (l, m) are larger—they extend further from the axis—than the $(0, 0)$ mode. Therefore, an aperture of suitable radius can intercept a large fraction of the energy in high-order modes while not being noticed by low-order modes. Just how well this works can be seen from Fig. 8.7: choose an ordinate $(1/\pi)(a/w)^2$ that gives a loss per pass of 0.5% for a $(0, 0)$ mode; it gives a loss per pass of at least 5% for all others.[34]

[34]Use the curves in Fig. 8.7 for a 'stable' cavity having concave mirrors. The curves give the diffraction loss incurred at one encounter with a mirror (or limiting aperture placed elsewhere). If the cavity is not symmetrical, having different limiting apertures at the two ends, different losses may occur on each transit.

The diffraction loss has such a sensitive dependence upon a/w that a desired diffraction loss is always achieved by adjustment, not by reading values from the graph. It is for this reason that we are content to give curves for square mirrors, rather than recomputing them for circular mirrors/apertures. The qualitative behaviour will be much the same, and quantitative conclusions (about precise values of a/w) are not being drawn.

For a near-hemispherical cavity, the light beam is widest at the concave mirror; close to that mirror is a good place to put the mode-selecting aperture to 'scrape the sides off the beam'. In fact, there is usually an obstacle or aperture already present, such as the walls of a discharge capillary. Small adjustments of cavity length (problem 8.14) can bring the spot size w to an appropriate fraction of the capillary radius a; in such a case mode selection is easy.

8.13 Mode matching

We sometimes wish to take the output from one laser cavity and feed it into a second cavity, exciting a single transverse mode in that second cavity. We need to 'match' the field distribution of the incident light to that of the recipient cavity mode.[35] A possible layout is indicated in Fig. 8.10. The light will 'fit' a transverse mode of the recipient cavity if it has its waist at the right place and if it has the right spot size there (equivalently, the right confocal parameter). In other words, we must supply light whose $q(z) = z - z_{\text{waist}} - ib$ matches the $q(z)$ of the cavity's own resonant modes—at some chosen z (any z will do). The lens's power and location must be chosen to achieve this, using the tools described in Chapter 7. The calculation of example 7.1 (p. 158) illustrates what may need to be done.

[35]We might for example be using the input light to pump a laser material, using the multiple reflections within the cavity to enhance the pump power. Or we might be using a non-linear optical material to generate a new frequency, again exploiting the cavity to enhance the driving field.

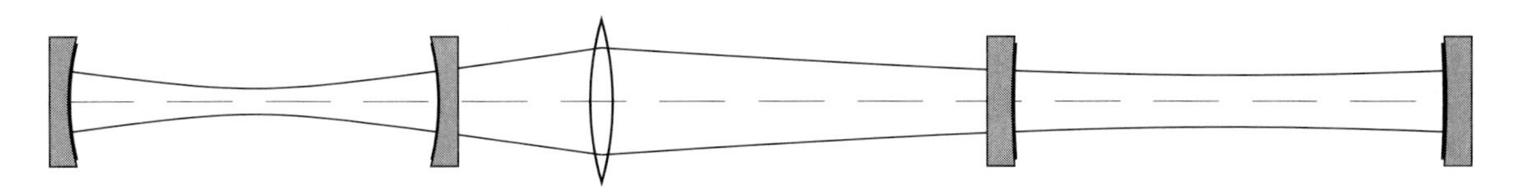

Fig. 8.10 Light is shown emerging from one mirror cavity, perhaps that of a laser that is generating the light. It is desired to focus the light beam into a second cavity so that it excites just one selected transverse mode in that cavity.

Problems

Problem 8.1 (c) The Gauss–Hermite wave solutions—hard
Show that the Gauss–Hermite expression eqn (8.1) is a solution of the scalar wave equation, to the same accuracy as applies to the U of eqn (7.21).

Problem 8.2 (c) The 'phase anomaly'
Consider a beam consisting of one of the eigenfunctions of eqn (8.1) propagating from $z = -\infty$ to $z = +\infty$. The beam starts wide, passes through a waist, and ends up wide again. This can be used as a model for what happens to any beam of light that passes through a focus. Indeed, it isn't only a model, because *any* such beam can be expressed as a sum over the complete set of (l, m) eigenfunctions (transverse modes).

The wave on-axis at z has phase $\omega t - kz + (l + m + 1)\alpha$. Over the full range of z, α changes from $-\pi/2$ to $\pi/2$, so the wave undergoes a phase change $(l + m + 1)\pi$ on top of what we'd expect from totalling optical paths. In the context of focused beams any such 'extra' phase is called a **phase anomaly**. For even-symmetry modes, meaning those with $(l+m)$ even, $(l + m + 1)\pi$ is an odd multiple of π. Conversely, $(l + m + 1)\pi$ is an even multiple of π for modes with $(l + m)$ odd.

The phase anomaly happens because our beam has a narrow section where it passes through its waist. This is consistent with problem 7.6 and sidenote 39 on p. 164, where we learned that an addition to the phase arises whenever a beam is narrow. Manifestations of the same physics also turn up in problem 8.9(4) as a perhaps-unexpected half wavelength; and eqn (8.12) contains a negative sign for even-symmetry modes.

The phase anomaly has been the subject of classic and complicated theoretical studies described in Born and Wolf (1999), §8.8.4. These studies show that there is a phase anomaly of π (modulo 2π) whenever a light beam passes through a focus.[36] Experiment agrees: Billet's split lens (Born and Wolf 1999, p. 296) shows that interference fringes have light and dark interchanged when light before a focus is interfered with light after the focus. But: there is no restriction here in theory or experiment to beams of light with even or odd symmetry.

So does the phase anomaly depend upon the evenness or oddness of $(l+m)$? Sort this out.

[36] This phase shift is sometimes referred to as the Gouy phase shift, after investigations dating back to the 1890s. Analysis using Gauss–Hermite functions (once you've sorted out the confusion I've fed you) yields the same conclusions as the reasoning in Born and Wolf and is much simpler.

Problem 8.3 (a) The condition for a low-loss cavity mode: pictorial
This problem investigates the low-loss condition in a picture-drawing way. We adopt the following provisional statement:

- we have a cavity eigenfunction, a low-loss mode, if a diagram like Fig. 8.1 *can be drawn* with a Gaussian beam between the mirrors.

(1) Apply $R = (b^2 + z^2)/z$ to a cavity whose mirrors are both 'concave towards the middle' and are labelled as in Fig. 8.1. Show[37] that a cavity like that in Fig 8.1 can be drawn only if the mirror separation $L < (R_1 + R_2)$, where R_1 and R_2 are the radii of curvature of the two mirrors.

(2) Using the same conditions as in part (1), show that if the centre of curvature of mirror M_2 is to the left of M_1 then the centre of curvature of M_1 is to the right of M_2; and conversely. So the two mirror centres are either both inside or both outside the cavity.

(3) Show that the results of parts (1) and (2) can be recast into a mathematical form identical to eqn (8.4).

(4) [Harder] Repeat the arguments above for the case where mirror M_1 is 'convex to the middle', lying to the right of the beam's waist.[38]

Comment: It happens that the reasoning of this problem has identified the full condition for there to be a low-loss cavity mode. And I think the physical picture used is very lucid. However, the conditions have been discovered piecemeal, and we could not know from this reasoning alone that we had found all the constraints. So it remains that there is a real need for the formal, but more complicated, method described in §8.5.

The picture-drawing criteria are also quite tricky to work with if one of the mirrors is convex. The algebraic way of deriving eqn (8.4) is again found to be the more secure route.

[37] *Hint*: Show that each mirror's centre lies on the far side of the beam's waist.

[38] *Hint*: Show that both mirror centres lie outside the cavity. The radius R_1 of the left-hand mirror must be put into the 'stability' condition as a negative quantity now. Show that within this convention $L > R_1 + R_2$.

Problem 8.4 (b) The condition for a low-loss cavity mode: formal

(1) Fill in all the mathematical steps that are only outlined in §8.5.[39]

(2) Look back to problem 7.10 and investigate how the reasoning there is affected if it is applied to a general (l, m) Gauss–Hermite beam instead

[39] *Suggestion*: Work out the product of the last two square matrices. The product of the first two need not be worked out separately because it can now be obtained by inspection. Only one more matrix multiplication remains to be done.

Since the matrix products can get a bit messy, it is a good idea to grind out $AD - BC$, and check that it is 1, as a way of waving antennae for mistakes.

of a Gaussian beam ($l = m = 0$). Show that all important results remain valid: knowledge of q is all we need to tell us the location of the beam's waist, its spot size there, its spot size anywhere else, its far-field angular divergence, and which way it is going.[40] And we can find the q at one place from that at another by using the same matrices as were introduced for the purpose in Chapter 7.

[40] *Hint*: The Hermite polynomials are real, so they do not affect the matching of phases on a boundary.

Problem 8.5 (c) All done with mirrors
Set up the calculation of §8.5 *without* unfolding the cavity into a lens train.

Problem 8.6 (b) Ray calculation for low loss
Return to the lens chain of Fig. 8.4(b) and send through it a nothing-special ray defined in plane z_1 by (y_1, θ_1). The ray travels to the right only, and through air except where it meets a lens, so the refractive index $n = 1$ throughout and will be dropped from the equations.

(1) Show that in plane z_2 we have

$$\begin{pmatrix} y_2 \\ \theta_2 \end{pmatrix} = \begin{pmatrix} A & B \\ C & D \end{pmatrix} \begin{pmatrix} y_1 \\ \theta_1 \end{pmatrix},$$

where A, B, C and D take the same values as in §8.5.

(2) Every time our ray travels through the lens-chain period $2L$, its (y, θ) gets multiplied by the square matrix. After a large number of periods, (y, θ) may have grown to some huge value, or it may have become minute (in which case tracing it in the reverse direction would make it huge), or it may remain finite; we want it to remain finite. How do we set a condition for this?

Our ray is propagating through a periodic structure: the lens chain repeats every $2L$. So we remember methods that are known to work on other periodic structures. The most familiar is probably the monatomic linear chain, used in solid-state physics to introduce the dispersion relation for phonons. Use the trick that worked there: find the eigenfunctions (y, θ) defined by the eigenvalue condition[41]

$$\begin{pmatrix} y_2 \\ \theta_2 \end{pmatrix} = \begin{pmatrix} A & B \\ C & D \end{pmatrix} \begin{pmatrix} y_1 \\ \theta_1 \end{pmatrix} = a \begin{pmatrix} y_1 \\ \theta_1 \end{pmatrix}. \tag{8.9}$$

[41] As with differential equations, this eigenvalue condition enables us to find a complete set of functions characteristic of the problem. The behaviour of a general ray is not being restricted, other than by the physics itself, because such a ray can have its (y, θ) written as a sum over the eigenfunction set.

Show that a satisfies

$$a^2 - a(A + D) + 1 = 0.$$

(3) Show that the product of the roots for a is 1. What are the possibilities for the two roots (real, complex, ...)? What behaviour would real a give? Show that we must have $(A + D)^2 < 4$; we are very close to recovering eqn (8.4).

Comment 1 Notice the very close similarity between this and the reasoning used on the monatomic linear chain: all periodic structures produce the same kind of mathematics, and submit to the same technique.

The monatomic linear chain propagates vibrational waves below a cut-off frequency only; it is a low-pass filter. So we are used to the idea that waves can be supported only if some condition is met. Similar things happen with electrical filters.[42]

[42] As with electrical transmission lines, electrical filters provide us with concepts and terminology that are useful in other areas of physics, hence the description of the monatomic linear chain as a 'low-pass filter'. For electrical filters a good reference is Bleaney and Bleaney (1976), Chapter 9.

Comment 2 Something very strange is going on! We know that the spread of a Gaussian beam is entirely caused by diffraction: this is physically obvious, but it is also evident from the presence of $k = (\pm)2\pi/\lambda$ in eqns (7.11) and (7.12). By contrast, any time we use a ray-optics picture, we are taking the limit $\lambda \to 0$. There seems to be a complete incompatibility between the treatments of §8.5 (diffraction fundamental) and problem 8.6 (no diffraction at all). Yet these apparently incompatible calculations give the same result. Is this comprehensible?

Refer back to eqn (8.5). When we approach one of the limiting conditions, $b = \sqrt{4-(A+D)^2}/2|C|$ becomes small; right on the limit it is zero. It is precisely when $b \to 0$ that Gaussian-beam optics becomes indistinguishable from ray optics; see, e.g., problems 7.8 and 7.11(3). We can after all be comfortable with finding the 'stability' condition using either the diffraction or ray pictures.

Comment 3 Arising from the last comment, notice that as $b \to 0$, the far-field angular divergence of the beam becomes very large: problem 7.6(9). This means that on at least one of the mirrors, the spot size w becomes very large as we approach the edge of the permitted regime. This again shows that a ray-optics limit is being approached.

Comment 4 The eigenvalue condition (8.9) says nothing about the absolute values of the quantities y_1, θ_1 constituting the eigenfunction. The ray calculation can give no idea of the size of the region that the light occupies in a 'stable' cavity. In fact, the *only* thing the ray calculation in problem 8.6 is good for is deriving condition (8.4); and even that needs quite a bit of thought before you can trust it.

(4) In part (3) we found that the eigenvalues a of eqn (8.9) are complex. But surely the (y, θ) for a ray must be real? What would it mean that two pairs of (y, θ) values are related by a complex multiplier? Sort this out.

Problem 8.7 (a) Use of the 'stability' chart

(1) Check that the shaded areas and their outlines have been correctly drawn in Fig. 8.5.

(2) A laser cavity is made from two mirrors whose radii are $R_1 = 1\,\text{m}$, $R_2 = 0.35\,\text{m}$. The separation L is adjustable. Make a copy of Fig. 8.5 and draw on it the locus traced out as L is changed. Identify the values of L for which the cavity starts and stops functioning correctly, and check that they agree with condition (8.4).

Problem 8.8 (a) The general symmetrical cavity
For this problem a mirror cavity has $R_1 = R_2 = R$.

(1) Show from eqn (8.4) that when $R_1 = R_2 = R$, the condition for a

low-loss mode reduces to $L < 2R$.

(2) Obtain the result of part (1) from the chart of Fig. 8.5.

(3) Derive the low-loss condition from scratch using the methods of §8.5. Notice that when $f_1 = f_2$ the lens train is a periodic structure with period L, not $2L$. So the eigenfunction condition is $q(z_1 + L) = q(z_1)$; you can obtain the necessary $ABCD$ by multiplying only two matrices.[43]

(4) Show from eqn (8.5) that the beam has its waist at the centre of the symmetrical cavity.

(5) Take the special case[44] where the cavity has $R_1 = R_2 = L$; we have a symmetrical confocal cavity. Show that a mode can fit symmetrically into this cavity with $b = L/2$. It will now be clear where the name 'confocal parameter' for b comes from.[45]

(6) Show that a symmetrical confocal cavity with mirror spacing L has $L\lambda = \pi w_{\text{mirror}}^2$. Use this to confirm the interpretation of $a^2/L\lambda$ given in the caption to Fig. 8.7 and in sidenote 24 on p. 180.

[43] *Comment*: The insight of part (3) is interesting. Something similar happens with other periodic structures when a newly introduced symmetry causes the period to be halved. A diatomic lattice in solid-state physics has acoustic and optical phonons, whose branches of the dispersion curve join up into a double-width Brillouin zone if the two atomic species are made identical. Bragg scattering from a crystal lattice loses reciprocal-lattice points if the lattice period is halved because the reciprocal-lattice period is doubled. And so on.

[44] *Hint*: For part (5) you *must* use the $ABCD$ matrix obtained in part (3), treating the period of the lens structure as L. If you use the matrix elements calculated in problem 8.4, you get a shock: $q = 0/0$. This means that q is indeterminate, a direct consequence of the degeneracy described in §8.9: many field distributions can be made by linearly combining degenerate eigenfunctions. A similar indeterminacy reappears elsewhere in this chapter, particularly in problem 8.12.

[45] The value of b can also be obtained very quickly by making use of eqn (7.11).

Problem 8.9 (a) Longitudinal modes

(1) A cavity mode, fully specified, has three eigenvalues: the l, m that identify the transverse mode, and a third p that identifies the longitudinal mode. Work out[46] the resonance frequency condition for an (l, m, p) mode in the general case where the cavity does not have anything special about its shape (except that it may be assumed 'stable'). Check that you agree with eqn (8.6).

(2) Take a cavity to have length $L = 0.5\,\text{m}$. Find the order of magnitude of the longitudinal-mode eigenvalue p. Find the separation in frequency of the longitudinal modes (the α phases can be ignored for the moment). Compare with the frequency for a wavelength of 633 nm. You should find that the mode spacing is *very* small compared to the absolute frequency.

(3) Knowing the result of part (2), make the approximation that all longitudinal modes with a given l, m (perhaps those oscillating within one spectral line) have the same values of $\alpha(z_1)$ and $\alpha(z_2)$. Show that the resonance frequencies belonging to fixed l, m form a 'comb' with regularly spaced 'teeth' along the frequency axis, and with spacing in frequency $c/2L$. Show that the effect of changing l and/or m is, in general, to make a new comb that is staggered relative to the previous one. (It is this characteristic that is important for laser operation.)

(4) Take the special case $R_1 = R_2 = L = 2b$, a symmetrical confocal cavity as in problem 8.8(5). Show that for this case the resonance condition is

$$2L/\lambda = p + \tfrac{1}{2}(l + m + 1)$$

so the modes form just two combs, each interlacing the other, with the frequencies given by eqns (8.7).

[46] *Warning*: It is possible to get the right answer by a wrong route. We are to take each mirror as imposing a boundary condition $U = 0$. Therefore, a wave arriving at a mirror and contributing amplitude $U(x, y)$ there generates a reflected wave with amplitude $-U(x, y)$. This happens at both mirrors and the two negative signs cancel out. Nevertheless, they should both be there in the reasoning.

Now that this point has been made, discussions will sometimes omit the double negative sign to avoid a distracting complication.

Problem 8.10 (c) A paradox: prolate spheroid

Consider a cavity that is symmetrical but otherwise of general shape. Make an ellipse whose minor axis coincides with the cavity axis, and whose proportions are such that its ends match our mirrors in location and curvature. Rotate the ellipse about the optical axis, to make an oblate spheroid. Make the spheroid conducting. We can think of our usual mirror cavity as two small pieces cut from the surface of this complete spheroid. This description repeats the comment in §8.3.

The spheroid forms a closed box within which we can imagine solving Maxwell's equations, or at least the scalar wave equation. The solutions are tabulated in Abramowitz and Stegun (1965), Chapter 21. The usual Gauss–Hermite solutions appear when $kL \gg 1$.

The problem for you to think about is this: The spheroidal box is a closed cavity, so it has a full set of honest-to-goodness eigenfunctions. Never mind that the eigenfunctions are messy; they exist. However, we are in danger of proving too much. Suppose we look instead at a prolate spheroid ($L > 2R$, stretched along the axis like a rugby ball). That too is a closed loss-free cavity possessing a full set of eigenfunctions. Yet, this is the case we've learned to call 'high-loss' or 'unstable'. So what is wrong with the prolate cavity?[47]

[47] An algebraic analysis is *not* required!

Problem 8.11 (b) The symmetrical confocal cavity
Take a symmetrical confocal cavity consisting of two mirrors with focal length f separated by $L = 2f$. 'Unfold' the cavity into a lens chain as in §8.5. All lenses are identical now. The z-axis passes through the centres of the original mirrors and through the foci of all the lenses.

(1) Launch a ray along the lens chain. The ray is described by (y_1, θ_1) and has nothing special about it. Show that after passing through $2L$ the ray's (y, θ) are the same as at the beginning but reversed in sign. Show that after $4L$ the ray returns it is original (y, θ). Sketch the path of this ray through the original cavity.[48]

[48] *Hint*: The matrix manipulations (and the physics) are easiest for this system if you start at a point half way between two lenses, rather than just after one lens. This trick helps only slightly with part (1), but it helps much more with the rest of the problem (the matrix is simpler—problem 7.3(6)).

(2) Launch a Gaussian beam through the lens chain. The beam is to propagate along the z-axis but has nothing special about its z_{waist} or its confocal parameter b. Its complex radius is q_1 at a point z_1 half way between two lenses. At location z_2, half way between the next pair of lenses (so $z_2 = z_1 + L$) the beam's q takes the value q_2. Show that $q_2 = -f^2/q_1$. Show that after travel through a further L the complex radius returns to its original value.

(3) Show that the beam of part (2), applied to the mirror cavity, does not in general 'fit over' its left–right path when travelling right–left. Write down the condition that it does fit over itself and show that it leads to $|q_1| = f$. Show that the beam's waist must then lie within the cavity but can have any value of z otherwise, and that once the waist location has been chosen the spot size is determined. Sketch such a beam.

(4) Consider a Gaussian beam whose waist lies at z_1 (half way between two lenses), but which does not meet the condition of part (3). Show that after travel through L along the lens chain it has its waist at z_2 but $b_2 = f^2/b_1$. Sketch such a wave travelling along the lens chain, and fitting into the mirror cavity. Compare with problem 7.11(4).

(5) Parts (2)–(4) show that a great variety of field distributions can be eigenfunctions of the symmetrical confocal cavity. Relate this finding to the degeneracy found in problem 8.9(4); there is also a connection with the failure to find a definite value for q according to the hint with problem 8.8(5).

(6) A paradox. The beams of parts (2)–(4) are eigenfunctions of the lens chain by the criterion of §8.5, but not according to that of problem 8.8(3). Are the beams of the present problem excluded by the calculation of problem 8.8, or does problem 8.8 not yield all possibilities? There's quite a lot to think about here.

Problem 8.12 (b) The symmetrical confocal cavity meets Fourier

Figure 8.11 shows a general mirror cavity. An electric-field amplitude $U_1(x_1, y_1)$ exists at (x_1, y_1, z_1) on the left-hand mirror surface[49] and undergoes diffraction to the right-hand surface, where a representative point is (x_2, y_2, z_2).

[49] The alert reader may object: $U = 0$ on each mirror because in the present chapter we take each mirror to be a perfect conductor. A more correct statement can easily be made based on the warning issued with problem 8.9(1).

(1) To start with, let the mirrors be separated by L, and let the radii of curvature of the mirrors take general values R_1, R_2 (positive if curved as drawn). Show that the distance ρ from (x_1, y_1, z_1) to (x_2, y_2, z_2) is, correct to second order in small quantities like x_1/L,

$$\rho = L - \frac{x_1 x_2}{L} - \frac{y_1 y_2}{L} + \left(\frac{R_1 - L}{2R_1}\right)\frac{x_1^2 + y_1^2}{L} + \left(\frac{R_2 - L}{2R_2}\right)\frac{x_2^2 + y_2^2}{L}. \quad (8.10)$$

(2) Write down the Kirchhoff integral that gives the amplitude $U_2(x_2, y_2)$ arriving at (x_2, y_2, z_2) on mirror 2 owing to amplitude $U_1(x_1, y_1)$ at (x_1, y_1, z_1) on mirror 1.

(3) For the rest of this problem take the symmetrical confocal case, $R_1 = R_2 = L$. Show that

$$U_2(x_2, y_2) = \frac{\mathrm{e}^{ikL}}{\mathrm{i}\lambda L}\int U_1(x_1, y_1)\exp(-\mathrm{i}k x_1 x_2/L - \mathrm{i}k y_1 y_2/L)\,\mathrm{d}x_1\,\mathrm{d}y_1. \quad (8.11)$$

(4) Equation (8.11) is a two-dimensional Fourier transform. We are accustomed to finding Fourier transforms only where there is a Fraunhofer case of diffraction. Figures 3.16 and 3.17 and problem 3.28 may help to show why we have Fraunhofer here.

(5) Write down the Kirchhoff integral that gives an expression for the amplitude at (x_1, y_1, z_1) caused by diffraction from Huygens sources $U_2(x_2, y_2)$ at (x_2, y_2, z_2) on mirror 2.

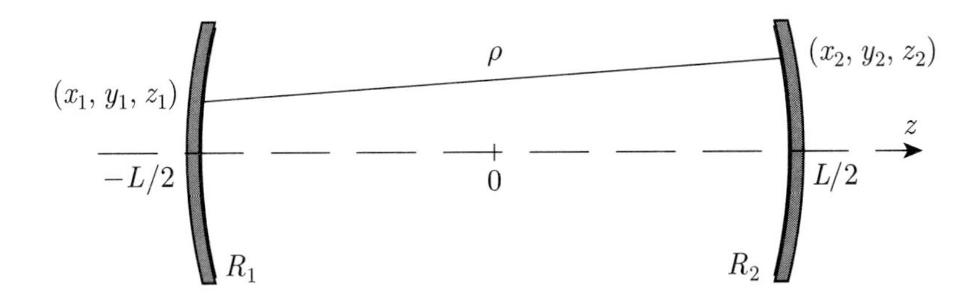

Fig. 8.11 The geometry for investigating diffraction from one mirror to the other of a cavity. The diagram defines quantities used in problems 8.12 and 8.15.

(6) Put parts (3) and (5) together to show that the cavity eigenfunction condition is[50]

$$U_1(x_1, y_1) = -\mathrm{e}^{\mathrm{i}k2L}\, U_1(-x_1, -y_1). \tag{8.12}$$

Since the amplitude returning to mirror 1 is required to match what was there originally in both magnitude and phase, condition (8.12) determines both the transverse and the longitudinal eigenvalue structure of the field.

(7) Show that eqn (8.12) has two separate classes of solution:

$$U(x_1, y_1) = U_\mathrm{s}(x_1, y_1) = +U_\mathrm{s}(-x_1, -y_1) \quad \text{with} \quad \mathrm{e}^{\mathrm{i}k2L} = -1,$$
$$U(x_1, y_1) = U_\mathrm{a}(x_1, y_1) = -U_\mathrm{a}(-x_1, -y_1) \quad \text{with} \quad \mathrm{e}^{\mathrm{i}k2L} = +1.$$

We derive again the property of eqn (8.7): all symmetric (meaning $l+m$ even) modes are degenerate, and all antisymmetric modes are degenerate, and the two sets of frequencies interlace. The minus sign in eqn (8.12) can be seen here to be the consequence of factors i in two Kirchhoff integrals. It has attracted comment previously in connection with eqns (8.7), see sidenote 28 on p. 180.

[50] Since the boundary condition at each mirror is $U = 0$, the source on mirror 2 should have been $-U(x_2, y_2)$. A second negative sign comes in when what is diffracted to mirror 1 is reflected from another conducting mirror. The two negative signs cancel, but ought to be included for honesty. Because these signs cancel, they are not responsible for the negative sign in eqn (8.12).

Problem 8.13 (b) Practicalities with a confocal Fabry–Perot

(1) A good way of using an ordinary plane-parallel Fabry–Perot is to scan it, while recording photoelectrically the intensity at the centre of the ring fringe pattern. The commonest method of scanning is to put an étalon in a box and change the pressure, and hence the refractive index, of the enclosed gas. The effect is to change the optical path between the plates through a small number of wavelengths. Could a similar technique be used to scan a confocal Fabry–Perot?

(2) An alternative way of scanning an étalon is to mount the plates against a piezoelectric spacer, and scan the plate separation through a small range by applying a voltage (a ramp or a sinewave) to the spacer. This arrangement sounds attractive, but it is hard to prevent translational motion from being accompanied by tilt—of an amount that is unacceptable for a plane-parallel Fabry–Perot. Is the confocal cavity equally sensitive to such errors?[51]

(3) Is there a reason why the confocal Fabry–Perot has not totally ousted the plane-parallel one?

[51] *Answer*: No. Indeed the confocal Fabry–Perot usually *is* scanned piezoelectrically. But the problem for you is to give a clear explanation as to why the sensitivity to tilt is acceptably reduced.

Problem 8.14 (b) Examples of cavity design

(1) Consider a symmetrical confocal cavity of length $L = 0.5\,\mathrm{m}$, used at wavelength $\lambda = 633\,\mathrm{nm}$. Assume a $(0, 0)$ mode to be excited, with its waist at the mid-point of the cavity. Find the spot size w for this mode: at the cavity centre; at one mirror.[52]

(2) Consider a cavity with one plane mirror and one spherical mirror of radius R, the mirrors separated by L. Show that the spot size for a $(0, 0)$ mode at the spherical mirror is w, where

$$w^2 = \frac{R\lambda}{\pi}\sqrt{\frac{L}{R-L}}. \tag{8.13}$$

[52] *Suggestion*: You don't need matrices or fancy analysis for any of this problem. Use eqns (7.11) and (7.12).

(3) Take $L = 0.5\,\mathrm{m}$, $\lambda = 633\,\mathrm{nm}$, and find the mirror radius R needed to make $w = 1\,\mathrm{mm}$. There are two possibilities for R. Take first the larger solution. Suppose that the spherical mirror is a disc of glass 15 mm in diameter; what is the depth of the hollow in its curved surface?

(4) Take the same numerical data as in part (3) and take the smaller of the two possibilities[53] for R. Find the value of $(R - L)$ that is required to give $w = 1\,\mathrm{mm}$, and that required to give $w = 2\,\mathrm{mm}$. You should find that they are not so very different.

(5) Show that when $(R - L)$ is small, $(R - L) \propto w^{-4}$.

[53]This gives a 'near-hemispherical' cavity.

Problem 8.15 (c+) The modes of a non-confocal cavity[54]
Problem 8.12 has shown us that a symmetrical confocal cavity has modes with a high degree of degeneracy: all modes with even values of $(l + m)$ are degenerate with each other, and all modes with odd values of $(l+m)$ are degenerate with each other. The result is that *any* even function of (x, y) is an acceptable eigenfunction solution; likewise for any odd function. A problem with so many solutions is as intractable as one with none.

We have mentioned in §8.9 that a laser built with a symmetrical confocal cavity has unpredictable behaviour because the light does not know which linear combination of Gauss–Hermite functions to choose. Degeneracy causes difficulties for the light as well as for the mathematics! The cure when designing a laser is to remove the degeneracy by shifting the cavity design away from confocal. Perhaps the same trick would tame the mathematics? This problem will show we've got the right idea.

Use the diagram of Fig. 8.11. Set the mirror radii equal: $R_1 = R_2 = R$, so the cavity is symmetrical, but do *not* set $R = L$.

(1) Look at eqn (8.10). The path ρ is to be used in an exponent so we should do everything we can to tidy it: a substitution is called for. Choose

$$\xi_1 = \sqrt{\frac{k}{L}}x_1, \quad \eta_1 = \sqrt{\frac{k}{L}}y_1, \quad \xi_2 = \sqrt{\frac{k}{L}}x_2, \quad \eta_2 = \sqrt{\frac{k}{L}}y_2, \quad A = \frac{R-L}{2R},$$

$$V_1(\xi_1, \eta_1) = U_1(x_1, y_1, z_1) = \text{amplitude on the left-hand mirror,}$$
$$V_2(\xi_2, \eta_2) = U_2(x_2, y_2, z_2) = \text{amplitude on the right-hand mirror.}$$

Show that the Kirchhoff diffraction integral simplifies (!) to

$$V_2(\xi_2, \eta_2) = \frac{\mathrm{e}^{\mathrm{i}kL}}{2\pi\mathrm{i}} \int V_1(\xi_1, \eta_1)$$
$$\times \exp\Big\{\mathrm{i}\Big[-\xi_1\xi_2 - \eta_1\eta_2 + A(\xi_1^2 + \eta_1^2) + A(\xi_2^2 + \eta_2^2)\Big]\Big\}\, \mathrm{d}\xi_1\, \mathrm{d}\eta_1.$$

(2) Refer back to Fig. 8.4 and problem 8.8(3). Our mirror cavity can be unfolded into a lens chain, whose period is L rather than $2L$. Therefore, we can identify the propagating modes (eigenfunctions) of the lens chain by requiring the amplitude just after lens 2 to be the same as that just after lens 1 apart from a multiplying factor:

$$V_2(\xi, \eta) = a\, V_1(\xi, \eta).$$

[54]The 'c+' grading indicates that this problem is mathematically fierce! The calculation is outlined here for one reason only: without it everything in this chapter would have been based on an 'it can be shown'.

Consider the effect of a round trip through the cavity (or propagation along the lens chain through $2L$). Show that requiring the amplitude to match itself in phase means $a^2 = 1$, so $a = \pm 1$. Hence,

$$V_1(\xi_2, \eta_2) = \pm \frac{e^{ikL}}{2\pi i} \int V_1(\xi_1, \eta_1) \times \exp\left\{ i\left[-\xi_1\xi_2 - \eta_1\eta_2 + A(\xi_1^2 + \eta_1^2 + \xi_2^2 + \eta_2^2) \right] \right\} d\xi_1 \, d\eta_1. \quad (8.14)$$

(3) An equation in which the unknown function is differentiated is a differential equation. One where the function is integrated is an integral equation. So eqn (8.14) is an integral equation for $V_1(\xi, \eta)$. There is a close relationship between integral equations and differential equations, and quite a number of properties carry over from one to the other.[55] So: if we can think of a technique that would crack a two-variable (i.e. partial) differential equation, it may crack eqn (8.14) too. Separate the variables!

[55] Of course: we chose to start this problem from a Kirchhoff integral, but that's just another way of dealing with the scalar wave equation.

Put $V_1(\xi_1, \eta_1) = X(\xi_1)\, Y(\eta_1)$, and show that eqn (8.14) reduces to

$$X(\xi_2) = \frac{\beta_x}{\sqrt{2\pi}} \exp(iA\xi_2^2) \int_{-\infty}^{\infty} X(\xi_1) \exp(iA\xi_1^2 - i\xi_1\xi_2) \, d\xi_1, \quad (8.15)$$

with a similar equation for $Y(\eta)$ with eigenvalue β_y, and with the eigenvalues related by $\beta_x^2 \beta_y^2 = -e^{ik2L}$.

(4) Equation (8.15) is the integral equation we have to solve. There is only one general method for tackling an integral equation: opportunism. Given squares in the exponent, it's sensible to think of a Gaussian as trial solution, so try

$$X(\xi) = X_0(\xi) = \exp(-\xi^2/2\sigma^2).$$

and show that it solves eqn (8.15) provided that

$$\sigma^{-4} = 1 - 4A^2, \quad (8.16)$$

$$\beta^2 = \beta_0^2 = 1/\sigma^2 - 2iA = -i\, e^{i2\alpha}, \quad (8.17)$$

$$\tan(2\alpha) = 1/(2\sigma^2 A). \quad (8.18)$$

We know that in an eigenfunction problem, the lowest eigenfunction never passes through zero (in quantum mechanics we say it has no nodes); the next has one node, the next two, and so on. What we found has to be the lowest eigenfunction, whatever that means here, since it has no nodes.

(5) How do we find the other eigenfunctions? The one solution we've found closely resembles the ground-state wave function of a quantum-mechanical harmonic oscillator, so we can make a guess and look for raising and lowering operators. Define

$$\hat{a}^+ = \sigma \frac{d}{d\xi} - \frac{\xi}{\sigma}.$$

Show that

$$\begin{aligned} &\text{if} \quad X_0(\xi) \text{ is an eigenfunction with eigenvalue } \beta_0, \\ &\text{then} \quad \hat{a}^+ X_0(\xi) \text{ is an eigenfunction with eigenvalue } \beta_1 = \beta_0\, \mathrm{e}^{\mathrm{i}2\alpha} \\ &\text{and } (\hat{a}^+)^l X_0(\xi) \text{ is an eigenfunction with eigenvalue } \beta_l = \beta_0\, \mathrm{e}^{\mathrm{i}2l\alpha}. \end{aligned}$$

(6) At this point we can look up our table of harmonic-oscillator wave functions, and verify that $\hat{a}^+$ is indeed the correct raising operator, and so

$$\begin{aligned} (\hat{a}^+)^l \exp(-\xi^2/2\sigma^2) &= \text{ the } l\text{th harmonic oscillator wave function} \\ &= \exp(-\xi^2/2\sigma^2)\, \mathrm{H}_l(\xi/\sigma) \end{aligned}$$

(with an arbitrary coefficient since eqn (8.14) is a homogeneous equation). Now work backwards through the substitutions and show that on either mirror ($x, y = x_1, y_1$ or x_2, y_2),

$$U(x,y) \propto \exp\{-(x^2+y^2)/w^2\}\, \mathrm{H}_l(\sqrt{2}\, x/w)\, \mathrm{H}_m(\sqrt{2}\, y/w), \tag{8.19}$$

$$w^2 = \frac{2L}{k\sqrt{(1-4A^2)}} = \frac{R\lambda}{\pi}\sqrt{\frac{L}{2R-L}}, \tag{8.20}$$

$$\tan\alpha = \sqrt{L/(2R-L)}, \qquad \text{with } 0 < \alpha < \pi/2, \tag{8.21}$$

$$k\, 2L = 4(l+m+1)\alpha + 2\pi p, \qquad \text{where } p = \text{any integer.} \tag{8.22}$$

(7) Equation (8.19) is the wave amplitude on the curved surface of one of the mirrors. Reconcile this with eqn (8.1), bearing in mind that eqn (8.1) gives U in a plane normal to the z-axis and not on a surface fitting the curvature of the wavefront.

(8) Reconcile eqn (8.20) with eqns (7.11) and (7.12), using $z = L/2$. Show that the α in eqns (8.18) and (8.21) is the same as the α in eqns (7.13) and (7.14), again evaluated at $z = L/2$. Compare eqn (8.21) with problem 8.14(2). Compare eqn (8.22) with eqn (8.6).

(9) Show from eqn (8.16) that σ^2 is either purely real or purely imaginary. Show that the solutions derived here are unacceptable if σ^2 is imaginary; therefore σ^2 is real. Use this to obtain a condition on A and hence the 'stability' condition for a symmetrical cavity: $0 < (1 - L/2R) < 1$.

Comment: The calculation of this problem does not quite give the amplitude of eqn (8.1), because it gives the value of U on a mirror surface only, not for general z. To cover this detail it is easiest to call upon the results of problem 8.1.

9 Coherence: qualitative

9.1 Introduction

Coherence is a much misunderstood subject, and even at the present level there may be things to unlearn.

Most of our discussion of light in this book has concentrated on a single Fourier component, a pure sinewave behaving like $\cos(\boldsymbol{k}\cdot\boldsymbol{r}-\omega t)$ or $\mathrm{e}^{\mathrm{i}(\boldsymbol{k}\cdot\boldsymbol{r}-\omega t)}$. Such a wave behaves in a totally predictable way, for all times t and at all locations $\boldsymbol{r}$. Look back to problem 1.12(1).

Real light is not exactly monochromatic, but contains a range of frequencies. A single spectral line covers a frequency range, its **linewidth**, typically a few gigahertz. Even the output from a laser contains a range of frequencies.

A wave consisting of several frequencies is not fully predictable: its behaviour contains *randomness. Incoherence*, when it happens, is a consequence of this randomness. Conversely, *coherence* is obtained when we succeed in having some order nonetheless.

Two separate waves may or may not be coherent with each other. We make an experimental test by superposing them and looking for 'stationary' interference: effects characteristic of an addition of wavetrains possessing a definite long-term phase relation between them.

The set of short statements given above indicates the direction that our discussion will take in the present chapter.

In advanced work, coherence is a quantitative concept, on a scale from 0 (complete incoherence) to 1 (complete coherence). The appropriate quantitative measures will be constructed in the next chapter. In the present chapter we concentrate on the two extreme limits,[1] as these provide the most help in getting the basic concepts right. And the treatment will be as simple as possible.

[1] Together with making more-or-less rough estimates of where the boundary lies between them.

9.2 Terminology

We shall need to use certain words in precise senses. In some cases there does not seem to be a standard terminology, and the reader is warned not to expect uniformity of usage between different books.

The electric field of a wave will be described in the usual way by the analytic signal $U(\boldsymbol{r},t)$. Because more than one frequency is assumed to be present we cannot write U simply as $U_0\,\mathrm{e}^{\mathrm{i}(\boldsymbol{k}\cdot\boldsymbol{r}-\omega t)}$, but must think of it as a sum over a number of frequencies, as in eqn (2.15). We shall call[2] $U(\boldsymbol{r},t)$ the **amplitude** of the wave at location $\boldsymbol{r}$ and instant of time t.

[2] In this chapter and the next we call U 'amplitude', rather than 'complex amplitude' which is too much of a mouthful. Note that 'amplitude' now refers to the actual value of the field, not just its peak (which would be $|U|$ for a sinewave). If we need a name for the peak field it will be called *peak amplitude*. Since $U(\boldsymbol{r},t)$ contains information about both magnitude and phase of the field, and is changed by changes of either, we may even refer to changes of 'amplitude' when it would be more correct to speak of changes of phase.

The whole wave $U(\boldsymbol{r}, t)$, at fixed $\boldsymbol{r}$, treated as a function of t defined for all t, will be called a **wavetrain** passing through $\boldsymbol{r}$. It is entire *wavetrains* that are or are not coherent with each other.[3]

- Two *wavetrains* are **coherent** if their random variations with time are statistically correlated (one with the other). In fact, 'coherent' and 'correlated' are identical in meaning in this context.
- In more detail: two *wavetrains* are *coherent* if they are fully correlated; they are **incoherent** if they are uncorrelated; and they are **partially coherent** if they have some correlation but are not fully correlated.

These statements are the most basic that we can give. We shall see just how appropriate they are in Chapter 10 where correlation is handled by statistical methods. However, at this stage those methods would risk burying the meaning under the algebra. Therefore, in the present chapter we first describe things in more qualitative terms. The following bulleted statements are not as formally correct as those above, but are intended to lead us in gently—towards the same concept of course.

- Imagine that somehow we could come to know the (complex) amplitude $U_1(t)$ of one wavetrain at some time t_1. A second *wavetrain* $U_2(t)$ is *coherent* with the first if we could then predict the value of $U_2(t_1)$ at the same time t_1—and this predictability exists for any and all choices of t_1, not just for some special instant.

Usually, we are interested in near-monochromatic light, such as a single spectral line, and then the wave possesses a phase. For this special case:

- Two *wavetrains* are *coherent* if they have a phase difference that is independent of time t.

Comment: The wavetrains we are considering have a variation with time that is at least partly random. Therefore, we must (in imagination if not in actuality) regularly update our information about the 'one' amplitude if we are to know what it is doing, let alone predict successfully what the 'second' amplitude is doing. This accounts for 'somehow come to know'.

Comment: 'Coherent' is a word that describes what two wavetrains have *in common*. It is used in rather the same way as 'similar', though that isn't quite the meaning. You can't have one similar thing, because 'similar' isn't that sort of adjective; if a thing is similar it is similar to something else. In the same way, one wavetrain on its own can't be 'coherent', it must be coherent with some other wavetrain (though that might perhaps be a shifted version of itself).[4]

It is, therefore, sloppy to say things like 'laser light is coherent'. We shall avoid statements like this until it is quite clear what is the shorthand being used.

If we have two ideally monochromatic waves of the same frequency they are of necessity fully coherent (with each other). Therefore, less-than-full coherence is always a consequence of having more than one frequency present. This statement is investigated in problem 9.1.

[3] To forestall a possible misunderstanding: A photon emitted by an atom can be thought of in classical terms as a **wave packet**. A wavetrain will usually be composed of many wave packets because it is defined over a much longer timescale than that of a wave packet—seconds rather than nanoseconds.

[4] *Comment*: To make sure we remember that 'coherent' (or 'incoherent') expresses what two wavetrains have in common, I shall sometimes say that two wavetrains are 'coherent *with each other*', though strictly the words 'with each other' are not necessary and could undermine the correct understanding of 'coherent' as already containing their meaning.

Table 9.1 Interference experiments are classified according to the way in which a beam of light is divided into two (or more) pieces that are later recombined. The two possibilities are 'division of wavefront' exemplified by the Young slits, and 'division of amplitude' exemplified by the Michelson interferometer. If fringes are to be observed, coherence requirements must be met, listed here for each case.

	division of wavefront	division of amplitude
examples	Young's slits Lloyd's mirror Rayleigh refractometer	Michelson interferometer Fabry–Perot Jamin refractometer
beam properties required	small angular spread (bandwidth $\Delta\nu$ unimportant)	small bandwidth (collimation unimportant)
beam properties restated as a coherence requirement	'transverse' (spatial) coherence required; sufficient coherence area	'longitudinal' (temporal) coherence required; sufficient { coherence length / coherence time
quantitative measure of coherence	cross-correlation function $\langle U(\boldsymbol{r}_1,t)^* \, U(\boldsymbol{r}_2,t+\tau)\rangle$	autocorrelation function $\langle U(\boldsymbol{r},t)^* \, U(\boldsymbol{r},t+\tau)\rangle$

The link between coherence and the ability to observe interference is explored in §9.4 and problem 9.2 (and later too of course). Problem 9.3 revises the elementary property: when waves are coherent and are superposed you add amplitudes; when they are incoherent you get the right answer by adding intensities.

A summary of the different types of interference experiment, and the requirements for them to succeed, is given in Table 9.1.

9.3 Young fringes: tolerance to frequency range

One of the statements given in Table 9.1 is: for observing the Young fringes successfully there is almost no requirement on the monochromaticity of the light used. This is demonstrated in problem 9.4. We are, therefore, able to dispose of a possible distraction, so that we can concentrate in §9.4 on the important matter: the light's collimation.

9.4 Young fringes: tolerance to collimation

Figure 9.1 shows a Young-slits apparatus, with lenses included to make the ray-optics geometry easier. Also, to make the geometry easier, hole S is a square of side w, and the 'slits' Σ_1, Σ_2 may be thought of as small pinholes. We'll refer to S as the 'source', since all light downstream from S originates from secondary sources within the open area of S, even though, of course, the actual source is the lamp behind S.

Light originating from the centre of hole S yields cosine-squared fringes

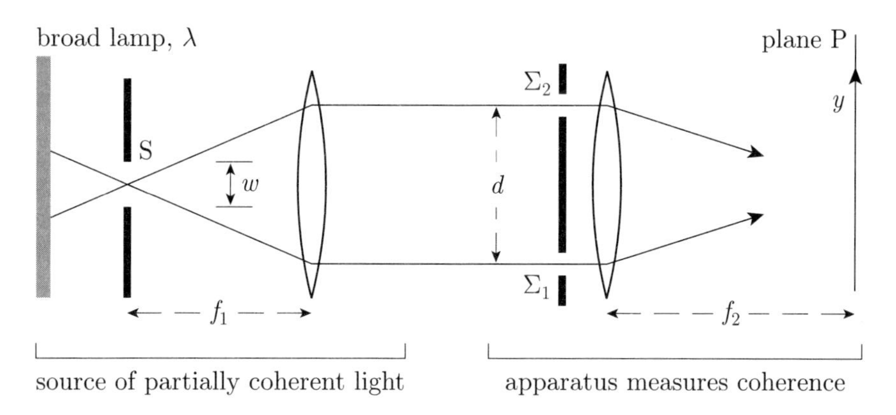

Fig. 9.1 A Young-slits apparatus with light entering through a square hole of width w. Light arrives at the slits in a range of angles w/f_1. It is this angular spread that has the effect of degrading the transverse coherence at the slits.

in plane P, with a separation (light–light or dark–dark) of $f_2\lambda/d$:

$$\text{intensity} \propto \cos^2(\tfrac{1}{2}kdy/f_2) \propto 1 + \cos(kdy/f_2).$$

Light originating from the bottom of hole S yields a similar pattern in plane P, but displaced upwards by $\frac{1}{2}w(f_2/f_1)$; this is simply the displacement of the ray-optics image (magnification f_2/f_1) of the displaced source point. Comparing the shift to the fringe spacing, we find that the intensity has had its cosine changed so that

$$\begin{aligned}\text{intensity} &\propto 1 + \cos\{(kd/f_2)(y - \tfrac{1}{2}wf_2/f_1)\}\\ &\propto 1 + \cos(kdy/f_2 - \delta\phi), \qquad \text{where} \quad \delta\phi = kd(\tfrac{1}{2}w)/f_1.\end{aligned}$$

Light from the top of S gives fringe displacement $-\frac{1}{2}w(f_2/f_1)$ and phase shift $+\delta\phi$; and in between in between.

Things are simpler if we think about angles rather than the linear displacements in plane P. Then

$$\text{range of phases in cosine} = 2\delta\phi = 2\pi\frac{\text{range } w/f_1 \text{ of arrival directions}}{\text{angular width } \lambda/d \text{ of one fringe}}. \tag{9.1}$$

Here the 'arrival directions' are the directions within which light is travelling when it arrives at the Young slits.[5]

The Young fringes are blurred away completely if the cosines contain phases covering a range 2π; equivalently, the range of arrival directions is equal to the angular width of a fringe. We then have $w = w_0$ where $w_0 = f_1\lambda/d$. In this condition, interference must still be happening, yet we have done something so clumsy that it is failing to manifest itself. The electromagnetic fields—the wavetrains—passing through the two slits Σ_1 and Σ_2 have been arranged to be *incoherent* (with each other).

Change viewpoint to thinking about the wavetrains (in the sense of §9.2) passing through slits Σ_1, Σ_2. Suppose that at some instant the phase of the wavetrain through Σ_1 is ϕ_1. Light coming from the bottom

[5] If it's not obvious how to find these directions, apply Fig. 1.3, reversed left–right, to the ray paths through the first lens. There are reasons why the lenses make the geometry easy ....

of S reaches slit Σ_2 after travelling an extra distance, compared with light reaching Σ_1, of $d(\frac{1}{2}w)/f_1$. Thus, it is phase delayed, relative to ϕ_1, by $kd(\frac{1}{2}w)/f_1 = \delta\phi$, the same $\delta\phi$ as above.[6] Taking light from all parts of hole S, we have a phase ϕ_2 at Σ_2 that could be anywhere in the range $\phi_1 \pm \delta\phi$. The apparatus investigates the wavetrains' coherence by superposing them in plane P and looking for fringes. The condition for there to be *no* fringes is equivalent to $2\delta\phi = 2\pi$: the two wavetrains have completely random phases relative to each other.[7]

It should be obvious now: smearing out the fringes involves the same physics as randomizing the wavetrains' phase difference; and complete smearing means randomizing $(\phi_2 - \phi_1)$ over the range 2π. Either of these ideas (smearing and phase-randomizing) expresses exactly what 'incoherent' means, and either can be used to fill out the meaning of the other. Conversely, if we make $w \ll w_0$, we return to having clearly visible fringes; and the wavetrains passing through the two slits have well-defined phases relative to each other;[8] the two wavetrains are now coherent with each other.[9]

Comment: In all of the above an assumption was made: that there is no phase relation between light originating from any two separate points within the source hole S. When that is the case, we add the intensities they contribute at plane P. That is, we sum *amplitudes* for light coming by two routes from any one source point in S, but add *intensities* for light coming from different source points.

Why was this assumption made? It was to be sure that fringes disappear for the smallest possible w_0. And conversely, if we want fringes to be seen clearly, then we should design to succeed given the worst that the sources in hole S could possibly do. And: it's simplest this way.

Comment: Another insight may be obtained by looking at the physics in yet another way. Figure 9.1 has been drawn so as to give the impression that light reaches Σ_1 and Σ_2 from entirely different parts of the lamp. If that were the case, there could be no coherence (between the wavetrains passing through Σ_1 and through Σ_2)—ever. But light from any given part of the lamp is diffracted through angles up to about λ/w after passing through slit S, so it *may* by able to reach both slits. At the point where we just fail to achieve any coherence at Σ_1 and Σ_2, $w = w_0$ and $\lambda/w = d/f_1$: the angular spread caused by diffraction at S is barely enough to bring light from the same part of the lamp onto both slits. In particular, light aiming towards Σ_1 fails completely to reach Σ_2 because Σ_2 is at a diffraction zero. We have discovered a particular case of the van Cittert–Zernike theorem (see §10.9 and problem 10.14).

[6] You are asked to check the geometry of this in problem 9.5. Problem 9.6 offers a kind of comprehension test.

[7] The reasoning here should clarify what was meant in §9.2 when we talked of 'somehow coming to know' the phase ϕ_1 of one wavetrain, and then asked whether that knowledge would permit us to predict the phase ϕ_2 of the other. We do not of course need to know ϕ_1; the point is that even if we did know it we would not thereby be enabled to deduce ϕ_2 when $2\delta\phi = 2\pi$.

[8] It happens here that the phase difference between the wavetrains is zero. But we could easily give it some other value by shifting source S up or down the page. Such a shift would cause the fringe pattern to move, but would have no effect upon its quality. We now have an illustration of the other introductory explanation of coherence given in §9.2: wavetrains are coherent if they have a constant phase difference.

[9] There is, of course, an intermediate case, where w is less than w_0, but not *very* much less. There is then *partial* coherence of the wavetrains passing through the slits; fringes exist but they lack perfect contrast between light and dark. To deal properly with partial coherence we require the correlation-function methods of the next chapter. For the present, only the extreme cases of full coherence and full incoherence need be understood—at least in any detail.

9.5 Coherence area

The ideas outlined in §9.4 have obvious application to the correct design of a working Young-slits apparatus. But they are more important for introducing a new concept: the **coherence area**. For this purpose, we now hold fixed the source size w, and therefore the range of directions

$\theta_s = w/f_1$ within which light travels after the first lens; a fixed arrangement prepares the light beam in the middle section of Fig. 9.1. The (transverse) coherence properties of that beam are to be investigated by sampling it with slits (or holes) whose spacing d may be varied. (This accords with the annotation of Fig. 9.1.) Let $d_0 = f_1\lambda/w = \lambda/\theta_s$. Then we shall say that the electromagnetic wavetrain passing through slit Σ_2:

- is *fully coherent* with that at Σ_1 when the fringes are present and sharp ($d \ll d_0$)
- is *partially coherent* with that at Σ_1 when fringes are present but indistinct ($d < d_0$)
- is *incoherent* with that at Σ_1 when fringes are absent ($d \gtrsim d_0$).

We'll say that when the Young fringes are present Σ_2 lies within the *coherence area* surrounding Σ_1. There will be a Σ_1–Σ_2 separation d_0 at which fringes disappear, and this gives the size of the coherence area.[10]

Comment: When there is at least partial coherence, hole Σ_2 may lie as much as d_0 above or below Σ_1, and (with a square source hole S) as much as d_0 away in either direction horizontally. Therefore, the actual area of coherence, by our definition, is $(2d_0)^2$. People tend to say 'coherence area' yet specify a linear dimension from middle to edge of that area. I'll permit myself to say that we have a coherence area of d_0 when that shouldn't cause confusion with the proper meaning of 'area'.[11]

Here are some statements within the new viewpoint.

(1) Suppose that light of mean wavelength λ arrives within an angle range θ_s at some plane of interest. Then, for the special case (assumed without comment up to now) of a beam arriving with equal strength at all angles within θ_s, $d_0 = \lambda/\theta_s$. This happens[12] for light from a uniform square or rectangular source, as in §9.4.

(2) In the special case of light coming from a uniform circular source, subtending an angle—angular diameter—of θ_s at the plane of interest, the coherence falls to zero at

$$\text{circular source: (radius of) coherence area } d_0 = 1.22\,\lambda/\theta_s. \quad (9.2)$$

(3) In the case of a uniform circular source, the coherence falls to 88% at $d = \lambda/(\pi\theta_s) = 0.32\lambda/\theta_s$, and this is conventionally used to define an area of 'almost full' coherence.[13]

(4) From statements (1) and (2), it should be clear that d_0 is always of order λ/θ_s, though there may be a numerical factor of order unity:

$$\text{(linear dimension of) coherence area} \approx \lambda/(\text{range of angles } \theta_s \text{ within which light is incident}). \quad (9.3)$$

9.6 The Michelson stellar interferometer

A star approximates a uniform circular-disc light source. Light from it arrives at us within a range of angles θ_s; this is the angular diameter of the star. The Michelson stellar interferometer exploits eqn (9.2): by

[10] This is somewhat oversimplified. The coherence, as properly quantified by a correlation function, has a peak centred on point Σ_1—a peak without a hard edge. Around Σ_1, the correlation falls, and reaches zero at $d = d_0$, but it does not then remain zero; it has some low outlying 'wings', as we'll see in problem 10.14. We'll use d_0 as an arbitrary but sensible measure of the coherence area. There is a similarity to the way that we assign a width (really an estimate of the FWHM) to a diffraction pattern by giving the distance from centre to first zero. Thus, d_0 is an estimate of the FWHM of the peak of the correlation as a function of the Σ_1–Σ_2 separation. Should we wish to specify an area over which the coherence is 'usefully large' (implying 'easily visible' fringes by some criterion), we'd use something somewhat smaller than d_0.

[11] There are glorious opportunities for different books' treatments to differ from each other by factors of 2 here.

[12] It is assumed, as in §9.4, that light originating from one point on the source has no phase relation to light originating from any other point, even for source points close together. More usefully: it is assumed that light travelling in any one direction is unrelated in phase to (incoherent with) light travelling in any other direction. This assumption is made throughout the present chapter, and is not stated every time, to avoid clutter.

[13] For an explanation of what it is that is quantified by the value 88%, see §10.8. For an idea of orders of magnitude, see problem 9.7.

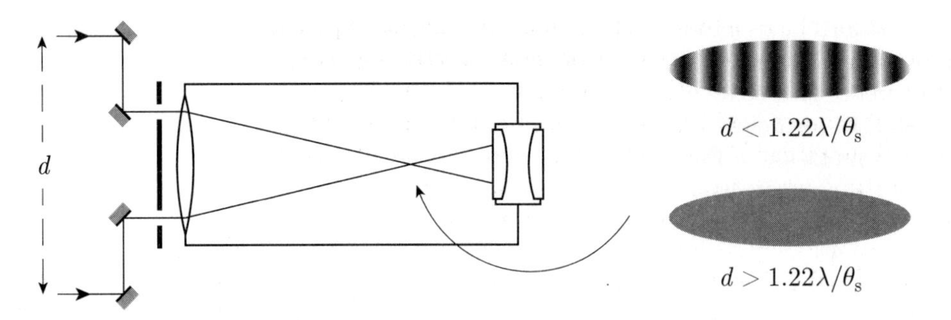

Fig. 9.2 A Michelson stellar interferometer can be thought of as an astronomical telescope, modified by the provision of the four mirrors shown in front of the objective. At least one of the outlying mirrors is movable so that distance d can be adjusted. An impression is given of the appearance of the fringe pattern in the focal plane of the telescope objective.

measuring the coherence area we can obtain a value for the angular diameter θ_s.

An arrangement that implements the stellar interferometer is shown in Fig. 9.2. By adjusting d to (just) blur out the fringes, the observer may measure the coherence area d_0 and hence estimate θ_s, the angular diameter of the star under investigation. The outlying mirrors perform a function identical to that of the slits Σ_1, Σ_2 of Fig. 9.1. Problem 9.8 asks you to investigate the proportions of the fringe pattern, to check the practicality of the device.

A modern stellar interferometer can look very different in its construction from the in-principle diagram of Fig. 9.2. See, e.g., Tango and Twiss (1980), Hajian and Armstrong (2001).

Atmospheric turbulence limits the resolution achievable with a 'conventional' astronomical telescope. It is shown in problem 9.8(2) that increase of a telescope's diameter above about 0.5 m fails to improve resolution, because light entering the telescope at opposite sides of the objective undergoes random relative phase changes of order 2π. It is the raison d'être (and the inventive beauty) of the stellar interferometer that it circumvents this limitation. At the same time, turbulence must affect a stellar interferometer too. If the mirror separation d is 5 m, we'd expect phase fluctuations of order $10 \times 2\pi$; equivalently, fringe movements of about ten fringe spacings.[14] The reason why Michelson was able to make measurements with his original interferometer is that the fringes could be detected by eye in spite of their random movements. In modern practice, the detection of fringes is automated, and the instrumentation must allow for the fringe movement; see problem 10.15.

[14] Starlight observed through the atmosphere has more or less uniform phase over patches of size r_0 (definition of r_0); more precisely, the r.m.s. variation of phase across r_0 is close to one radian. The accepted estimate of r_0 (for sites where telescopes are likely to be placed) is around 0.1 m (Fried and Mevers 1974). In 5 m, there are 50 such patches, and the expected fluctuation of relative phase over that distance is of order 50 rad (Tango and Twiss 1980), again representing about ten fringes. Better estimates can be made by taking account of the statistics of atmospheric fluctuations, using data in the works cited.

9.7 Aperture synthesis

The Michelson stellar interferometer achieves the resolution appropriate to a telescope of diameter d, while having no more than a couple of small mirrors at that separation. The extra resolution applies only in the plane of the diagram of Fig. 9.2.

Could we not extend the idea by using more mirrors until we acquire a proper image-forming system, while still using light structures attached to a small telescope? (A card pierced by an array of pinholes, placed over a telescope objective, would sacrifice intensity, but the resolution shouldn't be much degraded.) And why not fill in the spaces between the mirrors until you have a larger telescope ...? The reason why this fails is given in the last section: atmospheric turbulence precludes any improvement in resolution by such means.[15] But the idea is widely used in radio astronomy.[16] (Problem 3.26(1) shows how much the radio astronomer needs such an improvement in resolution.) Two or more radio aerials are used, and their output signals at radio frequency are conveyed to a central laboratory, combined and processed. Since the aim is to *interfere* the signals received, the radio-frequency signal-handling system must avoid introducing significant relative phase shifts into the signal paths. To achieve the ultimate in resolution, the aerials should be located on opposite sides of the Earth, or at least on different continents. In such cases, the maintenance of sufficient accuracy in the relative timing is an interesting technical problem. Readable accounts of radio interferometry may be found in Strom *et al.* (1975) and, for very long baseline work, in Readhead (1982).

[15] 'Adaptive optics' is a technique whereby we deform a mirror within an astronomical telescope in real time so as to cancel out the worst of atmospheric turbulence; at this level of sophistication, it *is* worth enlarging the telescope to improve resolution. None of what has been said here argues against building as large a telescope as possible to maximize the gathering of energy.

[16] The longer wavelength at radio frequencies means that atmosphere-induced phase shifts are less significant than at optical frequencies.

The radio aerials just described change their orientation relative to the stars as the Earth rotates; they can be thought of as tracing out portions of a huge receiving dish. Therefore, it is possible to achieve a proper imaging by processing the entirety of measurements taken over a day or so. The name 'aperture synthesis' refers to the synthesizing of signals that would have arrived within the aperture of the huge dish.

9.8 Longitudinal and transverse coherence

The discussion so far in this chapter has been concerned with what we shall now call **transverse** (sometimes rather imprecisely called 'spatial') coherence. It is time to introduce the other possibility, **longitudinal** (or **temporal**) coherence, and to explain the difference between them.

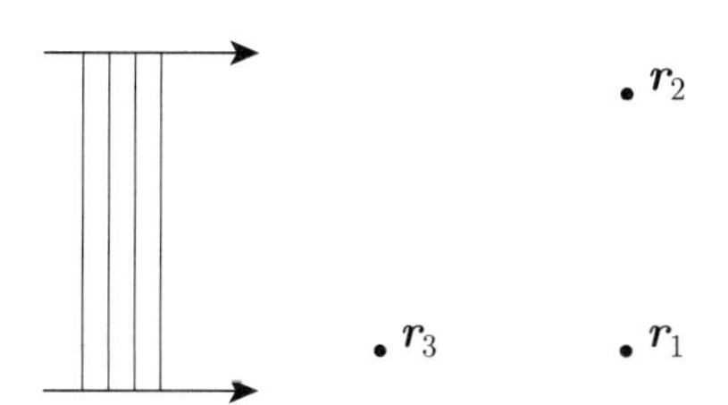

Fig. 9.3 A wave travels in general terms from left to right. It may well not be perfectly collimated, and it may well not be perfectly monochromatic. Transverse coherence concerns itself with the question: is wavetrain $U(\boldsymbol{r}_2, t)$ correlated with $U(\boldsymbol{r}_1, t)$; equivalently, could we predict the phase of the wave at $(\boldsymbol{r}_2, t)$ knowing that at $(\boldsymbol{r}_1, t)$ and do so for all values of t? Longitudinal coherence asks the question: is wavetrain $U(\boldsymbol{r}_3, t)$ correlated with $U(\boldsymbol{r}_1, t)$; equivalently, could we predict the phase at $(\boldsymbol{r}_3, t)$ knowing that at $(\boldsymbol{r}_1, t)$?

Figure 9.3 shows a wave travelling from left to right. We are dealing with practical cases, so this wave is neither ideally collimated nor ideally monochromatic. It, therefore, has some randomness in the behaviour of its amplitude $U(\boldsymbol{r}, t)$. The randomness is best thought of by addressing two questions: Within what limits could the amplitude (meaning complex amplitude U, including the phase information embedded within it) at $\boldsymbol{r}_2$ be predicted from knowledge of that at $\boldsymbol{r}_1$? And within what limits could the amplitude at $\boldsymbol{r}_3$ be predicted from the same knowledge?[17]

We already know the answer to the first question: we need to have $|\boldsymbol{r}_2 - \boldsymbol{r}_1| \ll$ (linear dimension of coherence area) $\approx \lambda/\theta_s$, where θ_s is the range of angles within which the light arrives. The next sections show that the second question has an answer of similar form but different content: we need to have $|\boldsymbol{r}_3 - \boldsymbol{r}_1| \ll c \times$ (coherence time) $\approx c/\Delta\nu$, where $\Delta\nu$ is the range of frequencies present in the light.

[17] From the discussion of §9.2 we should expect that this is a 'temporary' formulation, to be superseded by asking instead: to what extent is wavetrain $U(\boldsymbol{r}_3, t)$ correlated with wavetrain $U(\boldsymbol{r}_1, t)$?

It should be obvious .... A highly monochromatic light beam can be poorly collimated; and a white-light beam can be well collimated. So the two kinds of coherence are independent: either can be good or bad independently of the other; and each needs separate attention if we have a specification to meet.

9.9 Interference of two parallel plane waves

Longitudinal coherence concerns the properties of light containing a range of frequencies. We introduce the subject by thinking first about a case where there are only two frequencies present.

Two light beams are superposed by means of a half-reflecting mirror as shown in Fig. 9.4. After leaving the mirror, the amplitudes in one of the outputs sum to $U_1 + U_2$, where

$$U_1(z,t) = a_1 \exp\{\mathrm{i}(k_1 z - \omega_1 t - \alpha_1)\},$$
$$U_2(z,t) = a_2 \exp\{\mathrm{i}(k_2 z - \omega_2 t - \alpha_2)\}.$$

The intensity, as measured in a time interval T, is proportional to the time average of $U^* U$, as in eqn (2.35). We'll write it simply as $\langle U^* U\rangle$, leaving the averaging interval T to be specified later. Then at location z and time t we have (omitting the z to avoid clutter)

$$\begin{aligned}\text{intensity} &\propto \Big\langle \{U_1(t)^* + U_2(t)^*\}\{U_1(t) + U_2(t)\}\Big\rangle \\ &= \big\langle U_1(t)^* U_1(t)\big\rangle + \big\langle U_2(t)^* U_2(t)\big\rangle + 2\,\mathrm{Re}\,\big\langle U_1(t)^* U_2(t)\big\rangle \\ &= \underset{\text{intensity in beam 1}}{a_1^2} + \underset{\text{intensity in beam 2}}{a_2^2} + \underset{\text{interference term}}{2a_1 a_2 \big\langle \cos\{(\omega_1 - \omega_2)t - (\phi_1 - \phi_2)\}\big\rangle}. \qquad (9.4)\end{aligned}$$

Here $\phi_1 = k_1 z - \alpha_1$ and $\phi_2 = k_2 z - \alpha_2$.

The value we get for the interference term in eqn (9.4) depends on the

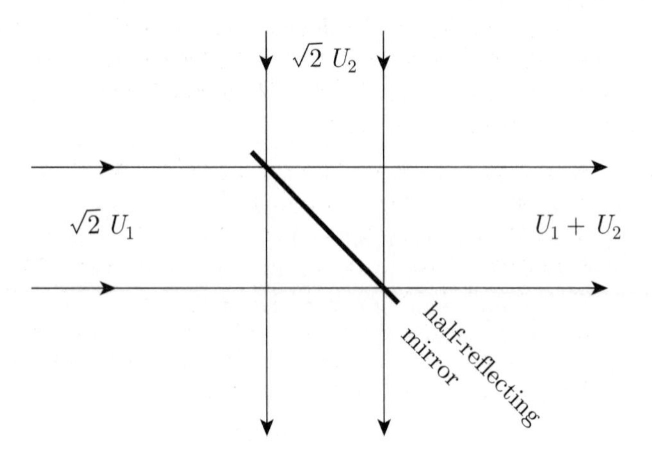

Fig. 9.4 Two waves are superposed by means of the part-reflecting mirror shown. In one of the two output beams the amplitude is $U_1 + U_2$. If the mirror is half-reflecting and loss-free the input amplitudes have magnitudes $\sqrt{2}\,|U_1|$ and $\sqrt{2}\,|U_2|$.

averaging time T:

$$\Big\langle \cos\{(\omega_1 - \omega_2)t - (\phi_1 - \phi_2)\}\Big\rangle \begin{cases} = 0 & \text{if } T \gg |\omega_1 - \omega_2|^{-1} \\ \neq 0 & \text{if } T \lesssim |\omega_1 - \omega_2|^{-1}. \end{cases} \tag{9.5}$$

We may see interference effects when $|\omega_1 - \omega_2| \lesssim (\text{averaging time})^{-1}$.

The interpretation of eqn (9.5) is straightforward, but it divides into a number of special cases. We state some such cases here, with representative numerical values, and ask you to check them in problems 9.9 and 9.10.

(1) Given two ideally monochromatic waves, whose frequencies differ[18] by less than $\sim$10 Hz, a human eye can easily observe interference in the form of 'beats' between the two waves. The conditions given are, of course, quite difficult to meet, but a point of principle is being made.

(2) Given two ideally monochromatic waves, whose frequencies differ by more than $\sim$70 Hz, a human eye cannot observe interference; interference is happening but this particular detector is too slow to detect beats.

(3) Semiconducting photodetectors are available with response times as small as $\sim 10^{-10}$ s. If such a detector is connected to a comparably fast oscilloscope, it can exhibit beats between two monochromatic waves so long as their frequency difference is not too much above 1 GHz. By contrast with item (1), the conditions envisaged here are quite easy to arrange, given two high-quality lasers.

(4) Consider now two beams of light that originate from two independent sources, say rather good gas-discharge lamps, radiating frequencies randomly distributed over a linewidth of 10^9 Hz. A photodiode with response time 10^{-10} s delivers an output current proportional to the intensity it receives. That current contains a zero-frequency component measuring the mean intensity. It also contains beat frequencies: the differences between all incident Fourier components, of each beam and of both beams, so it contains frequencies from zero to 2×10^9 Hz. (A detailed confirmation of this is requested in problem 9.10.)

(5) The two sources envisaged in item (4) produce interference effects when you use the right equipment to observe them. They do not, however, produce effects that can be seen by eye,[19] because almost all beat frequencies are too fast to be seen. The eye simply records the sum of the two intensities. This is in accord with the rule from problem 9.3: when the waves are incoherent, add intensities. But wait—if the waves are incoherent with each other, how could the photodiode of item (4) detect interference?

[18] It is a matter of common experience that an eye can detect intensity changes up to a frequency of order 30 Hz. A television screen, refreshed (in Europe) 25 times a second (albeit interlaced to give coarse coverage 50 times a second), gives flicker that can be perceived but is generally regarded as acceptable. A computer screen is usually written 70 or more times a second and this is fast enough to give no perceptible flicker.

[19] Items (2) and (5) in this list are responsible for an incorrect statement, commonly believed when the present author was a student: 'two independent light sources cannot produce interference'. Items (3) and (4) show why the statement, thus worded, is wrong.

If the statement quoted were right, it would apply in all frequency ranges. Yet, at radio frequencies, established technology relies on its being false. Most radio receivers interfere the incoming signal against a locally generated sinewave to produce a signal (then further processed) at the difference (beat) frequency. The difference frequency is commonly 10.7 MHz or 465 kHz, higher than an audio frequency, which gives the technique its name, 'superheterodyne' for supersonic [ultrasonic] heterodyne [mixing of frequencies from different sources]. An optical-frequency equivalent, perfectly practical—in fact, item (3) in the list, is called optical heterodyning.

It is also commonly said that 'a photon interferes only with itself'. The reader will be able to think of a counterexample to the most obvious understanding of this slogan. Of course, quantum interference is a remarkably complex subject; for an idea of the subtleties see, e.g., Hariharan and Sanders (1996).

9.10 Fast and slow detectors

The discussion in §9.9, reinforced by problems 9.9 and 9.10, leads to the following statement:

- *Any* two light beams, whether from independent sources or not, can cause interference.[20]

[20] The word 'can' is important. There are obvious cases where two light beams *do not* cause interference: when they have orthogonal polarizations; when they are 90° out of phase; when they do not overlap. The point of 'can' is to concentrate on what the light can do after any such impediments are removed: after the experimenter has made the beams overlap; has made the polarizations identical; has tried different relative phase shifts; and has chosen a detector with an appropriate speed of response ....

It is a condition for observing the interference that the detector be 'fast enough', i.e. its response time T must be such that

$$\textbf{fast detector}:\quad T \ll (\text{the interesting values of } |\omega_1 - \omega_2|)^{-1}.$$

This is, of course, a condition on the choice of apparatus, not on the light.

Now that the above point has been made, we shall henceforth look at a more restricted class of experiment where the detector's response time T is very long:

$$\textbf{slow detector}:\quad T \gg (\text{the interesting values of } |\omega_1 - \omega_2|)^{-1}.$$

With a slow detector, beat frequencies are not seen, and we concern ourselves with interference effects that are stationary only. The eye is almost always a slow detector for this purpose (but not, e.g., in item (1) of the list in §9.9).

We now give a 'new and improved' experimental test for coherence (applicable to both transverse and longitudinal cases):

- Two (or more) wavetrains are *coherent* (with each other) if their superposition gives *stationary* interference, verified by observation using a *slow* detector.

It was an unstated assumption in §§9.1–9.8 that we sought stationary interference using a slow detector.

A 'slow' detector has been defined in terms of the 'interesting' values of frequency differences, which is not the most precise of definitions as yet. We shall see that what 'slow' really means is that the detector's response time T should be long compared with the coherence time of the light under examination.

9.11 Coherence time and coherence length

Let a light wave contain frequencies in the range $\Delta\nu = (\omega_2 - \omega_1)/2\pi$, with $\Delta\nu \ll$ the light's mean frequency $\omega_0/2\pi$. We might be working with a single spectral line.[21]

[21] We are assuming, for simplicity, that the light's power spectrum is a 'top-hat' function of frequency within range $\Delta\nu$, and that the amplitudes of the different frequency components are random.

Imagine that we could somehow come to know the wave's phase at some representative time t. At a later time $t + \tau$, frequency $\omega_1/2\pi$ has acquired a phase advance of $\omega_1\tau$; frequency $\omega_2/2\pi$ has acquired a phase advance of $\omega_2\tau$. The wave as a whole has suffered phase changes that cover a range of $(\omega_2 - \omega_1)\tau$. The phase uncertainty, which is the same as the range of phases, increases with τ. It reaches 2π when τ has the value τ_c, which is called the **coherence time**. Then

$$\text{coherence time } \tau_c = 2\pi/(\omega_2 - \omega_1) = 1/\Delta\nu.$$

For time intervals τ much less than τ_c, the phase uncertainty remains small;[22] the wave's behaviour remains predictable (§9.2) from our knowledge of what it was doing at time t. Wavetrain $U(t+\tau)$ is coherent with wavetrain $U(t)$ provided $|\tau| \ll \tau_c$.

Conversely, over time intervals τ much greater than τ_c, the phase of the wave is completely unpredictable: $U(t+\tau)$ is incoherent with— uncorrelated with— $U(t)$ when $|\tau| \gg \tau_c$.

For a wave travelling towards $+z$, amplitude U is a function of $(z-ct)$, so $U(z-l,t) = U(z,t+\tau)$ when $l = c\tau$; everything said about time shifts τ can be re-expressed in terms of displacements $l = c\tau$. It is useful to be able to think about coherence in both of these ways.

The **coherence length** $l_c = c\tau_c$ is the distance that light travels, in the longitudinal (downstream) direction, within one coherence time. We can predict the phase of light at position $z - l$, knowing the phase at z, provided $|l| \lesssim l_c$. Equivalently, wavetrain $U(z-l,t)$ is coherent with wavetrain $U(z,t)$, provided $|l| \lesssim l_c$. In the extreme case[23] when $|l| \ll l_c$, the phase difference is constant for all t (or all z), though it depends upon l.

In summary, we have:[24]

$$\text{coherence time} = \tau_c \qquad \approx 1/\Delta\nu, \tag{9.6a}$$

$$\text{coherence length} = l_c = c\tau_c \approx c/\Delta\nu. \tag{9.6b}$$

Let us give a numerical illustration. Suppose a spectral line is centred on $\nu_0 = 5 \times 10^{14}$ Hz and has linewidth $\Delta\nu = 10^{10}$ Hz. The time τ_c over which the phase uncertainty grows to 2π is 5×10^4 periods of vibration. Even if we look at a time interval that is only 1% of the coherence time, it still contains 500 periods of oscillation. During this time, the wave looks like $V_0 \cos(\omega_0 t + \alpha)$ with constant V_0 and α: it is a 'clean' sinewave. Moreover, everything that has been said about times after t_0 can also be said about times before t_0, so the wave is a clean sinewave for several thousand periods.

Suppose we choose a new starting time t_1, many τ_c later than t_0. Around t_1, the wave is again a clean sinewave for some thousands of periods. But the phase of this sinewave bears no resemblance to the phase we had before: in the time between t_0 and t_1 the wave has 'forgotten' the phase it had.

In the case of transverse coherence, we devised an experimental test: we sampled a wave at two points (the $\boldsymbol{r}_1$ and $\boldsymbol{r}_2$ of Fig. 9.3) by placing Young slits there, and wavetrains passing through the slits were superposed to see if they would interfere (stationary interference using a slow detector). For longitudinal coherence we need an equivalent procedure. This is to interfere $U(t+\tau)$ with $U(t)$, using a Michelson interferometer to impose the relative time shift τ. In the notation of Fig. 9.3, $\tau = |\boldsymbol{r}_3 - \boldsymbol{r}_1|/c$. An experimental arrangement is described in §9.12.

[22] The idea that one can predict the future over short time intervals but not over long is familiar. Knowing what the weather patterns are doing today, you can give a rather good forecast for tomorrow. But today's data are useless for predicting the weather in a month's time. The same applies to 'retrodicting' yesterday's weather from today's if, for some reason, we don't know what it did. Everything being said is as applicable to negative τ as to positive. Hence the presence of $|\tau|$ and $|l|$ in what follows.

[23] Suppose we have a spectral line with width $\Delta\nu = 10^{10}$ Hz. The coherence time is 10^{-10} s and the coherence length 30 mm. The 'extreme case' being discussed is where l is much less than 30 mm but still may be many wavelengths. The phase difference has plenty of opportunity to vary with l. By 'extreme' we did not mean $l \ll \lambda$.

[24] Coherence time, like coherence area, is a 'soft-edged' quantity, unless we supplement the above explanations with a quantitative definition. (The 'top-hat' spectral line profile has served its purpose and is no longer being assumed.) This is why I've used $\lesssim$ rather than $<$ or $\leq$ and $\approx$ rather than $=$. Often it's sufficient to have an order-of-magnitude idea of the coherence time (or coherence length); correlation-function methods exist if we need to be more quantitative.

Comment: All the discussion above has concerned the phase of a wave. What about its size, its peak amplitude?[25] Is the peak amplitude, or the intensity, the same at time $t + \tau$ as it was at t? No, why should it be? After a time interval $\tau \gg \tau_c$ the wave has 'forgotten' everything that it was doing at time t. The peak amplitude must itself fluctuate, becoming unpredictable on timescale τ_c, the same as for the phase.[26] This idea will reappear in §§10.10–10.13, where we shall see that fluctuations of intensity can be exploited just as can fluctuations of phase.

[25] The V_0 of three paragraphs back.

[26] This statement applies to light from a non-laser source. The internal workings of a continuously operating laser give the output light very different statistical properties from those of 'ordinary' light.

Discussions of random wavetrains in optics can seem abstract and difficult to grasp because we can't display the wavetrains on an oscilloscope; oscilloscopes don't yet work at 10^{15} Hz. If only we could ask Mr Tompkins[27] to slow down the light by about 11 orders of magnitude .... Well, we can build quite a realistic model. A block diagram of a demonstration electronic circuit is shown in Fig. 9.5. The apparatus takes 'white' noise[28] and filters it down to narrow-band noise with a bandwidth that is (in the example given) a few percent of the centre frequency. The resulting waveform is displayed on an oscilloscope. Over a few cycles of oscillation the waveform resembles a clean sinewave. By comparing successive sweeps (conveniently superimposed by using repetitive triggering) with and without 'delayed-sweep' triggering, we can easily see that the waveform loses memory of what it was doing at the triggering time, and the memory loss is total after a time $\tau_c \sim 1/\Delta\nu$.

[27] The allusion is to Gamow (1965).

[28] We start from 'white' noise so that our starting waveform is unquestionably random.

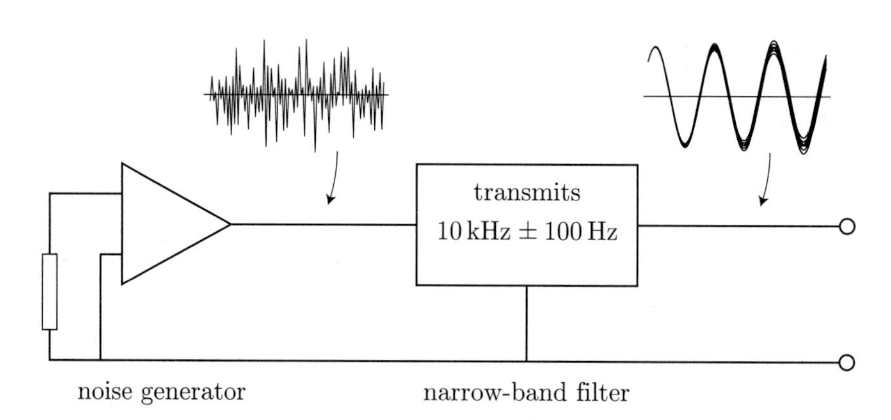

Fig. 9.5 An outline of an electronic circuit that generates narrow-band noise centred on a frequency of 10 kHz. An amplifier amplifies 'nothing', meaning Johnson noise in a resistor plus the amplifier's own added noise. This noise is 'white', from zero frequency to several megahertz, and can be seen to be so as 'grass' on an oscilloscope. Within its frequency range it is clearly random. The amplified noise is next passed through a filter centred on 10 kHz and with a bandwidth that may be in the range 100–1000 Hz. The output is still random, though now confined to a narrow range of frequencies; it is **narrow-band noise**.

The left-hand waveform gives an impression of the appearance of the white noise, as displayed by a single-sweep oscilloscope trace. If the narrow-band noise is displayed by an oscilloscope set for single sweep, the waveform looks like a clean sinewave. But a second trace does not exactly reproduce the first. The right-hand waveform gives an impression of the narrow-band noise, as seen using an ordinary repetitive display triggered at the left-hand end of the display. Successive traces gradually get out of step with each other. Because of fluctuations in the peak amplitude, an actual display is somewhat messier than that drawn.

If the oscilloscope is readjusted to display the narrow-band waveform on 'delayed sweep', starting 10 ms or so after the triggering event, successive traces written to the screen do not resemble each other. It is easy to be convinced that over a timescale of $1/\Delta\nu$, here a few milliseconds, the waveform becomes unpredictable.

The filter in the circuit of Fig. 9.5 is simplest if it uses an RLC circuit, or its active-filter equivalent, to give an 'ordinary' resonance. In spectroscopy we should call the resulting frequency dependence a Lorentzian line profile; it is the shape of a spectral line that is 'homogeneously broadened' (problems 10.6 and 10.7). Even though narrow-band noise is generated in the circuit by a mechanism different from that in (say) a gas discharge, the end result is surprisingly realistic.

9.12 A Michelson interferometer investigating longitudinal coherence

We show how eqns (9.6) can be investigated experimentally. For definiteness, we assume an input in the form of a simple model wavetrain.

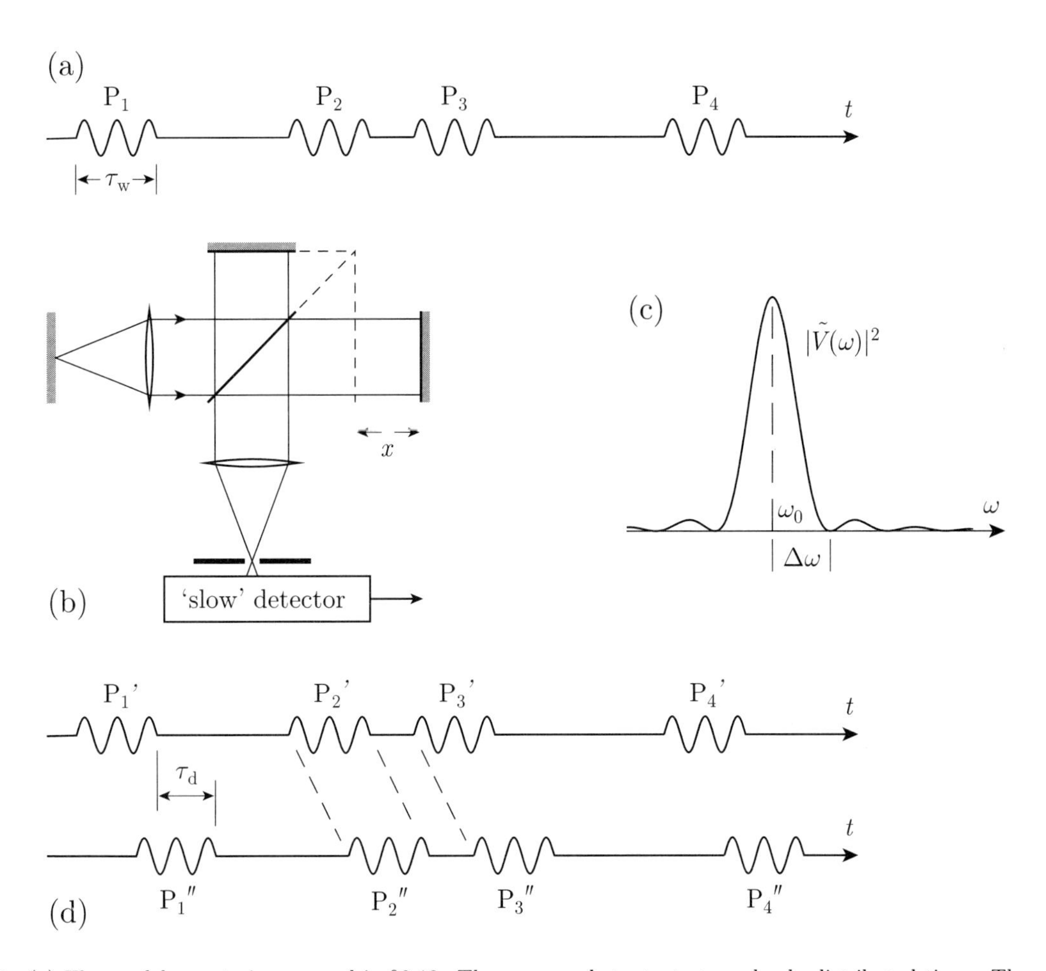

Fig. 9.6 (a) The model wavetrain assumed in §9.12. The wave packets start at randomly distributed times. The wavetrain of (a) is passed through a Michelson interferometer (b) that splits the wavetrain into two and recombines the pieces at the interferometer's output with a relative time shift $\tau_d = 2x/c$. The two pieces being recombined are shown at (d). The frequency spectrum possessed by the model wavetrain is shown in (c).

We imagine a beam of radiation—a wavetrain—to be constructed from building-block wave packets, starting at random times, as shown in Fig. 9.6(a). Each wave packet[29] is a sinewave of angular frequency ω_0, lasting for time τ_w. Using this model, we'll investigate the frequency spectrum of the light, and its (longitudinal) coherence properties.

[29] It is simplest for now to imagine that the wave packets are identical, all starting at the same point in a cycle (as drawn). This permits the use of the convolution theorem for finding the power spectrum $P(\omega)$. However, the same spectrum would be obtained if the starting phases were random—problem 10.5. To model the emission of photons by atoms it is more realistic to assume random phases.

First we find the frequency spectrum of the model wavetrain.

$$\begin{aligned} V(t) = \text{convolution of (single wave-packet waveform)} \\ \text{with (random array of } \delta\text{-functions).} \end{aligned}$$

Then by the convolution theorem,

$$\begin{aligned} \widetilde{V}(\omega) &= \text{Fourier transform of } V(t) \\ &= \{\widetilde{V}(\omega) \text{ for single wave-packet waveform}\} \\ &\quad \times \{\widetilde{V}(\omega) \text{ for random array of } \delta\text{-functions}\}. \end{aligned}$$

The power spectrum (problem 9.11(1)) works out to be

$$P(\omega) \propto \left(\frac{\sin\{\frac{1}{2}(\omega - \omega_0)\tau_w\}}{\frac{1}{2}(\omega - \omega_0)\tau_w} \right)^2 \times N, \tag{9.7}$$

where N is the number of wave packets per second.

The spectrum is shown in Fig. 9.6(c); it is centred on ω_0, and the width of the central peak (arbitrarily taken as the distance between centre and first zero) is $\Delta\omega = 2\pi/\tau_w$. Thus: the *linewidth* $\Delta\nu$ in this model is related to the *wave-packet duration* τ_w by $\Delta\nu = 1/\tau_w$.

Next we find the coherence length by applying the experimental test: for what conditions can we have (stationary) interference? We take the wavetrain and use it to illuminate a Michelson interferometer (light falling normally onto 'parallel' mirrors). The beam is split into two, and the pieces are recombined after one has been given a delay $\tau_d = 2x/c$ relative to the other, where x is the displacement of one mirror from the equal-path condition.

Figure. 9.6(d) shows the appearance of the output waveforms being superposed. Two portions, say P_2', P_2'' formed from one original wave packet P_2 overlap at the detector and interfere. The same is true of P_3', P_3'', and so on. The combined effect of such overlaps on the detector is the same for all wave packets because the two portions of a wave packet always have the same relative phase of $\omega_0 \tau_d$. Thus, interference is observed: reinforcement or cancellation depending on the phase difference between the waves (stationary interference with slow detector). All other contributions to the output (such as the overlap of P_2'' with P_3') time-average to zero because the wave packets have random start times. The outcome is shown in Fig. 9.7.

We see that the beams exiting from the two interferometer arms are:

fully coherent with each other when $|\tau_d| = 0$,
partially coherent with each other when $0 < |\tau_d| < \tau_w$,
totally incoherent with each other when $|\tau_d| > \tau_w$.

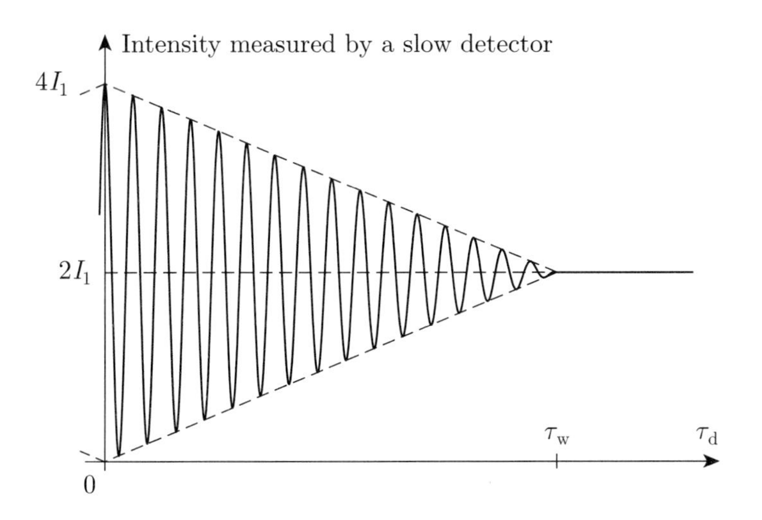

Fig. 9.7 A graph obtained by recording the output from a Michelson interferometer fed with the waveform of Fig. 9.6(a). Such a graph is called an **interferogram**, a name dating back to the time when such information was recorded on chart paper. The intensity reaching the output of the interferometer is recorded while the movable mirror scans the value of the path difference $2x$, imposing a relative time delay $\tau_d = 2x/c$. Interference is almost total when the time delay $\tau_d \ll \tau_w$, and diminishes to zero when τ_d reaches the duration τ_w of a wave packet. Intensity I_1 is the intensity measured if only one of the beams is allowed to reach the detector.

For the present model then

$$\text{coherence time} \quad = \tau_c \ = \tau_w \ = 1/\Delta\nu, \tag{9.8a}$$

$$\text{coherence length} = c\tau_c = c\tau_w = c/\Delta\nu. \tag{9.8b}$$

These results are fully in accord with eqns (9.6). In a way the agreement is too good, in that $(\tau_c\Delta\nu)$ has come out to be exactly 1. This was of course carefully arranged . . . .

Comment: The purpose of the present section is to demonstrate the appropriateness of the Michelson interferometer for measuring longitudinal coherence. The particular wave-packet model chosen for illustration has been incidental. Indeed the statements

$$\tau_c \times \Delta\nu \approx 1, \qquad l_c \times \Delta\bar{\nu} \approx 1$$

are of course just Fourier-transform uncertainty relations to put alongside others like eqn (4.6). We have enough familiarity with such relations not to need another model in support. And anyway, the reasoning of §9.11 is convincing and, though mentioning a model, is rather obviously not too model-dependent.[30]

[30] The wave-packet model could also be a source of unease: the linewidth $\Delta\nu$ is wholly caused by the finite duration τ_w of the wave packets; the spectral line is homogeneously broadened. We may be left wondering whether something different would happen with inhomogeneous broadening, such as Doppler broadening. Of course, the details *are* different, and we no longer have the non-overlap of wave-packet pieces to supply an 'obvious' reason for coherence to cease when $\tau_d \approx \tau_c = c/\Delta\nu$. However the phase-uncertainty reasoning of §9.11 remains secure, and with it relations (9.8).

9.13 Fringe visibility

We define the **visibility** of interference fringes as

$$\text{visibility} \equiv \frac{I_{\max} - I_{\min}}{I_{\max} + I_{\min}}. \tag{9.9}$$

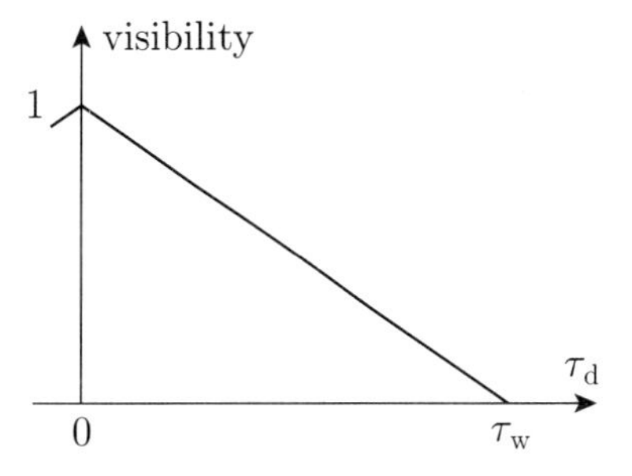

Fig. 9.8 The visibility of the fringes in the interferogram of Fig. 9.7. The definition of 'visibility' is given in §9.13.

It is assumed that we have a pattern consisting of many fringes, and visibility is defined for a part of the pattern where $I_{\max}$ and $I_{\min}$ are not varying too fast from one fringe to the next. In the model of §9.12, this condition is met (a real interferogram would have far more fringes than are drawn in Fig. 9.7), and the visibility falls with τ_d as shown in Fig. 9.8. We could now think of using visibility as a quantitative measure of coherence, since it is 1 when the coherence (of the two beams being interfered) is perfect, 0 when there is no coherence, and looks sensible in between. However, we'll see in Chapter 10 that it's better to go the whole hog and use correlation functions.

9.14 Orders of magnitude

Table 9.2 gives some representative values of the coherence time and coherence length for the light obtained from a variety of sources. It may be seen that the range of values achievable is extremely wide.

Table 9.2 The longitudinal coherence properties of the light coming from some representative light sources. The frequency range is $\Delta\nu$, the coherence time τ_c and the coherence length l_c.

source	λ/nm	$\nu/10^{14}$ Hz	$\Delta\nu$	τ_c/s	l_c
white light	400–600	5–7.5	2.5×10^{14} Hz	4×10^{-15}	1.2 μm (a few λ)
Hg lamp with single isotope	546.1	5.49	300 MHz	3×10^{-9}	1 m
ordinary He–Ne laser	632.8	4.74	1 GHz	10^{-9}	0.3 m
elaborately stabilized He–Ne laser	632.8	4.74	10 kHz	10^{-4}	30 km

9.15 Discussion

In some ways what I've *not* said is more important than what I have:

(a) Notice that I've carefully avoided ever saying that a light *source* (even a laser) possesses a coherence property. What has, or does not have, coherence is a light *beam* (or beams) within which we may compare two (or more) chosen wavetrains.

(b) To point the distinction: You can increase the coherence area of a light beam by improving its collimation (reducing the arrival-angle range θ_s); you can increase the coherence length of a light beam by passing the beam through a filter to reduce its bandwidth $\Delta\nu$. In neither case do you do anything to change the light source. And what you measure may not have much to do with what left the source.

(c) I've also tried to avoid ever saying that a light beam is 'coherent' without further qualification. Such a qualification might mean giving a coherence area or a coherence time. Or we might specify some other beam against which the first beam is to be compared.

(d) It is *just* possible to say that a laser beam is 'coherent' so far as its transverse coherence is concerned, if the beam possesses only a single transverse mode; its coherence area is then as wide as the beam itself. But even this is a sloppiness best avoided.

(e) Almost none of the discussion so far has involved lasers or laser light.[31] There is a good reason for this. *In*coherence is randomness, and non-laser light best illustrates randomness. Coherence is a non-randomness that we may have to impose (controlling or reducing θ_s or $\Delta\nu$), and, especially with non-laser light, it has to be worked for. It's when things don't come easily that you really have to understand what to do!

[31] Elementary descriptions of coherence often seem to give the impression that it is something to do with lasers, or something to do with coercing atoms to radiate in phase with each other ....

9.15.1 What of lasers?

As might be expected, this has to be answered in two parts, dealing separately with transverse coherence and longitudinal coherence.

It is not too hard to arrange that a laser oscillates, and emits radiation, in a single transverse mode. Possible measures by which this may be achieved have been mentioned in §8.12. Most commonly, and for obvious reasons, the laser is made to emit just the 00 transverse mode. The laser beam then has a uniform phase across the whole width of its wavefront, and its coherence area is as wide as the beam. It is usually this large coherence area that is being referred to if it is said that laser light is 'much more coherent' than non-laser light. Just how significant this is will become apparent when the mode structure of non-laser light is discussed quantitatively in Chapter 11.

So far as longitudinal coherence is concerned, it's useful to look at the two entries for lasers in Table 9.2. A common-or-garden gas laser gives out radiation within a frequency range that is rather similar to that from a good gas-discharge lamp. The coherence time is similar too, necessarily. So the laser is not inherently superior. It is possible to do better, but at the cost of working quite hard in the design and construction of the laser. A laser emitting a single longitudinal mode of its cavity emits a 'clean' sinewave, spoiled only[32] by fluctuations in the separation of the cavity mirrors (temperature changes, vibrations, ...). The achievable coherence time is entirely a function of how well the laser can be stabilized. A laser *can* therefore be 'much more coherent' than the best non-laser source, so far as coherence time (and of course coherence length) is concerned, but is not inevitably so.

[32] Well not quite only, because there are fundamental sources of fluctuations too: the random phases of spontaneously emitted photons cause the laser output to drift in phase. But fluctuations in the cavity length dominate in almost all practical cases, even after feedback stabilization has been applied.

9.15.2 The Young slits: another look

Look back to the discussion of the Young fringes in §9.4. Knowing about coherence time and coherence length, we can go over the ground again in new ways.

Consider first a case where the source slit S is wide ($w \gg f_1\lambda/d$); the Young-slits experiment should fail, but how? Over any time interval of order τ_c, the wave passing through either of slits Σ_1, Σ_2 is a 'clean' sinewave[33] and has some phase; the two waves have some phase difference, and combine to give a fringe pattern. As time advances, the phases,

[33] For simplicity it is being assumed that we are illuminating the apparatus with a fairly narrow spectral line. Then a range of frequencies is being input; the coherence time is not infinite. But the frequency range is not so large as to make description messy.

and the phase difference, of the two waves change over timescale τ_c; the fringes move to new positions. A 'fast' detector will see fringes in rapid motion; a 'slow' detector will see no fringes at all. In the whole of §9.4 it was assumed that the detector used was 'slow' in this sense.

We should also redescribe the opposite limit: where source slit S is made small enough ($w \ll f_1\lambda/d$) to give a coherence area spanning both Young slits. The wavetrain $V_1(t)$ passing through slit Σ_1 is much as before: a clean sinewave over hundreds to thousands of optical periods, but acquiring a new phase over timescale τ_c. But now the wave $V_2(t)$ passing through slit Σ_2 varies in the *same random way*, 'tracking' the random variations of V_1. 'Correlated'. This is what we arranged to happen by enlarging the coherence area.[34]

[34] We could, of course, arrange that $w \ll f_1\lambda/d$ by putting the slits closer together (reducing d), rather than reducing w. All that matters is the relative size of d and the coherence area.

The optical paths in a Young-slits apparatus should be thought about. Typically, those path differences are only a few wavelengths, say 5–10. For stationary fringes to be observed, the coherence length must be safely longer than this; and that is what we found in §9.3.

In the first comment of §9.4 an assumption was made about phase variations across the source slit S; we can now say that the coherence area there was assumed to be near-zero. See problem 9.5, parts (3) and (4).

9.15.3 Fast and slow detectors: another look

We can refine what it means for a detector to be 'fast' or 'slow'. A 'fast' detector must respond to the random changes in an interference pattern, so its averaging time (response time) $T \ll \tau_c$. Equivalently, it must have a frequency response that extends up to a few times $1/\tau_c$. A 'slow' detector is the opposite: it must average away the fluctuations over a timescale[35] of a few τ_c; its averaging time need not be much greater than this because any randomness in the light will show up after at most a few τ_c. Given that optical coherence times tend to be less than microseconds, a detector can respond to intensity variations at megahertz frequencies and still be 'slow' for almost all optical purposes.

[35] Therefore, 'fast' and 'slow' are not absolute concepts, but refer to frequency responses that must be chosen according to the frequency range of the light under examination.

9.15.4 Grating monochromator: another look

Consider a grating monochromator with light incident normally on the grating. Let the light consist of a single very short pulse, resembling that in problem 4.10. The output is a wavetrain of length $Nd\sin\theta$; the monochromator built it for us that way. By §9.12, the wavetrain has coherence length equal to its actual length $Nd\sin\theta$. We can now relate:

$$\begin{aligned}
&(Nd\sin\theta)\\
&\quad= (\text{coherence length of output wavetrain}) && \text{just found}\\
&\quad= c/(\text{frequency spread of output}) && \text{by eqn (9.6b)}\\
&\quad= c/(\text{frequency difference the grating can resolve}) && \text{by eqn (4.14)}\\
&\quad= (\text{longest path difference in spectrometer}) && \text{by eqn (4.6).}
\end{aligned}$$

Just so. It all fits.

9.15.5 Polarized and unpolarized light

Light travelling in the z-direction is polarized if there is a phase relationship between field components E_x and E_y, a phase difference that is constant with time. That is: the E_x and E_y wavetrains are coherent (with each other).

Light is 'unpolarized' or 'randomly polarized' if fields E_x and E_y have no long-term relationship between their phases. That is, the E_x and E_y wavetrains are incoherent.

The things we have said in this chapter about random wavetrains and coherence are rather well illustrated by polarized light. It is perfectly possible for light to possess a wide range of frequencies, even to be white, and yet to be polarized; you only have to look through polarizing sunglasses. In such a case, $E_x(t)$ varies randomly with time t; likewise $E_y(t)$ varies randomly with t; yet the two wavetrains have exactly similar random variations with time: $E_y(t) = (\text{constant}) \times E_x(t)$. If the light has some other polarization, such as circular, the two wavetrains have a phase difference that is non-zero but still time-independent. If we have any difficulty imagining two wavetrains that are random yet similar, this example should make it easier.

Problem 9.12 asks you to think about polarized and unpolarized light, and to describe it in the kind of way that was used for coherent and incoherent wavetrains in §9.15.2.

Problems

Problem 9.1 (a) Ideally monochromatic waves and coherence
Consider a plane wave whose amplitude is represented by the analytic signal $U(\boldsymbol{r}, t) = U_0\, \mathrm{e}^{\mathrm{i}(\boldsymbol{k} \cdot \boldsymbol{r} - \omega t)}$, where U_0 is constant.[36] This expression defines the wave amplitude at all times t and all locations $\boldsymbol{r}$—that is, it specifies the entire wavetrain. The only frequency present is $\omega/2\pi$, so the wave is exactly monochromatic.

(1) Show that $U_2 \equiv U(\boldsymbol{r}+\boldsymbol{s}, t) = U(\boldsymbol{r}, t)\, \mathrm{e}^{\mathrm{i}\boldsymbol{k} \cdot \boldsymbol{s}}$, in which we shall take $\boldsymbol{s}$ as constant. Show that wavetrain U_2 is fully coherent with wavetrain $U(\boldsymbol{r}, t)$.

(2) Show that $U_3 \equiv U(\boldsymbol{r}, t+\tau) = U(\boldsymbol{r}, t)\, \mathrm{e}^{-\mathrm{i}\omega\tau}$, in which we shall take τ as constant. Show that wavetrain U_3 is fully coherent[37] with $U(\boldsymbol{r}, t)$.

(3) Consider a wave, again exactly monochromatic, represented by complex amplitude $f(\boldsymbol{r})\, \mathrm{e}^{-\mathrm{i}\omega t}$, where $f(\boldsymbol{r})$ is some function of position that we need not specify. A second wave of the same frequency has complex amplitude $g(\boldsymbol{r})\, \mathrm{e}^{-\mathrm{i}\omega t}$. Investigate these two waves along the lines spelt out in parts (1) and (2), and show that similar conclusions apply.[38]

(4) In §9.2 it is stated that a near-monochromatic wave possesses a phase (albeit one that varies randomly with time over a timescale τ_c), but a wave with a broad frequency range does not. Why is this the case?

[36] We could work instead with the real amplitude $V = \mathrm{Re}\, U$, but U is mathematically more convenient.

[37] *Comment*: In these special cases there is no upper limit on $|\boldsymbol{s}|$ or $|\tau|$. An amplitude at *any* place or time can be predicted if we have obtained knowledge of the amplitude at some other place and time. This is a special feature of an ideally monochromatic wave. Ideas in problem 1.12 can be called upon to show that this had to happen.

[38] Of course, $f(\boldsymbol{r})$ and $g(\boldsymbol{r})$ are not completely general functions, because we must have solutions of the appropriate wave equation.

Problem 9.2 (a) Ideally monochromatic waves and interference

(1) Let two waves have amplitudes $A\,\mathrm{e}^{\mathrm{i}(\boldsymbol{k}_1\cdot\boldsymbol{r}-\omega t)}$, $B\,\mathrm{e}^{\mathrm{i}(\boldsymbol{k}_2\cdot\boldsymbol{r}-\omega t-\alpha)}$, where A and B are real so the phases are wholly displayed in the exponents. Arrange for these waves to be superposed at locations $\boldsymbol{r}$. Show that at $\boldsymbol{r}$

$$\begin{aligned} \text{amplitude} \quad U_{\text{total}} &= \mathrm{e}^{\mathrm{i}(\boldsymbol{k}_1\cdot\boldsymbol{r}-\omega t)}\{A + B\,\mathrm{e}^{\mathrm{i}(\boldsymbol{k}_2-\boldsymbol{k}_1)\cdot\boldsymbol{r}-\mathrm{i}\alpha}\},\\ \text{intensity} \quad \propto |U_{\text{total}}|^2 &= A^2 + B^2 + 2AB\cos\{(\boldsymbol{k}_2-\boldsymbol{k}_1)\cdot\boldsymbol{r}-\alpha\}. \end{aligned} \tag{9.10}$$

The terms A^2, B^2 represent the intensities of the individual waves. The cosine term is recognizable as interference: something additional (positive or negative), happening to the intensity, that depends upon the relative phases of the contributing waves.

(2) Take the case where the waves intersect at an angle, like the overlap on the downstream side of Young's slits. Show that fringes bisect the angle between the two wave directions.

(3) Use a half-reflecting mirror to make the two waves travel in the same direction and overlap. Show that interference effects can be identified if we find some way of making controlled variations in the phase difference α. We can do this in a Michelson interferometer if we move one of the interferometer mirrors.[39]

[39] We would more usually shift our gaze to a different part of the interference pattern, thereby observing ring fringes. But that requires light to be travelling in a range of directions, a condition excluded in the present problem.

Problem 9.3 (a) Add intensities when waves are incoherent

Consider two monochromatic waves with amplitudes $U_1 = A\,\mathrm{e}^{\mathrm{i}(\boldsymbol{k}\cdot\boldsymbol{r}-\omega t)}$, $U_2 = B\,\mathrm{e}^{\mathrm{i}(\boldsymbol{k}\cdot\boldsymbol{r}-\omega t-\alpha)}$, where A and B are real and the phase difference α is constant.

(1) Imagine an interference experiment, like that of problem 9.2(3), in which the waves are superimposed at $\boldsymbol{r}$. Show that we have

$$\begin{aligned} \text{amplitude} \quad U_{\text{total}} &= \mathrm{e}^{\mathrm{i}(\boldsymbol{k}\cdot\boldsymbol{r}-\omega t)}(A + B\,\mathrm{e}^{-\mathrm{i}\alpha}),\\ \text{intensity} \quad \propto |U_{\text{total}}|^2 &= A^2 + B^2 + 2AB\cos\alpha. \end{aligned}$$

(2) Now measure the intensity, not once but many times, each time with a different value of α, and with the αs distributed randomly over a range of 2π. Show that the average of all these measurements is

$$\langle |U_{\text{total}}|^2\rangle = A^2 + B^2.$$

Recognize that what we have just done is to take an *ensemble average*, in the sense in which this term is used in statistical mechanics. By taking an ensemble average, we have rendered invisible the terms containing the random phase α; equivalently, the interference effects have been averaged away. It is as if *intensities* should be added instead of *amplitudes*.[40]

[40] *Comment*: The wave amplitudes obey Maxwell's equations (1.3) and the scalar wave equation (1.24). These equations are *linear*, so *it is always correct to add amplitudes*. In part (2) we added amplitudes as always; but the cross-term $2AB\cos\alpha$ disappeared because of the averaging over random phases. The wording *as if* is carefully chosen ....

(3) Let's make things less abstract. Imagine that we do an interference experiment by superposing two wavetrains of light and observe the result by eye. One wave is monochromatic, $U_1 = A\,\mathrm{e}^{\mathrm{i}(\boldsymbol{k}\cdot\boldsymbol{r}-\omega t)}$; the other is $U_2 = B\,\mathrm{e}^{\mathrm{i}(\boldsymbol{k}\cdot\boldsymbol{r}-\omega t-\alpha)}$, where B is constant but α changes randomly[41] over a timescale of a microsecond or so (very long compared with an optical period but very short compared with the timescale an eye can

[41] The changes of α mean that this wave is modulated and cannot be strictly monochromatic. Compare with problem 1.12(1).

notice). Argue for yourself that each 'measurement' made by the eye is an average over an ensemble of about 10^5 different values of α. The ensemble average is now real: it's what we see.[42]

[42] The critical reader may object that what we observe isn't really an ensemble average, but a time average. However, for a *stationary random process*, which we have here, the two averages are the same. Compare this with what we do in statistical mechanics. A system in equilibrium has 'average' properties that we measure by taking a long-time average; but we calculate those properties using an ensemble average. The equality of the averages is related to the *ergodic hypothesis*.

Problem 9.4 (a) What frequency range can be tolerated with Young's fringes?
Refer to §9.3. Let a Young-slits apparatus have slits separated by d. Let the incoming light contain frequencies in range $\Delta\nu$ centred on ν_0. We wish to be able to see clearly ten fringes, five each side of the pattern's centre.

(1) Suppose that we set the following criterion for 'seeing clearly': in the vicinity of the fifth fringe, where $d\sin\theta = p\lambda_0 = pc/\nu_0$ with $p = 5$, the order of interference p shall be permitted to cover a range of $\frac{1}{2}$. Show that this permits $\Delta\nu/\nu_0 = 1/10$.

(2) The condition set in part (1) is unduly pessimistic. Show that the fringe contrast in the vicinity of the fifth bright fringe is zero[43] if the range permitted to p is 1. Show that this permits $\Delta\nu/\nu_0 = 1/5$.

[43] *Hint*: Remember the elementary way for locating the first zero in the diffraction pattern due a single slit. See problem 3.6.

(3) Argue that *some* fringes will be detectable even if the frequencies supplied cover a factor-2 range, equivalent to the whole visible spectrum.

(4) Discuss the extent to which the colour vision of the human eye helps when white light is input.

Problem 9.5 (a) Effect of arrival angles on the Young fringes

(1) Check in detail all the statements made in §9.4.

(2) In §9.4 it is stated that light from the bottom of source S has to travel further to Σ_2 than to Σ_1 by an extra $d(\frac{1}{2}w)/f_1$. Draw a ray diagram to show convincingly that this is so.

(3) In §9.4 the emphasis was on making the coherence area as large as possible, so we were content to use small-angle approximations. Suppose we wish instead to make a coherence area very small, then the range of arrival angles must be large. Rework the reasoning with light arriving at the Young slits within an angle range from $-\theta_1$ to $+\theta_1$. Show that the (linear dimension of the) coherence area is $\frac{1}{2}\lambda/\sin\theta_1$. Check that this correctly goes over into our previous expression when θ_1 is made small.

(4) Apply the result of part (3) to discuss the assumption, made explicit in the first comment in §9.4 and again at the end of §9.15.2, that the coherence area at source S is very small.

Problem 9.6 (b) Variations on the Young slits
Imagine changing the physical system of §9.4 as follows. The Young 'slits' remain small holes. Replace source hole S by a large-area piece of ground glass, so that the transverse-coherence condition $w/f_1 < \lambda/d$ is *not* met. Illuminate the ground glass with a laser oscillating in a single transverse mode, say the 00 mode—but not necessarily a single frequency. Can fringes be seen: with a fast detector; with a slow detector? How should you describe the action of the ground glass?

Would things change if Σ_1 and Σ_2 were elongated into slits?

Problem 9.7 (b) Coherence area for sunlight

Sunlight arrives at a point on the Earth in a cone of angular diameter $\theta_s = 0.5°$. Show that this gives 'almost fully coherent' illumination over a circular area of radius $37\lambda \approx 20\,\mu m$, while there is *some* coherence over a radius $140\lambda \approx 70\,\mu m$.

Problem 9.8 (b) The Michelson stellar interferometer

(1) In the stellar interferometer of Fig. 9.2, show that the following geometrical relations apply (the 'slits' are those in front of the telescope objective):

(a) Length/breadth of star 'image' = (slit length)/(slit width).

(b) Number of fringes appearing in star image
$= 2 \times$ (slit separation)/(slit width).

(c) Fringe spacing $= f\lambda/$(slit separation), which is independent of d.

There is, therefore, no difficulty in arranging the apparatus to give a fringe pattern that is comfortable to observe.

(2) Atmospheric turbulence limits the resolution of a visible-frequency astronomical telescope to ~0.25 arcsec. Work out the diameter[44] of a telescope which, if limited by diffraction only, resolves to 0.25 arcsec.

[44] This represents the diameter above which further increase of size gives no improvement in *resolution*. Do you agree with the value quoted in §9.6?

(3) Michelson used his stellar interferometer to measure the angular diameter of the star Betelgeuse, and found it to be 0.047 arcsec. Find the mirror separation he needed to conduct the measurement.

(4) Light forming the fringy image in the stellar interferometer has not passed through the whole of a circular lens, but through two rectangular slits. Is the factor 1.22 in the insets to Fig. 9.2 correct?

(5) How long is it worth making the slits and outlying mirrors in a stellar interferometer?

Problem 9.9 (b) Interference of two parallel plane waves

(1) In Fig. 9.4, the input beam amplitudes are marked as $\sqrt{2}\,U_1$ and $\sqrt{2}\,U_2$. Show that these expressions are correct for a loss-free half-reflecting mirror, except for possible phase factors.

(2) In connection with Fig. 9.4 it is stated that one of the output amplitudes is $U_1 + U_2$. Light energy must be dividing itself between the two outputs in such a way as to conserve energy overall. What property must the mirror have to guarantee energy conservation? What is the amplitude in the other output?

(3) Check all the enumerated statements in §9.9 following eqn (9.5).

(4) In item (3) of the list in §9.9, what happens to the signal if a piece of ground glass is placed between the half-reflecting mirror and the photodiode?

(5) A single discharge lamp, radiating frequencies randomly distributed over a bandwidth of 10^9 Hz, illuminates a Michelson interferometer. The mirrors are set 'parallel', and the optical paths in the interferometer arms differ by $l < 3\,\text{mm}$. Figure 9.4 applies to the recombination of the beams from the interferometer arms.

(a) Use a Fourier integral to describe the input amplitude as a sum over frequencies within the bandwidth supplied; use it to calculate the amplitude at the output.
(b) Show that the intensity at the interferometer output exhibits interference, and that the coherence of the two beams is almost perfect. What condition, given in the question, made the coherence so good?

(6) Part (5) shows that the two beams being interfered are fully coherent (with each other), in spite of the random structure within each beam. The beams' fields, though individually random, are highly correlated. The reason of course lies in our special method of preparation, starting with a single beam from a single lamp.[45]

[45] We have here some explanation as to how the erroneous statement, given in sidenote 19 on p. 203, arose: that interference requires light beams or photons to have been obtained from a single source. Interference can be engineered rather easily, even interference visible by eye, if you use beams of light derived from a single source; so it's easy to fall into thinking that a single source is somehow compulsory.

Problem 9.10 (c) Beats using two independent light sources

(1) Consider item (3) in the enumerated statements in §9.9, where two lasers give beats at the output from a fast photodiode. What happens if the two lasers have frequencies that drift slowly over a range of about 100 MHz?

(2) A discharge lamp, radiating frequencies randomly distributed over a bandwidth of 10^9 Hz, illuminates a photodiode whose response time is 10^{-10} s. Because the photodiode has a square-law response to the optical amplitude $V_{\rm in}(t)$, its current exhibits beats between frequencies present in the input; this is in accord with statement (4) of §9.9. Show that indeed the photocurrent contains Fourier components in the frequency range $\mathrm{d}\omega/2\pi$ centred on $\omega/2\pi$, whose amplitude is proportional to

$$\frac{\mathrm{d}\omega}{2\pi}\int_{-\infty}^{\infty}\widetilde{V}(\Omega)^*\,\widetilde{V}(\Omega+\omega)\frac{\mathrm{d}\Omega}{2\pi}, \quad \text{where} \quad V_{\rm in}(t)=\int_{-\infty}^{\infty}\widetilde{V}(\Omega)\,\mathrm{e}^{-\mathrm{i}\Omega t}\,\frac{\mathrm{d}\Omega}{2\pi}$$

and where $\widetilde{V}(\Omega)$ is defined by the last statement.

(3) [Harder] Show further that the electrical power, delivered by the photodiode to an external circuit, in the frequency range $\mathrm{d}\omega/2\pi$ is proportional to

$$\frac{\mathrm{d}\omega}{2\pi}\int_{0}^{\infty}P(\Omega)\,P(\Omega+\omega)\,\frac{\mathrm{d}\Omega}{2\pi},$$

where $P(\Omega)\,\mathrm{d}\Omega/2\pi$ is the incident optical power[46] in the frequency range $\mathrm{d}\Omega/2\pi$, related to $\widetilde{V}(\Omega)$ by eqn (2.22).

[46] Strictly, we have defined $P(\omega)$ in §2.8 to mean power per unit area as well as per frequency interval.

(4) Two discharge lamps, with power spectra $P_1(\Omega)$, $P_2(\Omega)$, provide the two light beams of §9.9. The bandwidths of these power spectra are both of the order of 10^9 Hz. A photodiode with response time $< 10^{-10}$ s measures the intensity in one exit beam. Show that the (electrical) power delivered by the photodiode in the frequency range $\mathrm{d}\omega/2\pi$ is now proportional to

$$\frac{\mathrm{d}\omega}{2\pi}\int_{0}^{\infty}\Big(P_1(\Omega)\,P_1(\Omega+\omega)+P_2(\Omega)\,P_2(\Omega+\omega) + P_1(\Omega)\,P_2(\Omega+\omega)+P_2(\Omega)\,P_1(\Omega+\omega)\Big)\frac{\mathrm{d}\Omega}{2\pi}.$$

Give a physical interpretation for each of the four terms in this expression.

Problem 9.11 (a) The wave-packet model of §9.12
Consider the tutorial model of repeated wave packets, discussed in §9.12.

(1) Fill in the missing steps in the calculation of the power spectrum $P(\omega)$. Sections 2.8 and 2.9.5 may help. Show that the expression for $P(\omega)$ is quoted incompletely in eqn (9.7). Obtain the full form, and show that the extra term is negligible if $\omega_0 \tau_w \gg 1$.

(2) Analyse the interference taking place at the output of the interferometer. Confirm that the overlap of 'different' wave packets makes no contribution to the time-averaged output.[47] Show that the 'envelope' of the interference pattern shown in Fig. 9.7 is indeed a linear function of τ_d.

[47] This is revisited, using more formal methods, in problem 10.5.

Problem 9.12 (a) Polarized and unpolarized light

(1) Randomly polarized light is obtained from a lamp emitting a fairly narrow spectral line, and travels in the z-direction. Describe the behaviour of fields E_x and E_y on timescales long and short compared with the coherence time. Model your description on that for the two wave-trains of §9.15.2.

(2) Suppose we have two narrow-band noise generators similar to that of Fig. 9.5, giving outputs with the same centre frequency and the same bandwidth. The two outputs are connected to the x- and y-inputs of an oscilloscope set up for an x–y display. Describe what is seen

(a) if the intensity of the oscilloscope is switched on for a few milliseconds only, and the observation is repeated every few seconds;
(b) if the oscilloscope display runs continuously.

(3) Suggest how the apparatus could be modified to give a representation of

(a) linearly polarized light,
(b) circularly polarized light,
(c) partially polarized light.

(4) In the arrangement by which the oscilloscope gives a display mimicking circularly polarized light, describe what is seen

(a) if the intensity of the oscilloscope is switched on for a few milliseconds only, and the observation is repeated every few seconds;
(b) if the oscilloscope display runs continuously.

Coherence: correlation functions

10

10.1 Introduction

Any light that isn't perfectly monochromatic (and that's every kind of light) necessarily has some degree of randomness. In Chapter 9 we described this randomness, how it originates and how (within limits) it may be controlled. However, there were obvious inadequacies: having no quantity to represent a 'degree of coherence',[1] we could not give a precise condition for coherence to disappear or for fringes to be so indistinct that they should be considered absent. Moreover, we found ourselves using awkward wordings: 'obtaining knowledge'; 'permits us to predict'; 'so clumsy that interference fails to manifest itself'; 'a wavetrain is the whole wave defined for all t'. The author of this book used words with more than usual care in Chapter 9, in an attempt at giving a correct impression and excluding common misconceptions; but the need for that care was itself an indication that something better was called for.

[1] Visibility was suggested, but problem 10.3 shows that it isn't ideal.

To give a quantitative description of randomness, we must work with statistical averages. Although the mathematics will sometimes look complicated, the physics will often be clearer than before.

The statistical average we need to introduce is the correlation function. This will provide us with

- a means of quantifying a 'degree of coherence' that ranges from 0 (total incoherence) to 1 (perfect coherence)
- a mathematical language that renders unnecessary the carefully crafted sentences of Chapter 9 and keeps our ideas on the rails
- a set of new and useful properties.

10.2 Correlation function: definition

Let $\phi(t)$ and $\psi(t)$ be two functions, in general complex, of time t, which are observed during a long time interval T. The correlation function of ϕ with ψ is defined as

$$\phi \star \psi = \left\langle \phi(t)^* \, \psi(t+\tau) \right\rangle \equiv \frac{1}{T} \int_{-T/2}^{T/2} \phi(t)^* \, \psi(t+\tau)\, \mathrm{d}t. \qquad (10.1)$$

The averaging indicated by $\langle\,\rangle$ is an average over time t, and the result is a function of the time shift τ.

If $\phi(t)$ and $\psi(t)$ are different functions of t, what we make from them is said to be a **cross-correlation** function. If they are the same, what we make is an **autocorrelation** function.

To understand what a correlation function means, think first about the simpler average $\langle \phi(t)^* \, \psi(t) \rangle$. If $\phi(t)$ and $\psi(t)$ take both positive and negative values, but tend to be positive together and negative together, their product will be mostly positive and the average will be positive (they are positively correlated). If the functions tend to have opposite signs, the average of their product will be negative (they are negatively correlated, anti-correlated). If $\phi(t)$ and $\psi(t)$ take both positive and negative values, and do so in dissimilar ways, their product will be as often positive as negative and the average of it will be small or zero (they are uncorrelated). The averaged product, therefore, is a rather good vehicle for giving a value to the degree of resemblance of the two functions.[2,3]

The correlation function $\langle \phi(t)^* \, \psi(t+\tau) \rangle$, including τ, quantifies the extent to which the functions $\phi(t)$ and $\psi(t+\tau)$ tend to resemble (be correlated with) each other when subjected to a relative time shift τ. 'Tends to' means that we are looking for a statistical resemblance, not a definite mathematical dependence; that's inevitable given that we shall be looking at a phenomenon with randomness.

The correlation function needed most often is the autocorrelation of a wavetrain $U(t)$ with itself:

$$\Gamma(\tau) \equiv \langle U(t)^* \, U(t+\tau) \rangle = \frac{1}{T} \int_{-T/2}^{T/2} U(t)^* \, U(t+\tau) \, dt. \tag{10.2}$$

In defining $\Gamma(\tau)$ in eqn (10.2), we have used the analytic signal $U(t)$, rather than the real $V(t)$. This is because we shall need $\Gamma(\tau)$ to describe the resemblance of $U(t+\tau)$ to $U(t)$ in phase as well as in magnitude.[4]

Which of the two amplitudes U in $\Gamma(\tau)$ carries the star (for complex conjugation) is a matter of arbitrary convention; make a choice and stick to it.

Compare eqn (10.1) with eqn (2.29), and check the distinction between a correlation and a convolution.

The comparison of $U(t)$ with $U(t+\tau)$ is implemented by integrating the product over a long time interval T. So the correlation function tells us the correlation (similarity) between the entirety of $U(t)$ and that of $U(t+\tau)$. It is *wavetrains* (as defined in §9.2) that are being compared. In Chapter 9 we had repeatedly to emphasize that coherence and correlation were properties of entire wavetrains (and not a property of, say, $U(t)$ for some specific t); here the notation contains all the reminder we need.

A similar improvement in clarity and conciseness comes from replacing phrases like 'somehow come to know ... permits us to predict' with 'is correlated with'. And we benefit from having a quantitative measure of the correlation as well.

Elementary properties of the correlation function $\Gamma(\tau)$ are worked out in problem 10.1, and will be made extensive use of in what follows.

[2] *Comment*: The reader familiar with statistics may notice: If we collect data on two quantities x and y, and we wish to know whether positive x tends to accompany positive y, or the reverse, we draw a scatter plot of points (x_i, y_i) and see if they lie non-randomly on the xy plane. The quantitative measure of correlation is built from an expression containing $\sum_i x_i \, y_i$, in which we see multiplication used in the same way, and for the same reason, as in $\langle \phi(t)^* \, \psi(t) \rangle$.

[3] *Comment*: There is a similarity here to the mathematics that we use in quantum mechanics when expanding a wave function $\psi(x)$ in terms of some basis set of functions $\phi_i(x)$:

$$\psi(x) = \sum_i a_i \, \phi_i(x),$$

where

$$a_i = \int \phi_i^*(x) \, \psi(x) \, dx.$$

The quantities a_i here are the expansion coefficients of wave function $\psi(x)$, and the integral is usually thought of as just the way that such coefficients happen to be evaluated. But some textbooks (e.g. Matthews (1974)) explain a_i as an 'overlap integral', which evaluates the degree of similarity between $\psi(x)$ and $\phi_i(x)$. If a particular coefficient $a_i = 0$, the function $\phi_i(x)$ is completely absent from the expansion of $\psi(x)$, so there is no resemblance between the functions. If $a_i = 1$, and necessarily all other coefficients are zero, then ϕ_i and ψ are so similar that they're identical.

[4] In this chapter we shall make frequent use of both U and V. The reader is warned to be attentive to which amplitude-quantity is being used, as we shall not interrupt the discussion to flag it on each occasion.

10.3 Autocorrelation and the Michelson interferometer

A Michelson interferometer is shown in Fig. 10.1 on p. 223. The optical paths taken by light through the two interferometer arms differ by $2x$, so light passing through the shorter arm is given a time advance, over that passing through the longer arm, of $\tau = 2x/c$. The amplitudes being superposed at the detector are, therefore, $U(t)$ and $U(t+\tau)$. The intensity measured by a slow detector is proportional to

$$\begin{aligned}&\big\langle\{U(t)+U(t+\tau)\}^*\{U(t)+U(t+\tau)\}\big\rangle\\&\quad=\big\langle|U(t)|^2\big\rangle+\big\langle|U(t+\tau)|^2\big\rangle+\big\langle U(t)^*\,U(t+\tau)\big\rangle+\big\langle U(t)\,U(t+\tau)^*\big\rangle\\&\quad=2\big\langle|U(t)|^2\big\rangle+2\,\mathrm{Re}\big\langle U(t)^*\,U(t+\tau)\big\rangle.\end{aligned}\tag{10.3}$$

The quantity $2\langle|U(t)|^2\rangle$ represents the summed intensities of the two wavetrains. So the other term, in which we recognize the autocorrelation function $\Gamma(\tau)$, must represent the addition caused by interference. The correlation function $\Gamma(\tau)$, therefore, relates very directly to interference effects that we can measure. You are asked in problem 10.1 to check the statements made in this section. We shall discuss the Michelson in more detail in §§10.5 and 10.7.

Experience in §9.12 has shown us that Michelson fringes have a contrast (visibility) that diminishes with increase of $|\tau|$ (to and beyond τ_c). Such a behaviour must likewise be present in the autocorrelation function $\Gamma(\tau)$ since it describes the same partial-coherence physics.

10.4 Normalized autocorrelation function

The correlation function $\langle U(t)^*\,U(t+\tau)\rangle$ takes values that are proportional to the intensity of the light, so it is not an ideal measure of the degree of similarity between wavetrains $U(t)$ and $U(t+\tau)$. It would be better to divide by a quantity proportional to intensity. We, therefore, define a **normalized correlation function** $\gamma(\tau)$ by

$$\gamma(\tau)\equiv\frac{\langle U(t)^*\,U(t+\tau)\rangle}{\langle U(t)^*\,U(t)\rangle}.\tag{10.4}$$

We shall see (it's worked out in problem 10.2) that $|\gamma(\tau)|$ quantifies the degree of statistical similarity between wavetrains $U(t)$ and $U(t+\tau)$ in just the way we would hope for: it ranges on a scale from 0 to 1. Also, the argument of the complex quantity $\gamma(\tau)$ tells us the (best-fitting) relative phase of $U(t)$ and $U(t+\tau)$. We shall now be able to make statements, about wavetrain $U(t)$, like the following:

$U(t)$ has 100% longitudinal coherence over time interval τ if $|\gamma(\tau)| = 1$
$U(t)$ has 88% longitudinal coherence over time interval τ if $|\gamma(\tau)| = 0.88$
$U(t)$ has no longitudinal coherence over time interval τ if $|\gamma(\tau)| = 0$.

Correlation, as measured by $|\gamma(\tau)|$, and coherence are now rather clearly seen to be two words for the same thing.[5]

[5] This fills out the meaning of the 'first and most basic' definition of coherence given in §9.2. In a similar way, a (differently defined) $\gamma(\tau)$ can be made to quantify transverse coherence, see §10.8.

10.5 Fringe visibility

We return to the Michelson interferometer, using it to illustrate the interference of partially coherent light waves.

Equation (10.3) rearranges in an obvious way into

$$\begin{aligned} \text{intensity} &\propto \langle |U(t)|^2 \rangle + \operatorname{Re} \langle U(t)^* \, U(t+\tau) \rangle \\ &= \langle |U(t)|^2 \rangle \{ 1 + \operatorname{Re} \gamma(\tau) \}. \end{aligned} \tag{10.5}$$

Equation (10.17) (obtained in problem 10.2) shows us that $\operatorname{Re}\gamma(\tau)$ is an oscillating function of τ with excursion $\pm|\gamma|$, and that $|\gamma|$ falls off slowly with τ on a scale connected with the coherence time τ_c of the radiation. (Compare also with Fig. 9.7.) Therefore, over a few fringes,

$$\text{visibility} = \frac{I_{\text{max}} - I_{\text{min}}}{I_{\text{max}} + I_{\text{min}}} = \frac{(1+|\gamma|) - (1-|\gamma|)}{(1+|\gamma|) + (1-|\gamma|)} = |\gamma(\tau)|. \tag{10.6}$$

We see that $|\gamma(\tau)|$ is the same measure of coherence (of wavetrain $U(t)$ compared with wavetrain $U(t+\tau)$) as the fringe visibility introduced in §9.13. However, we have taken a somewhat special case, because the two wavetrains being interfered are of equal intensity. Problem 10.3 investigates the case of wavetrains having unequal intensity; $|\gamma(\tau)|$ is then no longer the same as the visibility and it is $|\gamma(\tau)|$ that gives the better measure of coherence.

10.6 The Wiener–Khintchine theorem

We've seen in eqn (9.6a) that a long coherence time is associated with a small bandwidth, and vice versa. So it is unsurprising to find that our correlation functions are related to the spectrum of the radiation.

Let $V(t)$ be the real amplitude of a wavetrain. We define an autocorrelation function using this real amplitude by

$$\big\langle V(t)\, V(t+\tau) \big\rangle = \frac{1}{T} \int_{-T/2}^{T/2} V(t)\, V(t+\tau)\, \mathrm{d}t, \tag{10.7}$$

where T is some long time interval. The Wiener–Khintchine theorem tells us that

$$\int_{-\infty}^{\infty} \big\langle V(t)\, V(t+\tau) \big\rangle \, \mathrm{e}^{\mathrm{i}\omega\tau} \, \mathrm{d}\tau = \frac{|\widetilde{V}(\omega)|^2}{T} = \frac{Z_0}{2n} P(\omega), \tag{10.8}$$

where $P(\omega)$ is the power spectrum defined in eqn (2.22). In words: the power spectrum is proportional to the Fourier transform of the autocorrelation function. Problem 10.4 asks you to invent a proof of the Wiener–Khintchine theorem.

It often happens, in theoretical work, that $\langle V(t)\, V(t+\tau) \rangle$ is reasonably straightforward to calculate, so the Wiener–Khintchine theorem is the best route to finding the frequency spectrum for a **stationary random**

process. For example, it is also important for describing random noise in electrical circuits and elsewhere. The proof of the pudding is that it solves problems. Problems 10.5–10.7 exercise your ability to construct autocorrelation functions for physically interesting wavetrains and to infer the associated power spectra.[6] Problems 10.9 and 10.10 present the Wiener–Khintchine theorem in alternative ways.

[6]Problem 10.8 provides a consistency check.

10.7 Fourier transform spectroscopy

A Michelson interferometer, configured as a Fourier transform spectrometer, is shown in Fig. 10.1. The optical system is set up so that the optical path difference $2x\cos\theta$ closely approximates to $2x$; this is achieved by incorporating the fringe-forming lens and a pinhole at its focus just in front of the detector. The detector is 'slow' in the sense introduced in Chapter 9, meaning that it averages over a time long compared to the coherence time of the light. Since a coherence time is only of the order of nanoseconds, this does not preclude the detector from responding to changes of intensity as the fringe pattern is scanned by movement (order of milliseconds to seconds) of one of the interferometer mirrors.

Let electric-field amplitude $V(t)$ reach the detector at time t via the longer interferometer arm. The other part of the wavetrain, traversing the shorter arm, is advanced relatively by $\tau = 2x/c$ and contributes $V(t+\tau)$. The detector receives the superposed amplitude $V(t)+V(t+\tau)$ and (compare §10.3) records an intensity $I(\tau)$ given by

$$I(\tau) \propto \langle V(t)^2\rangle + \langle V(t)\,V(t+\tau)\rangle.$$

Setting $\tau = 0$ we find that, with the same proportionality constant,

$$\text{intensity for zero time-shift} = I(0) \propto 2\langle V(t)^2\rangle.$$

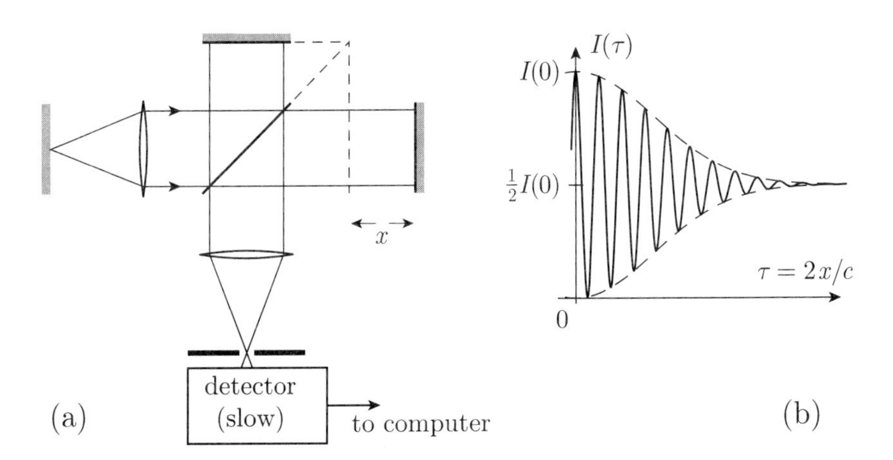

Fig. 10.1 (a) A Michelson interferometer set up to function as a Fourier-transform spectrometer. A fringe-forming lens and a pinhole in front of the detector select just the centre of the ring-fringe interference pattern. One mirror is moved so that the optical path difference $2x$ is scanned from zero to $2x_{\text{max}}$. (b) A sample interferogram such as might result from plotting the intensity at the detector against time delay $\tau = 2x/c$.

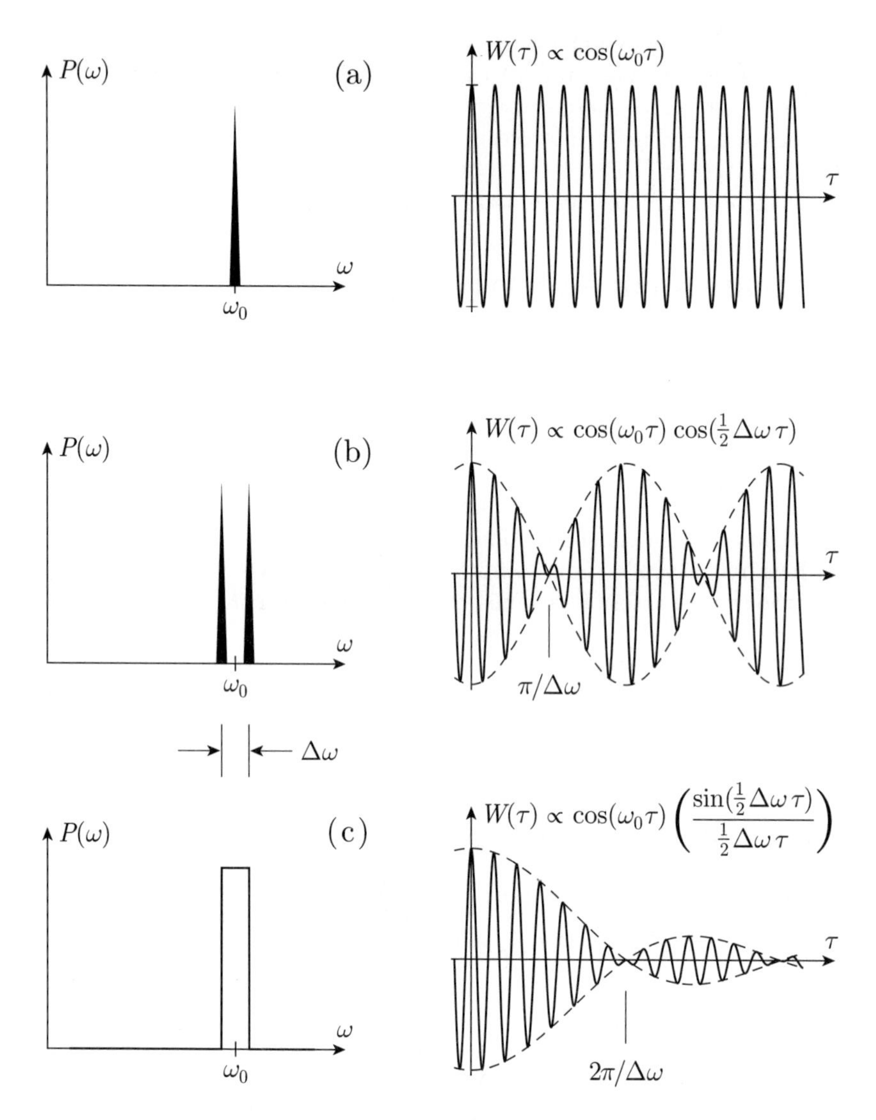

Fig. 10.2 Some examples of spectra and the associated interferograms obtained by passing the light through a Fourier-transform spectrometer. More examples are given in Fig. 10.3. In all cases in Figs. 10.2 and 10.3, except case (a) here, the ratio of centre frequency to FWHM, $\omega_0/\Delta\omega$, is 27/4, a ratio chosen to avoid artefacts that sometimes accompany a simpler ratio. In case (a) the frequency spectrum consists of a single δ-function at angular frequency ω_0. In case (b) the spectrum consists of two δ-functions separated by $\Delta\omega$. Problem 10.12 asks you to check all the details.

Then

$$W(\tau) \equiv I(\tau) - \tfrac{1}{2}I(0) \propto \big\langle V(t)\,V(t+\tau)\big\rangle. \tag{10.9}$$

We may now apply the Wiener–Khintchine theorem to this:[7]

$$P(\omega) \propto \int_{-\infty}^{\infty} W(\tau)\,\mathrm{e}^{\mathrm{i}\omega\tau}\,\mathrm{d}\tau \propto \int_{0}^{\infty} W(\tau)\,\cos(\omega\tau)\,\mathrm{d}\tau, \tag{10.10}$$

where the final expression is a valid alternative because $W(\tau)$ is an even function of τ.

[7] *Comment*: Result (10.10) can, of course, be derived by elementary methods, not relying on knowledge of the Wiener–Khintchine theorem, see problem 10.11.

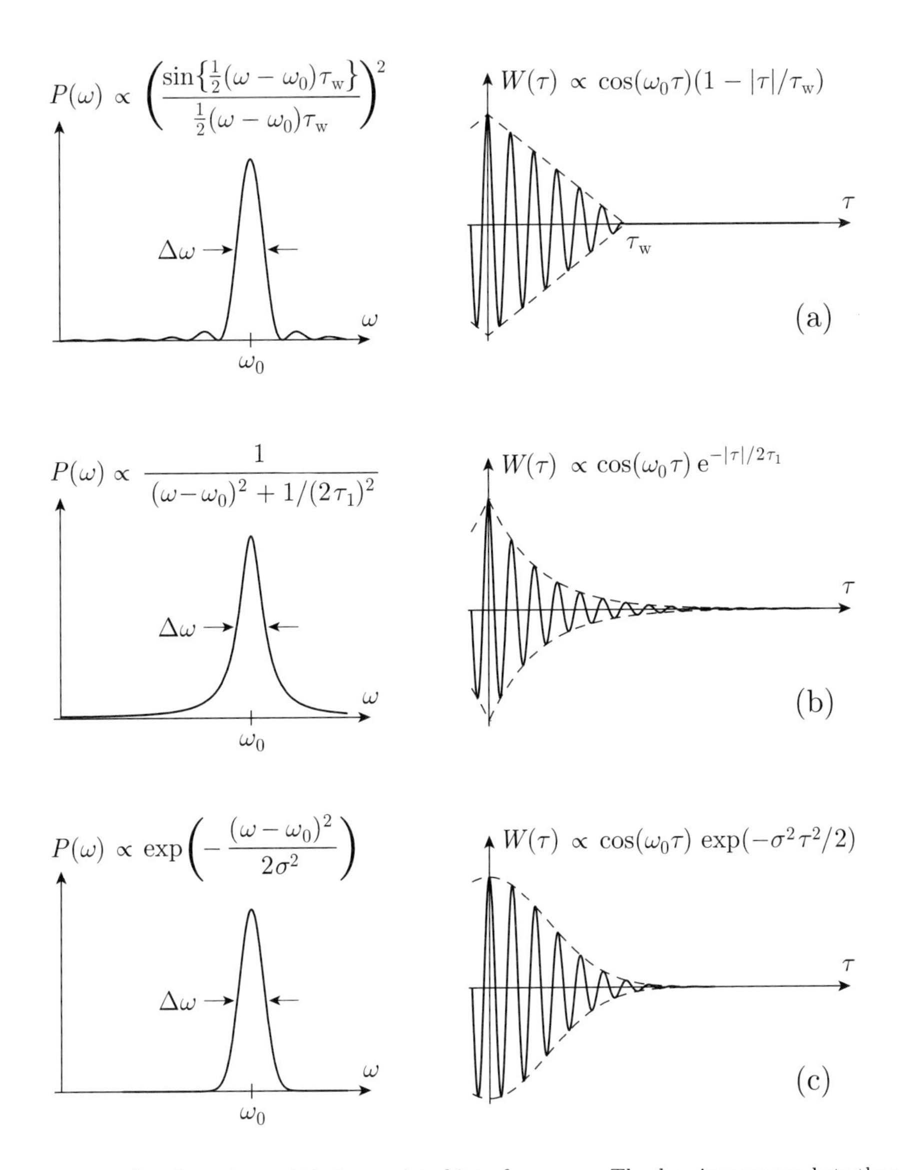

Fig. 10.3 Three more examples of spectra and their associated interferograms. The drawings are made to the same conventions and to the same scale as those of Fig. 10.2. In (a), τ_w is related to the FWHM $\Delta\omega$ by $\tau_w = 5.566/\Delta\omega$. In (b) $\tau_1 = 1/\Delta\omega$. In (c), the Gaussian parameter σ is related to the FWHM $\Delta\omega$ by $\sigma = \Delta\omega/\sqrt{8\ln 2}$.

It should now be clear how a spectrum can be deduced from measurements. The mirror spacing x is scanned and values of $I(\tau)$ are measured and fed into a computer. (The computer, therefore, holds within itself the equivalent of an interferogram.) Once a complete set of measurements has been collected, the Fourier transform is calculated, to give the spectrum $P(\omega)$ of the light.

The Fourier-transform technique has a number of advantages over other spectroscopic methods (at least in certain circumstances). Reasons why, and points of practical design, are discussed in Chapter 11.

Figures 10.2 and 10.3 catalogue a number of simple spectra and the associated interferograms, so that we have some model cases against which to compare experimental data. It should be stressed, however, that the Fourier-transform method does not rely on a person recognizing the shape of an interferogram, as the Fourier transform can be computed for a spectrum of arbitrary complexity.

10.8 Partial coherence: transverse

Correlation-function methods can be used to deal with cases of transverse coherence, in much the same way as they deal with longitudinal coherence. The advantages are similar too: we can quantify all degrees of coherence from none to full on a scale from 0 to 1; and we have a mathematical description that encapsulates the correct statistical ideas.

Figure 10.4 shows a wave (allowed to have some general shape) incident from the left on an aperture containing two pinholes. The pinholes 'sample' the optical field at points $\boldsymbol{r}_1$ and $\boldsymbol{r}_2$, and cause the sampled fields to overlap at places downstream. (Overlap may be because light was diffracted by passage through the small pinholes or because a lens is used as in Fig. 9.1.) The result is to superpose, at points like P, fields proportional to $U(\boldsymbol{r}_1, t)$ and $U(\boldsymbol{r}_2, t+\tau)$, where $\tau = (l_1 - l_2)/c$ is a time shift introduced by the difference of the two paths to P. We recognize that interference is happening by looking for fringes displayed in the plane through P, effectively scanning the value of τ as we shift our attention across the plane.

Our interest is now in the statistical resemblance of $U(\boldsymbol{r}_1, t)$ to $U(\boldsymbol{r}_2, t)$, and its dependence on the separation $|\boldsymbol{r}_2 - \boldsymbol{r}_1|$. Nevertheless, we have to vary τ as well, in order to explore the fringe pattern. It sounds complicated, but it's just the same as we found in Chapter 9.

To quantify the *resemblance* between $U(\boldsymbol{r}_1, t)$ and $U(\boldsymbol{r}_2, t)$, uncomplicated by dependence on intensities, we require a normalized cross-correlation function[8]

[8] The denominator is constructed, after the style of eqn (10.4), to make the result independent of the absolute intensity of *either* wavetrain.

$$\gamma_{12}(\tau) = \gamma(\boldsymbol{r}_1, \boldsymbol{r}_2, \tau) \equiv \frac{\left\langle U(\boldsymbol{r}_1, t)^* \, U(\boldsymbol{r}_2, t+\tau) \right\rangle}{\sqrt{\left\langle |U(\boldsymbol{r}_1, t)|^2 \right\rangle \left\langle |U(\boldsymbol{r}_2, t)|^2 \right\rangle}}. \tag{10.11}$$

This expression looks even more complicated than those encountered

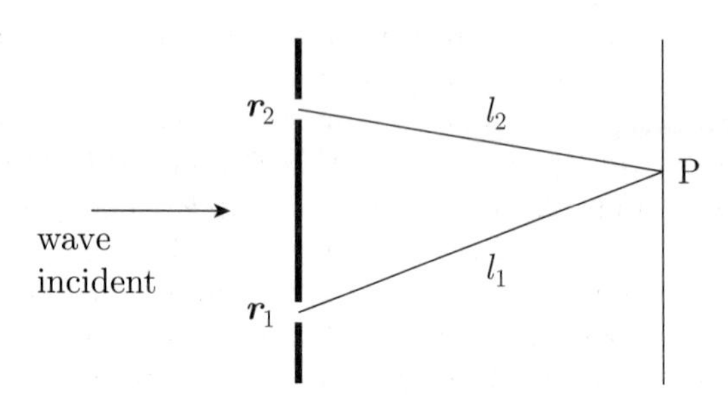

Fig. 10.4 Two pinholes form part of an 'apparatus for measuring transverse coherence' similar to the right-hand part of Fig. 9.1. Light travelling on from the pinholes is superposed in the plane through P, and may form fringes there. The contrast (visibility) of those fringes can be analysed to give the degree of coherence $|\gamma(\boldsymbol{r}_1, \boldsymbol{r}_2, \tau)|$ of the fields at $\boldsymbol{r}_1$ and $\boldsymbol{r}_2$.

previously, but we have just seen the need for it to contain all three of $\boldsymbol{r}_1$, $\boldsymbol{r}_2$ and τ. The 'degree of transverse coherence' between the fields at $\boldsymbol{r}_1$ and $\boldsymbol{r}_2$ is now quantified[9,10] by $|\gamma_{12}(\tau = 0)|$. A possible procedure for obtaining this quantity from experiment is outlined in problem 10.13.

We may look back to the qualitative discussions of transverse coherence in Chapter 9, and in particular to the itemized statements in §9.5. These statements may be made again in the language of correlation functions. The electromagnetic wavetrain passing through pinhole Σ_2 in Fig. 9.1:

- is *fully coherent* with the wavetrain passing through pinhole Σ_1 if $|\gamma_{12}(\tau = 0)| = 1$
- is *partially coherent* with that at Σ_1 if $0 < |\gamma_{12}(\tau = 0)| < 1$
- is *incoherent* with that at Σ_1 if $|\gamma_{12}(\tau = 0)| = 0$.

Moreover, this time following the enumerated statements in §9.5:

(1) For light arriving from a uniform square source subtending angle θ_s, $|\gamma_{12}(\tau = 0)|$ first falls to zero[11] when Σ_1 and Σ_2 are separated (transversely) by the coherence 'area' $d_0 = \lambda/\theta_s$.
(2) For light arriving from a uniform circular source subtending angle θ_s, $|\gamma_{12}(\tau = 0)|$ first falls to zero when Σ_1 and Σ_2 are separated (transversely) by $d_0 = 1.22\lambda/\theta_s$.
(3) For light arriving from a uniform circular source subtending angle θ_s, $|\gamma_{12}(\tau = 0)|$ falls to 88% when $d = \lambda/(\pi\theta_s)$, and this is conventionally used to define an 'area' (meaning radius d) within which there is 'almost full' coherence.

As with the longitudinal case, $|\gamma_{12}(\tau)|$ is the same as visibility when the two waves being interfered (at P in Fig. 10.4) have equal intensity. But, if the intensities are unequal, $|\gamma|$ and visibility differ, and it is $|\gamma_{12}(\tau)|$ that gives the better measure of coherence. See problem 10.13.

10.9 The van Cittert–Zernike theorem

This theorem helps with the calculation of the transverse correlation function $\gamma_{12}(\tau)$, and gives an insight into the way that transverse coherence behaves. This is a good time to work problem 10.14, as the theorem is most easily understood by first seeing an example.

We refer to Fig. 10.5. Light is to travel to the XY plane, via an aperture in the xy plane. Suppose first that light falling on the xy-plane aperture is fully coherent transversely, having come from a point source, and is so arranged that a ray-optics image of that source is formed in the XY plane at $\mathrm{P}(X, Y)$. Surrounding P there will be a diffraction pattern whose amplitude (normalized to 1 at P) is $F(\xi, \eta)$ at location $(X + \xi, Y + \eta)$. This sets the scene. Now instead illuminate the xy-plane aperture with light that is fully *incoherent* transversely over the area of the aperture.[12] Light reaching the XY plane will now have a degree of randomness. The correlation function $\gamma_{12}(\xi, \eta)$ tells us the statistical

[9] If the two pinholes are illuminated obliquely, $|\gamma_{12}(\tau)|$ will have its peak at a non-zero value of τ. To be correct even in such cases we should quantify coherence with $|\gamma_{12}(\tau)|_{\max}$ in place of $|\gamma_{12}(\tau = 0)|$. This complication will be ignored henceforth for simplicity.

[10] We should point out that the locations of the fringes—any shift from the symmetrical position—can give us the argument of the complex $\gamma_{12}(\tau)$, and therefore a best-fitting relative phase between $U(\boldsymbol{r}_1, t)$ and $U(\boldsymbol{r}_2, t)$. This point relates to that of the previous sidenote, because a phase shift is most likely to be the result of oblique illumination. However phase shifts are usually of less interest than the modulus $|\gamma|_{\max}$, and so we give them little emphasis here.

[11] 'First falls to zero' is right. For discussion see §10.9.

[12] The aperture is probably being used as a bright source shining light into a wide range of directions towards the right of the diagram.

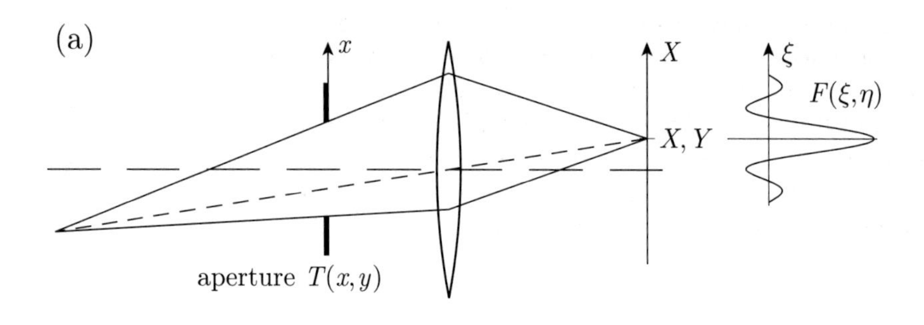

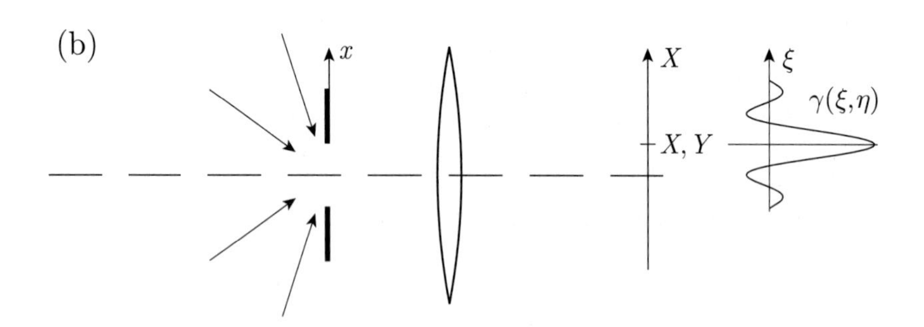

Fig. 10.5 Illustrating the van Cittert–Zernike theorem. In (a), light is arranged to form a Fraunhofer diffraction pattern $F(\xi, \eta)$ centred on (X, Y), having been restricted in width by an aperture with amplitude transmission $T(x, y)$. In (b) the same aperture is illuminated (transverse-) incoherently and the coherence properties of the transmitted light are sought in the XY plane. Correlation function $\gamma(\xi, \eta)$ measures the similarity between amplitudes $U(X + \xi, Y + \eta)$ and $U(X, Y)$. The van Cittert–Zernike theorem states that $\gamma(\xi, \eta) \propto F(\xi, \eta)$.

There is no need for there to be any lens between aperture and plane of observation (that shown is located in a nothing-particular way), and one has been included only to make the drawing deal with a rather general case.

similarity between the amplitude at $(X + \xi, Y + \eta)$ and that at (X, Y). The theorem says[13,14] that $\gamma_{12}(\xi, \eta) = F(\xi, \eta)$.

In the example of problem 10.14, we illuminate a slit of width w using light that is incoherent in the direction transverse to the slit. Light transmitted by the slit is gathered by a lens (focal length f), in a layout reminiscent of Fig. 3.4, and we examine the transverse coherence of the field in that lens's focal plane. The transverse-coherence correlation function $\gamma(\xi, \eta) = \text{sinc}(\frac{1}{2}kw\xi/f)$, where $\text{sinc}(x) = (\sin x)/x$. The similarity between this and the diffraction-pattern expression is obvious.[15]

The sinc function just quoted is derived in problem 10.14(6). It explains sidenote 10 on p. 199. The degree of coherence-correlation $|\gamma(\xi)|$ falls to zero when $\xi = d_0 = \lambda/(w/f)$, but then it rises again to go through a succession of diminishing peaks. If a pair of Young slits were placed at a separation ξ within one of these side peaks, fringes would be observable though with poor contrast. An arbitrary cut-off is conventionally made when assigning a width to the coherence area, and it is the distance to the first zero of $\gamma(\xi)$. The supposedly qualitative discussions of Chapter 9 were arranged to give expressions consistent with this cut-off.

[13] Notice that in case (b) of Fig 10.5 there is a fairly uniform intensity of light across a large area of the XY plane. So $\gamma(\xi, \eta)$ does not represent any fall-off of intensity (or amplitude) with ξ, η. It really does do 'what we came for', and describes the fall-off of correlation-resemblance-similarity of the fields.

[14] A derivation of the van Cittert–Zernike theorem can be found in Born and Wolf (1999), §10.4.2.

[15] The van Cittert–Zernike theorem finds application—and a more intuitive explanation in terms of transverse modes—in §11.5, in particular in sidenote 12 on p. 252.

A similar understanding applies to light received from a uniform circular source such as a star. The coherence 'area' $d_0 = 1.22\lambda/\theta_s$ is the separation of two points at which $|\gamma|$ falls to zero. The statements of §9.5, relating to cases like the Michelson stellar interferometer, are now seen to be quantitatively correct. We take the opportunity to mention problem 10.15, which shows how $|\gamma|$ can be obtained by processing measurements made in an automated way with a modern stellar interferometer.

10.10 Intensity correlation

It may help to refer back to the comment of §9.11, on p. 206, and to the electronic model of narrow-band noise in Fig. 9.5. If we have 'ordinarily random' light it 'forgets' what it has been doing on a timescale of the coherence time τ_c. This 'forgetting' should apply as much to the peak amplitude, and therefore the intensity, as it does to the phase. Therefore, we expect light to exhibit fluctuations of its intensity on timescale τ_c. It's natural to ask: can we observe these fluctuations, and do they provide an alternative route to measurement of τ_c —and, come to think of it, perhaps of coherence area as well? A surprise is that these apparently over-detailed questions turn out to be extraordinarily fruitful.

To discuss fluctuations of intensity we need new correlation functions. To start with we look at the 'longitudinal' case and define

$$\gamma^{(2)}(\tau) \equiv \frac{\big\langle U(t)^*\, U(t+\tau)^*\, U(t+\tau)\, U(t)\big\rangle}{\big\langle U(t)^*\, U(t)\big\rangle \big\langle U(t+\tau)^*\, U(t+\tau)\big\rangle} = \frac{\big\langle I(t)\, I(t+\tau)\big\rangle}{\big\langle I(t)\big\rangle^2}. \tag{10.12}$$

This quantifies the extent to which intensity $I(t+\tau)$ is correlated with $I(t)$. Correlation function $\gamma^{(2)}(\tau)$ is called a 'second-order' correlation function while our previous $\gamma(\tau)$, distinguished if necessary by being relabelled as $\gamma^{(1)}(\tau)$, is said to be 'first-order'.

The intensity $I(t)$ is always positive, so the correlation $\big\langle I(t)\, I(t+\tau)\big\rangle$ is necessarily positive too. We can get a better understanding of intensity correlation by thinking about the way that the light's intensity undergoes fluctuations[16] around its mean $\langle I(t)\rangle$. The interesting quantity is

$$\frac{\Big\langle \{I(t) - \big\langle I(t)\big\rangle\}\, \{I(t+\tau) - \big\langle I(t+\tau)\big\rangle\}\Big\rangle}{\big\langle I(t)\big\rangle^2} = \gamma^{(2)}(\tau) - 1 \tag{10.13}$$

(see problem 10.16). In view of special cases to be taken in the next section, we should point out that property (10.13) is general.

[16] *Comment*: The intensity I needs to be thought about here. It was defined in eqn (1.20) as an average over time, the intention being to remove 'lumpiness' of $\boldsymbol{E}(t) \times \boldsymbol{H}(t)$ taking place on timescale π/ω. The same averaging is implied here. However the average defining I should be thought of as taken over a time only just long enough to remove the 'lumps'. To see why, note that averaging over a time greater than τ_c would remove the fluctuations that we have 'come for'.

10.11 Chaotic light and laser light

Section 10.10 began with mention of 'ordinarily random' light. Clearly this needs explanation. What is meant is light emitted by atoms or molecules that radiate independently, such as atoms in an ordinary gas discharge, or the emitters of black-body radiation. In the present context this is called **chaotic** light.[17]

[17] The name is somewhat unfortunate, clashing as it does with the modern meaning of 'chaos' as an extreme sensitivity to initial conditions and the consequences thereof.

We may think of chaotic light in the following way. Atoms radiate photons (which we may choose to model as classical wave packets) at random times and with random phases. By chance, sometimes the light radiated within a short time contains more wave packets than average, sometimes fewer. It is these random variations that are responsible for the intensity fluctuations that we measure.

Chaotic light must be contrasted with the light emitted by a laser. The strong radiation field within the laser (having many photons per mode) causes new photons to be added in phase with (and otherwise strongly correlated with) those already present. Different photons are not radiated independently. The effect is to give the laser light very different statistical properties from those of chaotic light. In fact, the output from a c.w. (continuous-wave) laser resembles the nice clean sinewave that we obtain from an oscillator at radio frequencies: such radiation has only slight fluctuations of intensity and therefore has $\gamma^{(2)} = 1$.

A property of chaotic light (though not of c.w.-laser light) is that there is a connection between the second-order and first-order correlation functions:[18]

[18] Problem 10.17 shows that a condition must be met for measurements to yield this value of $\gamma^{(2)}$. The light detected must be that occupying only a single transverse mode.

$$\gamma^{(2)}(\tau) - 1 = |\gamma^{(1)}(\tau)|^2. \tag{10.14}$$

Here $\gamma^{(1)}(\tau)$ is the first-order (amplitude) correlation function, previously written simply as $\gamma(\tau)$. Equation (10.14) demonstrates the correctness of the surmise we made at the beginning of §10.10. Fluctuations of intensity (for chaotic light) manifest the same underlying physics as fluctuations of complex amplitude: measurements of intensity correlations can give information on $|\gamma^{(1)}(\tau)|^2$. This function of τ has a peak, whose width tells us the coherence time τ_c of the light.[19]

[19] The fact that we are able to obtain only the modulus of $\gamma^{(1)}(\tau)$ means we have less information that we might ideally wish: for a single spectral line of angular frequency ω_0, $\gamma^{(1)}(\tau)$ contains a factor $\exp(-i\omega_0 t)$. If we knew $\mathrm{Re}\,\gamma^{(1)}(\tau)$ we would have an oscillating function encoding the absolute frequency of the light, and we could feed it into eqn (10.8) to find the power spectrum. As it is, we are unable to deduce absolute frequencies in the light. However, if we know by other means that we have a single spectral line we are able—in principle at least—to learn a lot about the profile of that line from $|\gamma^{(1)}(\tau)|$.

10.12 The Hanbury Brown–Twiss experiment

This classic experiment demonstrated that intensity correlations can be measured and that they provide a valid tool for measuring coherence time or coherence area—for the case of light that is chaotic.[20]

[20] *Comment*: The name of the experiment should be punctuated correctly. The experimenters were one person called R. Hanbury Brown and a second called R. Q. Twiss. No comma after Hanbury, which would make them three people. And no hyphen between Hanbury and Brown because there isn't one.

For practical reasons (see below), Hanbury Brown and Twiss chose to measure coherence area, rather than coherence time. Their apparatus is outlined in Fig. 10.6. A mercury lamp illuminated a pinhole, and the field downstream from the pinhole was investigated using two photodetectors. A half-reflecting mirror made it possible for the detectors to have a 'sideways' relative displacement that was adjustable over a range that included zero. When the detectors were optically coincident, they received light carrying identical fluctuations of intensity, and a positive correlation was recorded at the output of an electronic multiplier ('correlator'). When one detector was moved so as to be outside the coherence area surrounding the other, the correlation disappeared. In this way the coherence area could be measured. The expected size of the coherence

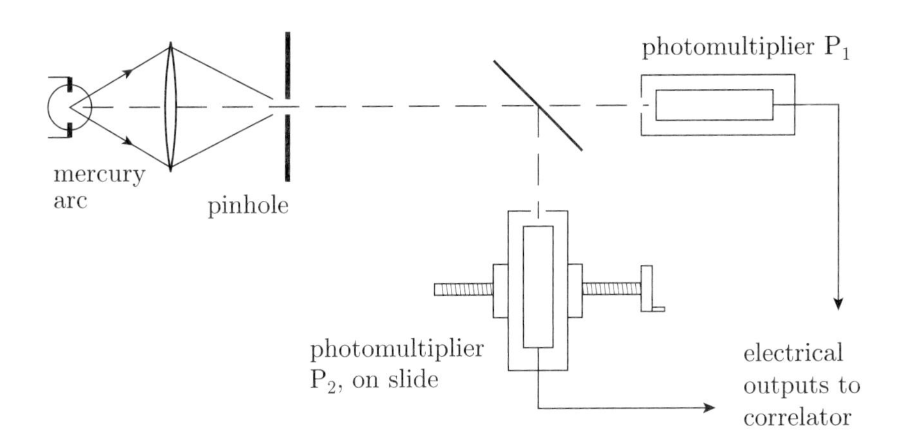

Fig. 10.6 The apparatus used in the experiment of Hanbury Brown and Twiss. Light from a powerful mercury lamp is focused onto a pinhole. From there it travels via a half-reflecting mirror to two photodetectors; Brown and Twiss used photomultipliers. The lateral position of one photodetector is adjustable, and the half-reflecting mirror makes it possible for the lateral separation (of one detector and the image of the other) to be reduced to zero. A square aperture in front of each photodetector defines the area over which light is collected, arranged to be just less than a coherence area for the illumination conditions. Each detector's electrical output is a.c.-coupled to the correlator (a multiplier), so that only fluctuations away from the mean intensity are transmitted to the correlator. The correlator generates an output voltage proportional to the time average of the product of the intensity fluctuations. The correlator's output falls to zero when the photomultipliers are given a lateral separation equal to (the linear dimension of) a coherence area.

area was, of course, known from the diameter of the pinhole and the apparatus geometry.

The mercury discharge lamp used by Hanbury Brown and Twiss gave out a spectral line with a linewidth of several terahertz. The intensity fluctuations likewise had Fourier components up to terahertz. The available detectors, amplifiers and multiplier did not have a response up to such high frequencies,[21] and this limitation very seriously diminished the fluctuation signals; the experiment had to be designed to work even so. In particular, the absence of high frequencies precluded any measurement of the light's coherence time ($\sim 10^{-12}$ s), and it was for this reason that the experimenters chose to investigate coherence area instead. More recent experiments using faster electronics (and narrower-bandwidth light) have been able to measure coherence time as well as coherence area.

Practical requirements in relation to coherence area are explored in problem 10.17. An estimate of the number of wave packets that overlapped is made in problem 10.18.

In the original Hanbury Brown–Twiss experiment, it was necessary for the lamp to be bright, in order for its intensity fluctuations to be detectable reliably above noise in the detection process.[22] Since a laser is very bright, there is an obvious temptation to think that the Hanbury Brown–Twiss experiment should be easy if done using a laser source. This is not so. As we have mentioned, c.w. laser light has $\gamma^{(2)}(\tau)$ close to 1, so there are no fluctuations of intensity to record.[23] We remind the reader again that eqn (10.14) does not apply to laser light.

[21] In the original experiment, the electronics responded up to about 40 MHz, and this restricted bandwidth had the effect of diminishing the signals by a factor of about 10^{-5}. We can say that the detection system was not really 'fast' in the sense of responding to frequencies up to $1/\tau_c$, but it just avoided being so 'slow' as to average away all trace of the fluctuations.

[22] When light (even constant-intensity light) is photodetected, there is randomness in the current of photoelectrons, an effect called shot noise. The 'wanted' signal must be detectable in the presence of shot noise, a condition which requires a greater intensity than is needed simply to give overlapping wave packets. We do not give the mathematical analysis here. A similar addition of shot noise explains why the form of eqns (10.34) requires correction.

[23] And any that do exist do not obey eqn (10.14), so they don't tell us about anything except the laser's imperfections.

10.13 Stellar diameters measured by intensity correlation

In §9.6 and problem 9.8 we mentioned that atmospheric turbulence sets a limit on the resolution obtainable with a ground-based optical telescope. The as-Michelson-invented-it stellar interferometer (and the modern refinements mentioned in §9.6) overcomes this difficulty, but only up to a point. It's inconvenient that the fringes move around as the optical paths above the telescope fluctuate, though problem 10.15 shows how measurements can be automated to cope with this.

Atmospheric turbulence has a less damaging effect on the intensity of light than on the phase.[24] It is, therefore, attractive to think of building a new kind of Michelson stellar interferometer that measures the correlation of intensity, and using the coherence area of that correlation to obtain the angular diameter of the star being observed.

Given that measurement of intensity correlations requires bright light, it may seem puzzling that intensity fluctuations can be observed and measured with starlight, which we are used to thinking of as dim. We need to be quite careful with 'dim'. Imagine first that we could achieve a resolved image of a nearby star, by using a very large telescope freed from problems with the atmosphere. Then we would be working with an extended source having a temperature of 6000 K or so, and the conditions would be no worse than those for sunlight. We would even need to aperture the image so as to utilize only one coherence area from within it! More simply, we could imagine using a smaller (but still very large) telescope whose objective spans only one coherence area of the incident light. Either recourse would collect as much optical power as is in-principle usable. With a telescope or more practical size, we suffer a penalty in energy collection (relative to optimum) equal to the ratio (area of telescope objective)/(coherence area of incident light). This is the meaning of 'dim' in the present context.[25] However, the light received occupies a wide optical bandwidth—it is 'white' black-body radiation—and this gives an advantage (relative to the original laboratory experiment which used a single broad spectral line) of 100 or so, enough to make the measurement workable.

An intensity interferometer consists of two separate telescopes, one or both mounted on rails so that the separation can be adjusted. Each collects light onto a photodetector that records the intensity being collected. The two resulting electrical signals are multiplied together in the same way as in the Hanbury Brown–Twiss experiment, and the coherence area is measured by separating the telescopes until the intensity-fluctuation correlation is brought to zero. There is no need for optical paths to be matched to wavelength accuracy, so apparatus requirements are not as critical as for the Michelson (phase-correlation) instrument.[26]

It gets better. Errors of optical path up to $l_c = c\tau_c$ are tolerable in principle *within* each telescope; we do not need the optical quality of a telescope as the term is usually understood. Successful measurements

[24] The atmosphere does produce intensity changes, known as 'scintillation', the 'twinkling' of starlight. The timescale for scintillation is milliseconds. Intensity variations on this timescale are removed by a high-pass filter in the electronics, with a cutoff frequency of (say) 1 MHz. The removal of these low frequencies is harmless to the intensity-correlation measurement (though it would wreck the original Michelson measurement).

[25] Suppose that the coherence area is 190 m across, close to the upper limit for practical measurement, and each telescope has an area of $30\,\text{m}^2$. Then the penalty in energy gathering is a factor of about $2 \times 30/(190)^2 \sim 10^{-3}$. These figures are not far from those for a practical apparatus, as described by Hanbury Brown (1974).

[26] Optical paths do need to be matched to within the effective value of the coherence length $l_c = c\tau_c$, so the need for careful adjustment has not gone away entirely. In particular, if one telescope is nearer to the star than the other by l (which could be several metres), a time delay l/c must be applied (optically or electrically) to equalize the effective values of the two paths to within l_c. Problem 10.19 asks you to think about the 'effective value' of l_c and its order of magnitude.

have been made using searchlight reflectors or reflectors made from a honeycomb pattern of individual facets (Hanbury Brown 1974). What sets the limit on optical quality is the requirement that the telescopes should be looking at a well-defined small area of sky.

Intensity interferometry can be more convenient than the original Michelson 'amplitude interferometry' for the reasons given above. It does, however, have a significant limitation: it gives adequate signal/noise ratio only for stars that are sufficiently bright. There is a significant range[27] of stellar magnitudes over which amplitude interferometry is usable but intensity interferometry is not.

A readable account of intensity-correlation astronomy is given by Hanbury Brown (1974).

[27] Tango and Twiss (1980) give stellar magnitude 2 as the limit for intensity interferometry and magnitude 9 for amplitude interferometry. Since 2.5 magnitudes represent a factor-10 ratio of intensities, the range from magnitude 2 to magnitude 9 represents a factor approaching 1000.

10.14 Classical and quantum optics

This book is primarily about classical (non-quantum) optics. Up until now quantum phenomena have been mentioned only occasionally, in connection with the detection of photons, photoelectrically and photographically. And we have used classical 'wave packets' to model photons without anxiety as to the validity of the description.

Things change quite dramatically when we deal with intensity correlations. The quantum nature of light, and of the photon detection process, become much more important.

A warning is contained in problem 10.21. A natural-broadened spectral line (lifetime-broadened) of ordinary spectral intensity,[28] with its light gathered in a single transverse mode, has 'sparse' wave packets, occupying only about 2% of the time. If such a wavetrain is detected with a photomultiplier, photoelectrons are emitted singly at random times—nothing remarkable.[29] Yet a naïve classical calculation in problem 10.21 gives a $\gamma^{(2)}(\tau = 0)$ of order 50 (even greater if the light is made dimmer). This failure makes it clear that the detection process—at the very least—must be treated quantum-mechanically, and included in any description of what is measurable.

[28] The intensity per unit of frequency range.

[29] It is of course not being suggested that natural-broadened light is commonly or easily obtained in the laboratory—pressure broadening and Doppler broadening ordinarily dominate with spectral lines, while black-body radiation (even if filtered) is another story. The point being made is that simple models (e.g. that of §9.12) build in something equivalent to natural broadening, and such models are no longer adequate; indeed they may be actively misleading.

We are not precluded from ever using classical models, even in present contexts. Problem 10.20 shows that a laser beam contains typically 10^9 or more photons per mode, so many that it is a 'classical stable wave' (a nice clean sinewave). A classical Maxwell-equations description is appropriate to almost all properties of a laser beam.[30] Problem 10.18 shows that the Hanbury Brown–Twiss experiment was designed to overlap so many wave packets at the detectors that a classical picture (of fluctuations as variations in the number of overlapping photon wave packets) was basically valid. What made this possible was the use of light covering a wide range of frequencies: very different from the natural-broadening conditions assumed in the last paragraph.

[30] Though not, we are now discovering, to ordinary chaotic light in a single spectral line of ordinary intensity. The fact that we have never had to worry about this until now is itself something requiring explanation.

These considerations raise an issue in the justification of eqn (10.14). If a *simple* wave-packet model is used, then the spectral intensity must be thousands of times greater than is realistic in order for enough wave packets to overlap. If the spectral intensity is not to be so high then the model should incorporate wave packets whose energy spreads over a widened range of frequencies, which would make it overcomplicated. Or we must grasp the problem of doing the job using quantum mechanics.

One line of quantum-mechanical attack is given by Glauber (1970). He gives a quantum-mechanical derivation of eqn (10.14) using a density-matrix approach. This in effect implements an ensemble-averaging of the kind that is familiar from statistical mechanics.

A rather different approach, and one that fits more neatly with what we have said above, is given by Mandel and Wolf (1995), especially their chapter 9. They show that for most purposes a semiclassical treatment is valid, in which the field amplitude of the light can still be treated in a classical manner, but photoelectron emission is linked to that field amplitude via quantum-mechanical rules. The result is an expression for the intensity correlation with much the same content as eqn (10.14) but with refinements to allow for the sensitivities and frequency responses of the detectors, and describing a correlation in the *photoelectrons*. To pursue either this or the density-matrix calculations here would carry us too far beyond the remit of the present book. We, therefore, content ourselves with providing in problem 10.22 a derivation of eqn (10.14) using a somewhat shameless model: using the same wave packets as in problem 10.6 but with the intensity increased to the point that many wave packets overlap within the wavetrain.

One question must be given some sort of answer before we leave the present topic. Why have quantum effects made themselves felt now when they did not matter previously? Or perhaps the question should be the other way round: why could we ignore quantum effects in all earlier work when—we now see—the (non-laser) light was not intense enough to be in the classical limit where 'one photon more or less makes no difference'? Why, to take a specific case, could we describe interference of dim light in a Michelson interferometer or in the Young slits?

An answer can be given by appeal to the Correspondence Principle. It is open to us to imagine that interference effects (Michelson or Young) are calculated using light that is well into the classical limit (many photons per mode, perhaps originating from a laser). Consider now what happens if we weaken the light. Maxwell's equations are linear, and the same applies to the Schrödinger equation and its relativistic cousins that describe photons. So the predictions of a classical calculation should simply scale with diminution of amplitude, and should remain valid for light that has few photons per mode. Correlation functions like $\Gamma(\tau)$ and $\gamma^{(1)}(\tau)$ are a part of this same description, so their use is justified also.

The Correspondence Principle does not help with intensity-fluctuation phenomena, because the 'interference effects' are no longer the result of linear superpositions of the field amplitudes. It is, therefore, not open to us to calculate with strong fields and then argue that weak fields will behave similarly except for a simple scaling.[31]

[31] We only need think of the precedent of density fluctuations in a dilute gas. If the number of molecules within a small volume is N, this number fluctuates randomly by $\sim\sqrt{N}$. So fluctuations of density are proportional, not to the mean density, but to its square root.

Now that quantum effects have been mentioned, we may contemplate a whole variety of new phenomena. One is 'photon antibunching'. It is possible to prepare quantum states of light which, when detected, yield photoelectrons that are less-than-randomly coincident; this is in contrast to the intensity correlations for chaotic light where the photons are more-than-randomly coincident (where the intensity happens to be high, photons are clustered). We can think of antibunched light as having fairly sparse photon wave packets that are prevented from being close to each other in time. We mention antibunching here just because it contrasts with the intensity correlations of chaotic light discussed earlier in the present chapter. But 'quantum optics' embraces a wide range of other non-classical phenomena: entangled states and quantum computing, for example. The reader should consult books on quantum optics for accounts of these exciting developments.

Problems

Problem 10.1 (a) Properties of the autocorrelation function

(1) Show that $\Gamma(\tau) = \Gamma(-\tau)^*$, and that both $\mathrm{Re}\,\Gamma(\tau)$ and $|\Gamma(\tau)|$ are even functions of τ. Show likewise that $\mathrm{Im}\,\Gamma(\tau)$ is an odd function of τ.

(2) Show that $\langle U(t)\,U(t+\tau)\rangle = 0$, which explains why we had to conjugate one of the amplitudes in forming $\Gamma(\tau)$. Look at the definition of $U(t)$ in eqn (2.15) and at problem 2.10(2).

(3) Show that

$$\langle V(t)\,V(t+\tau)\rangle = \langle V(t)\,V(t-\tau)\rangle = \tfrac{1}{2}\,\mathrm{Re}\,\Gamma(\tau), \tag{10.15}$$

and that it is an even function[32] of τ.

(4) Show that for a monochromatic wavetrain with complex amplitude $U = U_0\,\mathrm{e}^{-\mathrm{i}\omega t}$, $\Gamma(\tau) = U_0^*\,U_0\,\mathrm{e}^{-\mathrm{i}\omega\tau}$.

Comment: Compare parts (3) and (4). The correlation $\langle V(t)\,V(t+\tau)\rangle$ can be zero, for some particular value of τ:

either because there is no long-term similarity between wavetrains $V(t)$ and $V(t+\tau)$,

or because the two wavetrains are similar but are 90° out of phase.

A similar ambiguity does not afflict $\langle U(t)^*\,U(t+\tau)\rangle$. This fills out a possibly opaque explanation given in §10.2 for our preferring when possible to use a correlation function built from U rather than from V.

[32] The fact that $\Gamma(-\tau)$ is strongly related to $\Gamma(\tau)$ can be linked to sidenote 22 on p. 205, where we spoke (though using the example of weather forecasting) of 'retrodicting' $U(t-\tau)$ from $U(t)$ as being similar to the predicting of $U(t+\tau)$.

Problem 10.2 (a) Basic properties of $\gamma(\tau)$

(1) Show that $\gamma(\tau)$ has the same symmetry properties as were found for $\Gamma(\tau)$ in problem 10.1(1).

(2) Show that $\gamma(0) = 1$.

(3) Consider the monochromatic wave of problem 10.1(4). Show that

$$\gamma(\tau) = \mathrm{e}^{-\mathrm{i}\omega\tau} \qquad \text{and so} \qquad |\gamma(\tau)| = 1 \quad \text{for all } \tau.$$

These two cases show that situations known to possess complete longitudinal coherence do indeed have $|\gamma(\tau)| = 1$.

(4) We must make sure that $\gamma(\tau)$ has the property

$$0 \le |\gamma(\tau)| \le 1,$$

otherwise our hoped-for interpretation of it can't be sensible. We have a special case of the Schwarz inequality. Prove that $|\gamma(\tau)| \le 1$ by your own method or from the following suggestion. The statement

$$\left\langle \left| U(t) - e^{i\alpha}\, U(t+\tau) \right|^2 \right\rangle \ge 0$$

must be true for any choice of α, since the quantity being averaged is a square. Use this to show that $\mathrm{Re}\{e^{i\alpha}\,\gamma(\tau)\} \le 1$; and tidy up.

(5) Consider a nearly monochromatic wavetrain (without using a specific model for it such as that in §9.12), whose mean angular frequency is ω_0, and whose coherence time $\tau_c \gg 1/\omega_0$. Argue qualitatively that

$$\gamma(\tau) = \begin{cases} e^{-i\omega_0 \tau} & \text{when } |\tau| \ll \tau_c \\ 0 & \text{when } |\tau| \gg \tau_c. \end{cases} \tag{10.16}$$

Make a plausibility argument (perhaps guided by Figs. 10.2 and 10.3) that

$$\gamma(\tau) = e^{-i\omega_0\tau} \times \left(\begin{array}{l} \text{a function of } \tau \text{ that varies slowly from} \\ \text{a maximum of 1 at } \tau = 0 \text{ to zero when} \\ |\tau| \to \infty \text{, with a peak width of order } \tau_c \end{array} \right). \tag{10.17}$$

Problem 10.3 (b) Michelson interferometer, correlation and visibility

(1) Confirm the statement in §10.5 that the intensity at the output of a Michelson interferometer introducing a path difference $2x$ is proportional to

$$\left\langle |U(t)|^2 \right\rangle \{1 + \mathrm{Re}\,\gamma(\tau)\}, \tag{10.18}$$

where $\tau = 2x/c$.

(2) Look back to eqn (10.17) and to the likely orders of magnitude of the optical period $2\pi/\omega$ and the coherence time τ_c. Check the steps in §10.5 leading to eqn (10.6): the visibility of Michelson fringes is the same as $|\gamma(\tau)|$. Fringe visibility gives a good quantitative measure of the degree of coherence-correlation of the two wavetrains being compared. Unfortunately, this simplicity isn't always available, as the rest of this problem will show.

(3) Pass a collimated beam of light into a Michelson interferometer as previously, but adapt the interferometer to attenuate the amplitude of the light passing through one arm. Let the amplitudes travelling via the two arms be (after emerging from the half-reflecting mirror) $U_1(t)$ and $U_2(t+\tau)$, where $U_2(t)/U_1(t)$ is the constant factor by which U_2 has been weakened relative to U_1. Let the intensities of the two beams be I_1 and I_2.

Consider first the case where the input light is monochromatic with angular frequency ω. Show that the intensity leaving the interferometer is proportional to

$$I_1 + I_2 + 2\sqrt{I_1\, I_2}\cos(\omega\tau).$$

Show that when $I_2 \neq I_1$, there is no value of τ, not even near zero, which makes the output intensity zero, so the visibility must be less than 1 even when the coherence is perfect.[33] We'll now define a correlation function that does better.

[33] You are asked to obtain this statement from the mathematics. It is, however, obvious. If two waves have unequal strength, the weaker cannot completely cancel the stronger, even when they are in antiphase, so the waves cannot cancel to give complete darkness.

(4) Define a **normalized cross-correlation function** by

$$\gamma_{12}(\tau) \equiv \frac{\big\langle U_1(t)^*\, U_2(t+\tau)\big\rangle}{\sqrt{\langle |U_1(t)|^2\rangle\,\langle |U_2(t)|^2\rangle}}. \tag{10.19}$$

Modify the calculation of problem 10.2(4) to show that the new correlation function satisfies the basic requirement for a measure of coherence

$$0 \le |\gamma_{12}(\tau)| \le 1.$$

(5) The monochromatic radiation of part (3) is input to the interferometer. Show that for this case

$$\gamma_{12}(\tau) = \mathrm{e}^{-\mathrm{i}\omega\tau} \qquad \text{and so} \qquad |\gamma_{12}(\tau)| = 1;$$

$|\gamma_{12}(\tau)| = 1$ holds for a case that we know has full coherence.

(6) Pass a collimated beam of light into the interferometer, modified as in part (3), but now the light is non-monochromatic. Show that the output intensity is proportional to

$$I_1 + I_2 + 2\sqrt{I_1\, I_2}\,\mathrm{Re}\,\gamma_{12}(\tau).$$

Show further that, for a near-monochromatic input,

$$\text{visibility } V(\tau) = \frac{2\sqrt{I_1\, I_2}}{I_1 + I_2}\,|\gamma_{12}(\tau)|, \qquad |\gamma_{12}(\tau)| = \frac{I_1 + I_2}{2\sqrt{I_1\, I_2}}\,V(\tau). \tag{10.20}$$

(7) Show directly from the square root in definition (10.19) that $\gamma_{12}(\tau)$ is unaffected by changing the attenuation in the interferometer (provided of course that this is done without affecting the path difference or the light source). This, of course, explains the apparently intimidating square root in the definition of $\gamma_{12}(\tau)$.

Problem 10.4 (a) The Wiener–Khintchine theorem

(1) Make a brute-force proof; if you solved problem 2.18, a similar route will probably serve here.

(2) Show that $\widetilde{V}(-\omega) = \widetilde{V}(\omega)^*$ is the transform of $v(t) = V(-t)$. Show that

$$\big\langle V(t)\, V(t+\tau)\big\rangle = \big\langle V(t) V(t-\tau)\big\rangle = \big\langle V(t)\, v(\tau - t)\big\rangle$$

and use the convolution theorem. Compare with problem 2.21(3).

Problem 10.5 (b) Sinewave wave packets occurring at random times

Consider again the wave-packet model described in §9.12. Represent the wavetrain by

$$V(t) = \sum_i V_0 \cos\{\omega_0(t - t_i) + \alpha_i\}\, h(t - t_i), \tag{10.21}$$

where a 'top-hat' function $h(t)$ is defined by

$$h(t) = \begin{cases} 1 & \text{for } 0 < t < \tau_{\mathrm{w}} \\ 0 & \text{otherwise.} \end{cases}$$

In eqn (10.21), the ith wave packet starts at time t_i and has phase α_i. The start times t_i are randomly distributed in time. The wavetrain $V(t)$ has N wave packets per unit time and lasts for time T, so the sum over i includes NT wave packets.

(1) Show that

$$\langle V(t)\, V(t+\tau)\rangle = \begin{cases} \frac{1}{2}V_0^2\, N \cos(\omega_0 \tau)\{\tau_{\mathrm{w}} - |\tau|\} & \text{for } |\tau| < \tau_{\mathrm{w}} \\ 0 & \text{otherwise.} \end{cases}$$

Show that this result holds for the case where all the α_i take the same value, which is the assumption made in §9.12. Show that it also holds in the case where the α_i are random phase angles (independent of the t_i).

(2) Find the power spectrum using the Wiener–Khintchine theorem. Check with the result found in §9.12. If there seems to be an extra term check against problem 9.11(1).

(3) Obtain the fringe visibility that results from passing the given wavetrain through a Michelson interferometer. Show that the visibility is $1 - |\tau|/\tau_{\mathrm{w}}$ (provided this is positive), and check against Fig. 9.8.[34]

[34] *Comment*: Problem 9.11 asked you to work out $\langle V(t)\, V(t+\tau)\rangle$ without saying that was what it was.

(4) How would things be changed if the wavetrain were obtained by 'gating' a single continuous sinewave, that is, opening a shutter at random times t_i to let through portions of that sinewave?

Problem 10.6 (b) Natural (lifetime) broadening of a spectral line
Consider a large number of atoms, which are put into an excited state at random times, and then radiate on a single atomic transition (spectral line). Each emission generates a 'wave packet', which we shall describe classically although we are to think of it as modelling a photon. The only mechanism for broadening the spectral line is the finite radiative lifetime τ_{r} of the excited state ('natural' broadening).

(1) Show that a suitable form for the wavetrain's amplitude is

$$V(t) = \sum_i V_0 \exp\{-(t-t_i)/2\tau_{\mathrm{r}}\} \cos\{\omega_0(t-t_i) + \alpha_i\}\, u(t-t_i), \tag{10.22}$$

where $u(t)$ is the 'unit step function'

$$u(t) = \begin{cases} 1 & \text{for } t > 0 \\ 0 & \text{for } t < 0. \end{cases} \tag{10.23}$$

In eqn (10.22), the ith wave packet starts at time t_i and has phase α_i. The start times t_i are randomly distributed in time. The wavetrain $V(t)$ has N wave packets per unit time and lasts for time T, so the sum over i includes NT wave packets.

As with problem 10.5, the phase angles α_i may be taken to be all equal or to have a random distribution (uncorrelated with the t_i); both cases are to be investigated.

(2) Use the given wave packet in a model based on that of §9.12 and obtain $\langle V(t)\,V(t+\tau)\rangle$ for the radiation emitted. Show that

$$\big\langle V(t)\,V(t+\tau)\big\rangle = \tfrac{1}{2}V_0^2\,N\tau_\mathrm{r}\cos(\omega_0\tau)\exp\{-|\tau|/2\tau_\mathrm{r}\}. \tag{10.24}$$

(3) Use the Wiener–Khintchine theorem to find the power spectrum for the radiation. Show that it has the form[35]

$$P(\omega) \propto \frac{1}{(\omega-\omega_0)^2 + 1/(2\tau_\mathrm{r})^2}. \tag{10.25}$$

[35] *Hint*: You will have to approximate, by agreeing to look at angular frequencies ω for which $|\omega-\omega_0| \ll \omega_0$. The same approximation happened in problems 9.11 and 10.5.

Comment: We have here a classical model for the 'natural' broadening of a spectral line, meaning the broadening caused by the finite lifetime of the emitted wave packet. The frequency dependence in eqn (10.25) is called a **Lorentzian**[36] line profile. The discussion here is more realistic than that given in elementary treatments, in that the random emission of many wave packets is properly taken into account. Yet the mathematics, once the unfamiliarity of the presentation is overcome, is hardly more complicated.

[36] A Lorentzian profile has already been encountered in Chapter 5, §9.11, and in one of the graphs of Fig. 10.3.

Problem 10.7 (b) Pressure (collision) broadening of a spectral line
We'll use a model for 'soft' collisions. During emission of a wave packet, an atom or molecule suffers one or more collisions with other atoms. The duration of a collision is negligible,[37] so its effect on the radiation is (almost instantly) to randomize the phase α of the emitted wave packet. We'll assume that the probability that a collision happens in time interval $\mathrm{d}t$ is $\mathrm{d}t/\tau_\mathrm{coll}$, independent of anything other than $\mathrm{d}t$ (such as when the previous collision happened).

(1) Show that the probability of a wave packet retaining a given value of α for a time of at least t is $\exp(-t/\tau_\mathrm{coll})$. Show that the probability of a wave packet retaining a given value of α for time t and then having it randomized in the next $\mathrm{d}t$ is $\exp(-t/\tau_\mathrm{coll})\,\mathrm{d}t/\tau_\mathrm{coll}$. Show that τ_coll is the mean time between phase-changing collisions.[38]

(2) In gas kinetic theory it is often assumed that the probability of a collision happening in a length $\mathrm{d}x$ of an atom's path is $\mathrm{d}x/\lambda$ where λ is the mean free path. Is this compatible with the original assumption of this problem, that collisions happen equally often in every time interval $\mathrm{d}t$?

(3) Show that the effect of collisions on $\langle V(t)\,V(t+\tau)\rangle$ is to multiply it by $\exp\{-|\tau|/\tau_\mathrm{coll}\}$.

(4) Use the model of radiation emission given in problem 10.6, in which atoms emit randomly occurring wave packets with radiative lifetime τ_r. Show that when both lifetime and collision effects are included

$$\big\langle V(t)\,V(t+\tau)\big\rangle = \tfrac{1}{2}V_0^2\,N\tau_\mathrm{r}\cos(\omega_0\tau)\exp\{-|\tau|/2\tau_\mathrm{r}\}\exp\{-|\tau|/\tau_\mathrm{coll}\},$$

where N is the number of wave packets emitted per unit time.

[37] The 'duration of a collision' is a time interval within which the radiation has slightly altered frequency and therefore accumulates an altered phase. It may well last for several optical periods, but it is assumed to be small compared with the radiative lifetime τ_r, and also small compared to the time between collisions. This is of course saying that the gas within which the atom lives is not too dense.

[38] *Hint*: A similar argument is used in the kinetic theory of gases to find the statistical distribution of free paths. There is also a similarity to the distribution of lifetimes in radioactive decay.

(5) Use the Wiener–Khintchine theorem to show that the power spectrum of the radiation is[39]

$$P(\omega) \propto \frac{1}{(\omega - \omega_0)^2 + (1/2\tau_r + 1/\tau_{coll})^2}. \tag{10.26}$$

Line broadening by collisions is usually called **pressure broadening** since $1/\tau_{coll}$ is proportional to the number density of atoms.[40] Both 'natural' and pressure broadening are cases of **homogeneous** broadening, within the jargon of spectroscopy and laser physics.

[39] *Comment*: As with problem 10.6 and eqn (10.25), the result (10.26) can be obtained by more elementary methods. But it is more satisfactory to use an explicitly statistical argument to deal with a superposition of random wave packets. In fact, for the present problem, 'elementary' methods are extraordinarily messy .... By contrast, problem 10.7 is so straightforward that there was no temptation even to deal separately with collisions and with the finite radiative lifetime.

[40] The broadening is not dependent upon pressure alone, because broadening and pressure have different dependences on temperature. A name like 'collision broadening' would be better chosen.

Problem 10.8 (c) More on pressure broadening
Problems 10.6 and 10.7 show that a Lorentzian line profile results from either natural broadening, or pressure broadening, or both.

(1) When a spectral line suffers two broadening mechanisms it usually has a profile that is the convolution of the two contributing profiles. For example, Doppler broadening and natural broadening yield such a convolution (called a **Voigt profile**). Show this. See if you can construct a convincing argument that a convolution should describe the combination of natural and pressure broadening.[41,42]

(2) Show by brute force that the convolution of two Lorentzians is another Lorentzian, whose width is the sum of the contributing widths.[43,44]

[41] *Comment*: We can work the reasoning in reverse to show from eqn (10.26) and part (2) that the convolution of part (1) (convolving profiles for natural and pressure broadening) must be correct. So even if you can't justify the convolution directly, at least you know it's right.

[42] Problem 5.10 shows another application of this convolution: the instrumental line profile of a Fabry–Perot pattern introduces a third Lorentzian to be convolved in making the overall line profile.

[43] A non-Fourier method may require a contour integration. Surprisingly, it's easier—and far more transparent—to take the Fourier transform of each function in the convolution, multiply the transforms, and transform back again.

[44] *Comment*: The addition property of part (2) is special to Lorentzians, and does not apply to other line shapes. In particular, the result of convolving two Gaussians is another Gaussian, but the width is the square root of the sum of the squares of the contributing widths.

Problem 10.9 (b) Alternative Wiener–Khintchine theorems

(1) Prove that

$$P(\omega) = \frac{n}{Z_0} \int_{-\infty}^{\infty} \Gamma(\tau) \cos(\omega\tau)\, d\tau. \tag{10.27}$$

Do this two ways:

(a) by putting $V = \frac{1}{2}(U + U^*)$ into eqn (10.8),

(b) by making use of eqn (10.15).

(2) Adapt one of the arguments you used in problem 10.4 to show directly that

$$\frac{n}{2Z_0} \int_{-\infty}^{\infty} \Gamma(\tau)\, e^{i\omega\tau}\, d\tau = \begin{cases} P(\omega) & \text{for } \omega > 0 \\ 0 & \text{for } \omega < 0. \end{cases} \tag{10.28}$$

(3) Think hard about the two results just obtained, looking back at problem 2.10. There are some surprising subtleties to be unravelled.

(4) Repeat problem 10.6 starting from the complex wavetrain

$$U(t) = \sum_i V_0 \exp\{-(t - t_i)/2\tau_r\} \exp\{-i\omega_0(t - t_i) - i\alpha_i\}\, u(t - t_i)$$

and using eqn (10.28). The calculation is a little easier than before and there is no need to approximate. But wait: the result of problem 10.6 *was* approximate, yet now the same result seems to be exact: we've proved too much. Sort this out![45]

[45] *Comment*: A trap is revealed by the contradiction of part (4). And as traps go, this one is very easy to fall into. It explains why the Wiener–Khintchine theorem was set up in §10.6 using the $V(t)$ correlation function, rather than the more obvious $\Gamma(\tau)$.

Problem 10.10 (c) Yet another Wiener–Khintchine property
Consider light emitted in a single spectral line centred on angular frequency ω_0. Define a line-profile (lineshape) function $g(\omega - \omega_0)$, normalized so that

$$\int_0^\infty g(\omega - \omega_0)\frac{\mathrm{d}\omega}{2\pi} = 1. \tag{10.29}$$

The line-profile function contains the frequency dependence of the power spectrum $P(\omega)$, but normalized to take out the absolute intensity of the light.[46]

Show that the 'one-sided' power spectrum of eqn (10.28) is related to the normalized line-profile by

$$P(\omega) = \frac{n}{2Z_0}\Gamma(0)\, g(\omega - \omega_0).$$

Show further that the normalized line-profile function is related to the normalized autocorrelation function by[47]

$$\gamma^{(1)}(\tau) = \int_0^\infty g(\omega - \omega_0)\,\mathrm{e}^{-\mathrm{i}\omega\tau}\frac{\mathrm{d}\omega}{2\pi}; \qquad g(\omega - \omega_0) = \int_{-\infty}^\infty \gamma^{(1)}(\tau)\,\mathrm{e}^{\mathrm{i}\omega\tau}\,\mathrm{d}\tau;$$

$$\int_{-\infty}^\infty |\gamma^{(1)}(\tau)|^2\,\mathrm{d}\tau = \int_0^\infty g(\omega - \omega_0)^2\,\frac{\mathrm{d}\omega}{2\pi}.$$

[46] Notice that the lower limit of integration is zero. We are taking it that our spectral line contains frequencies that do not extend down to anywhere near zero; and $g(\omega - \omega_0)$ is defined not to contain any mirror-image frequencies centred on $\omega = -\omega_0$. Remember that the $P(\omega)$ of eqn (2.22) is an even function of ω and does have such a mirror-image peak.

[47] You may be tempted to use a rotating-wave approximation (see the solution to problem 10.9), but in fact these results come out exactly if you use eqn (10.28).

Problem 10.11 (a) Fourier transform spectroscopy
(1) Use elementary methods first. Supply to the interferometer a monochromatic wave of angular frequency ω. Show that the fraction of the energy passed to the output (assuming an ideal beam splitter) is $\frac{1}{2}(1 + \cos\omega\tau)$, where $\tau = 2x/c$ as in §10.7.
(2) Supply the interferometer with power $P(\omega)\,\mathrm{d}\omega/2\pi$ in the vicinity of angular frequency ω. Find the power transmitted to the output.
(3) Now supply the interferometer with power covering a range of frequencies. Show that the power[48] reaching the output is $I(\tau)$, where

$$I(\tau) = \tfrac{1}{2}I(0) + \frac{1}{2}\int_0^\infty P(\omega)\cos(\omega\tau)\,\frac{\mathrm{d}\omega}{2\pi}. \tag{10.30}$$

[48] In this book, $P(\omega)$ represents power per unit area, and so $I(\tau)$ should likewise be a 'per area' quantity. This detail is unimportant here.

(4) Hence show that

$$P(\omega) = 8\int_0^\infty W(\tau)\cos(\omega\tau)\,\mathrm{d}\tau, \tag{10.31}$$

where $W(\tau) \equiv I(\tau) - \frac{1}{2}I(0)$ as in eqn (10.9).
(5) Now do the job by the correlation-function route outlined in §10.7, checking the mathematical statements made there.
(6) Show that when $\tau \gg \tau_\mathrm{c}$, where τ_c is the coherence time of the radiation, $I(\tau) \to \frac{1}{2}I(0)$ and so $W(\tau) \to 0$. This step shows that $W(\tau)$ behaves in a way that presents no mathematical obstacle to the taking of the Fourier transform (integrals will converge).

Problem 10.12 (a) Examples of spectra and interferograms
Check all the quantitative information in the spectra and their interferograms of Figs. 10.2 and 10.3.

Problem 10.13 (b) Fringe visibility with the Young slits
(1) Show that the intensity observed at point P in Fig. 10.4 is proportional to

$$I_1 + I_2 + 2\sqrt{I_1\, I_2}\, \mathrm{Re}\, \gamma_{12}(\tau),$$

where $\tau = (l_1 - l_2)/c$, and I_1, I_2 are the intensities arriving at P via $\boldsymbol{r}_1$ and $\boldsymbol{r}_2$. (Obvious similarities to problem 10.3.)
(2) Show that the fringe visibility is $\{2\sqrt{I_1\, I_2}/(I_1 + I_2)\}\, |\gamma_{12}(\tau)|$. You may assume that the radiation is nearly monochromatic, so there is no significant fall of $|\gamma_{12}(\tau)|$ over the range of τ that needs to be explored in a measurement of the visibility. It will be clear now that $|\gamma_{12}(\tau)|$ can be obtained from experimental measurements of visibility, together with measurements of intensities I_1 and I_2.
(3) An adequate idea of the fringe visibility can be obtained by scanning across four or five fringes either side of centre. Over this distance, we want there to be no significant additional fall-off of fringe visibility owing to the finite coherence length of the radiation. Adapt the discussion of §9.3 to find the bandwidth permitted by this requirement.
(4) In a particular experimental arrangement, the collimation of the incident light is adjusted to make $|\gamma(\boldsymbol{r}_1, \boldsymbol{r}_2, 0)| = \frac{1}{2}$, and the intensities arriving at P via the two pinholes are equal. Obtain the fringe visibility for this case.
(5) The light arriving at the pinholes is the same as in part (4), but that transmitted through pinhole 2 is attenuated to half its original amplitude, without imposing any new phase shift. Show that this leaves $\gamma(\boldsymbol{r}_1, \boldsymbol{r}_2, \tau)$ unaltered but changes the fringe visibility to $\frac{2}{5}$.

Problem 10.14 (b) Young fringes with imperfect collimation
This problem re-examines the transverse coherence of §9.4. A square hole S of side w is illuminated from the left as shown in Fig. 10.7, and the light transmitted passes through a lens to plane Σ. We investigate the transverse coherence of the optical field produced in plane Σ.

Light illuminates hole S within angle range 2α in the plane of the diagram, while the diffracted light is to be examined in plane Σ and along the X-axis only. We take the light to be collimated in the 'other' direction, so there are no complicating phase variations with y.

(1) Show that the amplitude $U(X, t)$ of light diffracted at time t to location X in plane Σ is[49]

$$U(X, t) = \frac{w\, \mathrm{e}^{\mathrm{i}kd}}{\mathrm{i}\lambda f} \int_{-w/2}^{w/2} U(x, t)\, \mathrm{e}^{-\mathrm{i}kxX/f}\, \mathrm{d}x, \qquad (10.32)$$

where d is the optical path from the centre of S to the centre of Σ.

[49] *Comment*: The functions of time t are written in eqn (10.32) in a way that is not quite correct, because the notation doesn't properly indicate that $U(X, t)$ depends upon $U(x, t)$ at an earlier time. However, this physics, called 'retardation', has not been neglected: it is there in the exponential factors $\mathrm{e}^{\mathrm{i}kd}\, \mathrm{e}^{-\mathrm{i}kxX/f}$. In fact, we may notice that eqn (10.32) contains a somewhat curious mixture of functions of time t and angular frequency ω, since there is λ in the coefficient and k in the exponent. Invent for yourself an argument showing that this is harmless.

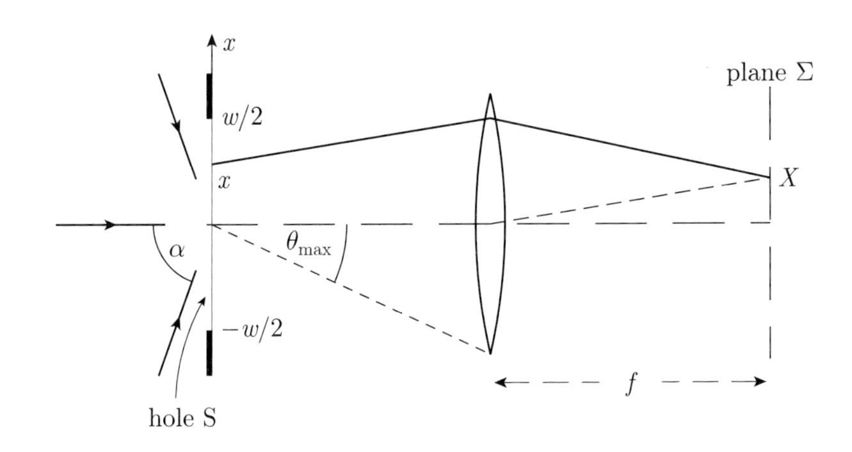

Fig. 10.7 The geometry used in problem 10.14.

(2) Show that the coherence area for light arriving at S has width $\lambda/2\alpha$ when α is small. For this, use the order-of-magnitude estimate of coherence area given in §9.5.

(3) Show that when α is not a small angle, the coherence area is more correctly given by $\lambda/(2\sin\alpha)$. Compare with problem 9.5.

(4) Show that the lens, thought of as part of a system forming an image somewhere to the right, can resolve detail in the plane of S down to a size of order $\lambda/\sin\theta_{\max}$, or possibly about half of this.[50]

(5) Hence justify the use of an 'incoherent limit' so far as $U(x,t)$ is concerned when $\alpha \gg \theta_{\max}$. This limit can be built into the algebra by taking $U(x,t)$ to have the property[51]

$$\left\langle U(x,t)^*\, U(x',t)\right\rangle = A\,\delta(x-x')$$

(same t), where A is independent of x.

(6) Show that on this assumption,[52,53,54]

$$\gamma(\xi) \equiv \frac{\left\langle U(X)^*\, U(X+\xi)\right\rangle}{\sqrt{\left\langle |U(X)|^2\right\rangle \left\langle |U(X+\xi)|^2\right\rangle}} = \frac{\sin(\frac{1}{2}kw\xi/f)}{(\frac{1}{2}kw\xi/f)}. \tag{10.33}$$

(7) Show that in the opposite limit, where S is illuminated by a plane wave with complete transverse coherence across S, $|\gamma(\xi)| = 1$.

Comment: Parts (6) and (7) show that the incoherent limit assumed in part (5) is a worst-case assumption so far as the transverse coherence in plane Σ is concerned. We should naturally make this assumption when designing apparatus aimed at guaranteeing a required degree of coherence in the light beam leaving the lens. Compare this statement with the first comment of §9.4 on p. 198; also the last paragraph of §9.15.2.

(8) The sinc function of eqn (10.33) has its first zero when $\frac{1}{2}kw\xi/f = \pi$. Show that this agrees with the first of the numbered statements in §9.5.

(9) Figure 10.7 has been drawn with the hole and lens separated by the lens's focal length f. Show that if this condition does not hold, the d

[50] Help in §12.6.

[51] A similar step is taken in the solution to problem 9.10, when thinking about field amplitudes of different frequency.

[52] *Comment*: The field $U(X,t)$ can't be calculated explicitly from eqn (10.32) because $U(x,t)$ is an unknown random function of both x and t. Nevertheless, the correlation function $\gamma(\xi)$ *can* be calculated. This illustrates nicely the power of correlation methods for dealing with tough problems. We had a foretaste in problem 4.16 where a random array of pinholes submitted to an attack that borrowed ideas like ensemble averages from statistical mechanics.

[53] *Comment*: Part (6) yields a $(\sin x)/x$ (or sinc) function, which is reminiscent of diffraction from a slit of width w. We have a rather strong pointer to the van Cittert–Zernike theorem.

[54] *Comment*: In part (6) the expression for $\gamma(\xi)$ should have its numerator more correctly written as $\langle U(X,t)^*\, U(X+\xi,t)\rangle$ to indicate that the field amplitudes being multiplied are taken at the same time t. And a similar clarification applies to the averages in the denominator.

of eqn (10.32) is dependent upon X but the derivation of eqn (10.33) is not disrupted.

Problem 10.15 (b) Michelson stellar interferometer: shifting fringes
In a Michelson stellar interferometer (set up for 'amplitude interferometry') the fringes shift randomly as a consequence of atmospheric turbulence. We ask how the interferometer can be instrumented so as to measure the fringe contrast.

Let the intensity of the fringe pattern, as a function of position x, be

$$I(x) = I_0\{1 + |\gamma|\cos(\kappa x + \alpha)\},$$

where $|\gamma|$ is the degree of coherence of the two beams being interfered (one via each of the paths drawn in Fig. 9.2) and κ is related to the wave vector k and the angle at which the two beams converge onto the fringe pattern. Phase angle α represents a phase difference caused by atmospheric turbulence, and it will be assumed to fluctuate randomly over several 2π.

Make two measurements of intensity, called I_1, I_2, in quick succession, in a time so short that α does not have time to change appreciably. Before I_2 is measured, an additional phase of $\pi/2$ is added to one of the interfering beams. Repeat such measurements, recording the results in a computer. Each pair of measurements will have a different α. Averaging the results is equivalent to taking an ensemble average (or time average) over the random values of α.

(1) Confirm that $|\gamma|$ is indeed the degree of coherence of the light arriving via the two paths. The reasoning in eqn (10.6) may help.

(2) Show that

$$\begin{aligned} \left\langle (I_1 - I_2)^2 \right\rangle &= I_0^2\,|\gamma|^2 \\ \left\langle I_1 + I_2 \right\rangle &= 2I_0 \\ |\gamma|^2 &= \frac{4\left\langle (I_1 - I_2)^2 \right\rangle}{\left\langle I_1 + I_2 \right\rangle^2}. \end{aligned} \tag{10.34}$$

Comment: Part (2) shows that the measurements can in principle yield a value for $|\gamma|$ without the need for a human to make observations by eye. It should be mentioned that the calculation here is very much 'in principle'. The intensities being measured are weak, and are measured by **photon counting**.[55] It is necessary to take account of the Poisson statistics in the counted pulses. When this is done, a correction must be applied to the form of eqn (10.34); see Tango and Twiss (1980), Appendix A.

[55] Photons cause photoelectric emission of electrons from the cathode of a photomultiplier. Each electron is accelerated to impact on a 'dynode' where it splashes out 'secondary' electrons by 'secondary emission'. The effect is that several (5–10) electrons now travel inside the photomultiplier instead of the original one: the current pulse has been amplified. The secondary electrons in turn are further multiplied at dynodes until a pulse of many (maybe 10^6) electrons is collected by a final anode and a current pulse is recorded in an external circuit. Pulses are counted and the counts are recorded in computer memory. This technique is known as photon counting, though the name is somewhat optimistic: only 5–30% of incident photons (depending on the wavelength) yield photoelectrons.

Problem 10.16 (a) Intensity fluctuations

(1) Take the statistical fluctuation of intensity (about its mean) as defined by the left-hand side of eqn (10.13). Show that it is equal to $\gamma^{(2)}(\tau) - 1$, where $\gamma^{(2)}(\tau)$ is defined in eqn (10.12).

(2) Take a 'classical stable wave', meaning what we have called a nice clean sinewave, $U = U_0\,\mathrm{e}^{-\mathrm{i}\omega t}$. Show that this wave has $\gamma^{(2)}(\tau) = 1$ and $\gamma^{(1)}(\tau) = \mathrm{e}^{-\mathrm{i}\omega\tau}$. Use this example to show that eqn (10.14) cannot be true in general. This is consistent with the statement in §10.11 that eqn (10.14) is restricted to the case of a chaotic light wave.

Problem 10.17 (b) Conditions for measuring $\gamma^{(2)}(\tau)$
Section 10.11 gives a brief description of chaotic light in terms of overlapping wave packets and fluctuations in their number.

(1) How might we collect the greatest number of wave packets at some detector that records intensity $I(t)$? We decide to focus the light onto the detector. Then we ask: how big an area can the detector have? Argue that it must not be larger than one coherence area: fluctuations of intensity on different coherence areas are independent and so will tend to cancel each other out. But then, the more strongly the light is focused, the more must the detector be limited in area. It is not clear whether focusing has conferred any advantage after all.

(2) Use eqn (11.13) to show that the cancellation found in part (1) is exact: *all* the advantage that we hoped to gain from focusing the light is exactly cancelled by the requirement to limit the detector's area. Conversely, if we try to collect more light by using a detector of large area the range of directions within which light may arrive is restricted. The étendue we may accept is no greater than $\lambda^2_{\text{vacuum}}$.

(3) Use eqn (11.9) to show that light reaching the detector must be selected so that it occupies just one transverse mode. This is another way of understanding the restriction found in part (2).

(4) In the experiment of Hanbury Brown and Twiss, light was admitted through a circular pinhole of diameter 0.19 mm. Each detector was 2.24 m away and had a square aperture in front of it, 5 mm on a side. The wavelength was 435.8 nm. Calculate the coherence area of the light approaching the aperture in front of the detector, and check that the aperture area was suitably chosen.

(5) Do the job another way, by calculating the étendue of the light beam accepted by the detector, comparing it with λ^2.

Problem 10.18 (b) Overlapping wave packets in Hanbury Brown and Twiss
In the experiment of Hanbury Brown and Twiss, the light source was a mercury lamp giving out a wavelength of 435.8 nm. According to Hanbury Brown (1974), p. 8, the distribution-over-states temperature T_{d} was 8000 K. The spectral line was subject to broadening that gave it a width $\Delta\nu$ of 8.5×10^{12} Hz.

Use the information given to show that the number of photons per mode for frequencies at the middle of the spectral line at 435.8 nm was about 1.6%. Using a plausible estimate of the radiative lifetime τ_{r} for the spectral line, estimate the mean number of overlapping wave packets in a single transverse mode.[56]

[56] Help in eqn (11.22).

Problem 10.19 (b) The effective coherence length for intensity correlation

Sidenote 26 on p. 232 says that the signal paths (optical plus electrical) in a stellar intensity interferometer must be matched to within a coherence length $c\tau_c$. Now the light being received is white, so its coherence length is only a micrometre or so. Matching paths to within a micrometre is not much easier than matching to within a wavelength. Is this technique really so much less critical than the original Michelson stellar interferometer?

Problem 10.20 (b) Number of photons per mode in a laser beam

Estimate the number of photons per mode[57] in the output from a helium–neon laser. Take the laser to give output power 1 mW at wavelength 633 nm, and to oscillate in one transverse mode and three longitudinal modes of its cavity. Each mode[58] has frequency width 1 MHz.

[57] *Hint*: Equation (11.10).

[58] The bandwidth of 1 MHz is something of a guesswork figure. The fundamental frequency range of a laser's output is inherently narrow: less than a hertz, caused by the addition, in random phases, of spontaneous-emission photons to the output. Practical lasers have wider bandwidths because of mirror vibrations; if these vibrations are controlled by negative-feedback correction, the laser's frequency width can be brought to a few kilohertz. This shows that the laser beam's photons are unlikely to be randomly distributed over (free-space) longitudinal modes within a bandwidth of $\delta\nu = 1\,\text{MHz}$: the estimated number of photons per mode should be increased accordingly. However, the main point being made here is that the number of photons per mode greatly exceeds the values in Table 11.1, whether by 11 orders of magnitude or by more does not matter too much.

In the above discussion, it is necessary to distinguish between the discrete longitudinal modes of the laser's own cavity (three of them) and the free-space longitudinal modes (forming a continuum within a very long quantization length) envisaged in eqn (11.10). It is tempting to think that these two kinds of longitudinal mode should be related in some straightforward way, perhaps having their occupations connected via the transmission coefficient of the output mirror. As we have seen, the relationship is not so simple.

Problem 10.21 (b) Value of $\gamma^{(2)}(\tau)$ for sparse classical wave packets

In this problem we show that classical models for intensity correlation can fail spectacularly.

Let a beam of light be modelled as in §9.12 and problem 10.5: it consists of identical wave packets of duration τ_w and intensity I_0, occurring randomly at a rate of N per unit time. Problem 11.2 shows that for this to be a realistic model for a natural-broadened spectral line of 'ordinary' intensity the wave packets must be 'sparse', meaning that $N\tau_w \ll 1$. If each wave packet represents a photon the likely order of magnitude of $N\tau_w$ is 2% (see Table 11.1).

Show that $\langle I(t)\rangle = I_0 N\tau_w$ and also that $\langle I(t)^2\rangle = I_0^2 N\tau_w$. Show that $\gamma^{(2)}(\tau = 0) = 1/(N\tau_w) \gg 1$. Show that this could well be of order 50.

Comment: The result just derived would be correct if each wave packet were made up of many ('bunched') photons, so that one photon more or less made little difference: the standard way of taking a classical limit. However, wave-packet models are often used—and have been in this chapter and Chapter 9—with each wave packet representing an individual photon. In that case the $\gamma^{(2)}$ obtained here is an obvious nonsense. If we were to shine the wavetrain onto a photomultiplier, we should receive a photoelectron pulse every so often, with negligible chance of two photoelectrons arriving together. Nothing in the statistics of the photoelectrons would resemble what we have calculated. It is forced on us that the description we give of light in the context of intensity correlation must be inherently quantum mechanical, at the least by inclusion of the physics of photoelectron detection.

Problem 10.22 (c) A shameless model for classical chaotic light

We pursue again the model of problem 10.6, this time permitting the light's intensity to reach the high values where classical physics should apply, even though that intensity is known to be too high to be realistic. The model is then used to derive eqn (10.14). The algebra is somewhat messy, which is why it is relegated to a problem graded (c).

Given the outcome of problem 10.17, we must restrict attention to light within a single transverse mode. Therefore, we can treat the problem as one-dimensional and consider that all wave packets being summed are part of a single beam travelling in one direction. Then amplitude $U(z,t)$ at position z and time t is a function of $(z-ct)$ and at a given z can be written as a function of t only.[59]

To make the mathematics tractable, we write the wavetrain in complex form (compare problem 10.9)[60]

$$U(t) = V_0 \sum_i \exp\{-\mathrm{i}\omega_0(t-t_i)-\mathrm{i}\alpha_i\} \exp\{-(t-t_i)/2\tau_\mathrm{r}\}\, u(t-t_i), \quad (10.35)$$

where there are N wave packets per unit time. The wave packets' starting times t_i are randomly distributed within the wavetrain of long duration T. The function $u(t-t_i)$ is a unit step defined as in eqn (10.23), so that the ith contribution to $U(t)$ is a sinewave that starts suddenly at time $t=t_i$ and then decays exponentially. We may think of the phase angles α_i as randomly distributed independently of the starting times t_i.

(1) Follow a route similar to that in problem 10.6 to show that

$$\left\langle U(t)^*\, U(t+\tau)\right\rangle = V_0^2\, N\tau_\mathrm{r}\, \exp(-\mathrm{i}\omega_0\tau) \exp\{-|\tau|/2\tau_\mathrm{r}\}. \qquad (10.36)$$

Check that this result is consistent with eqn (10.24).

(2) Take the special case $\tau = 0$ of the last result and show that

$$\left\langle U(t)^*\, U(t)\right\rangle = V_0^2\, N\tau_\mathrm{r}. \qquad (10.37)$$

Interpret this result in terms of the energy contained in each wave packet (initial amplitude V_0, radiative lifetime τ_r).

(3) Now we get to the serious stuff. Use expression (10.35) to show that

$$\begin{aligned}&\left\langle U(t)^*\, U(t+\tau)^*\, U(t+\tau)\, U(t)\right\rangle \\ &= V_0^4 \Biggl\langle \sum_i \mathrm{e}^{\mathrm{i}\alpha_i}\, \mathrm{e}^{-(t-t_i)/2\tau_\mathrm{r}}\, u(t-t_i) \times \sum_j \mathrm{e}^{\mathrm{i}\alpha_j}\, \mathrm{e}^{-(t+\tau-t_j)/2\tau_\mathrm{r}}\, u(t+\tau-t_j) \\ &\times \sum_k \mathrm{e}^{-\mathrm{i}\alpha_k}\, \mathrm{e}^{-(t+\tau-t_k)/2\tau_\mathrm{r}}\, u(t+\tau-t_k) \times \sum_l \mathrm{e}^{-\mathrm{i}\alpha_l}\, \mathrm{e}^{-(t-t_l)/2\tau_\mathrm{r}}\, u(t-t_l)\, \mathrm{e}^{-\mathrm{i}\omega_0(t_i+t_j-t_k-t_l)} \Biggr\rangle. \quad (10.38)\end{aligned}$$

(4) As before we take an ensemble average. Deal first with the starting times. Argue that there are three ways to make the ensemble average of the $\mathrm{e}^{-\mathrm{i}\omega_0(t_i+t_j-t_k-t_l)}$ terms non-zero:

$$\begin{aligned} i &= j = k = l \\ i &= k \neq j = l \\ i &= l \neq j = k. \end{aligned}$$

(5) Evaluate the three sums that result from simplifying eqn (10.38) using these possibilities.[61] Show that

[59] This may sound obvious, but an equivalent simplification did not apply to the calculation of eqn (10.33), for example. And the same simplification was not *necessary* (though assumed) in problem 10.6.

[60] See the notes on the solution to problem 10.9(4). We are using a 'rotating wave approximation' in which $U(t)$ slightly breaks the rules for the construction of an analytic signal.

[61] The first sum, over i only, can conveniently be averaged by taking a time average as was done in problem 10.6. A different procedure is better for the double sums: ensemble-average over the distribution of start times t_i and t_j. Since there are NT wave packets in time T the chance that t_i falls within $\mathrm{d}t_i$ is $N\,\mathrm{d}t_i$.

$$\gamma^{(2)}(\tau) = 1 + |\gamma^{(1)}(\tau)|^2 \left(1 + \frac{1}{2N\tau_r}\right), \tag{10.39}$$

Here $N\tau_r$ is the average number of wave packets that overlap each other in the wavetrain, or if we prefer, the probability that two wave packets overlap. Equation (10.39) agrees with eqn (10.14) in the limit $N\tau_r \to \infty$, which condition is equivalent to saying that many wave packets overlap. Thus we have a derivation—of sorts—of eqn (10.14).[62]

[62] A result similar to eqn (10.39), derived using a simpler model and showing a different dependence on $N\tau_r$, is given by Loudon (2000), Chapter 3.

(6) Show that eqn (10.39) also agrees with the outcome of problem 10.21 when we take the limit of small $N\tau_r$, apart from a factor 2 which is the result of using different pulse shapes in the two calculations.

Comment: The deficiencies of wave-packet models have been discussed in §10.14. The model of the present problem, taken literally, represents a natural-broadened spectral line. We know from problem 11.2(3) that the mean number of overlapping wave packets for such a wavetrain is the same as the number of photons per mode, while Table 11.1 shows that this is of order 2% for light from practical light sources.[63] The limiting case $N\tau_r \gg 1$ is not met by a factor of order 1000.

[63] Or course, it is difficult to make a natural-broadened light source, so 'practical' may be objected to.

Ideally we should replace the present model by one in which the number of overlapping wave packets is made large, without summing to an unreasonable spectral intensity, by making up the wavetrain from wave packets of many different frequencies. This would model a case of inhomogeneous broadening (or extreme pressure broadening). We shall not attempt such a model here. However, it may be surmised that the precise shape of the wave packets being summed is not critical, and so the present calculation may not be quite as inadequate as has been represented.

Comment: As mentioned in sidenote 60 (p. 247), the expression for $U(t)$ written in eqn (10.35) has slightly broken the rules for constructing a correct $U(t)$ in the same way as in problem 10.9(4). This approximation is quite unconnected with the largeness or smallness of $N\tau_r$. There is no escape route here that could remove the $1/(2N\tau_r)$ term from eqn (10.39).[64] Equation (10.14) really does rely on the taking of a limit where many wave packets overlap.

[64] It is *not* suggested here that any serious attention should be paid to the $1/(2N\tau_r)$ term in eqn (10.39). Problem 10.21 has shown that whenever this term is large enough to look significant we no longer have a valid classical model (one photon more or less makes little difference). 'If isn't negligible it's wrong.'

Optical practicalities: étendue, interferometry, fringe localization

11

11.1 Introduction

In this chapter we discuss several related topics. First we consider the flux of energy, mostly in the context of light from non-laser sources. An important geometrical factor is the étendue. It is the étendue that we must maximize when trying to get as much optical energy as possible into apparatus such as a spectrometer or an optical fibre. Optical instruments differ in their ability to collect light, and their relative merits in this regard are assessed by comparing their étendues.

We put this idea to use by analysing a Michelson interferometer used as a Fourier-transform spectrometer, and we discover that it is far better at light-gathering than a conventional (grating) spectrometer.

A property of interferometers of the Michelson type (achieving interference by division of amplitude) is that they generate **localized** fringes. We take the opportunity to explain localization, where the fringes lie, and how critical is the focusing on these fringes.

There is an insightful relationship between étendue and the number of transverse modes occupied by the electromagnetic field. There are further links with coherence area, with thermodynamics (entropy), and the understanding as to why laser light is so special.

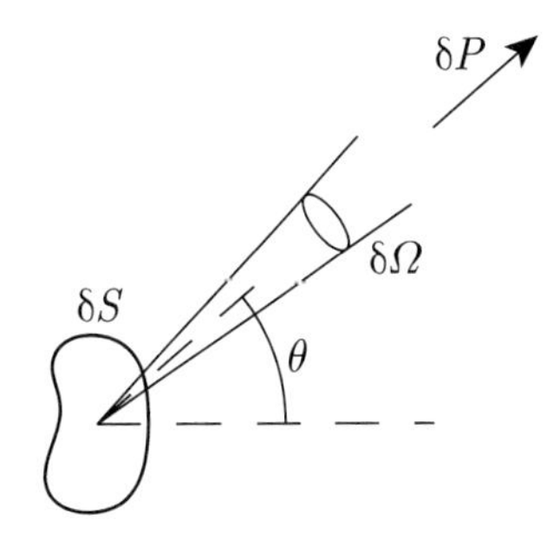

Fig. 11.1 Power δP is radiated from an element of surface area δS into an element of solid angle $\delta\Omega$.

11.2 Energy flow: étendue and radiance

Consider some area element δS that radiates optical power as shown in Fig. 11.1. The power δP that we can collect in a solid angle $\delta\Omega$ is proportional to $\delta\Omega$ and to $\delta S\cos\theta$ (the area projected along the line of sight). Given these dependences, we define a quantity **étendue** by[1]

$$\delta(\text{étendue}) \equiv n^2\,\delta S\,\delta\Omega\cos\theta. \tag{11.1}$$

The power δP collected is now given by[2]

$$\delta P = B\times n^2\,\delta S\,\delta\Omega\cos\theta. \tag{11.2}$$

If, as here, power δP is measured in watts, the coefficient B is called **radiance**;[3] if δP is measured in 'visual' units, weighted according to the

[1] The reason for including the factor n^2 (n is refractive index) will be made clear below.

[2] If B is independent of θ, the power δP is proportional to $\cos\theta$. This property is quite commonly encountered, and is called the Lambert cosine law.

[3] Old terminology dies hard, and B is still sometimes called *brightness*.

sensitivity to frequency of the human eye, then B is called **luminance**.

For a light source of finite area, radiating into a range of directions, eqn (11.2) can be integrated to give the total power P collected into apparatus with a defined étendue. In particular, if the source[4] has constant B (independent of locations within ΔS and directions within $\Delta\Omega$),

[4] Constant B means additionally that all parts of area ΔS radiate equally (apart from the $\cos\theta$ factor) into all parts of $\Delta\Omega$.

$$P = (\text{radiance } B) \times (\text{étendue}). \tag{11.3}$$

If we gather light radiated symmetrically into a cone of (non-small) semi-angle θ_{max}, we have (problem 11.5):

$$\text{étendue} = n^2\,\Delta S \int \cos\theta\,\mathrm{d}\Omega = \Delta S\,\pi(n\sin\theta_{\text{max}})^2, \tag{11.4}$$

in which we encounter the **numerical aperture** defined by

$$\text{numerical aperture} \equiv (n\sin\theta_{\text{max}}). \tag{11.5}$$

Although we have avoided approximating in the above, it often suffices to make small-angle approximations, and then eqn (11.1) integrates more simply to[5]

$$\text{small angles:} \qquad \text{étendue} \approx n^2\,\Delta S\,\Delta\Omega. \tag{11.6}$$

[5] Textbooks usually define étendue simply as $\Delta S\,\Delta\Omega$, with a small-angle approximation implied. The $\cos\theta$ factor of eqn (11.1) is, however, conventional in more careful treatments; see, e.g., Born and Wolf (1999), §§4.8.1 and 4.8.3. By contrast, the n^2 factor seems to be less usual, but its inclusion improves both the mathematics and the physics.

The presence of the numerical aperture in eqn (11.4), rather than $\sin\theta_{\text{max}}$ on its own, is just one of the neat consequences of putting n^2 into definitions (11.1) and (11.2). Numerical aperture will be encountered again in §§12.5.1 and 12.6.

Equation (11.3) may be applied to the collection of light into an instrument such a spectrometer. In such a case, ΔS refers to the radiating area from which light is usefully collected,[6] and a similar understanding applies to the directions included within $\Delta\Omega$.

[6] Area ΔS may be the surface of a light source, or an image of a light source, or sometimes some other area, possibly one 'along the way' in an optical system. However, areas 'along the way' don't radiate uniformly into $\Delta\Omega$ save in special cases. See problem 11.1(2). As mentioned in sidenote 4 (this page) we restrict attention to areas ΔS that do radiate uniformly.

If a source such as a lamp has a non-uniform radiance from different parts of its area, or radiates unequally into different directions, we can, of course, attempt to allow for the non-uniformity by integrating eqn (11.2):

$$P = \int B n^2 \cos\theta\,\mathrm{d}S\,\mathrm{d}\Omega.$$

However, non-uniformity is likely to be a symptom of a design fault—or an inappropriate choice of area for ΔS—and should be removed rather than allowed for.

When using a spectrometer, we shall wish to obtain strong signals on a photographic film or an electrical photodetector. Optimizing the power collection is to be achieved by maximizing the étendue of the optical equipment, since (as we shall see) radiance B is not something we can usually do much about. This motivates much of the discussion in §§11.3–11.10.

11.3 Conservation of étendue and radiance

When the étendue of an optical system is evaluated, using a properly chosen area according to the understandings in sidenote 6 (this page), it remains unchanged as the light is transformed by lenses, mirrors, or other optical components, or passes from one medium to another:

$$\text{étendue is an invariant}. \tag{11.7}$$

A further conservation rule follows from eqn (11.7): if an image is formed of a source, using lenses or mirrors,

$$(\text{radiance of image of source}) \leq (\text{radiance of original source}), \tag{11.8}$$

where equality holds when there is no loss of energy (**insertion loss**) in the imaging system. These two conservation rules are established in problem 11.1.

11.4 Longitudinal and transverse modes

A light beam can be decomposed into longitudinal and transverse modes, as was done in describing cavity modes in Chapter 8.

The étendue of a light beam is connected with the number of transverse modes that are populated by its photons. The relationship is:[7,8]

$$\text{(number of transverse modes occupied)} = \frac{\text{étendue}}{\lambda_{\text{vac}}^2}. \tag{11.9}$$

A derivation of eqn (11.9) is laid out in problem 11.3.

A wave occupying a single transverse mode has a photon flux, the number of photons per second crossing some plane, given by (problem 11.2)

$$\text{(photon flux)} = \text{(number of photons per mode)} \times \delta\nu, \tag{11.10}$$

where $\delta\nu$ is the range of frequencies occupied by the photons. The number of occupied longitudinal modes is related to $\delta\nu$, in a way that is presented in problem 11.2.

When light leaves an area ΔS into several longitudinal and transverse modes, the power $P(\nu)\,\delta\nu$ in frequency range $\delta\nu$ is easily seen to be

$$P(\nu)\,\delta\nu = \text{(number of photons per mode)} \left(\frac{\text{étendue}}{\lambda_{\text{vac}}^2}\right) h\nu\,\delta\nu. \tag{11.11}$$

We take the opportunity to obtain the radiance of a source radiating into frequency range $\delta\nu$:

$$B(\nu)\,\delta\nu = \frac{P(\nu)\,\delta\nu}{\text{étendue}} = \text{(number of photons per mode)}\,\frac{\nu^2}{c^2}\,h\nu\,\delta\nu. \tag{11.12}$$

Expressions (11.9), (11.11) and (11.12) are all to be doubled if two polarizations are excited.

[7]Equation (11.9) represents a purely geometrical relationship, so it applies to light of a single polarization. If we concern ourselves with unpolarized light, the number of occupied modes is doubled.

[8]A Gaussian beam occupies a single transverse mode. A direct check that it has étendue λ_{vac}^2 (with happy choices of definition for ΔS and $\Delta\Omega$) is made in problem 11.4.

11.5 Étendue and coherence area

Coherence area has been introduced in Chapter 9. The coherence area is related to the solid angle within which light arrives,[9] as is shown in Fig. 11.2: if angles are small, $\Delta S = \lambda^2/\Delta\Omega$.

Suppose that light is incident onto the area shown in Fig. 11.2, and arrives uniformly within solid angle $\Delta\Omega$. Then (small angles)

$$\begin{pmatrix}\text{étendue of light falling on}\\ \text{one coherence area } \Delta S\end{pmatrix} = n^2 \Delta S \Delta\Omega = n^2\lambda^2 = \lambda_{\text{vac}}^2. \tag{11.13}$$

This result is independent of the illuminating geometry and of the refractive index of the medium; everything has cancelled out.

Comparing eqns (11.9) and (11.13), we see that whenever we seek to collect light falling within a single coherence area we are in fact aiming to receive no more than a single transverse mode. A simple interpretation of this idea is explored in problem 11.7.[10]

[9]Small-angle expressions are used here for simplicity. An equivalent result is correct for large angles, as mentioned below.

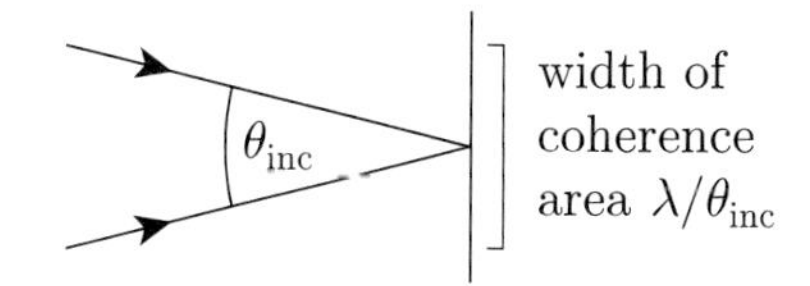

Fig. 11.2 This diagram shows a beam of light in cross section. It arrives at some plane within a range of angles θ_{inc} and in consequence the coherence area has width $\lambda/\theta_{\text{inc}}$. Something similar happens in directions at right-angles to the paper. Taking both directions together, the light arrives within solid angle $\Delta\Omega$ and yields a coherence area $\Delta S = \lambda^2/\Delta\Omega$.

[10]The derivation of expression (11.13) for coherence area has not taken detailed account of any particular geometry in the incident light beam; it is, therefore, less precise than the treatments in Chapters 9 and 10. However, there is an inevitable agreement in order of magnitude.

We have already seen an application of eqn (11.13). In the Hanbury Brown–Twiss experiment, it was necessary to observe fluctuations of intensity within a single coherence area. Problem 10.17 has shown that the experiments did precisely that.

Before we leave the present topic, we give a version of eqn (11.13) applicable to non-small angles. Let light be collected by a circular lens and focused into a cone with semi-angle $\theta_{\max}$, numerical aperture $(n \sin\theta_{\max})$. The coherence area in the focus is a circular patch of area ΔS and radius r where, by the inverse of eqn (11.4),[11]

$$\Delta S = \frac{\text{étendue}}{\pi(n\sin\theta_{\max})^2} = \frac{\lambda_{\text{vac}}^2}{\pi(n\sin\theta_{\max})^2}, \qquad r = \frac{\lambda_{\text{vac}}}{\pi(n\sin\theta_{\max})}. \tag{11.14}$$

Expressions (11.14) for ΔS and r also give us the dimensions of the area to which light can be focused if it occupies a single transverse mode.[12] If several transverse modes are excited (randomly, with nothing specially engineered), the achievable focus area will be increased by a factor equal to the number of occupied transverse modes. With this understanding, results (11.14) are of immediate relevance to the reading of a compact disc (problem 11.6) and the supply of power into an optical fibre (problem 14.2).[13]

11.6 Field modes and entropy

Consider a light beam that has some number of photons distributed among field modes, longitudinal and transverse. Like any other system of particles in quantum states, it must have entropy $k_B \ln W$, where W is the number of microstates in the macrostate and k_B is the Boltzmann constant. A chaotic (random, non-laser) light beam may occupy any of a large number of microstates that are macroscopically similar, so it has randomness and carries with it a non-zero entropy.[14]

The conservation laws for étendue and radiance stated in eqns (11.7) and (11.8) have a new interpretation: they tell us that a lens (or other optical component) doesn't change the number of field modes[15] occupied by photons; equivalently it doesn't alter the entropy.[16]

The prohibition on changing (in particular reducing) entropy-étendue applies with equal force to optical arrangements that don't form an image, e.g. where light is concentrated in a caustic.

By contrast with the above, a beam with only a single mode occupied can have no $k_B \ln W$ randomness, so it should not carry entropy.[17]

The idea that a (chaotic) light beam carries entropy justifies assumptions that were unexplained earlier in this book. It was assumed in Chapter 9 (particularly §9.15) that for random (non-laser, chaotic) light the only way to obtain a desired degree of coherence is to filter away all of the light that's doing things we don't want. This gloomy suspicion is now confirmed. Light in the 'wrong' field modes can't be rescued (somehow diverted into the 'right' modes) without diminishing the entropy and so violating the Second Law; it has to be discarded. In achieving

[11] A similar result can, of course, be obtained by using diffraction theory. The derivation given here is interesting for the rather different route that it follows. The reasoning is outlined in problem 11.5.

[12] This equality should 'feel' obvious. The van Cittert–Zernike theorem—which this is—should begin to feel obvious also ....

[13] Laser light is dangerous to eyes because it (usually) occupies a single transverse mode.

[14] As a light beam passes through some region of space, the photons that were present are swept out and replaced by others. So one microstate is succeeded by another, and what we observe over an ordinary timescale is a time average, equivalent to an ensemble average over microstates. All the conditions for $k_B \ln W$ to represent entropy are therefore met.

[15] *Aside*: There is a similarity to the Liouville theorem in classical particle mechanics (Mandl 1988, p. 208). A pretty discussion of the Liouville theorem is given by Penrose (1989), especially p. 235.

[16] I've made a bald statement for emphasis. More correctly: a lens can't reduce the number of occupied microstates while conserving the number of photons, because that would reduce W. Entropy can be increased if photons are scattered into additional directions. And the entropy of the light beam—though not that of the universe—can be reduced if photons are simply absorbed.

[17] We have $W = \prod_i \frac{(g_i + n_i - 1)!}{(g_i - 1)!\, n_i!} = 1$, whatever the value of n_i for the one state that's occupied.

Even in a single mode, photons can be 'loaded' into a variety of quantum states, see e.g., Loudon (2000), Chapter 5. Some randomness is therefore possible if the quantum state is changing. It seems best to treat such randomness as a quantum-mechanical phenomenon different in principle from entropy.

transverse coherence, we must throw away light from most of the source's area, or travelling in most directions, or both, until we have reduced the étendue to λ_{vac}^2. It's a high penalty. Likewise, to obtain a desired degree of longitudinal coherence (coherence time) we have to discard (by filtering) frequencies that are unwanted. Just how high these penalties are is a matter examined in problem 11.8.

11.7 Radiance of some optical sources

11.7.1 Radiance of a black body

The radiance $B(\nu)\,\delta\nu$ of a black body, for emission into frequency range $\delta\nu$, can be obtained at once from eqn (11.12):[18]

$$\text{radiance } B(\nu)\,\delta\nu = \frac{1}{e^{h\nu/k_{\rm B}T} - 1}\,\frac{2\nu^2}{c^2}\,h\nu\,\delta\nu. \tag{11.15}$$

Both polarizations are included.[19]

In eqn (11.15), the Planck distribution $(e^{h\nu/k_{\rm B}T} - 1)^{-1}$ is the value taken by the number of photons per mode when we have thermal equilibrium. Some representative numerical values are given in Table 11.1.[20]

[18] There is a 'Lambertian' dependence of the power radiated upon $\cos\theta$, which is not on display here because it has been included in the definition (11.2) of B.

[19] Radiance is here defined as in eqn (11.2), with a factor n^2. The fact that there is no n^2 on display in eqn (11.15)—it has cancelled—provides yet another reason why we favour defining things as in §11.2.

[20] These numerical values should be contrasted with the number of photons per mode in the output from a modest laser, as calculated in problem 10.20.

wavelength	$T = 2000\,\text{K}$	$T = 6000\,\text{K}$
435.8 nm	6.8×10^{-8}	0.41%
633 nm	1.2×10^{-5}	2.32%
780 nm	9.9×10^{-5}	4.85%
850 nm	2.1×10^{-4}	6.33%
1.3 µm	4.0×10^{-3}	18.8%

Table 11.1 The number of photons per mode for black-body radiation of two different temperatures.

11.7.2 Radiance of a gas-discharge lamp

The radiance of a black body represents a standard against which it is appropriate to compare the radiances of other light sources. In particular, we may ask whether a lamp emitting a line spectrum is or is not 'brighter' than a black body of comparable temperature—and what would be a 'comparable temperature'?

If we are to receive an intense line from a gas discharge, the discharge must be 'optically thick' at the middle frequencies of the spectral line. Otherwise, we could add more radiating gas 'behind' the discharge and we would be able to see extra light from that added gas through the original discharge. In an optically thick discharge, photons emitted in the middle of the discharge have a high chance of being absorbed by lower-state atoms before they can travel to the walls. There is frequent exchange of energy between atoms and photons, resulting in an approximation to local thermodynamic equilibrium within the bulk of the discharge; atoms acquire a Boltzmann distribution (and photons a Planck distribution) with a temperature $T_{\rm d}$. This temperature is called the 'distribution-over-states' temperature.

At the middle of a spectral line, the considerations just given mean that the intensity we receive approximates to that of a black body at temperature $T_{\rm d}$. For frequencies away from line centre, the atoms radiate less intensely, and they also absorb less intensely, so the mean free path for photons between absorptions is greater than at line centre. A point is reached where that mean free path is greater than the dimensions of the discharge, and the discharge then becomes 'optically thin'. The spectral line's emitted power falls with further frequency shift, following a profile that would have applied to the entire spectral line if the discharge had been thin at all frequencies.

As a rough rule of thumb, the distribution-over-states temperature for an intense discharge is about 6000 K; so a spectral line is at best about as bright (radiance, photons per mode) as sunlight of the same frequency.[21] A very clear and useful account of discharge physics is given by Wharmby (1997).

[21] Discharge physics is more complicated than this 'potted' account suggests. In particular, there is usually a region near the walls of the discharge where the gas is cooler. This cooler gas has its own T_d, lower than that for the bulk of the discharge. Photons near line centre come to equilibrium at this lower T_d on their way out while light a little away from line centre (outside the Doppler width for the cooler gas) does not. We have a case of 'self-reversal'. Such effects are described by Wharmby (1997).

11.7.3 Radiance of a light-emitting diode (LED)

Passage of a current through a suitably constructed diode results in a region of semiconductor where there is a surplus of electrons at the bottom of the conduction band and, likewise, a surplus of empty states at the top of the valence band.[22] Electrons in the conduction band come to thermal equilibrium with phonons and acquire a thermal distribution with a temperature equal to the lattice temperature (roughly room temperature). Relaxation to this distribution is fast, timescale about 10^{-11} s. The electrons then fall to empty states in the valence band with a radiative lifetime of order 1–10 ns. Once in the valence band they again thermalize (timescale 10^{-12} s) so that the states they entered at the top of the valence band remain empty.[23] These thermal effects mean that the photons emitted cover an energy range of roughly $2kT$.

[22] For a description of the basics of the device we refer the reader, e.g., to Fox (2001).

[23] The semiconductor is, therefore, optically thin, reabsorbing little of the emitted light. The timescales are given here because they are considerably shorter than those applying in a typical gas discharge, and result in a somewhat different way of thinking about what is happening.

The semiconductor almost always has a 'direct' band gap, so that the emission or absorption of photons is not accompanied by creation or absorption of phonons.

LEDs have a variety of constructions according to the purpose for which they are made. However, a diode intended for feeding an optical fibre might have an emitting area of 50 μm square radiating 1 mW at a wavelength of 850 nm into a numerical aperture of 1. Such an LED emits light with 3×10^{-2} photons per mode. We are again finding radiances roughly comparable to that for sunlight.

11.8 Étendue and interferometers

Any interferometer interferes light waves, obtained by dividing a beam of incident light into two (or more), either by division of wavefront or by division of amplitude. Consider first the case of division of wavefront, for which the paradigm is the Young slits. It's necessary to make the two divided beams coherent with each other, at least in one direction (the direction across the slits in a Young-slits case). To make discussion simple, imagine first that the beams must be coherent with each other in both transverse directions. In that case the light must be filtered, by narrowing its width and its range of directions, until it occupies only a single transverse mode. This filtering is known (problem 11.8) to result in only a small usable energy flow (for a non-laser source).

The penalty just introduced has arisen because the light beam has been made coherent in both transverse directions. We might expect that a Young-slits experiment, in which coherence is needed in only one direction, would be less disadvantageous: elongating the source from a pinhole to a slit as in Fig. 3.10 permits an increased flow of energy. This idea is explored in problem 11.9. The surprising result is that observation by eye is made more comfortable, but there is little improvement in visual brightness of the fringe pattern.

By contrast, an interferometer exploiting division of amplitude (the Michelson and its relatives) sets no requirement on transverse coherence. There is no need to perform any filtering down to one or a few transverse modes.[24] The bright fringes can be as bright as light received directly from the source without the interposition of the interferometer (if we ignore insertion loss).

[24] There is, of course, a requirement: that the frequency range of the light be small enough that the coherence length encompasses the intended path difference. However, this requirement is usually met by accepting the use of a limited range of path differences, rather than by filtering the light.

Suppose we are to design an interferometer to make an optical measurement, perhaps to measure the refractive index of a gas. We could adapt a Young-slits apparatus, and we would end up with something resembling a Rayleigh refractometer. We could adapt a Michelson interferometer and end up with something like a Jamin refractometer. Of these two approaches, the second is far more practical, in terms of the optical power available in the fringes.[25]

[25] This discussion is centred on the idea that we are going to use a gas-discharge lamp as light source. A laser gives high intensity in a single transverse mode, and a Rayleigh refractometer illuminated by a laser would work well. However, there is still no positive advantage in pursuing the wavefront-division route, so we would be wise to choose the greater versatility of the Jamin—which is also a slightly simpler instrument.

It might be objected that the 'textbook' description of the Rayleigh refractometer incorporates some elegant ideas, such as a set of fiducial fringes to assist measurement, which are not part of a 'textbook' Jamin. This, however, is to miss the point. Apparatus design is not set in stone, and there is nothing to prevent our hybridizing the designs to exploit the best features of both—or indeed of others. Excellent descriptions of the Rayleigh and Jamin designs may be found in Born and Wolf (1999).

The generalization is obvious:

- If you have any choice in the matter, design your interferometer to use division of amplitude.

11.9 Étendue and spectrometers

We consider here the light gathering by an optical apparatus such as a spectrometer: the larger we can make the étendue, the more energy we can collect and use.

In a spectroscopic instrument there is a trade-off between étendue and resolution. In the case of a grating spectrometer (problem 11.10), widening the entrance slit increases the étendue (by increasing area ΔS) but degrades the resolution, and conversely. A similar trade-off applies to other instruments, though for reasons that may not be quite so obvious.

Étendue further provides us with a means of comparing the merits of one instrument with another. We can ask: which is likely to be better for examining a weak source, a grating spectrometer or a Fourier-transform (Michelson-type) spectrometer? The answer (problems 11.14 and 11.15) may be surprising. A Michelson interferometer or a Fabry–Perot is a **whole-fringe instrument** possessing the properties (problem 11.15):

$$(\text{solid angle of acceptance}) \times (\text{chromatic resolving power}) = 2\pi \quad (11.16)$$

and

$$(\text{étendue}) \times (\text{chromatic resolving power}) = 2\pi \times (\text{area of a mirror}). \quad (11.17)$$

These properties represent a standard of comparison for the light gathering of a spectroscopic instrument. A Fourier-transform spectrometer is in line with this standard, while (it turns out) a comparable grating spectrometer falls short of it by a factor of order 200 (problem 11.14). This is the context for the ingenious measures described in §4.9.6: the grating spectrometer is coming from a long way behind, and there is correspondingly a great deal that might be gained from improving it.

11.10 A design study: a Fourier-transform spectrometer

The apparatus is a Michelson interferometer as described in §10.7. Our design study of it is pursued via problems 11.11–11.16, which discuss the light gathering (étendue) of the instrument, the resolution that can be achieved, and the trade-off between the two. The whole-fringe property of the interferometer, mentioned in §11.9, is derived in problem 11.15.

We have another agendum also in providing problems 11.11–11.16. There seems usually to be little time in a physics degree course for teaching the principles by which a piece of hardware (optical or otherwise) is designed. The author regards this as regrettable. It happens that the Fourier-transform spectrometer provides a rather good case study, where one can start from a blank sheet of paper and the relevant physics, and end up with a fully practical design for a piece of equipment.

11.11 Fringe localization

Any interferometer that exploits interference by division of amplitude (as in a Michelson interferometer) generates fringes that are **localized**. This may make the setting-up of the interferometer seem a less straightforward business than that for a division-of-wavefront arrangement such

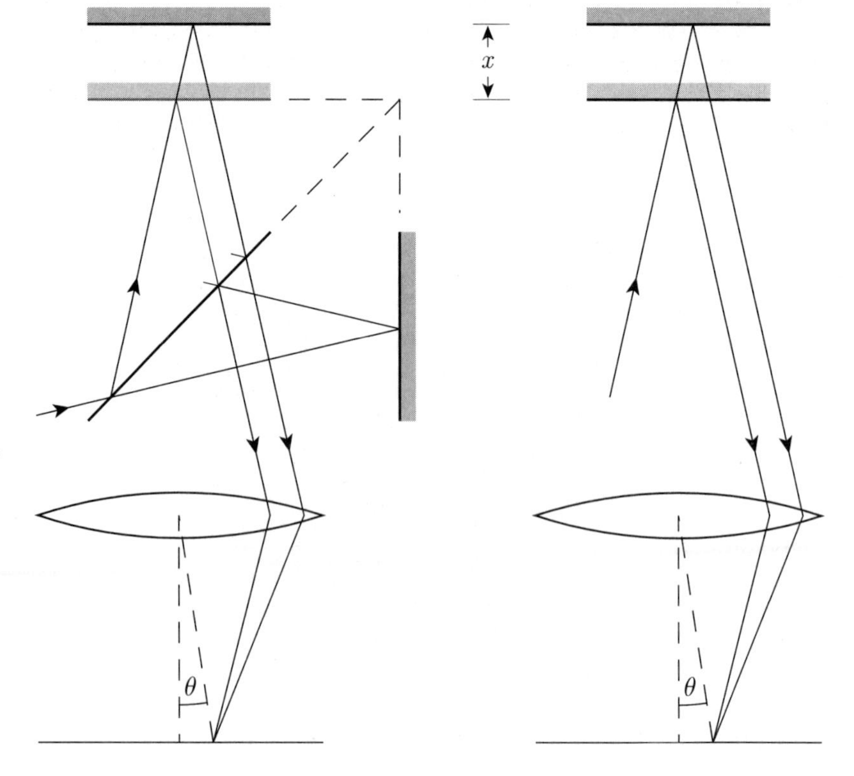

Fig. 11.3 Ray paths through a Michelson interferometer, compared with similar paths taken by light that is reflected at the two surfaces of a parallel-sided slab. The right-hand mirror of the Michelson, and the light rays to and from it, are shown mirror-imaged by the beam splitter.

To make the diagrams comparable, the 'slab' has been represented as two reflecting—or part-reflecting—surfaces with no refraction at the boundaries. The intention is to show that the interferometer is geometrically equivalent to the slab, without reproducing all the details of a real slab.

as Young's slits. However, the reasoning of §11.8 shows that there is a big advantage in energy gathering to be gained from using division of amplitude: localization, and the dealing with it, is a small price to pay. But, of course, we do have to understand localization: what it means; where the fringes are formed; and how critical is the focusing on them.

We shall restrict attention here to the case where a Michelson interferometer has its mirrors 'parallel'. Representative optical paths through the system are shown in Fig. 11.3. The same diagram shows rays reflected from a parallel-sided slab, to indicate that the ray geometry in the interferometer is the same as that for the simpler case of the slab.

For light following the paths shown, the optical path difference is $n\,2x\cos\theta$, where θ is the angle made by the ray paths *inside the slab*, x is the separation of the reflecting surfaces and n is the refractive index of the medium between the reflectors. The interference condition[26] is

$$\text{condition for dark fringe:} \qquad n\,2x\cos\theta = (p + \tfrac{1}{2})\lambda_{\text{vac}}. \qquad (11.18)$$

Given that the reflecting surfaces are parallel, x is being held constant. We also consider n and λ_{vac} to be constant. Then the phase difference between the two interfered light beams is controlled by $\cos\theta$ only.[27] To achieve good fringes, we must ensure that light ending up at a given place all has the same value of $\cos\theta$. What we need is shown in Fig. 1.3: a lens focused for infinity gathers all light travelling in a single direction and concentrates it in a single place.[28]

The analysis just given may seem rather abstract. It also gives us no idea of tolerances: the focal plane of the lens is the best place to look for fringes, but how critical is it to look just there? To investigate, we proceed by means of a series of diagrams. Figure 11.4 shows what happens if a pair of reflecting surfaces (which we'll call a slab) is illuminated by light from a monochromatic point source at S. Light reflected from the

[26] *Comment*: If we had a glass slab in air, there would be a phase change of π at one reflection or the other, and $(p + \frac{1}{2})$ would be replaced by p in eqn (11.18). However, our drawing of a slab is given to simplify the geometry while analysing the workings of a Michelson interferometer. The mirrors of the Michelson are likely to be identical to each other, so there should be no phase change of π here. (The physics of the beam splitter is another matter ....)

[27] This accounts for the description of these fringes as **fringes of equal inclination**.

[28] It may help to state that the interferometer with mirrors 'parallel' acts as an *angular filter*, with energy transmission dependent upon angle θ. We display this behaviour by following the interferometer by an *angular selector*, so we separate out light that has been filtered in the different ways.

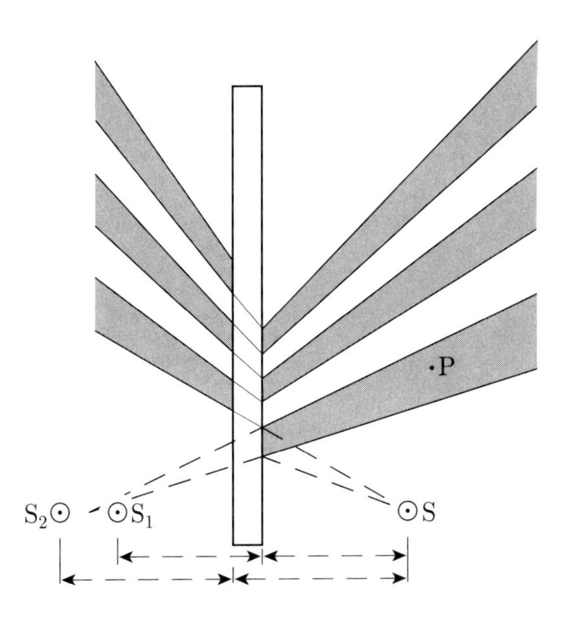

Fig. 11.4 A point source S illuminates a slab whose surfaces are part-reflecting, part-transmitting. Light travelling towards top right appears to have come from the two ray-optics images S_1, S_2 of the source. Equal-length arrows confirm the correctness of the image positions. Reflected light undergoes two-beam interference, with its energy redistributed into $\cos^2$ fringes; the grey shading indicates crudely where the intensity is less than average.

A slab acts as a low-grade Fabry–Perot: multiple reflections produce a succession of images of source S, of which only the brightest two are shown. At least one image to the right of the slab is needed to account for fringes in the light travelling towards top left.

The system has rotational symmetry about line S_2S.

right-hand surface appears to diverge from an image S_1 of the source, as far to the left of the reflector as the source is to the right of it. Light reflected from the other surface appears to diverge[29] from S_2. We have two (virtual-image) coherent sources S_1 and S_2 radiating spherical waves that overlap in the space to the right of the slab. When amplitudes are added at a point such as P, the result is light or dark, depending upon the value of the path difference $S_2P - S_1P$, which in turn depends upon angle θ according to eqn (11.18).[30] The shaded regions of the diagram indicate places where there is destructive interference and there is a beam of 'dark'. The fringes are, of course, $\cos^2$ fringes, so the intensity varies smoothly; the abrupt changes of shading are used in the figure to make the discussion that follows more stark.

[29] This is not quite correct for a real slab because distances are modified by the slab's refractive index.

[30] Equation (11.18) is approximate here because the two waves have slightly different values of θ.

The interference shown in Fig. 11.4 must, of course, conserve energy.[31] Therefore, where 'dark' travels one way away from the slab, there must be 'bright' travelling the other way. Lines on the diagram indicate how the fringes on the two sides of the slab 'fit between each other'.

[31] In a Michelson interferometer the mirrors do not transmit, so there are no fringes equivalent to those going to the left in Fig. 11.4. Energy must, of course, still be conserved, but the surplus or deficit appears elsewhere; see problem 11.19.

The arrangement shown in Fig. 11.4 yields fringes wherever there are two beams of light to be added together: the fringes are *non-localized.* This setup resembles a Young-slits arrangement, so far as the localization of the fringes is concerned. We shall see that this has happened because the light originated from a *point* source.

'Non-localized' may seem a strange term, since the fringes do have

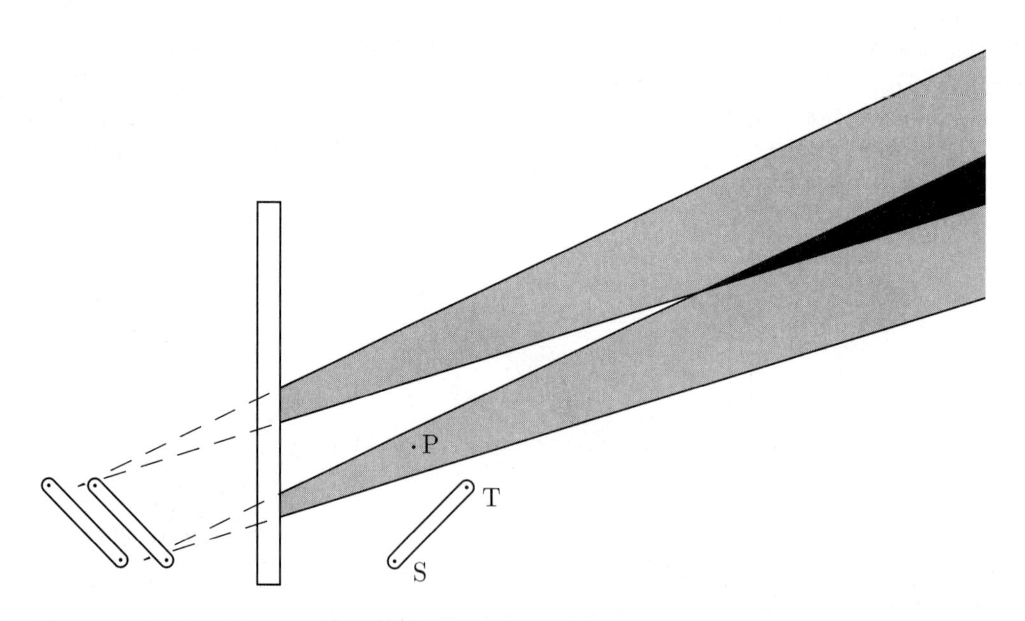

Fig. 11.5 The setup shown here is an elaboration of that shown in Fig. 11.4. The light source is now no longer a point, but is some luminous object ST. Source ST has been drawn in a nothing-special place and at a nothing-special orientation, in order to make it clear that the discussion is general. Light from S produces a set of fringes that radiate outwards from the images of S. Light from T produces a similar set of fringes, but displaced because they come from the images of T. Other similar fringes originate from places in between. A point such as P receives 'dark' from S but 'bright' from T; there can be no complete darkness there. But at large distances from the slab the fringes from S and T overlap with a displacement that becomes negligible compared with the width of a fringe: good fringes are seen, as is suggested by the black area.

locations; they do not exist over the whole of space. This is not the point. As we shall see shortly, there can be arrangements where interference happens, yet fringes are not seen: there is an additional condition to be met. 'Localized' fringes are those subject to such an additional condition; 'non-localized' fringes are those that are not.

Figure 11.5 moves us towards more practical conditions by showing what happens when light comes from an extended source. Light originating from point S produces a set of fringes, like those of Fig. 11.4, radiating out from the bottom end of the source–image pair. Light originating from point T does the same, but with everything shifted upwards and to the left. Light from points in between does something intermediate. Point P, to the right of the slab and fairly close to it, receives 'dark' from S but 'bright' from T, and little sign of any organized fringes can be seen at places similarly close to the slab.[32] However, we may see that the fringes overlap more and more as we go further away from the slab, and eventually they become so much wider than the separation of their sources that the overlap is total. The fringes are 'localized at infinity', in that they exist at sufficiently large distances from the slab—just such an 'additional' condition for seeing fringes as is claimed above.

The final step in our reasoning is made in Fig. 11.6. Rather than go to a large distance from the slab, we 'bring infinity closer' by using a lens. The 'beams of dark' are brought together so that their overlap is total in the lens's focal plane: the fringes are localized at that focal plane.[33]

Given that we have been made to concentrate attention on the lens's focal plane, we may view the lens and its focal plane as resembling a camera lens and film arranged to be 'focused for infinity'. This gives us another way of understanding 'localized at infinity'.

The reader will be able to imagine what happens to Fig. 11.6 if the source is made longer or shorter. If it is made shorter, the extreme

[32] It is possible to get into something of a tangle with irrelevancies. The reasoning given here proceeds by assuming that all points on the source radiate incoherently. That is: we assume that light from S never interferes destructively with light from T to yield 'dark' in the bright-expected areas. Would things be different if there were some significant degree of transverse coherence along the length of the source? It is for this reason that I have concentrated on where destructive interference sends 'dark'. If points S and T both send 'dark' to a given place, then it doesn't matter whether we should be adding amplitudes or intensities: the result is zero either way. What happens in the 'bright' areas is a separate issue; but that issue would have been with us in the absence of the slab ....

[33] To reconcile the jargon: The fringes are localized at the focal plane in the lens's image space; they are localized at infinity in its object space.

The reader is encouraged to confirm that the ray-optics rules of Fig. 1.3 have been rigorously applied in the preparation of Fig 11.6.

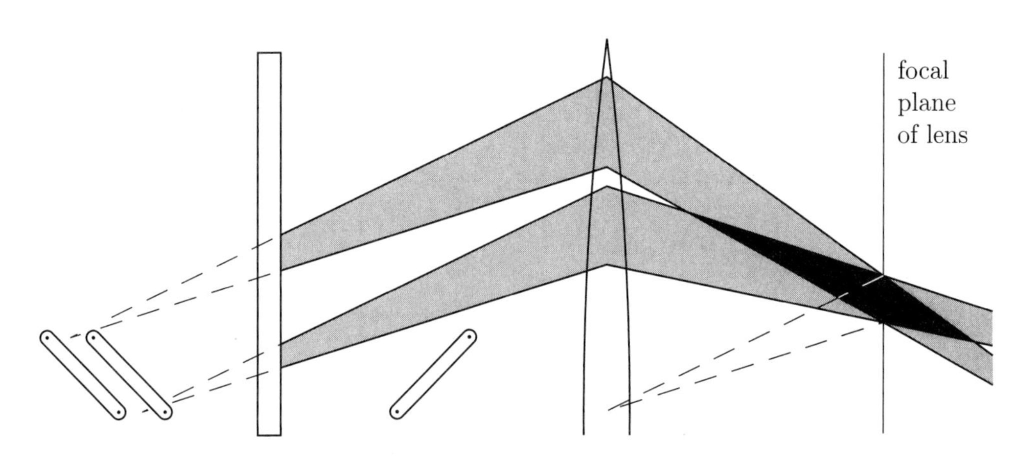

Fig. 11.6 The fringes of Fig. 11.5 are passed through a lens. Light from both ends of the source, and from points in between, is now focused so that all dark fringes come together at the lens's focal plane; and similarly all bright fringes come together (at other places) there. The best place to look for fringes is the focal plane. At the same time, the black diamond shape gives an idea of how far we can be from the focal plane and still see fringes of some sort; that is, it shows us the depth of focus.

beams of 'dark' get closer together, and the dark diamond of overlap gets longer. In the limiting case when the source is made a mere point, we are back to the non-localized fringes of Fig. 11.4. Conversely, as the source is made wider, the extreme beams of 'dark' get further apart; the diamond of overlap gets shorter and the focusing becomes more critical.

Given fixed values for slab thickness x, wavelength λ and refractive index n, the condition for a bright or dark fringe consists of a condition on θ only. Therefore, the fringes have axial symmetry about a direction normal to the reflecting surfaces (and passing through the centre of the lens if we use a lens). If we use an optical arrangement like that of Figs 5.2 or 5.8 or 11.6, the fringes are concentric circles centred on the lens axis. The reasoning of the present section fills gaps in the explanation of some diagrams that have appeared earlier in this book.

There is one other case of fringe localization that should be discussed: the case where the two reflecting surfaces form a thin wedge. The investigation of this case is left to the reader: problem 11.20.

Problems

Problem 11.1 (a) The conservation law for étendue
Figure 11.7 shows a ray-optics image formed when light is refracted at a curved surface separating two media; there is a similarity to Fig. 1.4. In the first instance, ΔS_1 is the area of some luminous source which radiates equally into all parts of solid angle $\Delta\Omega_1$. The image formed may be real or virtual, and is drawn real for simplicity.

(1) Use ray optics to show (small angles)[34] that

[34] Small angles are used here for simplicity. However, the connection linking étendue to field modes and to entropy makes it clear that a large-angle generalization must exist.

$$\frac{h_2}{h_1} = \frac{v}{u}\frac{n_1}{n_2}, \qquad \frac{\Delta S_2}{\Delta S_1} = \frac{v^2}{u^2}\frac{n_1^2}{n_2^2}, \qquad \frac{\Delta\Omega_2}{\Delta\Omega_1} = \frac{u^2}{v^2},$$

and hence

$$(\text{étendue})_2 = n_2^2\,\Delta S_2\,\Delta\Omega_2 = n_1^2\,\Delta S_1\,\Delta\Omega_1 = (\text{étendue})_1. \qquad (11.19)$$

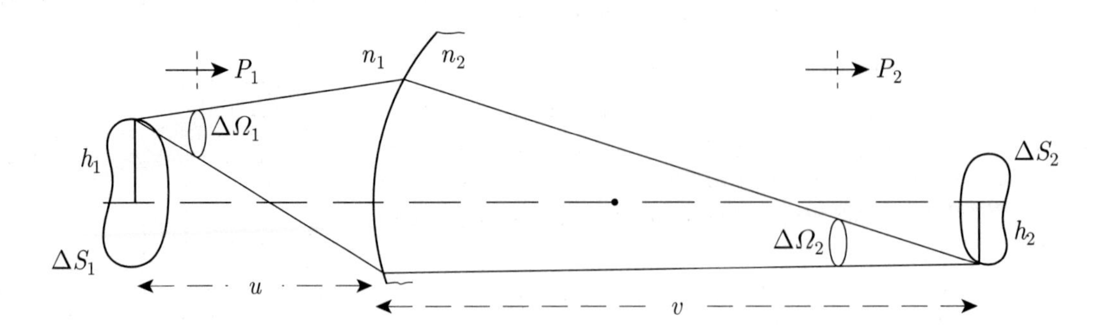

Fig. 11.7 The geometry for problem 11.1. Area element ΔS_1 is imaged to area ΔS_2. The solid angle collected is $\Delta\Omega_1$, and this results in an exit beam occupying solid angle $\Delta\Omega_2$.

(2) Show that the same value of étendue is obtained by using

$$n_2^2 \times \begin{pmatrix} \text{area of beam at} \\ \text{curved surface} \end{pmatrix} \times \begin{pmatrix} \text{solid angle into which} \\ \text{light leaves that surface} \end{pmatrix},$$

and that this works because there is no correlation between location and ray direction. Show that no other area 'along the way' through the optical system has such a lack of correlation, so étendue integrates to contain a simple area–solid-angle product only at the refracting surface or at an object or image.[35]

(3) In the diagram, P_1 and P_2 are the optical powers crossing the planes indicated. At best the system transmits all the power that's incident, and in practice it may lose a little.[36] Use the conservation of étendue to derive the 'conservation law' for radiance:

$$\frac{\text{radiance of image}}{\text{radiance of object}} = \frac{P_2}{P_1} \leq 1, \tag{11.20}$$

or, in words:

- an image formed by an optical system has (at best) a radiance equal to that of the original object.

(4) Argue that the conservation of étendue applies to the passage of light through any sequence of surfaces, and so applies (for example) to the case where a light beam is transformed by a glass lens in air.

[35] Equivalently, at a field stop or an aperture stop.

[36] Some energy may be absorbed, or reflected from the interface. We may say that the refracting surface introduces an 'insertion loss'.

Problem 11.2 (b) Longitudinal modes and their occupation

Consider a one-dimensional wave. It might be a Gaussian beam occupying a single transverse mode.[37] Let it occupy a very large length L, which is a 'quantization length', a length that we introduce simply to produce a countable number of longitudinal modes.[38]

(1) Apply periodic boundary conditions, and show that the number of longitudinal modes in frequency range $\delta\nu$ is $(L/v)\delta\nu$, where v is the speed of light (distinguished from c in case some medium is present). This is the number of (longitudinal) modes travelling in just one of the two possible directions.[39]

(2) Let a one-dimensional wavetrain have frequency range $\delta\nu$. Show that the number of photons per second that pass any fixed point within L, travelling in just one of the two possible directions, is[40]

$$(\text{photon flux}) = (\text{number } p \text{ of photons per mode}) \times \delta\nu. \tag{11.21}$$

(3) Why were 'periodic boundary conditions' desirable in part (1)?

(4) Let each of the photons of part (2) be represented as a wave packet having duration τ_r. Show that the total length of these wave packets within L is $p(L/v)\delta\nu\, v\tau_r$, where p is the number of photons per mode. Show further that

$$\frac{\text{total length of wave packets in wavetrain}}{\text{length of wavetrain}} = p \times \tau_r \delta\nu. \tag{11.22}$$

Interpret this as the average number of wave packets that overlap (if it's more than 1), or as the fraction of the time that is 'occupied' by light.

[37] We might equally well think of waves travelling along an electrical transmission line. Reasoning similar to that here is used in obtaining the thermal noise power travelling along such a line, and hence the thermal noise (Johnson noise, one-dimensional black-body radiation) radiated into the line by a resistor. See, e.g., Robinson (1974), §4.1.

[38] Compare with the large volume V that we introduce when deriving the density of states in statistical mechanics.

[39] Longitudinal and transverse modes are being distinguished in the same way as they were in Chapter 8. In all of this problem we consider a single polarization only.

[40] This result is easy to remember if we take an unconventional viewpoint and think that $\delta\nu$ modes pass per second, each occupied by p photons.

(5) Apply eqn (11.22) to a natural-broadened wavetrain, as modelled in §9.12 and problem 10.5: a sequence of identical wave packets each of duration τ_{r}. Show that eqn (11.22) simplifies for this case to

$$\text{fraction of wavetrain ‘occupied’} \approx p. \tag{11.23}$$

When applied to light of realistic intensity, this shows that photon wave packets are ‘sparse’, with only about 2% of the wavetrain occupied.[41]

[41] This numerical value is taken from Table 11.1. It is the basis of a number of arguments about intensity fluctuations in Chapter 10, especially in problem 10.21.

Problem 11.3 (b) Étendue and occupied transverse modes
Consider radiation occupying frequency interval $\delta\nu$, travelling in directions within solid angle $\delta\Omega$ and impinging on area ΔS at angle θ to the normal. We take $\delta\Omega = \sin\theta\,\delta\theta\,\delta\phi$ to encompass a small range of directions, so the light forms a near-collimated beam.

(1) We know from statistical mechanics that the number of modes per unit volume within frequency range $\delta\nu$ is $(4\pi\nu^2\,\delta\nu)/c^3$, where the usual factor 2 for polarizations is omitted as it will be dealt with separately. Argue that the expression must be changed to $(4\pi\nu^2\,\delta\nu)/v^3$ if the radiation travels in a medium in which the speed of light is $v = c/n$, where n is the refractive index.

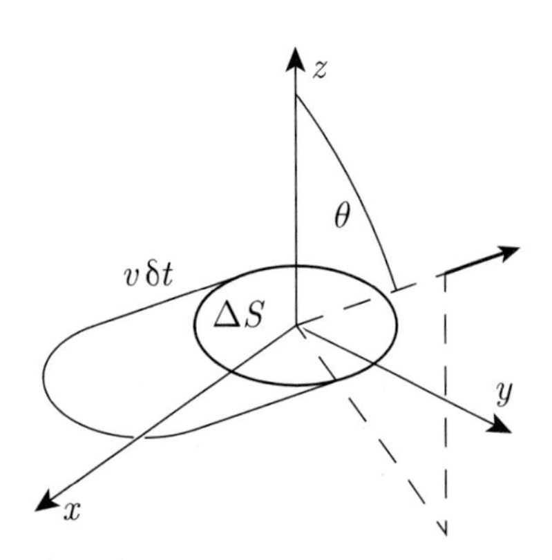

Fig. 11.8 Light arrives at area ΔS from within a cylinder making angle θ with the z-axis and of slant height $v\,\delta t$. All photons within the cylinder and travelling in direction θ meet or cross area ΔS in time δt.

(2) The expressions given in part (1) apply to radiation travelling in all directions. Argue that if radiation travels within solid-angle range $\delta\Omega$ it must occupy a fraction $\delta\Omega/4\pi$ of these modes, so the number of modes (per volume) it occupies within $\delta\nu$ and $\delta\Omega$ is $(\nu^2\,\delta\nu/v^3)\delta\Omega$.

(3) Next adapt a standard discussion in gas kinetic theory to work out the number of photons that impinge on area ΔS within time δt from within the solid angle $\delta\Omega$. Figure 11.8 may help. Those photons that impinge within time δt are those that occupy a cylinder whose base area is ΔS and slant height is $v\,\delta t$, making an angle θ to the normal to the surface. The volume of this cylinder is $\Delta S\, v\delta t\cos\theta$. Within this are p photons per mode in $(\nu^2\,\delta\nu/v^3)\delta\Omega$ modes per unit volume. Put this together, and show that the number of photons arriving per unit time is

$$\text{photon flux} = p \times \delta\nu \times \left(\frac{n^2\,\delta\Omega\,\Delta S\,\cos\theta}{\lambda_{\mathrm{vac}}^2}\right). \tag{11.24}$$

(4) The light beam occupies a small range of directions, so it can be described as composed of longitudinal and transverse modes, after the fashion[42] of the cavity modes in Chapter 8. Using eqn (11.21), identify $\delta\nu$ in eqn (11.24) as representing the number of longitudinal modes contributing to the photon flux.

[42] Any convenient set of eigenfunctions can be used to describe the transverse structure, not necessarily the Gauss–Hermite functions of eqn (8.1).

(5) Use the results of parts (3) and (4) to show that

$$(\text{number of transverse modes}) = \frac{n^2\,\delta\Omega\Delta S\cos\theta}{\lambda_{\mathrm{vac}}^2} = \frac{\text{étendue}}{\lambda_{\mathrm{vac}}^2}, \tag{11.25}$$

in agreement with eqn (11.9).

(6) Use eqn (11.24) to confirm eqns (11.11) and (11.12).[43]

[43] The present problem has concerned itself with the arrival of photons onto some surface element ΔS. The radiance of a black-body surface is obtained by appeal to the ‘principle of detailed balance’, requiring that what is radiated balances what is received.

Problem 11.4 (b) Étendue of a Gaussian beam

(1) Define the area ΔS of a Gaussian beam at its waist as a circle of radius w_0 where w_0 is the waist spot size defined in eqn (7.12). Take $\Delta\Omega$ as the (similarly defined) solid angle into which the beam diffracts in its far field. Show that with these definitions the étendue of a Gaussian beam comes out to be exactly λ_{vac}^2.

(2) What is wrong with the following argument? Take the same laser beam as in part (1) but take ΔS as the area of the beam at distance b (the confocal parameter) from the waist. Then ΔS is double what we had before, $\Delta\Omega$ is unchanged, so the étendue works out at $2\lambda_{\text{vac}}^2$. And an area ΔS farther from the beam waist would give an even larger value.

(3) [Harder] Generalize the discussion to the case of a Gauss–Hermite beam with transverse-mode eigenvalues l, m.

Problem 11.5 (a) Étendue and numerical aperture

(1) Confirm eqn (11.4).

(2) Combine eqns (11.4) and (11.9) to derive both parts of eqn (11.14).

Problem 11.6 (a) Reading a CD

The information track of a CD is read[44] by focusing a laser onto it through the body of the plastic disc. Show directly that refraction at the air–plastic interface has no effect on the numerical aperture, the étendue of the light beam, or the diffraction pattern formed at the focus (do not make a small-angle approximation).[45]

Does the refractive index affect the depth of focus?

[44] A fuller description of the optical system is given in §16.4.

[45] Thus, there is no advantage to be gained (and no disadvantage to be fought) by choosing any particular value for the refractive index of the plastic.

Problem 11.7 (a) Coherence area and transverse modes

A light beam consists of a mixture of the $(0,0)$ and $(1,0)$ transverse modes of eqn (8.1), with coefficients varying randomly with time. Show that the relative phases at locations $(x,y)=(-w,0)$ and $(x,y)=(w,0)$ are made unpredictable by the mixture.[46]

[46] This model shows very directly that a beam containing more than one transverse mode extends over more than one coherence area. Conversely, we see how it comes about that transverse coherence is associated with having only one transverse mode, as claimed in §11.5.

Problem 11.8 (a) The number of photons per mode

In black-body radiation, the number of photons per mode is given by the Planck distribution $(e^{h\nu/k_BT}-1)^{-1}$. 'Mode' means that both longitudinal and transverse characteristics are specified, as is the polarization.

(1) Show that the number of photons per mode, for black-body radiation, is greatest when the frequency is low and the temperature is high.

(2) Calculate the number of photons per mode in black-body radiation for the cases given in Table 11.1 and check the values given there.

(3) Find the power that can be extracted from a black-body lamp, at wavelength 633 nm, into étendue λ_{vac}^2 and in a frequency range of 1 GHz. Consider temperatures of (a) 2000 K and (b) 6000 K. Express the results in units of photons per second and in watts.

[*My answers*: (a) 2.3×10^4 photons s^{-1}, equivalent to 7.3×10^{-15} W; (b) 4.6×10^7 photons s^{-1}; 1.5×10^{-11} W. If polarized light is required these figures must be divided by 2.]

Comment: These figures show why laser light is so spectacularly more intense than black-body light. Even a modest laser, say a He–Ne radiating 1 mW into several longitudinal modes spanning 1 GHz, is brighter by a factor between 10^8 and 10^{12}.

Comment: The conclusions here give a more quantitative view of the discussion of laser coherence in §9.15.1.

(4) Discuss whether a laser source is essential, or whether something cheaper will do, for:

(a) the 780 nm radiation source used in reading a CD in a domestic CD player

(b) the 1.3 μm radiation source used for sending information along a single-transverse-mode optical fibre carrying 2 Gbit s^{-1}

(c) the 850 nm radiation used for sending information along a 'multimode' glass fibre which supports 1622 transverse modes[47]

(d) the radiation used in a 'laser printer' giving a resolution on the paper of 600 dots per inch

(e) for exposing a hologram whose area is 100 mm by 100 mm.

[47] This number of transverse modes is obtained, for a particular set of fibre characteristics, in problem 14.1.

Problem 11.9 (a) Étendue required for the Young slits
Investigate a Young-slits experiment in which the fringes are observed by eye. We want the fringes to be bright enough to see comfortably. For this it isn't the total optical power that's of interest, but the power per unit area on the observer's retina. Consider the cases

(1) where the source is so small that the illumination is transversely coherent along the length of the Young slits as well as across their width

(2) where the source is elongated into a slit after the fashion of Fig. 3.10, giving fringes in the usual pattern of stripes.[48]

[48] *Answer*: Surprisingly, case (2) offers hardly any increase of visual brightness, though it's more comfortable to observe.

Problem 11.10 (a) Energy throughput for a grating monochromator
Let a per-frequency radiance $B(\nu)$ be defined as in §11.7, so that a light source (not a laser) is described (small angles) by

$$\begin{pmatrix}\text{power radiated in frequency range } \delta\nu \\ \text{from area } \Delta S \text{ into solid angle } \Delta\Omega\end{pmatrix} = B(\nu)\,\delta\nu\,(n^2 \Delta S\, \Delta\Omega).$$

Imagine that we send its light into a grating monochromator. The power gathered by the instrument is[49]

$$\begin{pmatrix}\text{radiance of image of source} \\ \text{formed on entrance slit}\end{pmatrix} \times \begin{pmatrix}\text{area of ent-} \\ \text{rance slit}\end{pmatrix} \times \begin{pmatrix}\text{solid angle} \\ \text{collected}\end{pmatrix}.$$

[49] The spectrometer is in air so we drop the n^2 factor.

The radiance of the image at the entrance slit, formed by a condenser lens, is at best (eqn 11.8) equal to the radiance $B(\nu)\,\delta\nu$ of the light source, so the power we can gather (ignoring insertion loss) is

$$P = B(\nu)\,\delta\nu \times (\text{area of entrance slit}) \times (\text{solid angle collected}).$$

Comment: In practical laboratories, students are instructed to 'fill the entrance slit with light, and fill the grating with light'. The reason for this injunction is now clear: you lose out if the area sending light into the monochromator is less than the full area of the entrance slit; and you lose out if light isn't sent into the full available solid angle.[50]

[50] A similar point is made in §4.9.2 and problem 4.13, but the significance of the product (area) × (solid angle) should now be clearer.

We'll now apply these ideas to a monochromator whose grating is square of side Nd, used with angle of incidence α; we are using the same configuration as in problem 4.6. The collimator lens's focal length is f.
(1) Show that light leaving the instrument's entrance slit is collected within the solid angle $\Delta\Omega = (Nd/f)^2 \cos\alpha$.
(2) Let the entrance slit have width $(\Delta y)_{\text{slit}}$. Imagine that $(\Delta y)_{\text{slit}}$ is increased from zero; it begins to degrade the instrument's resolution when $(\Delta y)_{\text{slit}}$ reaches about $f\lambda/(Nd\cos\alpha)$. Compare with problem 4.6(6) and example 4.1 part (7) and eqn (4.10). To make the entrance slit much narrower than $f\lambda/(Nd\cos\alpha)$ just wastes energy that could be admitted to the monochromator and used. So we'll probably choose to make $(\Delta y)_{\text{slit}}$ about equal to $f\lambda/(Nd\cos\alpha)$. We might even make the slit considerably wider, if we don't mind degrading resolution in struggling for all the energy we can get. Either way, we'll have

$$(\Delta y)_{\text{slit}} \approx (\mathrm{d}y/\mathrm{d}\lambda)\,\delta\lambda, \tag{11.26}$$

where $\mathrm{d}y/\mathrm{d}\lambda = pf/(d\cos\alpha)$ is the grating dispersion (referred to the *entrance* plane), p is the order of the spectrum, and $\delta\lambda$ is the resolution we agree to accept.
(3) Let the length of the entrance slit be h. Now assemble together the factors introduced above to show that

$$\begin{aligned} P = {} & B(\nu) && \text{per-frequency-interval radiance of source}\\ & \times (\delta\nu)^2/\nu && \text{resolution demanded at frequency } \nu\\ & \times (Nd/f)^2 && 1/(f\text{-number})^2\text{; shape of apparatus}\\ & \times hf && (\text{linear dimension})^2\text{; larger apparatus permits larger slit area}\\ & \times \sin\alpha && (\approx p\lambda/d)\text{; maximize this for the best energy transmission.} \end{aligned} \tag{11.27}$$

The first two factors can't be controlled to any great extent. It's a good idea to have a small f-number, but we soon reach a limit set by aberrations in lenses or mirrors. After this, optimizing lies in doing the best we can with the last two factors. If all else fails we can increase the size of the entire spectrometer, but that has to be a last resort. The one thing we can optimize is $\sin\alpha$, making sure it isn't too far below 1. Since $\sin\alpha - \sin\theta = p\lambda/d$ and we are likely to aim to have $\theta \approx 0$, we have $\sin\alpha \approx \lambda(p/d)$. It follows that we must choose (p/d) with some care. There are reasons (unconnected with étendue, the possibility of overlapping orders) why it is best to choose p small, ideally $p = 1$. So we are pressed to choose *the most finely ruled grating that will work.*[51]
(4) Why does the analysis given here need further elaboration before it can be applied to a *spectrograph*, in which the spectrum is photographed?

Comment: We compare eqn (11.27) with corresponding expressions for other instruments in problem 11.14.

[51] 'That will work' is an allusion to the grating equation $d\sin\alpha \approx p\lambda$. Nothing will work if we choose values that make $p\lambda/d \approx \sin\alpha > 1$, either by making d too small or p too large. Additionally, a grating is usually required to record a *range* of wavelengths, and our choice of α must result in a range of θs that we can live with. Opportunities for dramatic improvement in the power P collected are not usually available unless we've been doing something silly.

Problem 11.11 (b) Resolution of Fourier-transform spectrometer[52]
An outline diagram of a Fourier-transform spectrometer is given in Fig. 10.1, and the important dimensions of it are identified in Fig. 11.9. Light reaching the detector is that which has passed through a small circular hole of diameter w at the focus of a lens of focal length f. The movable mirror on the right introduces a path difference $2x$ between the interferometer arms.

(1) Why is the limiting aperture located in front of the detector, rather than at the input?[53]

(2) Suppose that x is scanned from 0 to x_{max}. Show that the resolution, in terms of wavenumber $\bar{\nu} = 1/\lambda_{\text{vacuum}}$, can be expressed as an instrumental width $\Delta\bar{\nu}$, where

$$\Delta\bar{\nu} \approx 1/(4x_{\text{max}}), \qquad (11.28)$$

within a factor 2 or so.[54] Check this against the general statement about resolution in eqn (4.6).

(3) The resolution of a spectroscopic instrument is best described by defining an *instrumental line profile*. This is the apparent frequency spectrum of a monochromatic wave when examined with the given instrument.[55] Show that a pure sinewave of angular frequency ω_0 gives an apparent frequency distribution (instrumental line profile), after Fourier computation, of[56]

$$L(\omega - \omega_0) \propto \frac{\sin(\omega - \omega_0)T}{(\omega - \omega_0)T}, \qquad \text{where} \qquad T = 2x_{\text{max}}/c. \qquad (11.29)$$

[52] This problem, and the following seven, may be compared with a rather similar treatment in Thorne (1988), especially §7.9.

[53] An aperture is shown at the input, but it is optically a little wider than the exit hole (meaning it is wider than the real image of the exit hole formed 'backwards' by the mirrors and lenses), and it is there only to minimize the admission of stray light.

[54] *Hint*: Figures 10.2 and 10.3 give a number of examples upon which a discussion may be based if desired.

[55] An instrumental line profile has been encountered previously in problems 4.8 and 5.8. Instrumental width has been encountered, for the case of a grating spectrometer in eqn (4.8), and for the case of a Fabry–Perot in eqn (5.10).

[56] *Hint*: This result is approximate, and assumes that $\omega_0 T \gg 1$.

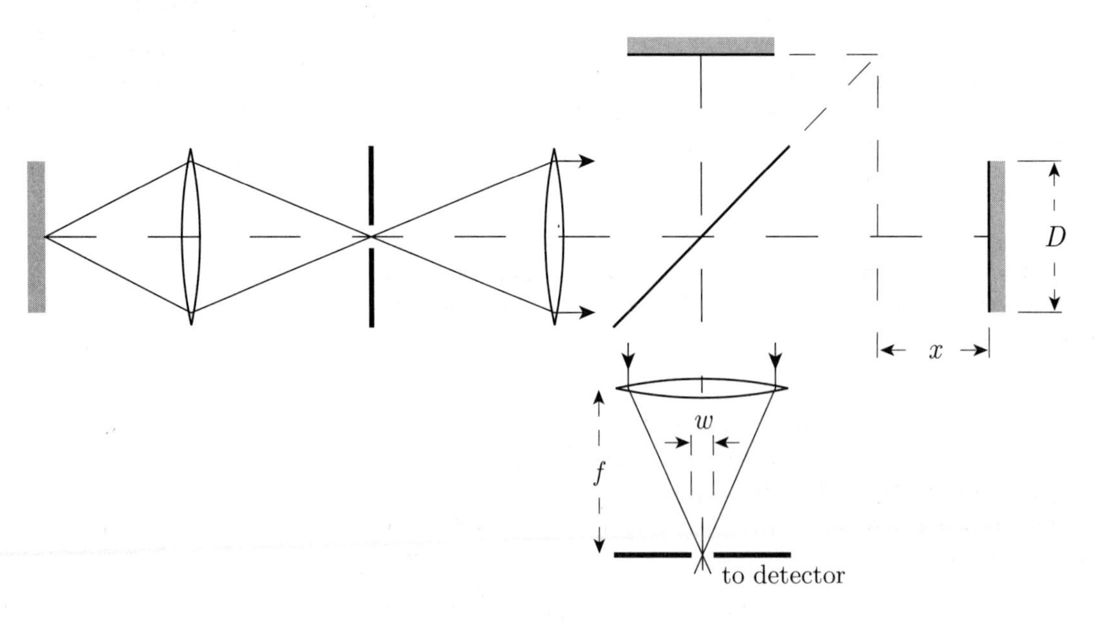

Fig. 11.9 A Michelson interferometer set up for use as a Fourier-transform spectrometer. The right-hand mirror is positioned to introduce an optical path difference of $2x$ between the two interferometer arms, distance x being adjustable over as large a range as may be necessary. Light passing through the interferometer has its range of directions limited by a small circular hole of diameter w in front of the detector.

(4) Draw sketch graphs of the functions $W(\tau)$ and $P(\omega)$ for this case and add them to the collection in §10.7.

(5) A spectral line with power spectrum $P(\omega)$ is analysed with the Fourier transform spectrometer. Consider each individual angular frequency component ω' received from the source, and work out what it contributes to the output. Show that the computed profile is

$$P_{\text{measured}}(\omega) \propto \int_{-\infty}^{\infty} P(\omega')\, L(\omega - \omega')\, \mathrm{d}\omega', \qquad (11.30)$$

which is the convolution of $P(\omega)$ with the instrumental line profile.

(6) Obtain the result of part (5) by a different route. Since we stop the scan at $x_{\max}$,

$$W_{\text{measured}}(\tau) = W_{\text{source}}(\tau) \times H(\tau), \qquad (11.31)$$

where $H(\tau)$ is a 'top hat' function

$$H(\tau) = \begin{cases} 1 & \text{for } -2x_{\max}/c < \tau < 2x_{\max}/c \\ 0 & \text{otherwise.} \end{cases} \qquad (11.32)$$

Now use the convolution theorem (backwards).

(7) It is useful to define an arbitrary criterion for the limit of resolution. Since $L(\omega - \omega_0)$ has the same form as the amplitude (not intensity, note) in a single-slit diffraction pattern, it makes sense to adapt the Rayleigh criterion. Show that with such a choice,[57]

$$\Delta\bar{\nu} = 1/(4x_{\max}). \qquad (11.33)$$

[57]This might look 'too good' in relation to eqn (4.6). Remember that $W(\tau)$ is an even function of τ, so by measuring from 0 to $x_{\max}$ we are in fact obtaining knowledge of $W(\tau)$ from $-x_{\max}$ to $x_{\max}$. This property has been relevant from the beginning of this question, in particular in setting the limits on $H(\tau)$ in eqn (11.32).

Problem 11.12 (c) Something is wrong!
The $L(\omega - \omega_0)$ of eqn (11.29) is the result of measuring the power spectrum $P(\omega)$ for a special case. It is negative for some values of $(\omega - \omega_0)$. But power can't be negative, so something must have gone wrong. What?

Problem 11.13 (b) Improving the instrumental line profile
The $L(\omega - \omega_0)$ calculated in problem 11.11 has obvious inconvenient properties:

- it is an oscillating function of $(\omega - \omega_0)T$, so it can yield negative values of $P_{\text{measured}}(\omega)$, which must be physical nonsense
- its positive excursions can create small peaks in $P_{\text{measured}}(\omega)$ that might be misinterpreted as weak spectral lines
- by falling slowly away from line centre, it could hide a quite distant, weaker, line.

(1) Show that the most negative value of $L(\omega - \omega_0)$ is 22% of its most positive value, so the oscillations of $L(\omega - \omega_0)$ are quite serious.

(2) Suggest a way in which these artefacts could be reduced or eliminated.[58]

[58]*Hint*: Think of similar problems. Problem 2.17. Apodizing.

Problem 11.14 (b) Light gathering: Fourier-transform and grating spectrometers compared

(1) Show that if the beam splitter in a Michelson interferometer divides the amplitude into the two arms unequally, the intensity falling on the detector is reduced, but the fringe visibility is unaffected.[59]

(2) Show that, if the resolution is not to be degraded from the value found in problem 11.11, the exit pinhole of the Fourier-transform instrument must have diameter w, where[60]

$$\frac{w}{f} \ll \left(\frac{2\lambda}{x_{\max}}\right)^{1/2}. \tag{11.34}$$

More realistically, the resolution may be limited both by the finite travel $x_{\max}$ of the movable mirror, and by the finite diameter w of the exit pinhole. We should construct an 'error budget' so that these two limitations result in a combined performance lying within a design requirement. A reasonable start is to allow the resolution to be degraded about equally by the two contributions. Then

$$\frac{w}{f} = \left(\frac{2\lambda}{x_{\max}}\right)^{1/2}, \tag{11.35}$$

and the overall resolution limit is changed to $\Delta\bar{\nu} = 1/(2x_{\max})$, twice the value obtained previously.

(3) Let the interferometer mirrors be square with side D. The étendue can be worked out by using the area $\pi w^2/4$ of the exit pinhole, taken with the solid angle D^2/f^2 within which light arrives at the pinhole. Show that the étendue is $\pi D^2\lambda/(2x_{\max}) = \pi D^2\lambda\Delta\bar{\nu}$. Notice that w and f have cancelled out, so there is nothing we can do to optimize the power by choosing favourable dimensions for the pinhole or the focal lengths of lenses.[61] The result obtained here should 'feel right', given the invariance of étendue.

(4) Let the source have radiance B. Show that the power reaching the detector in a bright fringe is (small angles)

$$\text{peak power at detector} = \pi B D^2 \lambda \Delta\bar{\nu}. \tag{11.36}$$

(5) Now make a comparison with corresponding quantities for a grating monochromator. Assume that the grating is used with incidence angle α, the same conditions as were assumed in problem 11.10. To make the instruments roughly comparable, we'll take the grating to be square with side $Nd = D$, and the entrance slit width will be set to give the same resolution $\Delta\bar{\nu}$ as is achieved with the Fourier-transform instrument.[62]

Use eqn (11.26) to show that the spectrometer's entrance slit has width $\Delta y = (pf_s/d\cos\alpha)\Delta\lambda = f_s\lambda\tan\alpha\,\Delta\bar{\nu}$, where f_s is the focal length of the monochromator's collimating lens. Show that the étendue is

$$\text{étendue of grating monochromator} = (Nd)^2(h\lambda/f_s)\sin\alpha\,\Delta\bar{\nu}. \tag{11.37}$$

Show that the power reaching the monochromator's detector is

$$\text{power reaching detector} = B(\nu)\,\delta\nu\,(Nd)^2(h\lambda/f_s)\sin\alpha\,\Delta\bar{\nu}. \tag{11.38}$$

[59] *Comment*: There are implications here for the effort we need to put into the quality of our equipment: the reflectivity of the beam splitter is not critical; spend money and effort elsewhere.

[60] For the precise numerical factor on the right see problem 11.16.

[61] *Comment*: This should agree with previous experience in problem 4.13, where all attempts at optimizing a condenser lens in front of a spectrometer cancelled out in a similar way.

[62] It is assumed that $4x_{\max} \le Nd\sin\alpha$. If this condition is not met, the instruments are not comparable because the interferometer achieves a resolution outside the capability of the grating.

(6) Assemble these results to show that the ratio

$$R \equiv \frac{\text{detector power in interferometer}}{\text{detector power in monochromator}} = \frac{B_{\text{total}}}{B(\nu)\,\delta\nu}\left(\pi \frac{f_{\text{s}}}{h}\frac{1}{\sin\alpha}\right). \tag{11.39}$$

(7) Show that the quantity in large brackets, $R_1 = \pi f_{\text{s}}/(h \sin\alpha)$, is the ratio of the two instruments' étendues. For a monochromator, the entrance slit length h cannot be made very large because of aberrations in the focusing optics (lenses or mirrors), so that $f_{\text{s}}/h \sim 40$. We can, therefore, estimate the geometrical factor, the ratio of étendues R_1, as about 200 in favour of the interferometer.

(8) Consider now the other factor in eqn (11.39). The numerator B_{total} is the radiance of the source totalled over the whole range of frequencies that is being accepted by the interferometer. The denominator is the radiance for frequencies that lie within the range $\delta\nu$ transmitted by the grating instrument. If the radiation were 'white', we should have $\delta\nu = c\Delta\bar{\nu}$ and

$$\frac{B_{\text{total}}}{B(\nu)\,\delta\nu} \approx \frac{\text{frequency range accepted by interferometer}}{\text{resolvable frequency difference}}, \tag{11.40}$$

which can easily be of order 10^4. Even if the radiation is far from white, say a line spectrum containing 100 lines, the ratio $B_{\text{total}}/B(\nu)\,\delta\nu$ will still be 100 or so. Altogether then,

$$R \approx 200 \times \frac{B_{\text{total}}}{B(\nu)\,\delta\nu}, \tag{11.41}$$

which is likely to be in the range 10^4 to 10^7.

Comment: This problem shows that the interferometer has a large superiority in energy gathering, composed of two factors

- a factor of order 200, of purely geometric origin (larger étendue), which permits the interferometer to accept more light energy for a given resolution
- a factor because the interferometer 'looks at all of the spectrum at once'.

Of course, it is not yet obvious what the accuracy will be after our computer has used the data to calculate a Fourier transform: we wonder how errors propagate through the calculation .... Nevertheless, we do win a great deal. The factor from étendue gives an unqualified advantage to the interferometer. The second factor requires more careful analysis, and may or may not favour the interferometer depending on conditions. There is a nice discussion in Thorne (1988), §7.9.

Problem 11.15 (b) The Fourier-transform spectrometer as a whole-fringe instrument

Problem 11.14 raises an obvious general question: what étendue is theoretically possible in a spectrographic instrument with a given resolution? Here we investigate.

(1) Show that the Fourier-transform spectrometer, with the design decisions of problem 11.14, has the properties:

$$\text{(solid angle within which light is accepted)} \times \text{(chromatic resolving power)} = \pi \tag{11.42a}$$

$$\text{(étendue)} \times \text{(chromatic resolving power)} = \pi \times \text{(area of a mirror)}. \tag{11.42b}$$

Here the solid angle referred to is $\pi w^2/(4f^2)$ within the notation of problem 11.14; and the chromatic resolving power is $\lambda/\Delta\lambda$ as in eqn (4.5).

(2) Explain in words why étendue has to be traded against resolution.

(3) Investigate whether a Fabry–Perot has a similar property.

Comment: Equation (11.42a) is similar to the property that defines a **whole-fringe instrument**:[63]

$$\text{(solid angle)} \times \text{(chromatic resolving power)} = 2\pi. \tag{11.43}$$

For discussion of this concept, see Jacquinot (1960). Jacquinot identifies a whole-fringe instrument as representing a useful standard of comparison so far as the gathering of light energy is concerned. We learn that a Fourier-transform spectrometer conforms to this standard.

Conversely, we learn that a grating spectrometer is *not* a whole-fringe instrument, and that it has by comparison a remarkably unfavourable étendue.

(4) Look at §8.10. The confocal Fabry–Perot accepts an even larger solid-angle range than does a normal Fabry–Perot. So does a whole-fringe instrument represent an unsurpassable optimum?

(5) Come to think of it, we could invent an improvement to a Fourier-transform spectrometer or a Fabry–Perot. Where the ring-fringes lie in front of the detector we could place an aperture that transmits several rings,[64] rather than just the central spot. Since all bright rings have the same area, we would gain a factor equal to the number of fringes transmitted. Why is this recourse rarely attempted?

[63] Equation (11.42a) seems to fall short of this by a factor 2. This is merely because we have allowed the finite pinhole diameter and the finite mirror travel both to degrade the resolution. Show this. The whole-fringe property of eqn (11.43) applies if all of the limitation on resolution comes from the pinhole. The same factor 2 accounts for a difference between eqns (11.42b) and (11.17).

[64] Not a large hole, which would wreck the resolution. A set of transparent annuli with black between them, rather like a Fresnel zone plate, arranged to match the sizes of the rings in the fringe pattern.

Problem 11.16 (c) Fourier-transform spectrometer: diameter of exit pinhole

A Fourier-transform spectrometer delivers light to its detector through a pinhole of diameter w. This diameter is now non-negligible.

(1) Show that a monochromatic input gives an intensity at the detector of $I(\tau)$, where $\tau = 2x/c$ and

$$W(\tau) = I(\tau) - \tfrac{1}{2}I(0) \propto \cos\{2kx\{1 - w^2/(16f^2)\}\} \times \frac{\sin\{kxw^2/(8f^2)\}}{kxw^2/(8f^2)}. \tag{11.44}$$

Interpret the two factors here.

(2) Find the instrumental line profile that results from the finite size of the source aperture. Show that it is a top-hat function with full width

$$\Delta\bar{\nu} = \frac{1}{\lambda}\frac{w^2}{8f^2}. \tag{11.45}$$

(3) Show that the $\Delta\bar{\nu}$ just calculated is the least spacing between resolvable spectral lines, if the resolution is limited entirely by the finite size of w.

(4) Problem 11.11(7) has shown that when the resolution is limited entirely by the finite travel of the moving mirror, the resolution is $\Delta\bar{\nu} = 1/(4x_{max})$. Show that, if we agree to make both limits on the resolution equally damaging, then we must choose w/f to have the value $\sqrt{2\lambda/x_{max}}$ that was claimed in eqn (11.35) of problem 11.14(2).

Problem 11.17 (b) The rate of scanning and digital sampling

A Fourier-transform spectrometer works by scanning one mirror between $x = 0$ and $x = x_{max}$. Discuss the allowable speed of the scanning and its relation to the response time of the 'slow' detector ('slow' defined as in §9.10).

During a scan, the intensity $I(x)$ reaching the detector is to be sampled at discrete values of x, digitized and fed to a computer for the Fourier analysis. What[65] is the greatest allowed separation Δx between samples?

[65] *Hint*: The idea needed is contained the **Nyquist sampling theorem**, which is well known in this kind of digital electronics. For example, it sets a lower limit on the sampling rate that must be used during the recording of sound that will end up impressed on a CD.

Problem 11.18 (a) Mechanical considerations

Consider a Fourier-transform spectrometer used at a wavelength of order 500 nm. Let the mirror diameter D be 50 mm.

(1) Estimate the tolerance on angular orientation of the movable mirror.

(2) Estimate the precision with which x must be known during its scan.

(3) Repeat these estimations for a wavelength of order 10 µm.

(4) Use your values to comment on the desirability of a Fourier-transform spectrometer as an instrument for the visible and for the infrared.[66]

[66] This merely scratches the surface. Problem 11.18 shows that a Fourier-transform instrument is easy to engineer for infrared use. But the really important thing about the infrared is how much we need the superior energy-gathering of the Fourier-transform technique. Reasons:

- Radiative lifetimes are long (roughly $\propto \omega^{-3}$) so spontaneous emission is infrequent, and spectroscopy must be done in absorption.
- Continuum sources for absorption spectroscopy are weak ($\propto \omega^2$ according to Rayleigh–Jeans).
- Detectors are notoriously insensitive once the photon energy $\hbar\omega$ gets too small to activate a quantum detection process.

Problem 11.19 (a) Energy conservation in slab interference

(1) Consider the equivalent of Fig. 11.4 for a Michelson interferometer. Light travelling to the right exhibits bright and dark fringes. In the case of the slab, energy conservation was assured by a second set of fringes, with light and dark interchanged, travelling to the left. But in the Michelson interferometer the mirrors are opaque and the 'transmitted' fringes are absent. This is drawn attention to in sidenote 31 on p. 258. Yet energy must still be conserved: what does not go into a dark fringe must go somewhere else. Where?

(2) A slab is illuminated as in Fig. 11.4, but its left-hand surface is perfectly reflecting. Discuss the conservation of energy in the resulting interference pattern.

Problem 11.20 (b) Localization of 'wedge' fringes

When light is reflected from two surfaces that are close together and whose spacing varies, we have 'fringes of equal thickness'. They are localized *at* the thin layer. Construct an explanation of this localization.

12 Image formation: diffraction theory

12.1 Introduction

In Chapter 1, we gave ray-optics descriptions of the process by which a lens or mirror forms an image of some object. Additionally, in Chapter 3, we learned how diffraction affects the resolution of such an optical system, resolution meaning its ability to form a fine-detailed image.

In these previous treatments, we assumed that all points on an object radiated incoherently (relative to each other), though this assumption was not always drawn attention to. Such an assumption is reasonable if the object is self-luminous, or if it is illuminated by light from 'all directions', thereby giving a coherence area of order λ^2 on the object. There is, however, an obvious opposite limit: where the object is illuminated with complete transverse coherence across its area. The physics of image formation must be different for this case, at least in its details. This is the central theme of the present chapter.

It might be thought that we are on the point of pursuing somewhat arcane detail. However, the understanding of coherent imaging has been extraordinarily fruitful, leading to a rich set of new optical discoveries and inventions.

12.2 Image formation with transversely coherent illumination: informal

We introduce the subject by imagining that a part-transparent object is illuminated by a perfectly (transverse-) coherent plane monochromatic wave. The coherence area extends over the whole illuminated area of the object. The object lies in plane O in Fig. 12.1, and the light arrives parallel to the axis from the left. There is a set of Huygens secondary sources in plane O, and definite phase relations exist (perhaps made complicated by the object, but still there) between all the secondary waves. Everything downstream from the object, including image formation by the lens in plane I, involves adding amplitudes (not intensities), because we have set things up that way.

A complex amplitude $U_o(x, y)$ is present in plane O, and the propagation of light downstream from O can be worked out using a Kirchhoff diffraction integral taken over $U_o(x, y)$. To understand what happens it's

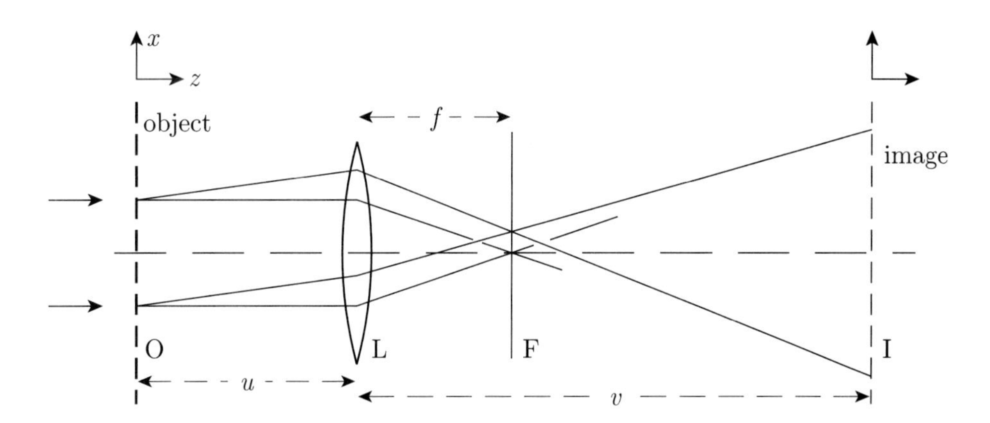

Fig. 12.1 A part-transparent object is illuminated from the left by a plane wave travelling parallel to the z-axis. The illumination is ideally transverse-coherent. The object's transmitted amplitude $U_o(x, y)$ is to be thought of as decomposed into sinewave (Fourier) components; the effects of two such components are shown. One is a constant; its light is transmitted unaffected, and is focused to a point on the axis of the system in plane F. The other component shown is proportional to $e^{i\beta x}$, which is what we would have had if there were no object but light arrived 'uphill' making angle θ with the axis, where $\sin\theta = \beta/k$; this light travels off ('continues') at angle θ and is concentrated to a focus, at plane F but above the axis. As light travels onwards from these foci (and others) it recombines in plane I to form a set of interference fringes that construct the image of the original object.

convenient to imagine $U_o(x, y)$ Fourier-analysed into a sum over sinewaves: functions like $e^{i\beta x}\, e^{i\gamma y}$. Each of the sinewaves follows its own distinctive history through the optical system.

To follow one of these histories, think of a sinewave component varying like $e^{i\beta x}$. We know from problem 3.19 how such a sinewave propagates: it travels away as a plane wave making angle θ to the z-axis, where $\sin\theta = \beta/k$. A representative such wave is sketched travelling 'uphill' in Fig. 12.1. The lens concentrates this wave to a point in the lens's 'back focal plane' F, at distance $f \tan\theta \approx f\beta/k$ (small angles) from the axis. The higher this sinewave's 'spatial frequency' $\beta/(2\pi)$, the farther is its focus from the axis. Thus each sinewave component of $U_o(x, y)$ goes to its own special place in plane F. This is precisely what is meant when we say that plane F displays the Fourier transform of $U_o(x, y)$.

Downstream from plane F, onward travel of the light results in the different sinewave components coming together in plane I to form an image of the object. The image is made up by interference of the (coherent remember) wave amplitudes coming from the object via plane F.

The interpretation just given can easily be confirmed in a student laboratory. We can use as object a 35 mm slide bearing a black-and-white grating pattern,[1] illuminate it with a small He–Ne laser, and see the grating orders as bright spots in lens L's focal plane F. To show that these grating orders are implicated in image formation, we can block selected spots and see that the image in plane I is modified. In particular, if all diffraction orders are blocked except the central order, the 'image' bears none of the grating lines. If the obstruction is moved so as just to admit one or both of the first orders, the grating lines reappear in the

[1] In such an experiment at Oxford, the slide bears a picture crossed by fine black lines, rather like the rastered picture displayed on a television screen. The picture assists with focusing the lenses and can itself be examined to investigate the effect of optical interventions.

'image'. If all odd-numbered orders are blocked, the 'image' shows lines, but double the correct number of them. And other tests are simple to invent and try out.

Problems 12.1 and 12.2 explore the ideas outlined here, aiming to reinforce the general approach, but also drawing attention to some over-simplifications that we have made in this introduction.

12.3 Image formation: ideal optical system

We put the ideas of §12.2 on a more formal footing with the aid of Fig. 12.2. The (scalar-but-complex) amplitude in plane O is $U_{\rm o}(x_{\rm o}, y_{\rm o})$, which may be some quite general function of $x_{\rm o}$ and $y_{\rm o}$. The amplitude in plane F can be calculated from a Kirchhoff integral (problem 12.3):[2]

[2] In contrast to problem 12.1, Cartesian coordinates are here being used rigorously.

$$U_{\rm F}(x_{\rm F}, y_{\rm F}) = \frac{{\rm e}^{{\rm i}kD_1}}{{\rm i}\lambda f_1} \int U_{\rm o}(x_{\rm o}, y_{\rm o}) \exp\{-{\rm i}k\, x_{\rm o}\, x_{\rm F}/f_1 - {\rm i}k\, y_{\rm o}\, y_{\rm F}/f_1\}\, {\rm d}x_{\rm o}\, {\rm d}y_{\rm o}, \quad (12.1)$$

where D_1 is the optical path from the origin of $(x_{\rm o}, y_{\rm o})$ to the origin of $(x_{\rm F}, y_{\rm F})$. Compare with eqn (3.28). Here the integration is taken over the whole $x_{\rm o} y_{\rm o}$ plane, and a small-angle approximation is used for simplicity.

The transformation from plane F to plane I is given by an integral of similar form:

$$U_{\rm i}(x_{\rm i}, y_{\rm i}) = \frac{{\rm e}^{{\rm i}kD_2}}{{\rm i}\lambda f_2} \int U_{\rm F}(x_{\rm F}, y_{\rm F}) \exp\{-{\rm i}k\, x_{\rm F}\, x_{\rm i}/f_2 - {\rm i}k\, y_{\rm F}\, y_{\rm i}/f_2\}\, {\rm d}x_{\rm F}\, {\rm d}y_{\rm F} \quad (12.2)$$

$$= -{\rm e}^{{\rm i}k(D_1+D_2)} \frac{f_1}{f_2}\, U_{\rm o}(-x_{\rm i} f_1/f_2, -y_{\rm i} f_1/f_2). \quad (12.3)$$

The last expression follows by an application of the Fourier inverse transformation, see problem 12.3. Equation (12.3) shows that the light

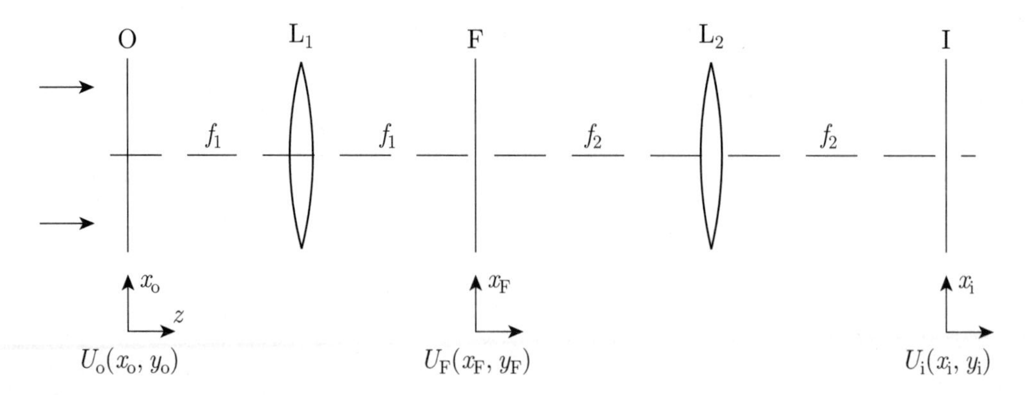

Fig. 12.2 This optical system is constructed to make mathematical analysis easy. An object lies in plane O and is illuminated as in Fig 12.1. Light diffracted from the object travels via lens L_1 and has its Fourier transform displayed in plane F. Light continuing on from there undergoes a second Fourier transformation to plane I. The double Fourier transformation results in a replica of the original object (apart from scale factors including a negative sign for inversion) in the image plane I.

amplitude at the image reproduces that at the object, except for inversion, magnification by f_2/f_1, a scaled amplitude that ensures energy conservation, and of course a phase factor connected with the distance the light has travelled.[3]

The optical arrangement given in Fig. 12.2 has been designed to make analysis as simple as possible: there is a Fourier-transform relationship ('exact' apart from scale factors and small-angle approximations) between the complex amplitude in plane O and that in plane F, and another between F and I. We have exploited (twice) the second of the two Fourier-transform arrangements shown in Fig. 3.19.

[3]The negative sign accompanying the phase factor $e^{ik(D_1+D_2)}$ is an instance of the 'phase anomaly' discussed in problem 8.2. A similar negative sign can also be seen in eqn (8.12) which results from mathematics very similar to that pursued here.

12.4 Image formation: imperfect optical system

The mathematics leading to eqn (12.3) is given to display the rôle of Fourier transformation in the formation of an image. Yet, much of the important physics will come from an understanding of why this presentation is inadequate.

In writing exponents like $-ikx_o x_F/f_1$, we have used a small-angle approximation for what was properly $-i\beta x_o = -i(k\sin\theta)x_o$, where θ is the angle through which the light leaving plane O is diffracted.[4] Now components $\exp(i\beta x_o)$ of $U_o(x_o, y_o)$ with $|\beta| > k$ do not give rise to light travelling away at any real angle θ; such components cannot be carried forward to contribute to the image. Moreover, the true upper limit on $|\beta|$ is less than k because light diffracted to large angles θ also fails to reach the image—for reasons we shall see. The optical system is a (spatial) low-pass filter: it builds up the image $U_i(x_i, y_i)$ from only the low Fourier components $\exp(i\beta x_o)\exp(i\gamma y_o)$ present at the object, having values of $|\beta|$, $|\gamma|$ below some cutoff.

[4]The small-angle approximation can, of course, be corrected, as is explored in problem 12.4. But this is a side-issue here.

All this leaves us with a task: to find out how the lens system filters the image and what image degradation results; to understand how the image quality can be optimized; and to see if new insights suggest further developments. As mentioned in the introduction to the present chapter, this programme has led to a blossoming of new ideas and techniques.

We shall return to making a formal development of the Fourier theory in §12.12. But we shall investigate the physics in a less formal way first, in §§12.5–12.9.

12.5 Microscope resolution: Abbe theory

Figure 12.3 shows a part-transparent object illuminated from the left and diffracting light into a lens which might perhaps be the objective lens of a microscope. Light diffracted through angle θ carries *information*: its (complex) amplitude is caused by, and conveys information about, a sinewave component $\exp(i\beta x_o)$ in the field transmitted by the object. The shorter is the period $2\pi/\beta$ of the sinewave the larger is the angle θ;

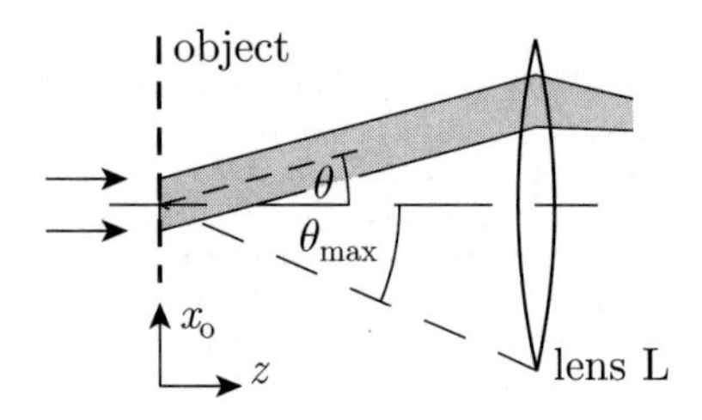

Fig. 12.3 The part-transparent object of Fig. 12.1 is illuminated by a plane wave parallel to the z-axis. The transmitted light is diffracted into a variety of directions, one of which is shown. Diffracted light can pass through the lens provided $\theta \leq \theta_{\max}$. This restriction (according to the discussion in §12.5.1) sets an upper limit on the spatial frequencies that are passed on to the image.

fine detail in the object is conveyed by light travelling at large angles θ to the axis. We want such large-angle light to reach the image and reproduce the detail there. With large angles in prospect, small-angle approximations won't do any more.

12.5.1 Abbe theory: introduction

We start by giving an unsatisfactory explanation, because the reader may have encountered it before, and it may make a familiar introduction to a more careful treatment. In Fig. 12.3, light is accepted into lens L provided that it is diffracted through an angle θ with $|\theta| < \theta_{\max}$. Light diffracted through angle $\theta_{\max}$ originates from a sinewave component with $\beta = \beta_{\max} = k \sin\theta_{\max}$ and period $d = \lambda/\sin\theta_{\max}$. We may say that the optical system resolves structure down to linear dimension $d_{\min}$, where[5]

$$\beta_{\max} = k\sin\theta_{\max}, \qquad d_{\min} = \frac{\lambda}{\sin\theta_{\max}} = \frac{\lambda_{\text{vac}}}{(n\sin\theta_{\max})}. \tag{12.4}$$

The quantity $(n\sin\theta_{\max})$ is called the **numerical aperture.**[6] We shall see that the result (12.4) is correct, though we prefer to account for it in a different way.

12.5.2 Abbe theory: explanation

Let us first see why all is not well with Fig. 12.3. Suppose that the object, which may conveniently be thought of at this stage as a grating, has width w. The grating order shown, a beam of width $w\cos\theta$, does not represent a single Fourier component, because it is of finite width. A Fourier component properly defined behaves like $\exp(\mathrm{i}\beta x_o)$ over the whole (x_o, y_o) plane.[7]

Figure 12.4 shows a single Fourier component $\exp(\mathrm{i}\beta x_o)$. This component propagates as a beam of *infinite* width travelling at angle θ to the axis; we draw that part of it that falls within the aperture of the lens. (Only one lens is drawn, as in Fig. 12.1, so that there is only one limiting aperture in the system.) A beam like this can always enter the lens, for any $\theta < \pi/2$, and not only for angles $\theta < \theta_{\max}$. It is not the case then that high Fourier components fail to enter the lens. They *do* fail to contribute to the image—it turns out—but they are lost later on in the optical system.

Figure 12.4 traces the selected Fourier component[8] through the lens to the image plane I. When the light arrives at I it covers an area that is the image of the object-plane area from which it came. There it overlaps with the zeroth order, and together they form the interference pattern that builds (their contribution to) the image of the original object.

Now consider what happens as we increase the space-frequency β to beyond $\beta_{\max}$. The Fourier component diffracted from $\exp(\mathrm{i}\beta x_o)$ makes a larger angle θ with the z-axis as shown in Fig. 12.5. That part of the diffracted beam collected by lens L now originates wholly from below the z-axis, so where it arrives at the image plane it falls wholly above the

[5] This idea was used in problem 10.14(4).

[6] We have encountered numerical aperture before, in a different context, in eqn (11.5), where it was connected with the ability of an optical system to gather energy.

[7] The reader may be reminded of the top-hat pulse of Fig. 2.2(a). When this is expressed as a Fourier integral, every contributing sinewave extends from $t = -\infty$ to $t = +\infty$. This is not paradoxical, because the job of the Fourier components is to add up correctly to the original $V(t)$, both within the width τ_w of the pulse, and outside that width where those components must sum to zero.

[8] It may help to be reminded of problem 2.16 and example 3.1.

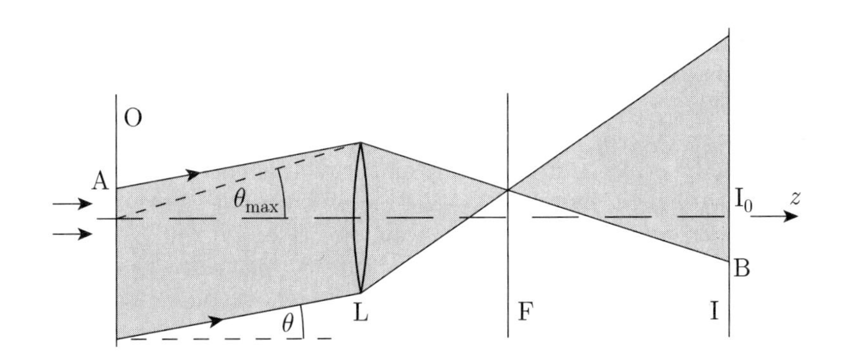

Fig. 12.4 A single Fourier component $\exp(i\beta x_o)$ is diffracted to travel away from the object as a plane wave at angle θ to the z-axis. We draw that part of the (infinitely wide) wave that passes through lens L; it contributes a magnified version of $\exp(i\beta x_o)$ at plane I. If this wave overlaps the zeroth order (diffracted to $\theta = 0$) around I_0, the two waves interfere and give rise to the fringes that constitute a visible image. If there is no overlap, these fringes are absent. The condition for overlap is $|\theta| \le \theta_{\rm max}$.

z-axis. It no longer can interfere with the zeroth order at places around I_0 where the image is required.[9] Conversely, to arrive at and below I_0 and so contribute to the image, light must have left the object plane from places above the axis, and therefore at an angle less than $\theta_{\rm max}$. The image is built up only from light with $|\theta| \le \theta_{\rm max}$ and $|\beta| \le k \sin\theta_{\rm max}$: eqn (12.4) turns out to be correct, though not for the reason that first led to its being written.[10]

We may bring the Abbe theory to life by means of a simple example. Suppose we use as object a transmission diffraction grating with rulings spaced by d. When the grating is illuminated normally, light travels away as a set of plane waves in the various orders of the grating.

- Suppose that all grating orders other than the zeroth are lost. (It may help to think of the optical system of Fig. 12.2 with obstacles placed in plane F blocking the discarded grating orders, even though the loss may take place elsewhere.) Then all information about the periodic structure of the grating is absent from the image. Equivalently, there are no superposed waves that could interfere to produce fringes. An image that should have been stripy is simply 'clear'.[11]
- Next imagine the system modified so that it transmits the first orders, as well as the zeroth order. Then the image is what we would have had if the grating generated only these three orders: a grating like that of example 3.1. The image has a sinusoidal variation[12] of amplitude across it, of the form (constant $+ \cos\beta_i x$).
- Let additional grating orders be transmitted. Each new order contributes a new sinewave to the image. The image acquires more sinewaves, more information, and is made to resemble more closely the original object (apart from magnification and inversion). The physics here resembles what is taught in most elementary presentations of Fourier series, when the student is shown the Fourier components of a square wave being added one by one, progressively giving a better approximation.

Some implications of eqn (12.4) for microscope design are obvious. To achieve the highest possible resolution, we need the highest possible

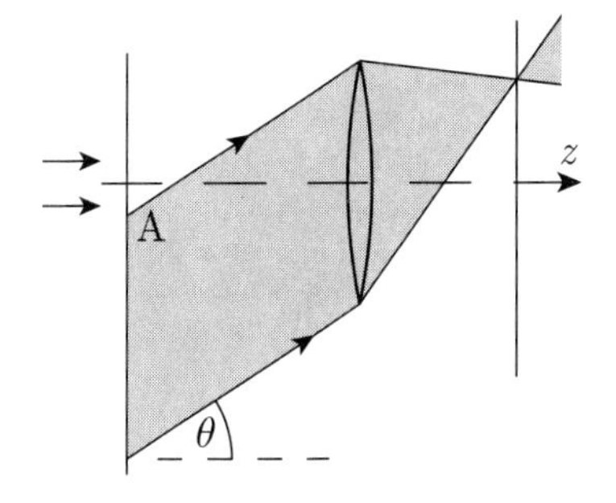

Fig. 12.5 Here the Fourier component $\exp(i\beta x_o)$ has $\beta > k\sin\theta_{\rm max}$. Unlike the case shown in Fig. 12.4, the diffracted light cannot contribute to an image around I_0, interfering with the zeroth order there, because it all ends up away from the axis in the image plane: the image B of A lies above the axis, and the rest of the beam even further above.

[9] In Fig. 12.4, point B is the image of point A. By the ray-optics rules for locating an image, a straight line from A to B passes through the centre of the lens. (It does not matter that the light considered does not follow this path.) Therefore, A and B must lie on opposite sides of the z-axis.

[10] The ideas given here are based on Walton (1986).

[11] We are repeating the outline description of a demonstration experiment given at the end of §12.2.

[12] This β_i is related to the β of the light that left the object by $\beta_i = (f_1/f_2)\beta$.

numerical aperture. And because angles are not small, designing the objective lens is a non-trivial task: aberrations must not add significantly to the diffraction spread of the image.[13]

We should record one warning before thinking about optimizing the design of a microscope. It was mentioned in §12.2 that a grating may be imaged in a way that doubles the number of lines. This rather startling effect may be achieved by removing all the odd-numbered orders of diffraction, so that the light 'thinks' the second order is the first, and so on. The doubled number of lines is an example of 'false detail'.[14] Filtering in the objective's focal plane cannot create Fourier components 'from nothing', but it can rearrange the existing components to add up to something misleading. Sometimes filtering can be advantageous, as we shall see in the next few sections, but we have to be aware that, applied without understanding or proper control, it can result in artefacts.

12.6 Improving the basic microscope

Once something is understood, it's often easy to see how to do better. Two developments will be described here.

First, we relax the condition that the object must be illuminated by collimated light travelling parallel to the axis. Imagine first that light is supplied to the object as a collimated beam,[15] but at oblique angle $\alpha_{\max}$. Figure 12.6(a) shows a first order of diffraction passing—just—through the objective, at the same time as the zeroth order. The angle of diffraction is now greater than before, so the light has been diffracted by a finer sinewave. When this light is transmitted to the image, it displays the finer detail.

To exploit this idea in a more practical way, we use a **condenser lens**. Let the condenser supply light within a cone of directions with semi-angle $\alpha_{\max}$, and let the objective accept light within a cone of semi-angle $\theta_{\max}$. Then detail of linear dimension $d = d_{\min}$ is just resolved if both a zeroth order and a first order of diffraction just make their way into the objective. We should expect[16]

$$d_{\min} = \begin{cases} \lambda/(\sin\alpha_{\max} + \sin\theta_{\max}) & \text{for } \alpha_{\max} \le \theta_{\max} \\ \lambda/(2\sin\theta_{\max}) & \text{for } \alpha_{\max} > \theta_{\max}. \end{cases} \tag{12.5}$$

A sensible choice is usually to make $\alpha_{\max} \approx \theta_{\max}$, and then

$$d_{\min} = \frac{\lambda_{\text{vac}}/2}{(n\sin\theta_{\max})} = \frac{\lambda_{\text{vac}}/2}{\text{numerical aperture}}. \tag{12.6}$$

The condenser achieves an improvement in resolution of about a factor 2.

This chapter started out with a statement that we were to investigate the effect of (transversely) coherent illumination on an imaging system. The effect of supplying light within a range of arrival directions is to supply light with a restricted coherence area. We may wonder how this affects things. Problem 12.6 shows that the illumination can be thought of as just-incoherent for the case where $\theta_{\max} = \alpha_{\max}$. Nevertheless, the

[13] The critical reader may—should—be uncomfortable. It is large-angle diffraction that impresses the required information on the light entering the microscope. Yet, our Chapter-3 theory of diffraction is then at the limits of its validity range: at the least, we should wonder whether to include an obliquity factor, and, at worst, we might need to abandon a scalar theory and deal properly with vector electric and magnetic fields. This accusation has to be correct. At the same time, the physical insights we have presented are compelling, and their essentials can be expected to survive into any corrected theory: large-angle diffraction is caused by high spatial frequencies at the object, and large-angle interference recreates those spatial frequencies at the image. We claim, therefore, that the essentials of the discussion here must be a reliable guide to the physics.

[14] When Abbe put forward his theory of microscope resolution, he received objections from microscopists who said 'but I can see detail smaller than that', to which the reply was 'yes, but it's false detail; it's not there in the object'.

[15] We revert to the reasoning of §12.5.1 here for simplicity. The reader is invited, in problem 12.5, to re-present the argument after the fashion of §12.5.2.

[16] If $\alpha_{\max} > \theta_{\max}$, some light arrives at angles $\alpha > \theta_{\max}$ and does not send its zeroth order into the objective. If a first order from this is diffracted to angle $\theta < \theta_{\max}$ and enters the objective, conditions are similar to those of dark-ground illumination. This diffracted light contributes to the image—because dark ground works. But problem 12.6 shows that it does not help with resolving small detail. The outcome is that $d_{\min}$ is not made significantly smaller by increase of $\alpha_{\max}$ beyond $\theta_{\max}$, hence the second of the two cases itemized in eqn (12.5).

(a)
order 0
α_{max}
order 1
object
objective L_1

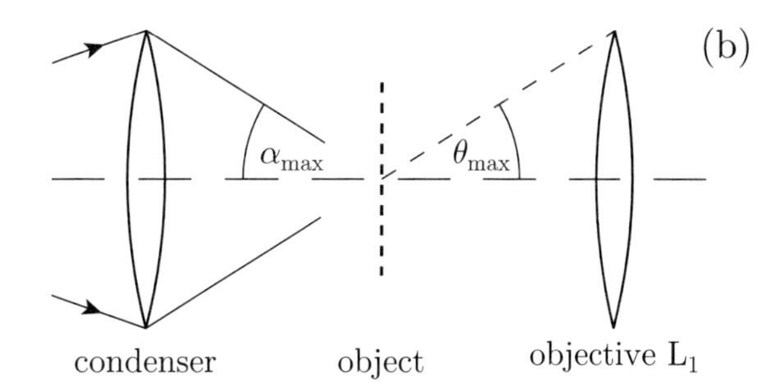

Fig. 12.6 (a) The resolution of a microscope may be enhanced by supplying collimated light at an oblique angle, so that the zeroth order and one side order of diffraction both scrape through the objective lens. (b) A more practical implementation. Light is supplied to the object within a range of directions by using the condenser lens shown. It is again the case that *some* of the light diffracted from fine detail in the object manages to get through the objective accompanied by its zeroth order.

route by which we have obtained the resolution limit is most conveniently seen as a development building on coherent-illumination ideas.

The second development that is suggested by our analysis comes from noticing the presence of the refractive index in eqns (12.4) and (12.6). If we could increase the refractive index, we could reduce the resolution distance. This idea leads to the use of **oil immersion**. We may fill the space between the object and the objective lens[17] (perhaps the condenser too?) with oil with refractive index n, and the resolution is improved by a factor n. A reasonable value of n is $\sim$1.5. A factor 1.5 does not sound much, but it is an opportunity not to be missed when we need to examine tiny detail.[18]

With all microscope designs, we have to think about the conditions of use of the completed instrument. In the first instance, a microscope is usually intended to be used with a human eye as the detector of the final image. Problem 12.7 examines the implications of this, and shows that the magnification and the numerical aperture of a microscope should be thought about together. There is no point in pursuing high magnification if the only outcome is a larger fuzzy image. Conversely, there is no point in resolving detail and then under-magnifying it so that it's too small to see.

[17] Compare with problem 11.6, which shows that a refractive index makes no difference to the available resolution when laser light reads information from a CD. Why is the present case different?

[18] The objective lens should be designed either for use in air or for use with oil, so that aberrations are minimized for the anticipated conditions of use.

12.7 Phase contrast

The central insight we've gained from §§12.3 and 12.4 is that *information* about an object is on display, in Fourier-transform form, in the objective lens's back focal plane. Also we mentioned (in §12.2) that obstacles placed in that plane can be used to verify the correctness of the theory. This leads us to ask: are there more creative things we can do in the focal plane, perhaps to apply some other sort of filtering (which is more interesting than it sounds)? This is the beginning of several developments, of which phase contrast is one.

Microscopes are probably used in biological and medical laboratories more than anywhere else. Objects of interest to biologists are usually

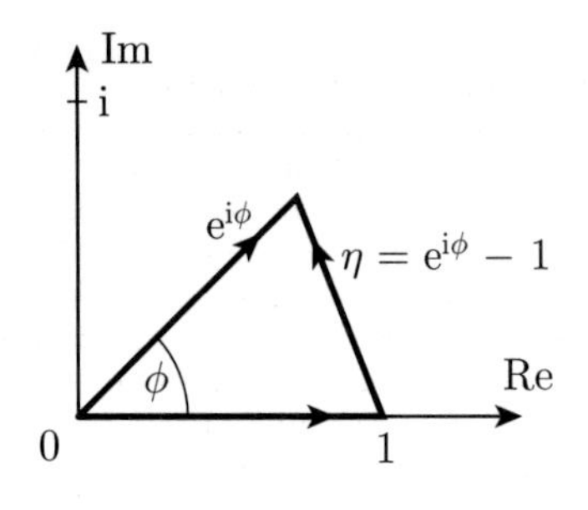

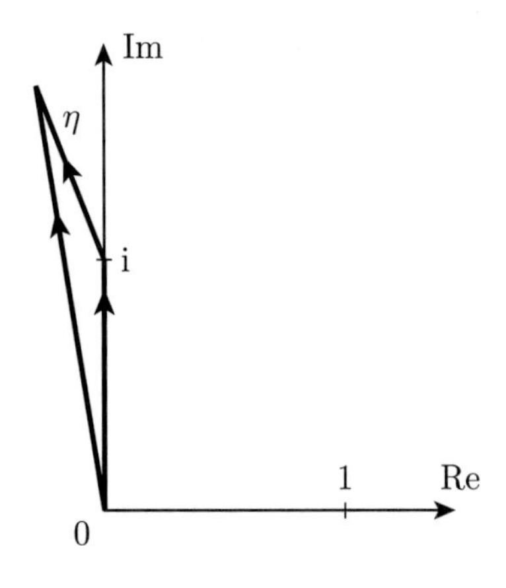

Fig. 12.7 The complex amplitudes involved in phase contrast are here shown as phasors—complex numbers drawn in the complex plane (Argand diagram). The amplitude transmitted by the object is $U_p(x, y) = U_0 e^{i\phi(x,y)}$, where $\phi(x, y)$ may have some general dependence on x, y. Complex number $e^{i\phi}$ is written as $1 + \eta(x, y)$, and the upper diagram shows '1' and 'η' drawn as phasors summing to $e^{i\phi}$. In the phase-contrast system, the '1' is phase-shifted by $\pm\pi/2$ by the action the phase plate. The lower diagram shows the outcome of phase-delaying by $\pi/2$. The image is now formed by adding this 'i' to the unfiltered $\eta(x, y)$, with the result that the total amplitude has magnitude affected by the value of ϕ.

Unlike the discussion in the text, these diagrams do not assume that ϕ is a small angle.

transparent, rather than light-and-dark, and a transparent object is almost invisible when looked at through a conventional microscope. However, the object often does have inhomogeneities of refractive index, or of physical thickness, so it causes changes of optical path; equivalently, it causes changes in the *phase* of the transmitted light. If only we could do something to 'decode' these phase changes into intensity changes ....

Example 3.1 and problem 4.3 offer us two models on which to test ideas. One discusses an 'amplitude grating', the other a 'phase grating'. We may compare the following two transmitted amplitudes, similar to those in the two cited precedents:

$$\text{amplitude grating:} \quad U_a(x, y) = U_0 \{1 + A\cos(\beta x)\},$$
$$\text{phase grating:} \quad U_p(x, y) = U_0 \{1 + iA\cos(\beta x)\},$$

where the constant A is real and (for simplicity) small ($\ll 1$). In both cases, there is a zeroth order of diffraction, generated by the '1' term, that represents the mean amplitude of the light, so is concerned mainly with the average intensity of the light leaving the object. And in both cases there are ± 1 orders of diffraction, coming from the cosine, diffracted to angle θ, where $\sin\theta = \pm\beta/k = \pm\lambda/d$ with $d = 2\pi/\beta$ being the grating spacing. If light in these three orders continues on (unmodified) to form an image, that image reproduces the object (apart from scale factors), giving an 'amplitude image' or a 'phase image'.

What is immediately noticeable about the expressions for U_a and U_p is how similar they are. Couldn't we somehow convert one to the other? The clue is that the '1' term and the $\cos(\beta x)$ terms go to different places in the objective lens's focal plane, so they can be manipulated in different ways. We think of putting a 'phase plate' in the focal plane, designed to impose a phase shift of $\pi/2$ (give or take a sign) on the zeroth order only—light passing through the point on the axis at F in Fig. 12.1. The field leaving plane F now mimics U_a (from the amplitude object) instead of reproducing U_p (from the phase object). An optical system that implements this trick achieves **phase contrast**.

The above reasoning has been constructed to deal with a grating object. It should, however, be clear that it applies to a phase object of any form,[19] since the value of β is not at all critical: a U_p equal to a sum of $e^{i\beta x} e^{i\gamma y}$ terms will be processed in the same way. Figure 12.7 shows the amplitude $U_p(x, y)/U_0$ as a general complex quantity plotted in the complex plane, and shows the effect of dissecting out the '1' part, phase shifting it, and reassembling the total.

Figure 12.8 shows a phase-contrast system conforming to the description above. It is examined in problem 12.8, which tests the feasibility of the arrangement and also investigates its limitations.

The apparatus of Fig. 12.8 is an in-principle optical system. It would be appropriate if the object were illuminated by a laser,[20] but that is not usually what the user wants. A diagram of a more practical configuration, suitable for use with a white-light source, can be found, for example, in Hecht (2002), §13.2.4.

[19] Well, almost any. See problem 12.8.

[20] It is entirely sensible to use a laser when demonstrating the principle.

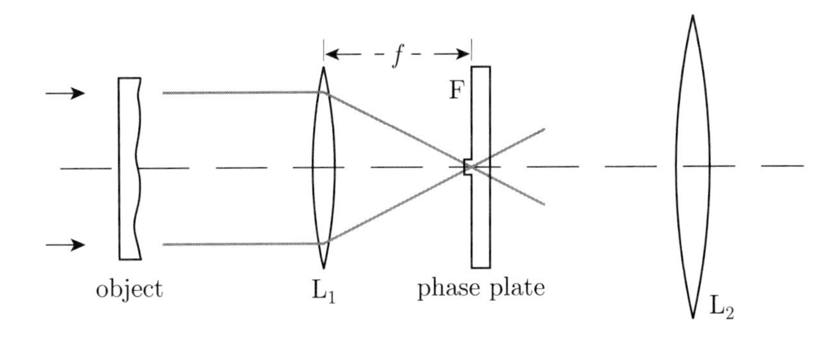

Fig. 12.8 A phase-contrast system. The object is illuminated by a collimated beam of light travelling parallel to the axis, and transmits amplitude $U_p(x, y)$. The zeroth order of diffraction (indicated by grey rays) is focused onto the phase plate's lump at the centre of focal plane F. Light diffracted by other Fourier components (not shown) travels to other areas of the phase plate. The final image formed by lens L_2 displays the original phase variations 'decoded' into intensity variations.

12.8 Dark-ground illumination

Phase contrast is not the only technique that can be used for making phase objects visible; it is always given pride of place in textbook accounts because it is so intellectually satisfying. In particular, phase contrast converts small phase changes linearly into changes of amplitude, so the images produced are quantitatively interpretable. However, the idea of processing the zeroth order differently from others can be pursued in other less precise ways. The simplest is to remove the zeroth order entirely. In the absence of phase variations in the object, the observer would see nothing but 'dark' through the microscope, hence the name 'dark ground' or 'dark field'. When phase variations are present, some light is diffracted into the objective; bright regions are seen, especially at edges in the object. The imaging has something of the high-pass-filter characteristic of phase contrast (problem 12.8(2)), but what is seen is related to the original phase variations in a less interpretable way.

Dark ground is usually implemented[21] by modifying the design of a microscope's condenser system so as to illuminate the object at large angles only. Examples of such designs are to be found in Longhurst (1973), pp. 346–7.

[21] A demonstration of the principle can be made by taking the apparatus of Fig. 12.8 and replacing the phase plate with a plate bearing an opaque spot at the centre. This spot absorbs the zeroth order of diffraction while allowing other orders through to the image.

12.9 Schlieren

Yet another way of rendering phase variations visible is to remove all orders of diffraction from one side of the objective lens's focal plane by means of a knife-edge. This technique is known as **Schlieren**.[22] (The zeroth order can be just retained or just rejected, according to taste.) Schlieren is often the method of choice for investigating fluid-flow phenomena such as take place in a high-speed wind tunnel. (The method requires significant changes of refractive index, caused by density changes, such as exist in shock waves; it is less suitable for a low-speed wind tunnel where the changes are smaller.)[23] Diagrams of two different Schlieren setups may be found in Hecht (2002), §13.2.5.

[22] As with other German nouns (Bremsstrahlung . . .), the word retains its capital letter in the middle of a sentence.

[23] A very similar optical system, a knife-edge at the focus of a lens or mirror, is used as a critical test for spherical aberration; it is called the Foucault knife-edge test. The lenses or mirrors used for Schlieren must therefore be rather well corrected for spherical aberration. If mirrors are used as the focusing elements they need to be 'off-axis paraboloids'.

12.10 Apodizing

This topic is not strictly a development of Abbe theory, but it fits nicely here where we are thinking about Fourier components and their effect on an image.[24]

[24] The word 'apodize' comes from Greek roots meaning 'to remove the feet', the 'feet' being the low-intensity parts of the diffraction pattern away from the central peak, otherwise called the 'wings'!

Consider a telescope looking at a star. If there are no aberrations, the image is a Fraunhofer diffraction pattern, with shape and size determined by the objective lens or mirror. The amplitude in the diffraction pattern falls off[25] only as 1/radius, because the amplitude $U(x, y)$ leaving the objective has step discontinuities at the lens edge. There is the possibility that a faint second star, even one well separated from the first according to the Rayleigh criterion, may be hidden under the 'wings' of the first star's diffraction pattern. If we could remove the step discontinuities, the diffracted amplitude would fall more rapidly in its wings, and the faint star might become detectable.

[25] 1/radius for a square objective, $1/(\text{radius})^{3/2}$ for a circular objective.

We may implement this idea by putting a 'shaded' plate over the objective, with a transmission profile somewhat like that shown in Fig. 12.9. The central peak of the diffraction pattern is widened, of course, but that price can be worth paying in exchange for getting a greatly reduced intensity in the wings. There is no simple optimization rule here: the design depends on how much one wants to reduce the near wings, or the far wings, or preserve a fairly narrow central peak. Quantitative designs are, therefore, made by numerical computation; see Jacquinot and Roizen-Dossier (1964). A model can be found in problem 12.9. A design

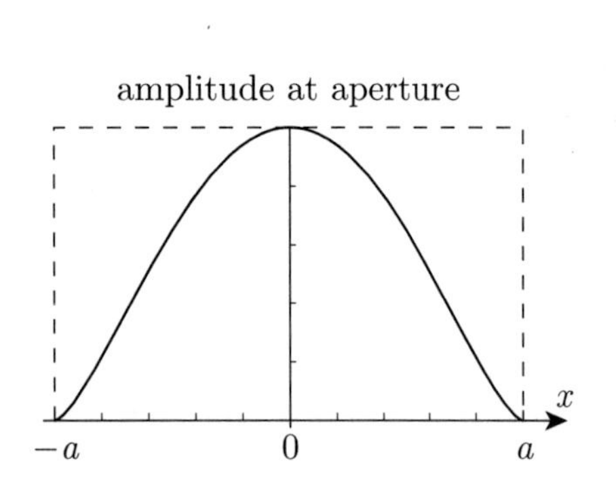

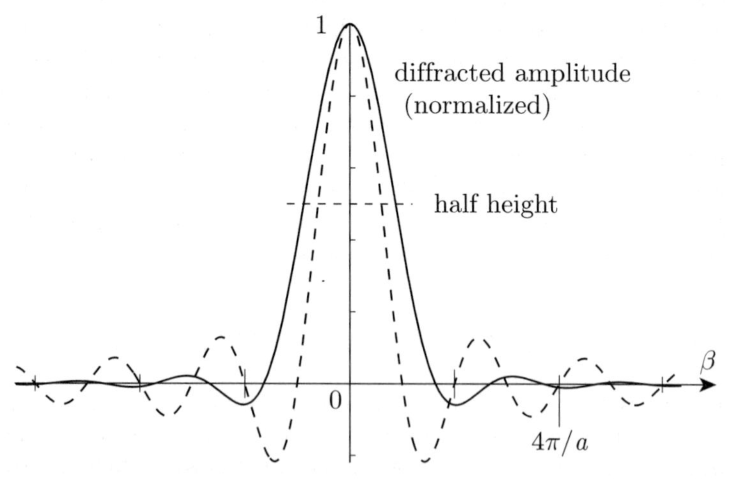

Fig. 12.9 An example of apodizing. The left-hand graph shows the amplitude transmitted by a 'shaded' one-dimensional aperture (solid curve) compared with that for a slit of the same width $2a$ (broken curve). The Fraunhofer diffraction patterns for the two cases are shown at the right, with $\beta = k \sin\theta$ as the independent variable. The apodized aperture gives a peak that is wider at the half-height, but has much weaker 'feet' away from pattern centre.

The shading function here is $(1 - x^2/a^2)^{3/2}$, chosen from among the examples given by Jacquinot and Roizen-Dossier (1964). The energy transmitted by the shaded aperture is, of course, less than that for the unshaded slit; the normalization of the intensity curves has concealed this loss, which is a factor 0.46. Since the diffraction pattern is widened, the intensity at the centre is reduced more severely, by a factor 0.35.

study, in connection with detecting planets close to a star, is given by Nisenson and Papaliolios (2001).

The idea of apodizing is fruitful in contexts other than that given. One such case has been encountered in problem 11.13, where there is a need to remove artefacts from the spectrum obtained with a Fourier-transform spectrometer. The measured values of $W(\tau) = I(\tau) - \frac{1}{2}I(0)$ are multiplied by a 'shading function' within the computer as part of the Fourier-transforming process (no need to do anything optically).

Another context in which apodizing (or at least the tailoring of an amplitude distribution) is encountered is in the design of communication satellites. The satellite radiates its signals down to a 'footprint' area to which it provides its service. The distribution of energy across that footprint is optimized by applying an aperture function to the transmitting aerial structure. The same idea in reverse applies to the reception of signals by the satellite from the ground, where a reciprocity property means that a satellite aerial receives best from areas to which it transmits best.[26]

[26] I am indebted to Professor D. J. Edwards and Dr K. Hungerford for information on this.

12.11 Holography

We discuss holography in Chapter 13. However, we should describe the reasoning that led Gabor to invent holography, because holography is another invention spawned by the ideas given in the present chapter.

Refer again to Fig. 12.1. The electromagnetic field in plane F contains all the information needed for constructing the image in plane I. Could we somehow record the field in plane F, perhaps photographically, and 'replay' it later? There is a difficulty, because a photographic film records intensity, not amplitude, so if we did only what's described we would destroy an essential part of the information. But we are familiar with optical arrangements where phase differences are made visible: interferometry. The arrangement of Fig. 12.1 can be modified to supply a separate 'reference wave' at plane F, so as to produce an interference pattern there. When this pattern is photographed, the photograph *does* contain enough information to make replay possible.

12.12 The point spread function

In this section we return to the mathematical analysis of §12.3 and build into it an allowance for the imperfections described in §12.4. The mathematics is presented in as simple a form as possible, but even so it can look somewhat heavy.

Refer to Fig. 12.2. The image formed in plane I cannot be an exact copy of the object. This may be because the optical system is imperfect and has discarded some spatial frequencies, or it may be because we have intervened to insert some filter, probably in plane F. A single mathematical treatment can deal with both kinds of image modification.

We return then to eqn (12.2) and modify it to build in a filtering function applied, for simplicity, in plane F.

$$U_{\mathrm{i}}(x_{\mathrm{i}}, y_{\mathrm{i}}) = \frac{\mathrm{e}^{\mathrm{i}kD_2}}{\mathrm{i}\lambda f_2} \int U_{\mathrm{F}}(x_{\mathrm{F}}, y_{\mathrm{F}})\, H(x_{\mathrm{F}}, y_{\mathrm{F}}) \times \exp\{-\mathrm{i}k\, x_{\mathrm{F}}\, x_{\mathrm{i}}/f_2 - \mathrm{i}k\, y_{\mathrm{F}}\, y_{\mathrm{i}}/f_2\}\, \mathrm{d}x_{\mathrm{F}}\, \mathrm{d}y_{\mathrm{F}}. \quad (12.7)$$

Here $U(x_{\mathrm{F}}, y_{\mathrm{F}})$ is the amplitude diffracted to plane F from $U_{\mathrm{o}}(x_{\mathrm{o}}, y_{\mathrm{o}})$ at the object, assumed to be given by eqn (12.1) with no allowance for imperfections.[27] Then all filtering or degradation in the whole optical system is attributed[28] to the filtering action of $H(x_{\mathrm{F}}, y_{\mathrm{F}})$. If $H(x_{\mathrm{F}}, y_{\mathrm{F}})$ reproduces the usual imperfections of an image, then H transmits low spatial frequencies fully but removes those high spatial frequencies that we know to be lost.

[27] As mentioned, the mathematical treatment here is made no more complicated than it has to be. We, therefore, have resorted again to a small-angle approximation, as may be seen by the linear dependence on x_{i}, y_{i} in the exponents.

[28] It is *as if* the field amplitude just to the left of F were $U_{\mathrm{F}}(x_{\mathrm{F}}, y_{\mathrm{F}})$, while that just to the right were $U_{\mathrm{F}}(x_{\mathrm{F}}, y_{\mathrm{F}})\, H(x_{\mathrm{F}}, y_{\mathrm{F}})$; and the rest of the optical system were perfect as assumed in §12.3. The point of *as if* is that there may be no actual filter placed in plane F, but H represents the effect of losses and imperfections however caused.

The product $U_{\mathrm{F}}H$ in eqn (12.7) suggests a reverse application of the convolution theorem. The (imperfect) image $U_{\mathrm{i}}(x_{\mathrm{i}}, y_{\mathrm{i}})$ must be a convolution,[29] of the 'ideal' image $U(x_{\mathrm{i}}, y_{\mathrm{i}})_{\mathrm{ideal}}$ given by eqn (12.3), taken with another function $h(x_{\mathrm{i}}, y_{\mathrm{i}})$, the **point spread function**, that is the Fourier transform of $H(x_{\mathrm{F}}, y_{\mathrm{F}})$. The image amplitude $U_{\mathrm{i}}(x_{\mathrm{i}}, y_{\mathrm{i}})$ has the form

[29] While this should be easy to see in general terms, the details need some rather messy tidying up. The algebra has been relegated to problem 12.10.

$$U_{\mathrm{i}}(x_{\mathrm{i}}, y_{\mathrm{i}}) = \int U_{\mathrm{i}}(x'_{\mathrm{i}}, y'_{\mathrm{i}})_{\mathrm{ideal}}\, h(x_{\mathrm{i}} - x'_{\mathrm{i}}, y_{\mathrm{i}} - y'_{\mathrm{i}})\, \mathrm{d}x'_{\mathrm{i}}\, \mathrm{d}y'_{\mathrm{i}}. \quad (12.8)$$

Problem 12.10(4) shows that $h(x, y)$ has the interpretation that we would expect from its name: it gives the mathematical form for the spread of amplitude around a would-be point image at $(0, 0)$.

The idea of a point spread function is not new to us. All of the two-dimensional Fraunhofer diffraction patterns given in Chapter 3 represent the spread of light around a would-be point image; so the diffracted amplitudes we calculated are the point spread functions for those cases.

Comment: The point spread function can be put to use in justifying the discussion given here. The reader may have been reluctant to accept the proposition at the beginning of this section: that any filtering or degradation, however caused, can be represented by an 'equivalent' filtering function H in the focal plane F. However, any optical system must possess a point spread function. And an $H(x_{\mathrm{F}}, y_{\mathrm{F}})$ can always be constructed to be the Fourier transform of that point spread function. The *as if* of sidenote 28 (this page) is justified.

12.13 Optical transfer function; modulation transfer function

We give a brief mention here to the optical transfer function and modulation transfer function, because they provide the ways in which optical-system performance is most commonly quantified, and they are connected to the point spread function.

First we must make explicit a shift of viewpoint from that of §§12.2–12.9. There we were interested in obtaining information about a small

object, so it was appropriate to express everything in terms of coordinates (x_o, y_o) and space (angular) frequencies (β, γ) in the object. Now we are more interested in the image, so we work with coordinates (x_i, y_i) and space (angular) frequencies (ξ, η). Then a sinewave[30] present in the image is a term in U_i like $\exp(-i\xi x_i - i\eta y_i)$. If the optical system resembles that of Fig. 12.2 then the new variables and the old are related by $\xi = k x_F/f_2 = (f_1/f_2)\beta$, $\eta = k y_F/f_2 = (f_1/f_2)\gamma$. The viewpoint given here was already implicit in the discussion of §12.12, but the symbols ξ, η did not need to be mentioned except in the supporting problem, problem 12.10.

[30] The image is inverted so, however we set things up, there are some negative signs doing their best to be troublesome. Here we switch the convention with Fourier exponentials, so that $e^{i\beta x}$ in the object becomes $e^{-i\xi x}$ in the image, and positive β goes with positive ξ.

Consider an optical system forming an image of an 'ordinary' object. We are thinking of a telescope looking at the sky, or a television camera looking at some scene illuminated by sunlight. Light coming from different points on the object will be incoherent, and we are concerned once more with incoherent imaging.

Perhaps surprisingly, there is a link with coherent imaging, and it is via the point spread function. If our optical system is used to image a single bright point, the image (amplitude) centred on (X, Y) is (proportional to) the point spread function $h(x_i - X, y_i - Y)$. Then[31] the image (intensity) is proportional to $|h(x_i - X, y_i - Y)|^2$. A second bright point, incoherent with the first, will add its own intensity function $|h|^2$ centred on its own image location. And by extension, a complicated image is formed by summing over many points using $|h|^2$ as the point spread function for intensity.

[31] Because there is only a single source point here, there is no distinction between coherent and incoherent illumination. This is why we are able to switch between coherent and incoherent cases while retaining the form of $|h(x, y)|^2$.

The **optical transfer function** (OTF) $K(\xi, \eta)$ is the Fourier transform (with variables ξ, η defined above) of $|h|^2$. It, therefore, does for incoherent imaging what $H(x_F, y_F)$ does for coherent imaging:[32] it describes how the optical system filters or modifies space frequencies that ought to contribute to the image.[33]

[32] The different Fourier components of intensity are not laid out in the lens's focal plane, so we cannot think of $K(\xi, \eta)$ as a filter, even a pretend-filter, doing its stuff in that plane.

[33] Problem 12.11 investigates the optical transfer function K and its relationship with the amplitude filtering function H.

In general, the optical transfer function K is a complex quantity, because it accounts for phase errors as well as the attenuation or removal of space frequencies. This means that it contains within itself the possibility of describing image distortions. There are cases where small amounts of distortion are unimportant in comparison to the resolution of detail. Then we use the **modulation transfer function** (MTF), which is $|K|$.

The statements just made need explanation. Suppose we use as object a sinewave grating, somewhat resembling that of example 3.1 (but with a cosine term in the intensity). The *intensity* of the ideal image (a scaled version of the object intensity) might have the form

$$I(x_i)_{\text{ideal}} = I_0\{1 + \cos(\xi x_i)\} = \tfrac{1}{2} I_0\{2 + e^{-i\xi x_i} + e^{i\xi x_i}\}, \qquad (12.9)$$

and the intensity of the actual, imperfect image is then[34]

$$I(x)_{\text{actual}} = \tfrac{1}{2} I_0\{2 + K(\xi)\, e^{-i\xi x_i} + K^*(\xi)\, e^{i\xi x_i}\}. \qquad (12.10)$$

Depending on the space frequency of the grating, the image may be a sinewave of good contrast ($|K| = 1$), a sinewave of poor contrast ($|K| < 1$), or a 'clear field' ($K = 0$). And if K is complex, the sinewave

[34] The reader may perhaps be startled: complex quantities in the description of an *intensity*? Surely only field amplitudes are complex; intensities are real. Well, no. Field amplitudes are real too. Fields often vary sinusoidally with position or time, so complex exponentials are natural for describing them. But that's nothing to do with the fact that they are fields, rather than something else. Here an intensity varies sinusoidally, so complex exponentials are convenient, for all the usual reasons. Of course, the intensity of eqn (12.9) must be real, and it is. But the same is true of amplitudes: eqn (2.16).

components are phase shifted, which means that they are in the wrong place: the image is distorted. As mentioned, the modulation transfer function is the value of $|K(\xi)|$, in which any phase shifts are ignored.

Measurement of the modulation transfer function is quite straightforward. In one method, a strip is made whose (intensity) transmission function has the form $\frac{1}{2}\{1 + \cos(\beta x)\}$, but where β slowly increases from one end of the strip to the other. The strip is imaged through the optical system under test, and the $\cos(\beta x)$ modulation is measured by a photodetector at a fixed location in the image space as the strip is translated.[35] The measured modulation gives $|K(\xi)|$, where ξ is related to β via the system's magnification as above.

[35] The OTF (and, therefore, also the MTF) depend on where we measure within the image. This is a complication that we have not attempted to build into the mathematics because it clutters things for little more insight than can be conveyed in words. However we should explain what the complications are. A lens, even a poor-quality lens, has axial symmetry, so we should describe its performance in cylindrical (r, ϕ) coordinates. The ability of the lens to form a sharp image tends to deteriorate as we move away from the lens's axis, so the OTF decreases with radius r. Furthermore, lines (or sinewaves) running round the axis, or radially out from the axis, suffer differently. The optical transfer 'function' is in fact two functions of r, one for sinewaves varying as $\exp(i\xi' r)$ and another for sinewaves like $\exp\{i\eta'\phi\}$.

Modern optical design, of lenses in particular, is based firmly on the ideas given here. Rays can be traced, by computer, through a proposed design, and used to calculate the optical transfer function or modulation transfer function for selected locations in the image plane. The process can be repeated with different designs until an optimum is reached. After a lens has been made, quality-control measurements (of OTF or MTF) check the lens's conformity with intention. At the calculation stage, rays are traced by applying Snell's law directly to each refraction. This computer-oriented procedure can be contrasted with a 'traditional' heavy algebraic analysis which writes $\sin\theta \approx \theta - \theta^3/6 + \cdots$, uses this approximation in Snell's law of refraction, and groups the outcomes of the θ^3 terms into five quantities called Seidel sums; the design process then tries to control the values of those sums.[36]

[36] The Seidel sums retain a descriptive usefulness, because they are identified with the five monochromatic aberrations: spherical aberration, coma, astigmatism, curvature of field, and distortion. An optical-transfer-function approach does not consider these defects separately, but simply tries to minimize the overall image degradation.

We do not pursue lens design at all in this book. The reason is that modern design is really a form of computer-aided trial-and-error that does not lend itself to analysis at the level we are aiming for. The reader is warned that lens design, though highly developed, is a non-trivial exercise, and it should not be assumed that every lens gives a performance limited only by diffraction!

Problems

Problem 12.1 (a) A workout on the ideas in §12.2
Use Fig. 12.1.

(1) Use the lens formula (1.31) to find the location of plane I.

(2) Show from ray optics that any object in plane O is imaged in plane I with linear magnification[37] $-v/u$. Show also from ray optics that the two beams diverging from the object area drawn in plane O overlap each other in plane I. We have said in §12.2 that the image is formed by interference of these two beams (and others); interference is possible because the beams overlap fully.

[37] I use an awkward mixture of the two sign conventions of Fig. 1.2 here. It's convenient to refer to object and image distances as u and v, both positive as drawn. But in anticipation of a switch to Cartesian coordinates I make the height of the image positive when it lies at positive x; hence the negative sign in $-v/u$.

(3) Apply in plane O an input amplitude $U_1 = U_0(1 + e^{i\beta x})\,e^{-i\omega t}$, prepared by shining in from the left a beam along the axis and a second beam, coherent with the first, at angle θ where $\beta = k\sin\theta$. There is no actual 'object' in plane O now, only the given field distribution (is there any difference?[38]); so these beams continue on their way unaltered.

[38] This question is discussed in §3.10.

Show that a set of straight-line interference fringes crosses plane O with spacing $d_o = 2\pi/\beta$. Problem 9.2 may help.

(4) Find the angle at which these two beams cross again in plane I. Show that their interference there creates straight-line fringes of spacing $d_i = d_o(v/u)$, in agreement with the expected ray-optics magnification.

(5) Now input to plane O an amplitude $U_2 = U_0\{1 + \cos(\beta x)\}\,e^{-i\omega t}$. Such an amplitude might be prepared by intersecting three incident plane waves. Show[39] that the image of this at plane I has amplitude $(u/v)U_0\{1+\cos(\beta x u/v)\}\,e^{-i\omega t}$, apart from a phase factor describing the time taken by light to go from O to I.

(6) Describe, in the same terms as we have been using, what happens when:

(a) a part-transparent object with amplitude transmission factor

$$T(x) = \tfrac{1}{2}\{1 + \cos(\beta x)\}$$

is illuminated normally by a plane monochromatic wave

(b) a similar arrangement uses a $T(x)$ containing other sinewave components as well as $\cos(\beta x)$.

(7) Use example 3.1 and problem 3.20 to describe the way in which U_1 and U_2 are broken up into individual sinewaves varying with x, and how each sinewave goes to its own place in plane F, so that the Fourier transform of the original amplitude $U(x, y)$ is displayed there. Follow the history of the same waves from plane F to plane I, and explain how the original sinewaves are recreated (with scale factor v/u) in plane I. A formal description of this process (of Fourier analysis and Fourier synthesis) is given in §12.3, but it should be possible to trace the essentials starting from no more than the introduction of §12.2.

[39] *Suggestion*: For the present problem don't cover the page with diffraction integrals; handwave each factor in turn using simple geometry and parts (3) and (4) above.

Problem 12.2 (b) Getting things right

Problem 12.1 contains insights that we'll build on, but in one respect it's seriously oversimplified: the phases are wrong. Can it really be that travel from O to F and from F to I both perform Fourier transformations when one stage contains a lens and the other doesn't? And what's so special about the distance FI?[40] Problem 3.8 may help.

Show that the optical layout of Fig. 12.2 prepares a Fourier transformation without phase discrepancies (though with scale factors) as light travels from O to F, and another between F and I.[41]

[40] A Kirchhoff integral carrying fields from F to I does not look dramatically different from one carrying fields from F to any other plane. And it doesn't at all meet the Fraunhofer–Fourier conditions of §§3.13 and 3.17.

[41] This repeats reasoning in Fig. 3.19(b) and problem 3.28.

Problem 12.3 (b) Image formation by a double Fourier transformation

(1) Use Fig. 12.2. Obtain eqn (12.1) by assembling the Kirchhoff integral that carries light from O to F. Check against eqn (3.28).

(2) Obtain eqn (12.3) by writing a second Kirchhoff integral, and then using the 'forward' and 'backward' Fourier transformations.[42]

(3) Interpret the presence of $-x_i f_1/f_2$ and $-y_i f_1/f_2$ in eqn (12.3) as showing that the image is inverted and magnified by the factor f_2/f_1.

(4) Show that ray-optics imaging of a point object accounts for the factor $-f_1/f_2$ of part (3). Show that a field $\propto \exp(i\beta x_o)$ leaving plane O

[42] This is in principle straightforward, but involves some messy chasing of constants.

is brought to a focus in plane F, becomes a plane wave again after lens L_2, and gives a field at I varying as $\exp\{-i(f_1/f_2)\beta x_i\}$. Why the negative sign in the exponent? Draw this wave, perhaps after the fashion of Fig. 12.4.[43] Compare with problem 12.1(3).

[43] Lens L_2 (Fig. 12.2) must be large enough that its outline doesn't obstruct this wanted Fourier component.

Problem 12.4 (b) Image formation with non-small angles
Let light be incident normally on a part-transparent object as shown in Fig. 12.2. A component $\exp(i\beta x_o)$ in the transmitted amplitude is diffracted to travel away at angle θ_o, where $\sin\theta_o = \beta/k = \lambda/d_o$ with d_o being the period of the sinewave: $d_o = 2\pi/\beta$. Small-angle approximations are not now to be made. Show that at the image plane this component is required to produce the sinewave $\exp(-i\xi x_i)$, where $\xi = (f_1/f_2)\beta$. Show that to produce this the light must arrive at angle θ_i to the axis, where $\sin\theta_i = -\xi/k = -(f_1/f_2)\sin\theta_o$. This is what must result when the small-angle approximations of §12.3 are corrected.[44]

[44] Notice that a proof, or at least a strong indication, of the Abbe sine condition is not far away. The Abbe sine condition is usually obtained by requiring errors of optical path (from object to image) to vanish to first order in an expansion in terms of x_o, y_o. Here we have an interestingly different route.

Problem 12.5 (b) Abbe theory with a condenser
The diagrams of §12.6 introduce the advantages of a condenser lens by using the unsatisfactory physical picture of §12.5.1. Draw your own diagrams, modelled on the discussion of §12.5.2, to show how the reasoning may be put into a better Fourier-component language. Use your diagrams to re-explain the functioning of the condenser given in §12.6.

Problem 12.6 (b) Illumination with a condenser
Take the optical system of Fig. 12.6(b), with the condenser supplying light within a cone of directions whose semi-angle is α_{max}. The microscope objective collects light within a cone of semi-angle θ_{max}, which for now we treat as chosen independently from the value of α_{max}.

(1) Using étendue, show that the coherence area is a circle of diameter[45]

$$d_{coh} = \frac{2}{\pi}\frac{\lambda_{vac}}{n\sin\alpha_{max}}.$$

[45] Remember eqn (11.14). A more conventional route would be to make a large-angle modification of §10.8 to yield

$$d_{coh} = \frac{1.22}{2}\frac{\lambda_{vac}/n}{\sin\alpha_{max}}.$$

This differs by only an unexciting numerical factor.

(2) Consider the case when $\alpha_{max} \ll \theta_{max}$. The reasoning leading to eqn (12.4) should apply. Show that

$$\frac{\text{resolution distance } d_{min}}{\text{diameter } d_{coh} \text{ of coherence area}} = \frac{\pi}{2}\frac{\sin\alpha_{max}}{\sin\theta_{max}} \ll 1.$$

This is the case discussed in the text in sections before §12.6.

(3) Consider a case where the condenser supplies light in the same angle range as is collected by the objective lens: $\alpha_{max} = \theta_{max}$. Then light arriving at the 'extreme' angle α_{max} can behave as in Fig. 12.6(a) and deliver both its zeroth order and its first order into the objective. For this light the linear dimension just resolved is $d_{min} = \lambda/(2\sin\theta_{max})$. Show that

$$\frac{\text{resolution distance } d_{min}}{\text{diameter } d_{coh} \text{ of coherence area}} = \frac{\pi}{4}.$$

The illumination for this case is 'just incoherent'.

(4) Show that when $\alpha_{\max} \gg \theta_{\max}$, the coherence area is small compared with the resolution distance. The illumination has been made transversely incoherent. This is the limit considered in problem 10.14, for example. Argue that in this limit we should be able to discuss the image formation by using the 'incoherent limit' that was assumed throughout Chapter 3. Adapt eqn (3.20b) to the case of non-small angles and show that[46] $d_{\min} = 1.22(\lambda_{\text{vac}}/n)/(2\sin\theta_{\max})$. Apart from the factor 1.22, this is in agreement with the second case itemized in eqn (12.5).

(5) Think about the last case in an Abbe-theory way.

[46]The factor 1.22 here makes it look as though the resolution has been worsened by 22% by increase of $\alpha_{\max}$ beyond $\theta_{\max}$. This is most unlikely. We have to remember that we have been setting arbitrary criteria for resolution that are order-of-magnitude right, but probably shouldn't be trusted at a 20% level.

Problem 12.7 (b) Worthwhile magnification for an optical microscope
A microscope forms its 'final' image (the last one before the eye's retina) at a distance D from the eye. We start by pretending that D is set equal to the 'least distance of distinct vision' D_{ldv}, taken as 250 mm. The magnification M is then

$$M = \frac{\text{linear dimension of final (virtual) image distant } D_{\text{ldv}} \text{ from eye}}{\text{corresponding linear dimension of object}}. \tag{12.11}$$

(1) Show that the size of the image *on the observer's retina* is much the same wherever the 'final' image is formed between D_{ldv} and infinity.[47] It is, therefore, harmless to pretend that the final image is formed at D_{ldv}, though the knowledgeable user prefers the greater comfort of forming it at infinity.

(2) Let the microscope's exit pupil (eye ring)[48] have diameter d_e. Choose d_e to match the diameter of the observer's eye pupil, around 5 mm. The numerical aperture of the microscope's objective is $(n\sin\theta_{\max})$. Show that the magnification under these conditions[49] is

$$M = \frac{2D_{\text{ldv}}}{d_e} \times (n\sin\theta_{\max}) \approx \frac{500\,\text{mm}}{5\,\text{mm}} \times (n\sin\theta_{\max}) = 100\,(n\sin\theta_{\max})$$
$$\sim 80 \text{ for an objective in air, or} \sim 120 \text{ for one in oil.}$$

(3) Take the microscope's resolution to be limited only by diffraction (and to take the value appropriate to illumination via a condenser): $d_{\min} = (\lambda/2)/(\text{numerical aperture})$. Show that

$$\text{linear size of resolvable detail in 'final' image} = D_{\text{ldv}}\lambda/d_e. \tag{12.12}$$

Show that this is also the resolution available if that image is observed at distance D_{ldv} through a perfect (= diffraction-limited) eye[50] with pupil diameter d_e.

(4) Show that if d_e is greater than the eye pupil's diameter, the eye fails to profit from the resolution provided so expensively by the microscope. Equivalently, light passing through the outer parts of the microscope objective is 'wasted', and so the effective value of $(n\sin\theta_{\max})$ is reduced.

(5) Show that if d_e is less than the eye pupil's diameter, the smallest resolved detail in the 'final' image is larger than what the eye itself could expect to resolve, so the image looks blurred. We have generated 'empty magnification'.

[47]This repeats arguments given in problem 1.16. The magnification M is best defined as an angular magnification, and this quantity is hardly affected by making different choices for D.

[48]The eye ring has been encountered in problem 3.18, in connection with a telescope. The eye ring of a microscope is defined in a similar way: it is the real image of the objective, formed by the eyepiece, and it is the best place to locate the observer's eye pupil.

[49]Permit $n\sin\theta_{\max}$ to be non-small with, possibly, $n \neq 1$. You may wish to use the Abbe sine condition, a result of geometrical optics; see, e.g., Born and Wolf (1999), p. 179. However, the job can be done quite tidily by using the conservation of étendue.

[50]A similar result is derived for a telescope in problem 3.18.

(6) Something has gone wrong! Microscopes commonly have magnifications much greater than the values calculated in part (2), up to $\sim$1000; 'empty magnification' up to about a factor 8 is quite usual. This is just to give greater comfort for the user.

Show that 'empty magnification' of a factor 8 can be obtained only at the expense of under-filling the eye pupil's area by a factor $8^2 = 64$. The power received, per unit area of the observer's retina, is reduced by this same factor (compared with what it could be if the eye pupil were filled)—the étendue is reduced—so empty magnification carries a high penalty in reduced image intensity.

(7) How are the conclusions of this problem changed if the instrument forms a real image on the photodetecting area of a television camera?

Problem 12.8 (b) Details of phase contrast

(1) Consider the phase-contrast system of Fig. 12.8. Given that the phase plate has refractive index n, what is the height that the 'lump' must have?[51]

[51] *Hint:* It is *not* $\lambda/4$.

(2) Suppose that the lump on the phase plate is a disc of radius r. Show that phase contrast works for spatial (angular) frequencies β satisfying $\beta > kr/f_{\text{objective}}$, but not for smaller frequencies. When using a phase-contrast microscope, you have to remember that the image is observed through a high-pass filter.[52]

[52] Once you know about this filtering, you see it everywhere: it puts a light-and-dark 'halo' round the edges of (the image of) an object. A good comparison of phase contrast with 'bright field', displayed for the same object, may be found in Klein and Furtak (1986), Fig. 7.66; the 'halo' effect is very evident.

(3) Is there any choice of radius r that can prevent this filtering effect?

Problem 12.9 (b) A model for apodizing

Work out the diffraction pattern caused by a slit that transmits

$$U(x,y) = \tfrac{1}{2}U_0\,\mathrm{e}^{-\mathrm{i}\omega t}\{1+\cos(2\pi x/a)\}$$

for $-a/2 \le x \le a/2$ and $-b/2 \le y \le b/2$. Assume normal incidence. Compare with the pattern obtained when $U(x,y)$ is replaced by $U_0\,\mathrm{e}^{-\mathrm{i}\omega t}$. Check that the results agree with statements in problem 2.17.

Problem 12.10 (c) The point spread function

This problem deals with the algebra leading to the point spread function of §12.12. Rename the unfiltered image function of eqn (12.3) as $U_{\mathrm{i}}(x_{\mathrm{i}}, y_{\mathrm{i}})_{\text{ideal}}$. Then eqn (12.2) may be rewritten as

$$U_{\mathrm{i}}(x_{\mathrm{i}}, y_{\mathrm{i}})_{\text{ideal}} = \frac{\mathrm{e}^{\mathrm{i}kD_2}}{\mathrm{i}\lambda f_2}\int U_{\mathrm{F}}(x_{\mathrm{F}}, y_{\mathrm{F}}) \times \exp\{-\mathrm{i}k\,x_{\mathrm{F}}\,x_{\mathrm{i}}/f_2 - \mathrm{i}k\,y_{\mathrm{F}}\,y_{\mathrm{i}}/f_2\}\,\mathrm{d}x_{\mathrm{F}}\,\mathrm{d}y_{\mathrm{F}}. \quad (12.13)$$

To remove the worst of the complications in this, define a set of scaled variables by

$$\xi = kx_{\mathrm{F}}/f_2, \qquad \eta = ky_{\mathrm{F}}/f_2, \qquad \widetilde{U}_{\mathrm{i}}(\xi,\eta) = -\mathrm{i}\lambda f_2\,\mathrm{e}^{\mathrm{i}kD_2}U_{\mathrm{F}}(x_{\mathrm{F}}, y_{\mathrm{F}}).$$

(1) Show that with these substitutions,

$$U_{\mathrm{i}}(x_{\mathrm{i}}, y_{\mathrm{i}})_{\text{ideal}} = \int \widetilde{U}_{\mathrm{i}}(\xi,\eta)\,\exp\{-\mathrm{i}\xi\,x_{\mathrm{i}} - \mathrm{i}\eta\,y_{\mathrm{i}}\}\frac{\mathrm{d}\xi}{2\pi}\frac{\mathrm{d}\eta}{2\pi},$$

so that $\widetilde{U}_\text{i}(\xi, \eta)$ is the Fourier transform (strict sense) of $U_\text{i}(x_\text{i}, y_\text{i})_\text{ideal}$.

(2) In terms of the scaled variables ξ, η, define a new version $\widetilde{h}(\xi, \eta)$ of the filtering function $H(x_\text{F}, y_\text{F})$ by

$$\widetilde{h}(\xi, \eta) = H(x_\text{F}, y_\text{F}), \tag{12.14}$$

and then $\widetilde{h}(\xi, \eta)$ is the transform of function $h(x_\text{i}, y_\text{i})$ defined by

$$h(x_\text{i}, y_\text{i}) = \int \widetilde{h}(\xi, \eta) \exp\{-\text{i}\xi x_\text{i} - \text{i}\eta y_\text{i}\} \frac{\text{d}\xi}{2\pi} \frac{\text{d}\eta}{2\pi}. \tag{12.15}$$

(3) Show now that the non-ideal image amplitude of eqn (12.7) is

$$\begin{aligned} U_\text{i}(x_\text{i}, y_\text{i}) &= \frac{\text{e}^{\text{i}kD_2}}{\text{i}\lambda f_2} \int U_\text{F}(x_\text{F}, y_\text{F})\, H(x_\text{F}, y_\text{F}) \\ &\qquad\qquad \times \exp\{-\text{i}k\, x_\text{F}\, x_\text{i}/f_2 - \text{i}k\, y_\text{F}\, y_\text{i}/f_2\}\, \text{d}x_\text{F}\, \text{d}y_\text{F} \\ &= \int \widetilde{U}_\text{i}(\xi, \eta)\, \widetilde{h}(\xi, \eta) \exp\{-\text{i}\xi x_\text{i} - \text{i}\eta y_\text{i}\} \frac{\text{d}\xi}{2\pi} \frac{\text{d}\eta}{2\pi} \\ &= \int U_\text{i}(x_\text{i}', y_\text{i}')_\text{ideal}\, h(x_\text{i} - x_\text{i}', y_\text{i} - y_\text{i}')\, \text{d}x_\text{i}'\, \text{d}y_\text{i}', \end{aligned} \tag{12.16}$$

by the convolution theorem, used backwards.

(4) Obtain a physical interpretation of h as follows. Consider what happens if we have a point source in the object plane, giving amplitude $U_\text{o}(x_\text{o}, y_\text{o}) \propto \delta(x_\text{o} + X f_1/f_2)\, \delta(y_\text{o} + Y f_1/f_2)$. Show that the ideal image of this is

$$U_\text{i}(x_\text{i}, y_\text{i})_\text{ideal} \propto \delta(x_\text{i} - X)\, \delta(y_\text{i} - Y)$$

and that the non-ideal image amplitude is

$$U_\text{i}(x_\text{i}, y_\text{i}) \propto h(x_\text{i} - X, y_\text{i} - Y). \tag{12.17}$$

Thus, $h(x, y)$ describes the spread of an image that 'ought' to be a point at $(0, 0)$.

Problem 12.11 (b) The optical transfer function

(1) Show that the optical transfer function is the autocorrelation function of $H(x_\text{F}, y_\text{F})$, equivalently $\widetilde{h}(\xi, \eta)$, with itself.

(2) The result of part (1) should raise a question. If we have a peaky function and we square it, the outcome is a narrower peak. If we have a peaky function and we form its autocorrelation function, the result is (almost always) a wider peak. Then the optical transfer function, representing the optical system's transmission of intensity, extends to higher space frequencies than its amplitude transfer function H. Is this surprising, or worrying?

13 Holography

13.1 Introduction

Holography[1] is perhaps the most dramatic application of diffraction physics. Although it was invented and demonstrated by Gabor well before the invention of the laser,[2] it is the opportunity to use lasers for recording and (usually) replay that has made most of the applications possible.

[1] The word means 'writing the whole', meaning that the whole wavefront is recorded.

[2] Gabor first published his invention of holography in Gabor (1948); the first laser was made in 1960.

Figure 13.1 shows in outline the processes that are carried out in making a hologram and in replaying it. We illuminate some object with a broad beam of light from a laser. Light scattered by the object

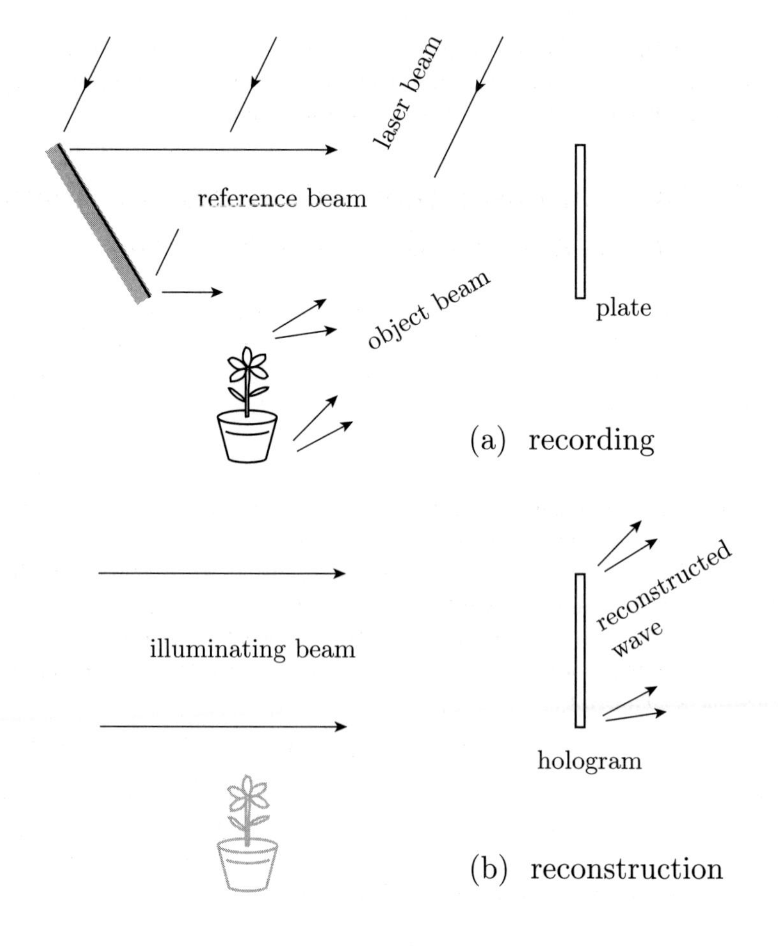

Fig. 13.1 (a) An arrangement for the recording of a hologram. A laser beam illuminates the object; light scattered by the object is called the object beam, and some of it falls on the photographic plate. At the same time, some of the laser light is reflected by a mirror directly onto the plate, forming the reference beam. The two beams form an interference-fringe pattern where they overlap, and that pattern is recorded in the photographic emulsion on the plate. (b) The developed photograph constitutes the required hologram. When this hologram is illuminated by the illuminating beam (here shown identical in its geometry to the reference beam), the photographed fringe pattern acts as a diffraction grating, diffraction from which re-forms the wave that formerly radiated from the object.

radiates away and forms the **object beam**. Another part of the original laser beam forms the **reference beam**. These beams overlap to form an interference pattern (somewhat complicated), which we record on a photographic plate or film. Preparation of the hologram is completed by 'developing' the photographic image so that the record of it is made permanent.

To replay the hologram, we illuminate it with the **illuminating beam**, which for the present relates to the plate in the same way as did the original reference beam. The light and dark regions of the plate (bearing a record of the original interference pattern) act as a complicated kind of diffraction grating. The Huygens sources just downstream from the hologram contain a 'reconstruction' of the original object beam's amplitude there; and in consequence the diffracted beam contains a reconstruction of the original object beam, or rather what it would have been downstream of the plate had the plate not intercepted it. Thus we see an image of the original object, in its original place, and, since the whole wave radiating from the object is reconstructed, in three-dimensional detail.[3]

[3]This description explains the alternative name for holography: *wavefront reconstruction.*

Of course, the statements just made require justification, and we need to know what conditions must be met. A formal proof that holography works is given in §13.5. However, most understanding can best be obtained by thinking about the interference fringes, and the subsequent diffraction from their photographic record. The next few sections deal with special cases in order to draw out these insights.

13.2 Special case: plane-wave object beam and plane-wave reference beam

Let the reference beam and object beam both be plane waves as shown in Fig. 13.2. The waves arrive at angles α, δ to the plate normal. Interference between these two beams makes a fringe pattern where they overlap. In the xy plane, the fringes have spacing d, where (see problem 13.1)

$$d(\sin\delta - \sin\alpha) = \lambda. \qquad (13.1)$$

Imagine that we photograph these fringes and develop the photograph. The resulting hologram-photograph bears a set of light and dark stripes separated by d; it resembles a diffraction grating. We illuminate the hologram with the illuminating beam, which like the original reference beam is incident at angle α to the normal. Light transmitted by the plate forms a set of diffraction-grating orders: plane waves travelling at angles θ_p to the plate normal, with

$$d(\sin\theta_p - \sin\alpha) = p\lambda, \qquad (13.2)$$

where p is an integer representing the order of the interference.[4] The diffracted beam of first order ($p = 1$) has $\theta_1 = \delta$ and recreates the (continuation of the) object beam. Thus, we achieve (among other things) the hoped-for reconstruction of the object beam.

[4]Equation (13.2) resembles the diffraction-grating equation (4.3), but with the terms in the bracket in reverse order. The sign in eqn (4.3) was chosen because a diffraction grating is most often used in practice with light incident at angle α somewhat larger than the angle θ_p of departure. No such custom—nor yet its opposite—applies generally to work with holograms, and the sign in eqn (13.2) is that which the author is most comfortable with. What here is called order $p = 1$ would, in Chapter 4, have been called order $p = -1$.

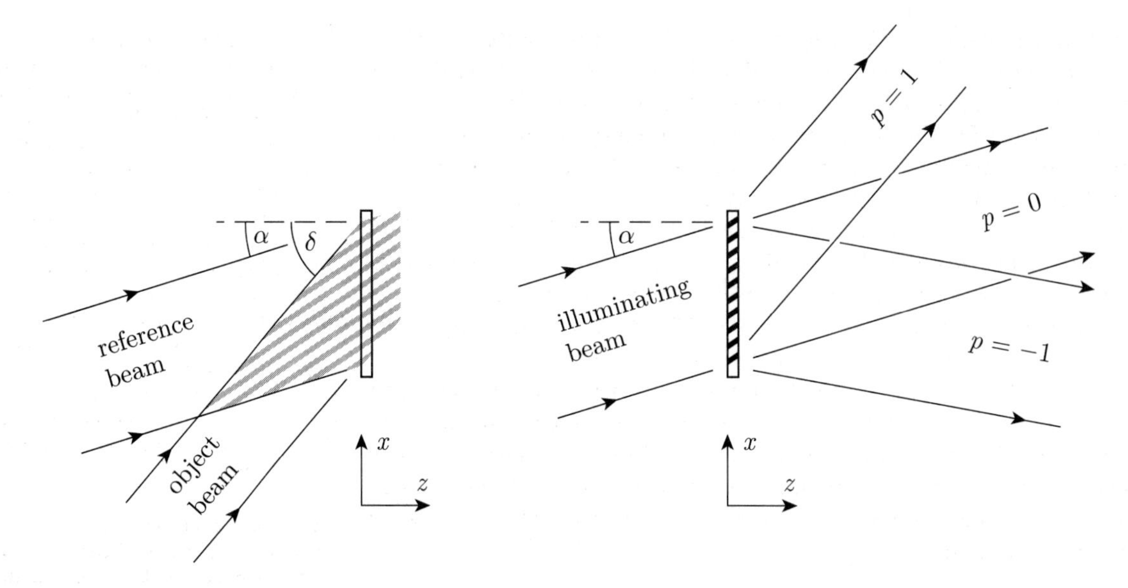

Fig. 13.2 In this example, two coherent plane waves are made to interfere and a photographic plate records their interference fringes (left diagram). The right-hand diagram shows the developed plate being illuminated by the 'illuminating beam', identical to the original reference beam. Diffraction order $p = 1$ reproduces the (continuation of the) object beam. There is also a straight-through beam $p = 0$, and the 'conjugate beam' $p = -1$. Other diffraction orders may exist, but are not shown.

Comment 1 This example displays the essential physics plainly: we photograph interference fringes, and use the photograph as a diffraction grating.

Comment 2 Reasonable angles α, δ give rise to fringe spacings d of only one or two optical wavelengths. The photographic emulsion used for holography has to have a high spatial resolution in order to record such fine-detail fringes—a much higher resolution than is needed for ordinary snapshots.

The $p = 1$ order reproducing the reconstructed object beam is known as the **primary beam**. It is, of course, the primary beam that is the 'wanted' output from the replay process. The $p = -1$ order is the **conjugate beam**.[5] There may well be ± 2 ... orders, but good layout[6] can prevent them from being a nuisance. We shall have to find out how the light in the unwanted orders can be made non-distracting: problem 13.2.

[5]We use terminology introduced by Smith (1969) that deservedly has become standard.

[6]Good layout means using quite large angles, hence the relevance of comment 2 above. The now-conventional off-axis layout is due to Leith and Upatnieks (1962).

13.3 The intensity of the reference beam

A hologram is required to hold within the photograph all necessary information about the object beam, which means both the amplitude and the phase. A photograph records intensity only, thereby destroying information about the phase of the light that it received. This is why we cannot make a hologram by photographing the object beam on its own. The reference beam is there to interfere with the object beam, with the resulting fringe pattern encoding (at least some) information about the object beam's phase.[7] This raises a question as to the relative intensity of the reference and object beams: How do we choose an intensity for the reference beam, and does a workable choice exist at all?[8]

What matters for the production of good-quality fringes is the ratio of reference intensity to object-beam intensity, $I_{\text{ref}}/I_{\text{obj}}$. The acceptable values for this ratio are investigated in problem 13.3, where it is shown that fringes of acceptable contrast can be obtained with values of $I_{\text{ref}}/I_{\text{obj}}$ between about 10^{-2} and 10^2. Other considerations militate against the use of a low intensity for the reference beam, and the usual practical recipe is to aim for $I_{\text{ref}}/I_{\text{obj}} \sim 10$. However, the important point for the present is that this ratio is not at all critical.

Section 13.2 has shown that the optical geometry of reference and object beams is not a critical matter in getting holography to work. Now we learn that the optical intensities, absolute and relative, are not critical either. We are undertaking a kind of feasibility study, such as might be done before trying out a new idea, and the whole process is looking very hopeful.

13.4 The response of a photographic emulsion

In §13.5 we shall give the formal reasoning that shows holography to work. Before that argument can be carried through properly, we need to know something about the response of a photographic emulsion when it is 'exposed' to light and the resulting image is developed.

The photographic process can be simplified to the three main steps outlined in Fig. 13.3.

(1) A photographic plate or film[9] is first exposed to light of intensity I_{exp} for time t_{exp}. Photons in that light cause photochemical damage to silver bromide crystals, producing a **latent image**.
(2) The exposed plate is placed in 'developer', which is a bath of chemicals whose effect is to reduce each damaged silver bromide crystal

[7] The difficulty of destroyed phase information is well known in other contexts. The X-ray crystallographer, trying to use X-ray diffraction to understand the electron density within a protein molecule, needs to perform an inverse Fourier transformation on the scattered X-ray field. But what his X-ray diffractometer gives him is a recording of intensity without phase information; and without that phase information the inverse transformation can't be done. One trick for making phase information recoverable is to embed a heavy atom within the protein, so that X-rays strongly scattered from the heavy atom provide the equivalent of a reference beam.

[8] Remember that the object beam is the outcome of diffraction/scattering, so it must have substantially non-uniform intensity. Interference with the reference beam must give adequate fringes even in the presence of this variation.

[9] The active ingredient in photography is an emulsion of silver bromide in gelatin, deposited on a glass plate or on a piece of flexible film. We are most familiar, in everyday photography, with the case where the 'backing' is flexible film. However, glass plates are the medium of choice for holography because of the need for mechanical stability. The habit in this context is therefore to speak of 'plates' even when a film might in fact be used. One also speaks of the 'plate' as being made into the hologram, though it is of course the emulsion on the plate that is doing the work for us.

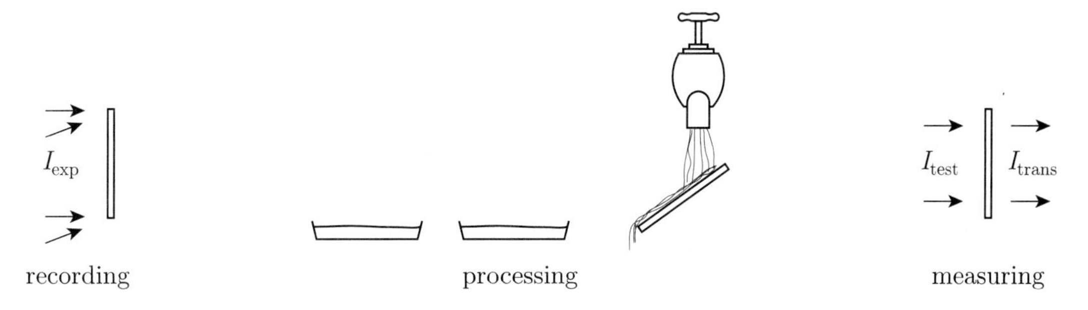

Fig. 13.3 The steps in the preparation and testing of a photograph such as might be used in preparing a hologram. First, the photographic emulsion on a plate is exposed to the interference fringe pattern formed by the reference and object beams. Then the exposed plate is processed (developed, fixed, washed and dried). The third step shown is to measure the transmission of the developed photograph, and is given to define the quantities I_{test}, I_{trans} used in Fig. 13.4.

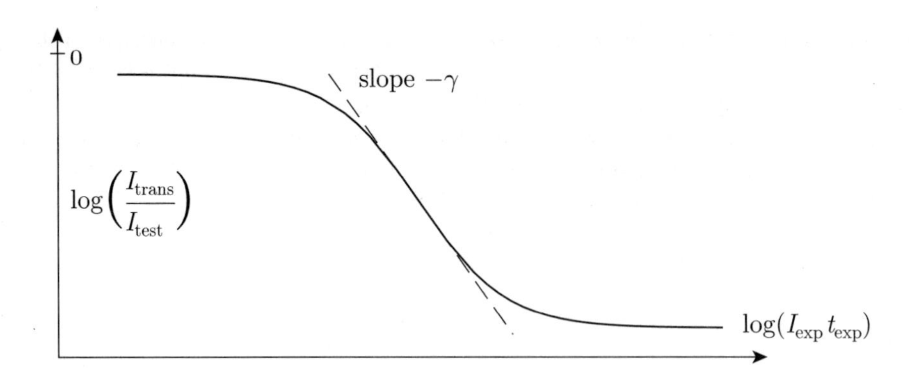

Fig. 13.4 The characteristics of a developed photographic emulsion. The quantities I_{trans} and I_{test} on the vertical axis are explained in Fig. 13.3: their ratio gives the transmission coefficient of the plate (and their inverse ratio gives the 'blackening'). The horizontal axis gives $I_{\text{exp}}t_{\text{exp}}$, the 'exposure'. On this log–log plot, the roughly linear section (usually covering about two decades on each axis) in the middle has a slope $-\gamma$, which defines the contrast parameter γ of the plate.

to a grain of metallic silver. Next, all remaining silver bromide is dissolved from the emulsion by a bath of 'fixer', after which the emulsion has no further sensitivity to light. Then the reagents and waste products are washed away, leaving just gelatin and the developed grains of silver that constitute the **negative image**.

(3) Finally, we must find out what we have done. We illuminate the completed photograph with light of intensity I_{test} and observe the intensity I_{trans} that is transmitted. The success of the photographic process is indicated by the extent to which I_{trans} is dependent upon the **exposure** $I_{\text{exp}}t_{\text{exp}}$. The graph in Fig. 13.4 shows a typical such dependence.

The graph of Fig. 13.4 is plotted, as is customary, with logarithmic scales on both axes, and is known as the Hurter–Driffield plot.[10] It shows three regions: on the extreme left so few photons[11] have been received during exposure that the blackening is independent of exposure; on the extreme right the plate is completely blackened by the exposure (and subsequent development) so that an even greater exposure could make it no blacker. The useful range is in the middle, where there is an approximately linear slope whose gradient is conventionally called $-\gamma$. The quantity γ is a measure of the 'contrastiness' of the final photograph: in ordinary photography a low γ (< 1) gives a grey and wishy-washy picture; a high γ ($\gg 1$) produces soot-and-whitewash. We shall see that for holography the value of γ is not critical; far less critical in fact than is the case with 'ordinary' photography.

On the steep part of the Hurter–Driffield curve we have approximately

$$I_{\text{trans}}/I_{\text{test}} = (\text{constant}) \times (I_{\text{exp}}t_{\text{exp}})^{-\gamma}. \tag{13.3}$$

Equation (13.3) gives the intensity transmission factor of the developed plate, in terms of the energy received during exposure. The amplitude transmission coefficient is the square root of this:[12]

$$U_{\text{trans}}/U_{\text{test}} = (\text{constant}) \times (I_{\text{exp}}t_{\text{exp}})^{-\gamma/2}. \tag{13.4}$$

[10] Cognoscenti may object that my graph is upside down compared to the usual Hurter–Driffield plot. The conventional plot has 'optical density' up the vertical axis, where optical density is $-\log_{10}(I_{\text{trans}}/I_{\text{test}})$. The presentation here contains the same physics and makes it unnecessary to be concerned with optical density.

[11] The blackening of the emulsion after development is a function of the 'exposure' (which is of course proportional to the number of photons received), except at very low intensities where the excitation produced by one photon can decay before a second photon arrives to lock in the latent image.

[12] The above description of photography is deliberately simplified, and may well be embarrassing to experts. In particular, I have considered only the production of a photographic 'negative' which is blackest where most light was received. To make an ordinary photographic print we undertake a second photographic process, making a 'positive', so that the final print is lightest where the exposure of the negative was greatest. There also exist 'reversal' processes, particularly for the making of colour slides, where the γ is engineered to be < 0 so that the making of a separate positive is rendered unnecessary. In the context of holography there is no benefit from having $\gamma < 0$, so the process leading to the production of an ordinary negative is all that we need to understand here.

13.5 The theory of holography

After the feasibility studies of §§ 13.2 and 13.3, we give a formal demonstration that the holographic process works.[13] The reference beam and object beam are shone onto the photographic plate as shown in Fig. 13.5.

[13]The treatment given here is loosely based on Stroke (1969).

The reference beam has complex amplitude (at the plate)

$$U_{\rm ref}(x, y) = U_{\rm r}\, {\rm e}^{{\rm i}\rho(x,y)}\, {\rm e}^{-{\rm i}\omega t},$$

where $U_{\rm r}$ is real so that all phase variation is explicitly contained in $\rho(x, y)$. The one constraint being laid on the reference beam is that its intensity is uniform over the plate; equivalently, $U_{\rm r}$ is constant. However, $\rho(x, y)$ is allowed to depend on position, so the reference beam is not restricted to having any special form, such as a plane wave or a spherical wave.

The object beam has complex amplitude

$$U_{\rm obj}(x, y) = U_{\rm o}(x, y)\, {\rm e}^{{\rm i}\phi(x,y)}\, {\rm e}^{-{\rm i}\omega t},$$

in which $U_{\rm o}(x, y)$ is again real, and both it and $\phi(x, y)$ can depend upon position, so this expression represents a quite general (though monochromatic) wave.

The complex amplitude falling on the photographic plate at position (x, y) is

$$U_{\rm exp}(x, y) = (U_{\rm r}\, {\rm e}^{{\rm i}\rho} + U_{\rm o}\, {\rm e}^{{\rm i}\phi}){\rm e}^{-{\rm i}\omega t}. \qquad (13.5)$$

The intensity exposing the plate at position (x, y) is therefore

$$I_{\rm exp} \propto |U_{\rm exp}(x, y)|^2 = U_{\rm r}^2 + U_{\rm r}\, U_{\rm o}\, {\rm e}^{{\rm i}(\phi-\rho)} + U_{\rm r}\, U_{\rm o}\, {\rm e}^{{\rm i}(\rho-\phi)} + U_{\rm o}^2. \qquad (13.6)$$

We expose the plate to this intensity for a suitably chosen exposure time $t_{\rm exp}$, then process the exposed plate chemically to a contrast factor γ, thereby completing the preparation of the hologram.

The amplitude transmission coefficient of the exposed and developed

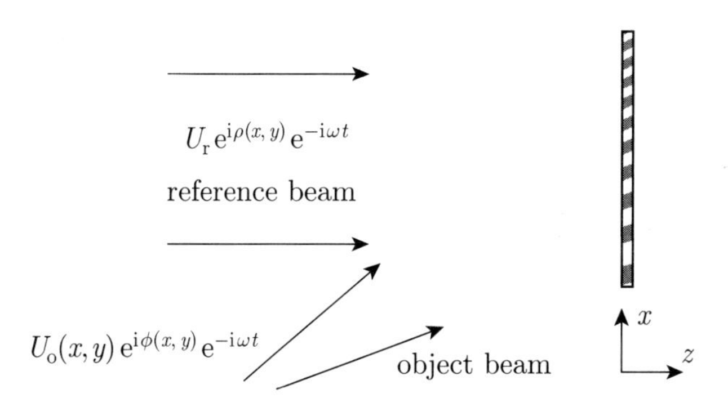

Fig. 13.5 Reference and object beams illuminate a photographic plate in the xy plane. The reference beam has constant (real) amplitude $U_{\rm r}$ but can have any reasonable dependence of phase $\rho(x, y)$ on position. The object beam has (real) amplitude $U_{\rm o}(x, y)$ and phase $\phi(x, y)$ that may both depend upon position.

plate at position (x, y) is, from eqn (13.4),

$$\begin{aligned} T(x,y) &\propto I_{\text{exp}}(x,y)^{-\gamma/2} \\ &\propto U_{\text{r}}^{-\gamma}\left\{1+\frac{U_{\text{o}}}{U_{\text{r}}}\text{e}^{\text{i}(\phi-\rho)}+\frac{U_{\text{o}}}{U_{\text{r}}}\text{e}^{\text{i}(\rho-\phi)}+\frac{U_{\text{o}}^2}{U_{\text{r}}^2}\right\}^{-\gamma/2} \\ &= U_{\text{r}}^{-\gamma}\left\{1-\frac{\gamma}{2}\frac{U_{\text{o}}}{U_{\text{r}}}\text{e}^{\text{i}(\phi-\rho)}-\frac{\gamma}{2}\frac{U_{\text{o}}}{U_{\text{r}}}\text{e}^{\text{i}(\rho-\phi)}-\frac{\gamma}{2}\frac{U_{\text{o}}^2}{U_{\text{r}}^2}\right. \\ &\qquad\qquad\left.+\text{ terms in }\text{e}^{2\text{i}(\phi-\rho)},\ \ldots\right\}. \end{aligned}$$

Finally, we illuminate the developed hologram with an 'illuminating beam' that is identical to the original reference beam $U_{\text{r}}\,\text{e}^{\text{i}\rho}\,\text{e}^{-\text{i}\omega t}$. The amplitude transmitted through the plate at (x, y) is therefore

$$\begin{aligned} &U_{\text{recons}}(x,y) \\ &= U_{\text{r}}\,\text{e}^{\text{i}\rho}\,\text{e}^{-\text{i}\omega t}\,T(x,y) \\ &\propto U_{\text{r}}\,\text{e}^{\text{i}\rho} && \text{illuminating beam (attenuated)} \\ &\quad-(\gamma/2)\,U_{\text{o}}\,\text{e}^{\text{i}\phi} && \text{object beam reconstructed} \\ &\quad-(\gamma/2)\,U_{\text{o}}\,\text{e}^{\text{i}(2\rho-\phi)} && \text{conjugate beam} \\ &\quad+\text{terms like }U_{\text{o}}^2\,\text{e}^{\text{i}\rho} && \text{forward scattering around illuminating beam} \\ &\quad+\text{terms like }U_{\text{o}}^2\,\text{e}^{\text{i}(2\phi-\rho)} && \text{second and higher orders of diffraction} \\ &\quad+\text{higher terms.} && (13.7) \end{aligned}$$

One of the terms in this can at once be identified as a reconstruction of the amplitude $U_{\text{obj}}(x, y)$ at x, y of the object beam. We have a general demonstration that the process of recording and reconstruction works for an arbitrarily complicated object beam.

Given a term $\propto U_{\text{obj}}(x, y)$, in the xy plane, we can imagine calculating from it the Kirchhoff-integral diffracted amplitude at places downstream. This amplitude is the same (proportionality constant apart) as what we would have had if the original object beam had passed unobstructed through the xy plane. We really have reconstructed both the object beam's wavefront and the whole of that beam's continuation.

Comment 1 The final proportionality sign conceals a factor $U_{\text{r}}^{-\gamma}$, so the object beam is properly reconstructed only if $U_{\text{r}} = \text{constant}$; reference and illuminating beams should be given uniform intensity across the plate. But a gradual variation of U_{r} often doesn't matter much; an ordinary picture lit more strongly at one side doesn't look too peculiar.

Comment 2 Annotations beside eqn (13.7) identify the individual terms, using the results of problems 13.1(6) and 13.4. Problem 13.2 shows that the terms additional to the object-beam reconstruction can be made to cause negligible confusion, if we have the sense to arrange that their waves travel off in uninteresting directions.

Comment 3 We may see rather clearly the need for the reference beam at the recording stage, by noticing that all of the needful terms disappear if we remove the U_{r} term from eqn (13.5).

Comment 4 An ordinary photograph has its contrast deliberately increased from that in the original scene (with $\gamma \sim 1.2$–1.5) for aesthetic reasons. Notice that the holographic reconstruction has exactly the same 'contrast' as the original object beam,[14] whatever the value of γ.

[14] This is one reason why the γ of the hologram is not critical.

Comment 5 Things would be especially simple mathematically if we could devise a way of making the photographic plate have $\gamma = -2$. However, this is not easy to do, and simplicity of calculation is not accompanied by any special practical merit.[15]

[15] This type of hologram tends to live more in examination questions than in the real world.

Comment 6 Figures 13.1 and 13.2 show special cases: the reference and illuminating beams are plane. That is not a necessary arrangement. Spherical waves are often used in practice. In fact, any reasonable wave, of near-uniform intensity, can be used so long as it is used for both reference and illuminating beams (and even that condition can be relaxed somewhat—see §13.8 and problem 13.6). The calculation of the present section was set up to draw attention to this generality: the reference-beam phase $\rho(x, y)$ was left completely unspecified.

13.6 Formation of an image

The discussion of §13.5 shows that wavefront reconstruction works generally. Nevertheless, the algebra may not give us much 'feel' for the way in which an image of the original object is made to appear. We need another tutorial example. One is given in problem 13.5. In that problem, a point scatterer gives rise to spherical waves that interfere with a plane reference wave, giving a set of circular fringes. When those fringes are photographed, the result is a diffraction grating with circular elements that behaves like a Fresnel zone plate. Putting the hologram into a plane illuminating beam results in two spherical waves, one (primary beam) radiating from the location of the original object, the other (conjugate beam) radiating towards a second image of that object, which is called the conjugate image.

The reasoning of §13.5 shows that the recording and reconstruction just described work for any object beam. Therefore, there can be any number of point scatterers generating spherical waves as described, each forming its own set of circular fringes, and each set of (photographed) fringes acts as its own zone-plate grating, undisrupted by the presence of the others. Figure 13.6 shows the recording and replay of a hologram of a simple three-dimensional object.

The layout of Fig. 13.6 displays a special case: where the reference and illuminating beams are plane waves normal to the photographic plate. In this case, the primary and conjugate images stand in mirror-image relation to each other, with the plane of the plate half-way between them. Although this is a special case, it displays a general property: the conjugate image is depth-reversed. It is for this reason that we almost always arrange the replay layout so as to discard the conjugate image, while exploiting the primary image.

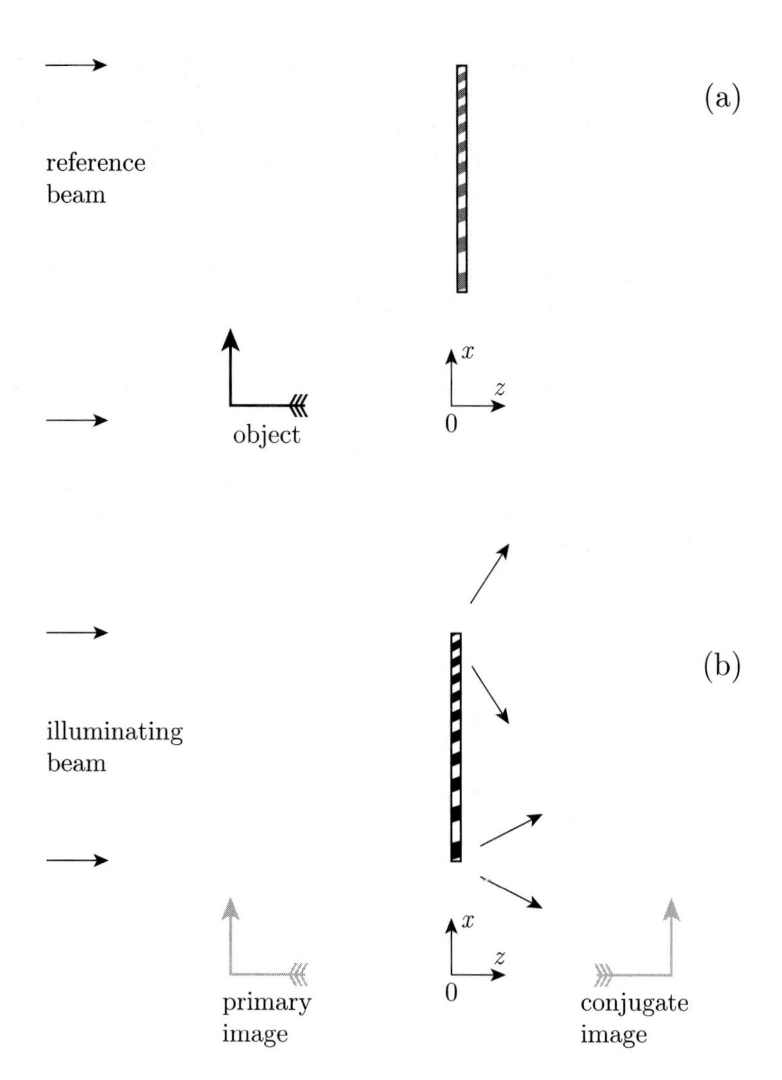

Fig. 13.6 The recording and replay of a simple three-dimensional object. In (a) the object scatters some light from a laser that also supplies the reference beam. In (b) the illuminating beam is shone onto the developed hologram, and diffraction makes light radiate from the primary image and towards the conjugate image. With the simple geometry used here (plane reference and illuminating beams both incident normally on the plate), primary and conjugate images stand in mirror-image relation to the hologram. It may be seen that the conjugate image is depth-reversed.

13.7 What if we break a hologram in half?

It is often stated that a hologram differs from an 'ordinary' picture in the following way: if you break the hologram in half (non-destructively by covering it with a card!), the whole image is still there.[16] This is often the case. But you must lose *something* by removing half the hologram.

There are two cases that are easily discussed.

- Suppose we view a hologram reconstruction, by eye, with a layout similar to that of Fig. 13.1. The reconstructed primary image is seen in three dimensions, as though through a window whose outline is the outline of the hologram plate. If we cover half of the hologram with a card, we halve the area of the window, and we restrict the observer's opportunity to look round foreground objects by moving his eye. Information is lost, and it's easy to believe that the loss is half that originally present.

[16]The statement isn't always true. Holograms are often incorporated into banknotes, credit cards and the like as security measures. In these cases, the reconstructed image often lies close to the plane of the hologram itself. If then you cover half the area of the hologram, you *do* lose half the picture.

- Suppose we form a new image of the primary image, using a camera whose lens is large enough to receive light from the whole area of the hologram. If half the area of the hologram is covered, the resolution of the image formed by the lens is worsened by a factor 2. Again, information is lost, in proportion to the area obstructed.

These two examples show what we expected. Obstructing part of the hologram area may not remove completely a part of the reconstructed image, but a penalty *is* exacted as a reduction of available information.

13.8 Replay with changed optical geometry

In descriptions of the holographic process so far, the illuminating beam (replaying a completed hologram) has exactly reproduced the reference beam (in its geometry, perhaps not its intensity) that was used in preparing the hologram. This constraint can be relaxed—up to a point.

Figure 13.7 shows what happens by means of a tutorial example resembling that of problem 13.5: a plane reference beam works with a

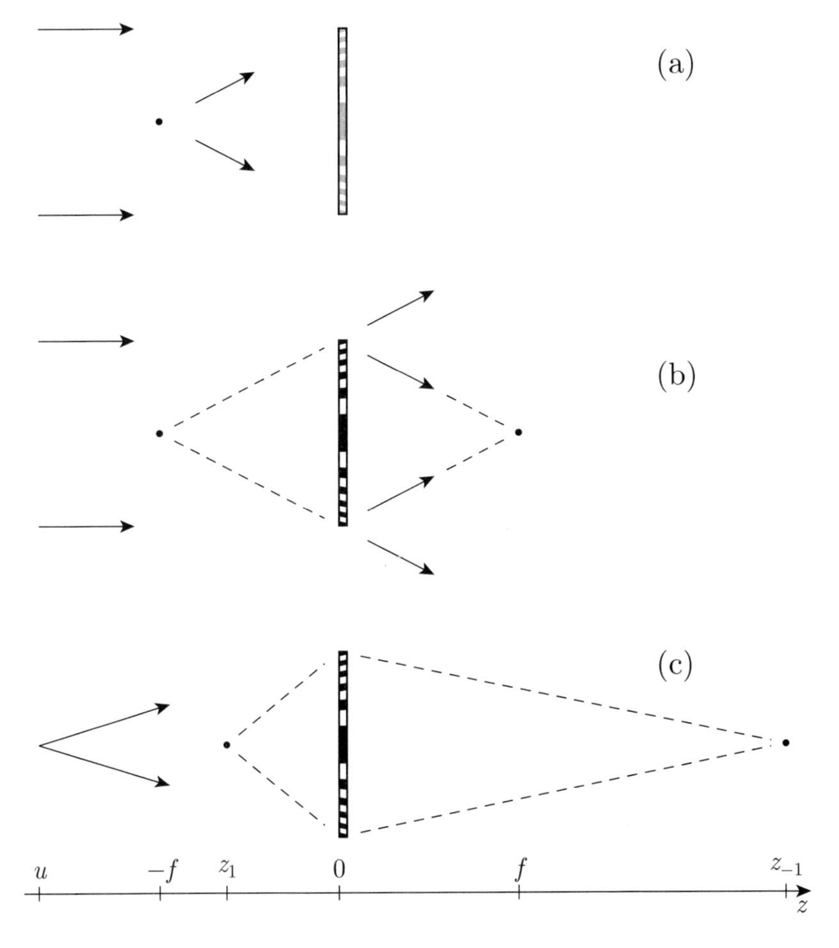

Fig. 13.7 (a) A point object scatters light from a laser which also provides a plane reference beam normal to the photographic plate. A spherical wave radiated from the object interferes with the plane reference wave, and the resulting circular fringes are recorded.
(b) The developed hologram is placed in an illuminating beam identical with the original reference beam. Two diffracted waves (in particular) are generated, one radiating from the location of the original object (primary beam), and one radiating towards a second point (conjugate beam). In the special-case layout drawn, the two point images lie at distances $z = \pm f$ on the z-axis with the hologram at $z = 0$. (The symbol f is used for these distances because it is the focal length of the Fresnel zone plate formed by the recorded fringe pattern.)
(c) An illuminating beam of different geometry is used with the same hologram as before. It diverges from a point at z-coordinate u, which as drawn is negative. The action of the hologram as a Fresnel zone plate is now to give diffracted waves diverging from a point at z_1 and converging towards a point at z_{-1}. Formulae for z_1 and z_{-1} are worked out in problem 13.6.

point scatterer whose scattered wave provides the object beam. The two beams interfere to give a set of circular fringes on a photographic plate placed normal to the reference beam. If the hologram is replayed using an illuminating beam identical to the original reference beam, the primary beam diverges from a point at the location of the original scatterer. If, instead, the hologram is replayed using a diverging spherical wave as illuminating beam, primary and conjugate beams are still generated, but the points from which they diverge (or to which they converge) are displaced. The locations of these displaced images are investigated in problem 13.6.

There are limits on how far the geometry can be changed. One limit comes from the limitations of a Fresnel zone plate: when used away from the original geometry it produces aberrations. But a more serious difficulty comes from the three-dimensional structure of the recorded fringes, as explained in the next section.

13.9 The effect of a thick photographic emulsion

Up until now, we have described a hologram as if it was formed within a photographic emulsion of zero thickness. In fact, a photographic emulsion has a thickness that may amount to several optical wavelengths.

In the diagrams of holography in this chapter, we have tried to give an impression of the three-dimensional structure of the photographed fringes. In particular, in Fig. 13.2 the fringes are shown bisecting the angle between reference and object beams, and these slanting fringes are recorded within the thickness of the photographic emulsion. In the reconstruction process, the primary beam is the result of mirror-reflection from each of these three-dimensional fringes. Constructive interference of waves reflected from parallel planes means that the light is obeying the Bragg condition of X-ray crystallography. Conversely, the conjugate beam does *not* obey the Bragg condition, and light scattered from some places within the emulsion interferes destructively with that from others. With an ideally thin emulsion, this destructive interference is absent, and the conjugate beam carries as much energy as the primary beam. With a thick emulsion, the conjugate beam is partly suppressed.[17]

[17] As are higher orders of diffraction, of course.

There are implications for the replay with changed optical geometry discussed in §13.8. The further we depart from the geometry used in preparing the hologram, the larger are the departures from the Bragg condition. This is often the main factor limiting the opportunity for choosing a replay geometry.

A very thick emulsion imposes a severe constraint, via the Bragg condition, on the reconstruction process. Not only must the geometry be correct, so must the wavelength. This property is exploited in making coloured holograms: the hologram is illuminated with white light, and only the correct wavelength is diffracted by each set of fringes.[18]

[18] This property was known and used long before the invention of holography. A process known as Lippmann photography is described in Born and Wolf (1999), p. 312. In the Lippmann arrangement, an image is formed (by a lens) on a photographic plate in the usual way, but a reflector immediately behind the emulsion causes standing waves: interference fringes run roughly parallel to the plane of the plate, with a spacing determined by the optical wavelength. The photograph records an image of some object illuminated by white light, and each colour produces fringes appropriate to its own wavelength. On replay with white light, only the right colour is reflected by each part of the photograph, because only it meets the Bragg condition. The picture produced is coloured, although the photographic emulsion is 'black-and-white'. The colour rendition is said still to be more accurate than that produced by 'colour film', because it is the physics of the Bragg reflection that produces the colour.

'Lippmann plates' (plates bearing an emulsion designed for the Lippmann process) are 'slow' by the standards of everyday photography, because of the need to record fringes spaced by a half-wavelength; and the viewing angle is critical. For these reasons Lippmann photography has never been much more than an interesting curiosity. Many of the earliest holograms were, however, made using Lippmann plates, because of the similar need to record very fine fringes.

The Bragg condition is used in a more extreme form when holograms are recorded in a bulk three-dimensional medium, with the aim of making a mass-storage device for holding computer information. But this is another story.

13.10 Phase holograms

Photographing interference fringes, according to the description of photography in §13.4, produces deposits of silver within the photographic emulsion. This silver is in the form of very fine particles, and is effectively 'black'. A hologram made this way absorbs a significant fraction of the light used to illuminate it. It would be better if we could avoid this loss of energy.

We have met something similar in connection with diffraction gratings. A 'reflection grating', even if not blazed, makes more efficient use of the available light than a part-absorbing transmission grating. The important characteristic of a reflection grating is that it introduces periodic changes of phase instead of periodic changes of absorption.

It is, therefore, worth asking whether a hologram can be made that has interference fringes recorded as phase changes instead of 'blackening' changes. There are several ways in which this may be done.

One way to make a phase hologram is to prepare the hologram first in the way described in §13.4. Then an extra process is undertaken in which all the silver is dissolved away. The optical thickness of the gelatine retains an impression of the image where the silver was, and so the outcome is the holographic equivalent of a 'phase grating'.

Another possibility is to record a hologram on photoresist.[19] After exposure to the fringe pattern, the surface of the photoresist is chemically etched until it is grooved to the required depth. The hologram can be used in transmission, or (more probably) the surface may be coated with aluminium to make a reflection hologram.[20] Neither process for making a phase hologram lends itself readily to the equivalent of 'blazing', but even without it the improvement is substantial.

[19] Photoresist is described in §4.8.

[20] This is, above all, the process used in making holographic diffraction gratings, as described in §4.8. The reason why gratings fabricated in this way are called 'holographic' should now be clear.

Once a hologram has been made with photoresist, a sequence of operations can be undertaken that resemble those that are used in the manufacture of a CD. A CD starts out as an exposed layer of photoresist on glass, and from there the manufacturer prepares various copies—and copies of copies—until he has a hard metal die that can impress its shape onto multiple plastic discs in an injection-moulding process. A similar set of moves with holography yields a die that can emboss a surface of plastic or metal for use on a banknote or credit card.

We mention again the preparation of 'volume holograms' for the bulk storage of information. These likewise are phase holograms.

13.11 Gabor's holograms

The original holograms prepared by their inventor Gabor were made in an apparatus that loosely resembled that of Fig. 13.7(a), with the reference beam supplied by a point source somewhat behind an on-axis object. The reconstruction layout then resembled that of Fig. 13.7(b): primary and conjugate images were one behind the other, and each was spoiled by the other's out-of-focus presence. Many books[21] present photographs of Gabor holograms and the reconstructions from them; they can seem rather unimpressive beside what is routinely done now.

[21]For example, Born and Wolf (1999), p. 509.

There were several reasons why Gabor used his in-line optical layout. The main one was the then-necessary use of non-laser light with restricted coherence area and coherence length. But another reason was the purpose for which Gabor invented holography. He wished to prepare a hologram with electrons in an electron microscope and to replay it with visible light. Such a two-step process could (it was hoped) overcome the defects of electron-microscope lenses, because the optical illuminating beam could mimic the electron reference beam—including its aberrated shape.[22] His demonstration holograms were made and replayed with visible light, but this was simply to establish the feasibility of the process.

[22]Remember that the reconstructed image is perfect if the illuminating beam mimics the reference beam, even if neither has any specially simple shape.

Holography has never taken off as an adjunct to electron microscopy (or X-ray microscopy to which it could also be adapted). Electron-optical lenses have been developed beyond the quality which in Gabor's day seemed to have hit a fundamental limit, so that holographically assisted microscopy does little that can't be done more easily in other ways.[23]

[23]The discussion here does nothing to diminish Gabor's achievement in inventing holography. It often happens that something invented for one purpose turns out to be much more useful for another. And, as we have mentioned, optical holography became a really serviceable technology with the availability of the laser and the invention (another very significant inventive step) of the off-axis layout of Fig. 13.1 (Leith and Upatnieks 1962).

13.12 Practicalities

The making of a hologram is generally somewhat more critical than the replay, so we concentrate on the recording process to start with.

The fringes to be recorded have a spacing that is only an optical wavelength or so. The photographic emulsion used must have a resolution that is fine enough to record these fringes clearly. It is sometimes possible to make things less critical by reducing the angle between reference and object beams (angle $\delta - \alpha$ in Fig. 13.2), and taking advantage of the resulting coarser fringes by using a coarser (which means faster) emulsion.

The fringes must not move during recording (from vibration, thermal expansion, air currents, drift of laser frequency, ...). Fine-grained plates are 'slow', so the rigidity problem is exacerbated by the need for long exposures. This was a really serious problem in the early days of holography (exposure times of tens of seconds), but with high-powered lasers (such as argon) and improved emulsions it is usually possible to use exposure times of less than a second.

It should not need saying that the laser providing the reference beam and illuminating the object must have transverse coherence over the whole of the area being used. One should also remember that most lasers have a limited coherence length (unless they have a single longitudinal mode and are frequency-stabilized). An attempt should be made at avoiding very long path differences.

We have already said that the reference beam should be a little more intense (say a factor 10) than the object beam.

The description so far has assumed that we use a single-wavelength[24] laser, which means that the hologram is recorded in a single colour, and its replay can convey only one colour. It is possible to record using a laser such as krypton that gives out enough frequencies to simulate white light. However, the replay will be intelligible only if a thick emulsion is used so that the Bragg condition causes the correct wavelengths to be selected.

[24]The jargon of laser physics describes a laser as 'single wavelength' if its radiation lies within a single spectral line (even if composed of several longitudinal modes), and as 'single frequency' if it is in a single longitudinal mode.

We now turn to the replay process. If a single wavelength is to be used, it is generally easiest to use a laser: a small He–Ne emitting 1 mW gives an image that is quite bright enough to see clearly. However, a non-laser source subtending a small angle at the hologram may sometimes be acceptable, though the image will be smeared through about the same angle range as that of the illuminating light. A white-light source, even if small, usually gives rather poor results (except with a thick emulsion), because each wavelength is diffracted through a different angle from the same piece of hologram, and the reconstructed image is made up from a mess of coloured images of different sizes or at different distances.

An exception to the above is the case where a hologram is made to record an object (probably a real image of the actual object) that is very close to the hologram itself. The result is still a hologram (rather than an ordinary picture): it may, e.g., be made with reference beam and object beam intersecting at an angle, so it has a fringe pattern across it as well as a real image of the object. In this case, replay with white light results in a strongly coloured image, but those colours that are seen at one time reconstruct an image whose smearing is not objectionable. A hologram of this type is to be seen on any credit card.

13.13 Applications of holography

Here we have space to touch on only one or two topics that have not already been encountered in passing.

Holography can be used in various ways to make visible small displacements, distortions or vibrations of some body. One possibility is to expose a hologram of an object, distort the object slightly and then take a second exposure on the same photographic emulsion. When the hologram is developed and replayed, there are superimposed reconstructions of the wavefronts from the object in its two configurations, and there is interference between the two reconstructed wavefronts. Fringes give 'contours' measuring the displacement that the object underwent between the two exposures.

A variation on this procedure is to make a single-exposure hologram, process it, and return it to the place where it was made with the original object still in place. The illumination geometry is kept as it was during exposure. If the object is now deformed, there is interference between the light that it scatters and the primary-reconstruction beam, again giving fringes that measure the displacement between 'before' and 'after' positions. This arrangement is usually said to give 'live fringes' because the object deformation and the resulting fringe pattern can be examined in 'real time'.

A third possibility is to make a hologram of a vibrating object such as a violin, using an exposure that is long compared with the period of vibration. Where the movements of the object are more than a fraction of a wavelength, the fringes (interference between object beam and reference beam) move so much during exposure that no holographic recording results. But at nodal lines, where the object is not moving, a recording is made in the ordinary way. On replay the hologram shows an image of just the nodal regions of the object, a display resembling the Chladni patterns traditionally made by scattering a fine powder on a vibrating surface.

For details of these and other ingenious applications of holography we have to refer the reader to more technical works.

Problems

Problem 13.1 (a) The detail of §13.2

(1) Draw the optical paths and their differences which give rise to interference fringes with the spacing d given by eqn (13.1). Draw the optical paths and their differences which cause the pth order of diffraction from a developed hologram to depart at angle θ_p as given by eqn (13.2). Draw these optical paths and their differences for the case $p = 1$, and make it completely obvious on your drawing that $\theta_1 = \delta$. The fact that, in the special case discussed in §13.2, the primary beam reconstructs the object beam must be made to appear obvious, or you haven't fully understood the interference–diffraction physics.

(2) Show that in eqn (13.2) the case $p = 0$ represents an output which is just the illuminating beam, transmitted without modification to its shape.[25]

[25] This diffracted beam cannot quite be identical to the illuminating beam, because it has reduced intensity: some of the incident energy has been diverted into other orders of diffraction, and some has been absorbed. It does, however, have the same *shape*.

(3) In eqn (13.7) we need to identify the various orders of diffraction within a mathematical expression for the field amplitude transmitted during reconstruction. The cases $p = 0$ and $p = 1$ should now be recognizable, but order $p = -1$ has a less obvious form. We obtain the required interpretation by using the special-case optical layout of Fig. 13.2. The photographic emulsion lies in the xy plane. Both reference and illuminating beams are plane waves incident at angle α to the z-direction, and the object beam is another plane wave incident at angle δ. Show that the complex amplitude of the reference beam at the xy plane is $U_{\text{ref}} = U_{\text{r}}\, \mathrm{e}^{\mathrm{i}\rho(x,y)}\, \mathrm{e}^{-\mathrm{i}\omega t}$, where U_{r} is real and $\rho(x, y) = kx\sin\alpha$. Show that the object beam has complex amplitude $U_{\text{obj}} = U_{\text{o}}\, \mathrm{e}^{\mathrm{i}\phi(x,y)}\, \mathrm{e}^{-\mathrm{i}\omega t}$ with $\phi(x, y) = kx\sin\delta$.

(4) Let the pth order of diffraction from the hologram have amplitude $U_p = (\text{real}) \exp\{\mathrm{i}\psi_p(x, y)\}\, \mathrm{e}^{-\mathrm{i}\omega t}$ in the xy plane. Show that in the special-case geometry of Fig. 13.2, $\psi_p = kx\sin\theta_p$.

(5) Still using the same geometry, show from eqns (13.1) and (13.2) that

$$\psi_p = \rho + p(\phi - \rho). \tag{13.8}$$

(6) Show from this that an amplitude in the xy plane

$$\begin{array}{lll} \text{like } e^{i\phi} & \text{belongs to the } \textit{primary beam} & p = 1, \\ e^{i\rho} & \qquad\qquad\qquad\quad \textit{illuminating beam} & p = 0, \\ e^{i(2\rho-\phi)} & \qquad\qquad\qquad\quad \textit{conjugate beam} & p = -1. \end{array} \tag{13.9}$$

(7) The results just obtained, in particular statement (13.9), have been derived for a special case. Invent an argument to justify their correctness for any reasonable arrangement of the optics.

(8) The fringes formed at the recording stage have a $\cos^2$ intensity variation. A $\cos^2$ diffraction grating gives diffraction orders $p = 0, \pm 1$ only (example 3.1). Why may we have higher orders of diffraction here?

Problem 13.2 (b) How muddling are the several orders of diffraction? Return to Fig. 13.1. Identify (measure them!) the largest and smallest of the angles δ that object waves make with the plate normal. Find the consequent range of fringe spacings recorded on the hologram. Find the ranges occupied by angles θ_p for the cases $p = -1, 0, 1, 2$ in the reconstruction stage. Draw the extreme rays for $p = 2$ and show that the 'image' they form is well clear of the $p = 1$ image.

[*My answers*: For the case $p = -1$: $-51° < \theta_p < -17°$. For $p = 0$: $\theta_p = 0$. For $p = 1$: $17° < \theta_p < 51°$. For $p = 2$: $36° < \theta_p < 90°$.]

Comment: There is little difficulty in making these diffracted beams completely non-distracting. So the $p \neq 1$ orders of diffraction are a nuisance only in that

- we have to design their nuisance value away by choosing appropriate angles at the recording stage
- they divert some energy in the reconstruction process which we should prefer to have in the primary ($p = 1$) beam, reconstructing the object beam.

Problem 13.3 (a) Relative intensities of reference and object beams
In §13.2 we saw that the whole process of holography consists in recording a fringe pattern and using the recorded pattern as a diffraction grating. The fringes must have reasonable contrast if they are to be recorded successfully.

(1) Show that the visibility V of the fringes is given by

$$V \equiv \frac{I_{\max} - I_{\min}}{I_{\max} + I_{\min}} = \frac{2\sqrt{R}}{1 + R}, \qquad \text{where} \qquad R = \frac{I_{\text{ref}}}{I_{\text{obj}}}. \tag{13.10}$$

(2) Show that V is unchanged if we replace R with $1/R$. Is this what you would expect?[26]

[26] Something similar happened in problem 6.2(1).

(3) Notice how slowly V varies with R: An R as large as 100 yields a V that is still as large as 0.2; equivalently, $I_{\text{max}}/I_{\text{min}} = 1.5$, a perfectly serviceable fringe contrast.[27]

[27] This insensitivity of V to R is one of those ideas that keeps popping up in unexpected places in physics and elsewhere: whenever we 'beat' one signal against another. It is a good part of the reason why a superheterodyne radio works. Likewise, it is vital to the success of phase-locked loops in control electronics and engineering. It is even relevant to the audibility of the 'beats' that are used in the tuning of a musical instrument.

Comment: It helps to turn round the statements above so that they refer to the acceptable object-beam intensity. This problem has shown that we can get fringes of usable contrast with an object-beam intensity anywhere between $10^{-2} I_{\text{ref}}$ and $10^{2} I_{\text{ref}}$. Even if we discard the possibility of allowing the object beam to be anywhere more intense than the reference beam, we still have a dynamic range of 100, which is not at all bad.[28] In a practical case, we must recognize that the object-beam intensity will vary with position over quite a wide range, and we undertook this problem to see whether recording of such a variable-intensity beam could be possible. It has turned out to be almost completely unproblematic.

[28] The dynamic range of a photographic film is only about 100—meaning that in Fig. 13.4 the extreme values of $I_{\text{trans}}/I_{\text{test}}$ have a ratio of order 100. In the jargon of photography, the greatest optical density is ~ 2.

Given that the object beam is stronger in some places than others, we could guess that it carries most information where it is most intense. This encourages us to make sure of getting good-quality fringes where the object beam is strong. There is good sense in the usual recommendation to make $I_{\text{ref}}/I_{\text{obj}} \sim 10$.

Problem 13.4 (b) Forward scattering around the illuminating beam
Refer to eqn (13.7). One term, $U_{\text{o}}^2\, \text{e}^{\text{i}\rho}$, is annotated as representing 'forward scattering around the illuminating beam'. We might at first think that this term gives a mere addition to the transmitted illuminating beam, but it does not because $U_{\text{o}}(x, y)$ is not constant.

(1) Return to §13.2 and apply two plane object beams at angles δ_1, δ_2 to the plate normal. Show that in addition to fringes given by eqn (13.1), these beams give rise to a third fringe system by interfering with each other.

(2) Use methods drawn from problem 13.2 to estimate the smallest distance (in the xy plane) over which U_{o}^2 may vary, and hence the largest angle through which the $U_{\text{o}}^2\, \text{e}^{\text{i}\rho}$ beam spreads downstream from the hologram. Any angles required may be estimated by measurement from Fig. 13.1.

(3) Show from eqn (13.6) that the unwanted fringes have intensity proportional to U_{obj}^2, while the wanted fringes have intensity proportional to $U_{\text{ref}} U_{\text{obj}}$. Thus, we can do something to minimize the nuisance value of the unwanted fringes by making the reference beam more intense than the object beam.[29]

[29] This shows us another reason why the reference beam should be made somewhat more intense than the object beam. We can see it as refining the understanding in comment 3 on p. 298, which pointed out the need for the reference beam to be used at all.

(4) Show that the minimization of part (3) also minimizes the second and higher orders of diffraction at the hologram.

Problem 13.5 (a) Special case: holographic imaging of a point
A broad laser beam provides the reference beam for the arrangement shown in Fig. 13.8. The laser also illuminates a small scatterer at P, which scatters a spherical object wave onto the photographic plate.

(1) Show that the interference fringes formed at the photographic plate

are circles centred on O (chosen to be the origin of coordinates) which lies at the foot of the perpendicular from P to the plate.

(2) Find the radius of the pth dark fringe.

(3) Take a small patch on the plate, well away from the z-axis, the patch being chosen large enough to contain several fringes, but small enough that those fringes have nearly constant spacing. Treat the patch (on the exposed and developed hologram) as a small diffraction grating. Illuminate the patch with the original laser beam alone (i.e. the illuminating beam is identical to the original reference beam). Show that the first order of diffraction from the grating leaves at exactly the angle at which the object beam arrived.[30]

(4) With the same geometry as that in part (3), show that the $p = -1$ order of diffraction (the conjugate beam) travels towards a point P′ on the z-axis where OP′ = PO. Draw this.

(5) Argue that all small patches of (replayed) hologram send light away from P and towards P′, so the hologram generates two images (virtual at P, real at P′) of the original point object.

(6) Show that the (photographed) fringe pattern of part (1) constitutes a Fresnel zone plate,[31] whose rings are centred at O. Given the focusing properties of a zone plate, we have a way of seeing why it is that the hologram forms images as it does.

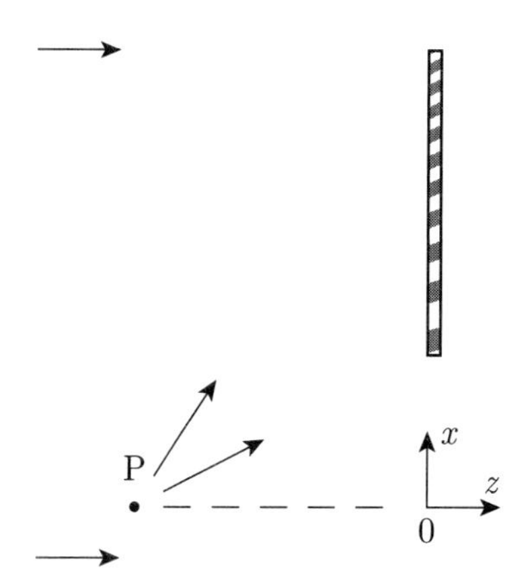

Fig. 13.8 A plane wave incident from the left provides the reference wave here. A part of that wave is scattered from a point at P, and is radiated away from P as a spherical wave. Interference occurs where the plane wave and spherical wave overlap, and results in a family of circular fringes, centred on O, that are recorded on the photographic plate.

Problem 13.6 (b) Use of a new geometry for the illuminating beam
Use Fig. 13.7. Locations on the z-axis are specified using a strict Cartesian convention for attaching signs.

(1) A Fresnel zone plate receives light from an object at a finite distance, rather than from infinity. Use the diffraction physics of sidenote 31 (this page) to work out the equation corresponding to eqn (1.32) that relates object and image distances. Show that the focal length of the Fresnel zone plate is the f of Fig. 13.7; but the 'lens' is simultaneously converging and diverging, so it would be better to say that its focal length is $\pm f$.

(2) Consider the case shown in Fig. 13.7(c), with a point source at $z = u$ (negative as drawn) illuminating the Fresnel zone plate. Show that

$$\frac{1}{z_1} = \frac{-1}{f} + \frac{1}{u}, \qquad \frac{1}{z_{-1}} = \frac{1}{f} + \frac{1}{u}. \tag{13.11}$$

(3) Show that when $u < -f$ or $u > f$, the primary image is virtual ($z_1 < 0$) while the conjugate image is real ($z_{-1} > 0$).

(4) Show that when $-f < u < 0$, both images are virtual ($z_1,\, z_{-1} < 0$).

(5) Show[32] that when $0 < u < f$ both images are real ($z_1,\, z_{-1} > 0$).

(6) Section 13.9 discusses the effect of a thick emulsion on the replay process. Estimate roughly the angular tolerance on the illuminating beam's direction (taking it that the reference beam was a plane wave incident normally) when the emulsion has thickness 10 μm.

Problem 13.7 (c) Gabor's electron–optical microscope
An electron beam (de Broglie wavelength λ_1) and scatterer are used in

[30] *Hint*: Strip the reasoning to its simplest form, until the result looks inescapable, not just a coincidence.

[31] For a reminder of the definition of Fresnel zones, see §3.13 and problem 3.21(7). A zone plate has alternate zones transparent and opaque. If a plane wave is incident on a zone plate, then a point downstream can be found on the axis such that all the transparent zones contribute an amplitude there in the same phase, while all in-between zones that would have contributed in a cancelling phase have been removed. A zone plate, therefore, has focusing properties. But, like our hologram, it has two foci, one for each of the first orders of diffraction.

[32] *Comment*: This problem shows that primary and conjugate images can each be real or virtual, depending on the geometry chosen. There are confusions in some of the literature where 'virtual' means 'primary' and 'real' means 'conjugate', even when they're not! It is this confusion, in particular, that was tidied up by Smith (1969).

the geometry of Fig. 13.7(b) to make a hologram. Electron-interference fringes are recorded on a photographic plate.

(1) We wish to replay by using light of wavelength λ_2. To this end we make a copy hologram that is a replica of the original, but magnified in the ratio of the wavelengths.[33] Show that both lateral and longitudinal magnification of the resulting image are equal to λ_2/λ_1.

(2) Discuss the possibility of replaying by a route that doesn't involve making a magnified replica of the original electron-prepared hologram.[34]

[33] This gives a replay geometry that reproduces the recording geometry and thereby eliminates aberrations.

[34] Problem 13.8 may help.

Problem 13.8 (e) General properties of any imaging system[35]
An imaging system imposes a **transformation** or **mapping**: in forming an image it maps points (in the object space $x_o y_o z_o$) into points (in the image space $x_i\, y_i\, z_i$); and it maps straight lines (which might be rays) in the object space onto other straight lines in the image space. The most general transformation with these properties is:

$$x_i = \frac{a_1 x_o + b_1 y_o + c_1 z_o + d_1}{A x_o + B y_o + C z_o + 1}, \qquad y_i = \frac{a_2 x_o + b_2 y_o + c_2 z_o + d_2}{A x_o + B y_o + C z_o + 1},$$

$$z_i = \frac{a_3 x_o + b_3 y_o + c_3 z_o + d_3}{A x_o + B y_o + C z_o + 1},$$

in which all denominators are the same but the numerators are different. The final constant in the denominators has been set equal to 1 because otherwise we have too many constants.

(1) Choose to define axes z_o, z_i in the two spaces so that the transformation will have axial symmetry about these axes. Then $x_o = y_o = 0$ implies $x_i = y_i = 0$. Show that $c_1 = c_2 = 0$, $d_1 = d_2 = 0$.

(2) A 180° rotation carrying x_o into $-x_o$ and y_o into $-y_o$ should carry x_i into $-x_i$ and y_i to $-y_i$ with no change of z_o or z_i. Show that $A = B = 0$ and $a_3 = b_3 = 0$. Discuss what condition we have imposed on the directions of the axes in the two spaces.

(3) Now that I've started you off, think of arguments that make $b_2 = a_1$ and $a_2 = -b_1$ and permit us to choose $a_2 = 0$.

(4) Choose to locate the origins[36] of both z_o and z_i so that $z_o = 0$ implies $z_i = 0$ and $x_i/x_o = y_i/y_o = 1$. Show that $a_1 = 1$ and $d_3 = 0$.

(5) Define[37] focal lengths f_1, f_3 so that $z_o = -\infty$ implies $z_i = f_3$ and $z_o = -f_1$ implies $z_i = +\infty$. Show that $C = 1/f_1$ and $c_3 = f_3/f_1$.

(6) Show that

$$\frac{f_3}{z_i} - \frac{f_1}{z_o} = 1, \qquad \text{and} \qquad \frac{x_i}{x_o} = \frac{y_i}{y_o} = \frac{f_1\, z_i}{f_3\, z_o}.$$

Compare these equations with eqn (1.33). It is demonstrated that any imaging system (not necessarily optical, it could be an electron microscope or some complicated holographic arrangement) must obey an object–image relationship that is identical in form to that for refraction at a spherical surface.

(7) Show that the longitudinal magnification $\delta z_i/\delta z_o$ is related to the square of the lateral magnification $\delta x_i/\delta x_o$.

[35] The reasoning in this problem is loosely based on Joos (1934). Joos derives a Newton-style formula that is equivalent to our result.

[36] It should be clear from the algebra that such values of z_o and z_i exist so no questions are being begged here. We are defining 'planes of unit magnification', otherwise known as *principal planes*. For a thick lens, the object and image distances and the associated focal lengths are all measured from the principal planes, see §1.9.4.

[37] The image-side focal length is now called f_3 for uniformity with problem 7.3.

14 Optical fibres

14.1 Introduction

The main application of optical fibres (now) is in optical communication. Telephone traffic between exchanges is almost all carried by optical fibre, and in many towns and cities[1] a fibre passes every house entrance.

When a signal is to be conveyed, it must be encoded in some way (**modulated**) onto light or some other wave. There is a range of frequencies, the **bandwidth** $\Delta\nu$, that is required for transmission of the signal; and that bandwidth is proportional to the rate at which information is to be transmitted. Conversely, the rate at which information may be carried along an optical fibre is determined by the range of frequencies that it is practical to transmit along the fibre, given the properties of the fibre and of the associated light sources and detectors.

Examples from more familiar contexts may fill out this idea. An amplitude-modulated radio broadcast requires a bandwidth $\Delta\nu \sim 9\,\text{kHz}$, centred on the **carrier frequency** ν_0 ($\nu_0 \sim 200\,\text{kHz}$ in the long waveband, $\nu_0 \sim 1\,\text{MHz}$ in the medium waveband). A frequency-modulated (FM) radio signal is somewhat more greedy for bandwidth and uses $\Delta\nu \sim 150\,\text{kHz}$ with a carrier frequency of order 100 MHz. A colour television[2] broadcast requires $\Delta\nu \sim 3\,\text{MHz}$ with a carrier frequency of order 700 MHz. These are examples where the raw signal is not digitized but is encoded onto the carrier in analogue form.

Digital signals usually require more bandwidth than analogue, though there are compensations in opportunities for reduced noise and the correction of errors. An example is a telephone conversation, which in analogue form contains audio frequencies up to 3 or 4 kHz. In digital form[3] the same conversation might be sampled 8000 times per second to 8-bit accuracy, giving a required bit-rate of $64\,\text{kbit}\,\text{s}^{-1}$. This in turn requires Fourier components up to at least 32 kHz if the individual binary digits are to be received cleanly.[4]

Telephone signals don't need particularly high quality, which is why we're satisfied with fitting signals into a $64\,\text{kbit}\,\text{s}^{-1}$ 'channel'. By contrast, a music CD uses a bit-rate of $4.3218\,\text{Mbit}\,\text{s}^{-1}$, for reasons set out in §16.3. But we usually don't transmit several of these signals together via a telephone link, so it doesn't matter that the CD uses a relatively high bit-rate.

If you need to transmit lots of information (a few thousand telephone conversations, say), you need a correspondingly high bandwidth.[5] It's easiest to achieve this if the carrier frequency $\nu_0 \gg$ the bandwidth $\Delta\nu$

[1] Oxford is such a city, having been 'fibred' in 1996–8. Additionally, the ethernet connections between the university buildings are made by fibre, only the last ~50 m to each computer being served by copper wire.

[2] The reader unfamiliar with these orders of magnitude, or unfamiliar with the concept of 'carrier wave plus sidebands', should consult a book on basic electronics.

[3] The reason for taking samples at a frequency as high as 8 kHz lies in the Nyquist sampling theorem, already encountered in problem 11.17. Our thumbnail calculation explains why a channel bandwidth of $64\,\text{kbit}\,\text{s}^{-1}$ was chosen as the industry standard. However, the encoding used in practice conforms to standards set by the Consultative Committee for International Telephone and Telegraph (CCITT). These standards permit digitizing in a more ambitious way, provided that subsequent 'compression' results in the signal fitting into a $64\,\text{kbit}\,\text{s}^{-1}$ channel.

[4] This is on the worst-case assumption that the signal may contain bit sequences of the form 01010101..., which has a period of two time-slots.

[5] Introductory descriptions often suggest that a high carrier frequency is required, when in fact the need is for high *bandwidth*.

required, so that $\Delta\nu$ is a small fraction of ν_0. Thus optical waves (carrier frequency $\sim 10^{15}$ Hz) are more promising for wideband communication than microwaves ($\nu_0 \sim 10^{10}$ Hz) or lower-frequency electromagnetic waves.[6]

[6]This does not stop microwaves from being very widely used, as may be seen where communication towers (including the Post Office Tower in London) send signals to each other using dish aerials that are focusing waves with a wavelength of the order of centimetres, fairly obviously. Nevertheless, optical fibre is increasingly superseding these links.

We have mentioned 64 kbit s^{-1} as the industry standard for telephone signals. Video signals need more, and a lot of attention is paid to using clever tricks for reducing the bit-rate without incurring too much degradation of the picture quality (MPEG standards,[7] for example). Even when all this has been done, you still need cables or fibres capable of high bit-rate transmission in order to carry the traffic. For transmission along high-frequency electrical (coaxial) cables, a typical European standard rate is 139.364 Mbit s^{-1}, equivalent to 1920 simultaneous telephone conversations.[8] This is to be compared with a bandwidth of order 2 GHz for a single-mode optical fibre. However, there are limitations to the performance of optical fibre, set by the underlying physics, which is one reason why this chapter does not end here.

[7]Motion Picture Experts Group. The MPEG standards are encoding schemes for 'compressing' a video signal into a reduced number of bits per second with minimized effects on picture quality.

[8]The reader alert enough to check these figures will find that 2177 telephone conversations ought to be accommodated. There is, however, an 'overhead' required for organizing the signals.

14.2 Fibre optics: basics

The beam of light travelling along a fibre is narrow, only a few wavelengths across. If such a beam were launched in free space, it would spread rapidly by diffraction. The fibre must be designed to have a focusing (or waveguiding) action that counters the tendency of the light to spread.

Waveguiding is achieved by making an optical fibre with a refractive index that is higher at the middle than at the outside. There are two extreme ways of achieving this:

- a step-index fibre, which has a **core** of uniform refractive index n_1 surrounded by a **cladding** of lower index n_2, and with a sudden transition between the two indices
- a graded-index fibre, which has a smooth variation of refractive index (often roughly quadratic with radius) over the core region, and again a lower-index cladding.[9]

[9]The reason why a cladding is still needed appears in problem 14.3.

With either construction, the variation of refractive index serves to confine the light. With a step-index fibre, we may think of the light as undergoing total internal reflection at the core–cladding boundary, so that it is continually reflected back into the core (problem 14.1). Alternatively, and more usefully, we may see the core and cladding as the inner and outer components of a dielectric waveguide along which a confined travelling wave propagates. A graded-index fibre has a more obvious focusing effect, a bit like a very thick converging lens.

Fibres divide into two types in another way that is more important when considering their applicability. Light waves may travel along the fibre in several possible *transverse modes*, somewhat similar to the transverse modes that are the possible eigenfunctions for a mirror cavity (Chapter 8). A fibre may be designed to support transmission of several transverse modes (**multimode** fibre), or it may be designed to support only one transverse mode (**single-mode** fibre).[10]

A multimode fibre often has a core diameter of 62.5 μm surrounded by cladding to a diameter of 125 μm. With 'light' of wavelength 850 nm the number of transverse modes is of order 1600 (see problem 14.1). Such a fibre is used for non-critical applications, and is step-index. A single-mode fibre usually has a graded-index core of diameter about 9 μm, surrounded by a lower-index cladding, again to a total diameter of 125 μm. In both cases, a complete 'cable' contains several fibres (if only to provide 'spares'), a protective plastic sheath round each fibre, and outside that various protective and strengthening layers making a complete structure that is a few millimetres thick, and is robust enough to be handled, pulled through ducts, lowered to the ocean floor ....[11]

[10] Longitudinal modes are not relevant now, as we shan't be interested in standing waves fitting integer numbers of half-wavelengths along the length of a fibre. I shall though often refer to 'transverse modes' as a reminder when discussing the possible distributions of $\boldsymbol{E}$-field over the cross section of the fibre.

[11] If a cable has to be more than ∼100 km long, it has to incorporate 'repeaters' at which the optical signal is amplified (directly or by demodulation, electronic amplification and remodulation). In such a case the cable must also carry electrical connections to power the repeaters.

14.3 Transverse modes

We can understand the existence of transverse modes by using a graded-index fibre as a model.[12] When the fibre has a large radius and a rather gentle variation of refractive index with radial coordinate, a number of transverse modes can propagate. Those modes that don't extend too far from the axis closely resemble the Gauss–Hermite modes already encountered in Chapter 8.

We shall assume that the relative permittivity of the fibre varies with radius r like[13]

$$\varepsilon_{\mathrm{r}} = \varepsilon(1 - \beta^2 r^2). \tag{14.1}$$

We further assume that a scalar wave equation can be used to describe a scalar amplitude V standing for E_x or E_y; the algebra is handled in problem 14.3. The scalar wave equation is

$$\nabla^2 V = \frac{\varepsilon(1-\beta^2 r^2)}{c^2}\frac{\partial^2 V}{\partial t^2} = \frac{1-\beta^2 r^2}{v^2}\frac{\partial^2 V}{\partial t^2}, \quad \text{where} \quad v^2 = \frac{c^2}{\varepsilon}. \tag{14.2}$$

When this equation is solved by separating the variables in x, y, z, the functions of x and y satisfy differential equations identical in form to those describing a quantum-mechanical harmonic oscillator. Solutions have the form

$$V \propto \exp\{-(x^2+y^2)/w^2\}\, \mathrm{H}_l(\sqrt{2}x/w)\, \mathrm{H}_m(\sqrt{2}y/w)\, \mathrm{e}^{\mathrm{i}(kz-\omega t)}, \tag{14.3}$$

where the H_l and H_m are Hermite polynomials. The spot size w is given by

$$w^2 = \frac{2v}{\beta\omega} = \frac{2c}{\beta\omega\sqrt{\varepsilon}}, \tag{14.4}$$

and the propagation constant k is given by

$$k^2 = \frac{\omega^2}{v^2} - 2\frac{\beta\omega}{v}(l+m+1). \tag{14.5}$$

The expressions given here are very similar to those obtained in Chapter 8 for the field that propagates along a chain of lenses, or bounces

[12] This model is used here to make a physicist comfortable with the idea of transverse modes. To keep the mathematics tractable, the model has had to be given parameters that make it far from representative of real-world fibres, as we shall see.

[13] The symbol β should not be confused with the Fourier-transform variable introduced in §3.8.

between two mirrors. The similarity is not an accident, and it solves a mystery that has been with us since Chapter 8: why do the eigenfunctions for a lens chain resemble those for a quantum-mechanical harmonic oscillator? Until now the mathematics has shown no obvious resemblance. We may now see that a graded-index medium, having a quadratic variation of relative permittivity with radius r, is the 'continuum limit' of Chapter 8's lens chain.[14] Compare this with a similar situation in solid-state physics, where a uniform elastic string is the 'continuum limit'[15] of a balls-and-springs monatomic chain. We recover the continuum limit for the monatomic chain when the sound wavelength is much greater than the atomic spacing. In the present case, we expect to recover the graded-index limit when the lenses of a lens chain are weakly focusing and close together so that the light beam does not diffract much between one lens and the next.[16]

[14] The continuum limit is taken formally in problem 14.4.

[15] Here, a long-wavelength limit is being taken. We may contrast this with the transition between wave and ray optics, where a short-wavelength limit is taken.

[16] At the least, the discovery of this continuum limit would justify using Gaussian (or even Gauss–Hermite or Gauss–Laguerre) functions as trial solutions in problem 8.15.

Equation (14.5) shows that the propagation constant k is different for different modes. The reason for this lies in the variation of the wave amplitude with coordinates perpendicular to the fibre axis. Sidenote 39 on p. 164 shows why such variations change k when they are put into the wave equation. The dependence of $1/v_{\text{phase}} = k/\omega$ upon l, m results in intermodal dispersion, discussed below; the dependence of k/ω upon ω results in intramodal dispersion.[17]

[17] *Inter-* = 'between' or 'among'; *intra-* = 'within'.

Although expressions (14.3) and (8.1) are similar, eqn (14.3) does not contain the dependence upon complex radius q that accounts for the diffraction of the Gauss–Hermite beam; likewise, the phase factor $e^{-i(l+m)\alpha}$ is absent here. A physically equivalent phase shift is, however, present in the reduced k of eqn (14.5).

We quote also the result of solving eqn (14.2) by separating variables in cylindrical coordinates r, ϕ, z instead of in Cartesians x, y, z. A scalar field is sought in the form

$$V = R(r)\, e^{is\phi}\, e^{i(kz-\omega t)},$$

and the differential equation for $R(r)$ is

$$\frac{d^2R}{dr^2} + \frac{1}{r}\frac{dR}{dr} + \left(\frac{\omega^2}{v^2} - k^2 - \frac{\omega^2}{v^2}\beta^2 r^2 - \frac{s^2}{r^2}\right) R = 0. \tag{14.6}$$

The solution is mathematically less straightforward than that given earlier, and the result is expressed in terms of a Laguerre polynomial $L_n^{(s)}(2r^2/w^2)$:

$$V \propto \exp(-r^2/w^2)\, r^s\, L_n^{(s)}(2r^2/w^2)\, e^{\pm is\phi}\, e^{i(kz-\omega t)}, \tag{14.7}$$

where s is an integer describing the variation of the field around the fibre axis, and n is an integer[18] describing the variation of amplitude with distance r from the fibre axis. It will be no surprise that the V of eqn (14.7) closely resembles a similar expression obtained by solving the scalar wave equation in r, ϕ, z coordinates in the context of Gaussian beams: eqn (8.2) on p. 174. The propagation constant for the field

[18] Not to be confused with refractive index.

eigenfunction (eqn 14.7) is k where

$$k^2 = \frac{\omega^2}{v^2} - \frac{2\beta\omega}{v}(2n + s + 1). \tag{14.8}$$

We know that the (r, ϕ) solutions form a complete set, as do the (x, y) solutions, so each set must be expressible in terms of linear combinations of the other; comparison of eqn (14.8) with eqn (14.5) shows that such linear combinations must have $(2n + s) = (l + m)$. This is in line with a similar property given in sidenote 10 on p. 174.

We can get an idea of the appearance of the fibre modes by looking first at Fig. 8.3. There are nodal surfaces, l of them in the y-direction coming from the functions of x, and m in the other direction. This behaviour is in line with our experience with other eigenfunctions, such as hydrogen-atom wave functions, where there are nodal surfaces in the r, θ, ϕ directions each associated with one of the atomic quantum numbers. The Gauss–Laguerre modes are of course more appropriate to the axial symmetry of a fibre. From the precedents we know what to expect:[19] nodal surfaces (cylinders) in the radial direction (n of them not counting a zero at $r = 0$ that happens when $s \neq 0$) and in the ϕ direction (planes, s of them if we linearly combine $e^{is\phi}$ with $e^{-is\phi}$). These features are general properties of eigenfunctions, and so must carry over to all realistic cases; they are not artefacts of our graded-index model.

[19] You are asked to make sketches of these modes in problem 14.5.

The graded-index model is discussed here mainly because of the connection that it makes with the lens chain of Fig. 8.4(b). Unfortunately, the model has rather scant connection with the optical fibres of technology. A single-mode fibre, made with a graded-index core, has fields that extend outside the core into the cladding where a different mathematical description is needed; in addition, the dimensions are so small compared with an optical wavelength that a scalar-wave treatment is invalid. The field eigenfunctions for a real graded-index fibre can of course be calculated from Maxwell's equations, but the mathematics is ugly.

The case of a step-index fibre may seem cleaner (though unconnected with our graded-index model). The (now multimode) fields can be worked out in full from Maxwell's equations; the mathematics contains Bessel functions. A number of books give detailed treatments of the mathematics, e.g. Yariv (1997) and Davis (1996), and give diagrams (agreeing with the description above) showing the shapes of different field modes. We do not reproduce the rather heavy calculations here.

In a step-index fibre the electromagnetic wave is confined by the refractive-index difference between the core (n_1) and cladding (n_2), the structure forming a dielectric waveguide. The wave propagates like $e^{i(kz-\omega t)}$ with $\omega n_2/c < k < \omega n_1/c$. This means that the electric and magnetic fields decay, roughly exponentially, with distance into the cladding, while having an oscillatory variation with radius inside the core.[20] Fields in the cladding decay on a distance scale of a few optical wavelengths, so nothing reaches the outer surface of the cladding (and therefore no energy leaks out); this is what it means that the wave is 'waveguided' by the refractive-index difference.

[20] Compare with the behaviour of the wave function of a quantum particle confined within a finite-depth square potential well. We are used to sketching wave functions for such a particle, making them bend towards the axis inside the well and away from the axis outside. The fields in the fibre behave similarly.

The lowest mode (and in a single-mode fibre the only mode), known in the trade as the HE_{11} mode, is one in which the $\boldsymbol{E}$-field runs across the fibre, having a profile that is *roughly* like a Gaussian 00 mode in Chapter 8.[21] Attempts at feeding light energy into a fibre consist in presenting the end of the fibre with a field distribution approximating the lowest-mode's shape. Higher modes are then excited in addition, to the extent that our prepared field distribution does not exactly match that of the fibre's chosen mode.[22]

We mention just one other special case. There exist modes in which the electric field has its lines of $\boldsymbol{E}$ forming circles around the fibre axis. Such a mode has $\boldsymbol{E}$ entirely transverse to the direction of wave propagation, so in the jargon of §7.10 it is a transverse-electric (TE) mode. The $\boldsymbol{B}$-field lines thread the circles, so $\boldsymbol{B}$ has a longitudinal component (as well as radial); the mode is not transverse-magnetic. In a similar way, there are modes in which the magnetic field has lines of $\boldsymbol{B}$ running round circles, and these are TM modes.[23] There are no modes that are exactly TEM.[24] The TE and TM modes are not of any special practical importance, and they are mentioned here to justify statements made in sidenote 29 on p. 159.

In the case of a step-index fibre with refractive indices n_1 (core) and n_2 (cladding) and with core radius a, it is conventional to define a 'normalized frequency' $\mathcal{V}$ by

$$\text{normalized frequency } \mathcal{V} \equiv \frac{2\pi a}{\lambda_{\text{vac}}} \sqrt{n_1^2 - n_2^2}. \tag{14.9}$$

The number of transverse modes that a fibre can support is dependent upon $\mathcal{V}$. The lowest (HE_{11}) mode[25] propagates whatever the value of $\mathcal{V}$. If $\mathcal{V} < 2.405$ (the first zero of Bessel function $J_0(x)$) only the HE_{11} mode propagates. If $\mathcal{V}$ is increased through 2.405 a TE mode joins the game, and at larger values yet more modes propagate. Therefore the condition for a fibre to be single-mode is $\mathcal{V} < 2.405$. Other (though qualitatively similar) conditions hold for fibres whose construction is other than step-index.[26]

[21] When the relative permittivity is not constant, whether graded or stepped, lines of $\boldsymbol{E}$ are not continuous but lines of $\boldsymbol{D}$ are. If lines of $\boldsymbol{D}$ are drawn, they resemble those sketched in Fig. 7.7. That is, the $\boldsymbol{E}$- and $\boldsymbol{D}$-fields for the HE_{11} mode are not wholly transverse to the fibre axis, but have longitudinal components as well.

[22] We can imagine making an eigenfunction expansion of the imposed field in terms of the eigenfunctions for propagation (i.e. the modes) of the fibre.

[23] The existence of these TE and TM modes is general, and does not rely on the fibre having any particular refractive-index profile.

[24] In a 'weakly guided' approximation, where refractive indices n_1 (core) and n_2 (cladding) are only slightly different, the electromagnetic fields are almost transverse and we then have what are called 'linearly polarized modes'. However we emphasize that the name 'linearly polarized' does not mean that the field is exactly TEM.

[25] Strictly two modes, as there are two polarizations.

[26] In the case of a multimode fibre supporting many modes (strictly *multi-*), the number of transverse modes is related in a simple way to the normalized frequency. See problem 14.1(2).

14.4 Dispersion

The optical signals transmitted along a fibre are always digital, in the form of zeros (no light) and ones (light present); the light travels as a sequence of pulses. The rate at which signals can be transmitted depends on how short the pulses can be made. But a pulse lengthens as it travels because of **dispersion**. If the pulses received have lengthened to the point of overlapping, the signals they carry will be undecipherable.

The effect of dispersion gets worse as a fibre is made longer. Therefore, the bit-rate that a fibre can transmit is inversely proportional to its length. A given fibre might then be rated with a **length–bandwidth product** of (say) $500\,\text{Mbit s}^{-1}\,\text{km}$, meaning that a 1 km length can carry $500\,\text{Mbit s}^{-1}$, but a 2 km length must be downrated to $250\,\text{Mbit s}^{-1}$, and so on.

14.4.1 Material dispersion

There are three reasons for dispersion. Perhaps the easiest to understand is **material dispersion**, that arising from the frequency dependence of the fibre's refractive index.[27] Different Fourier components of the signal travel at different speeds and smear out the pulses. The variation of group velocity with frequency is investigated in problem 14.6. It is shown there that the variation of group velocity with frequency is proportional to $\mathrm{d}^2n/\mathrm{d}\lambda^2$, where λ is the vacuum wavelength and n is refractive index.

It happens that silica (SiO_2) has $\mathrm{d}^2n/\mathrm{d}\lambda^2 = 0$ at a wavelength of $\sim 1.3\,\mu\mathrm{m}$. Material dispersion can be near-eliminated if we operate at this wavelength. This is the reason why many optical-fibre systems operate at $1.3\,\mu\mathrm{m}$: surprisingly (perhaps), it is easier to design a laser to give out a frequency suited to the fibre than to design a fibre material suited to a given light source.

[27] The core and cladding both consist of SiO_2, and have different refractive indices only because one or both has been slightly 'doped' with impurities. We can, therefore, treat them as having a common refractive index for present purposes.

14.4.2 Intermodal and intramodal dispersion

Intermodal dispersion arises because different transverse modes[28] travel with different values of the propagation constant k, even at the same optical frequency $\omega/2\pi$. This phenomenon is illustrated by the model fibre discussed in §14.3 and may be seen in eqn (14.5) or eqn (14.8): k depends upon the mode numbers l and m (or n and s). A pulse that is coupled into the fibre in such a way as to excite several transverse modes will change shape and lengthen as it travels because those modes get out of step with each other. Intermodal dispersion is of course a problem only with multimode fibres in which several modes are excited at the same time.

[28] The word 'dispersion' is here being extended beyond the meaning it was given in §1.7, when it referred to frequency dependence only.

In a multimode fibre, intermodal dispersion can be minimized by arranging that the fibre has only a small difference between core (n_1) and cladding (n_2) refractive indices, since k is 'sandwiched' between $\omega n_1/c$ and $\omega n_2/c$. The fibre must then have quite a large (relative to λ) core diameter in order to support the required number of transverse modes, as is shown by eqn (14.12). Given a fibre with these proportions, there is little to be gained by giving the core a graded refractive index (with the aim of exercising some control over the variation of k). We can see why a multimode fibre is normally step-index.[29]

[29] So the model of §14.3 does not apply. Although eqn (14.5) does not hold, the step-index expression for k has a somewhat similar dependence upon geometry and mode number.

Intramodal dispersion is dispersion (in the frequency-dependence sense of the word) that exists even within a single transverse mode. We may again use eqn (14.5) to illustrate. Since both terms on the right depend on ω, but in different ways, the group velocity $v_\mathrm{g} = \mathrm{d}\omega/\mathrm{d}k$ depends upon ω. Different Fourier components of a pulse travel with different group velocities; the effect is similar to material dispersion, though the physical cause is different. Intramodal dispersion is most serious with a single-mode fibre, since a small core radius is required for suppressing all modes except one, and this enhances the geometry-dependence of k.

Suppose a fibre carries signals at a bit-rate of 2 GHz; then a bit sequence like 0101010 has a fundamental Fourier component at 1 GHz.

Allowing for the individual bits to be reasonably square pulses, we might expect frequency components to be present in a range of perhaps 2–5 GHz. An even larger frequency range may be present if the laser's frequency shifts as it is switched on and off. Both intramodal and material dispersion become more serious as the frequency range is increased. Ingenious measures can be taken to reduce the problem: for example, we can choose the optical frequency so that intramodal and material dispersion cancel against each other. Another possibility is to 'chirp' the radiation source (Yariv 1997), so that the fastest Fourier components of a pulse are emitted later than the slowest, the timing being such that they catch up at the far end of the fibre.[30]

[30] A further idea is to use non-linearity in the $\boldsymbol{D}(\boldsymbol{E})$ relation of the glass to work against the frequency dependence. Pulses that are kept compact in this way are called solitons. Solitons have attracted attention for a long time as an ingenious way of controlling pulse lengthening, but they have been slow to reach the practical-engineering stage. The reason is that the non-linearity gets less effective as a pulse loses energy. With a long fibre, the energy of a pulse may fall by a factor 10^{-3} over the length of the fibre, thereby making the non-linearity ineffective over much of the fibre's length; yet it is with long fibres that the soliton mechanism is most needed.

14.5 Multimode fibres

Multimode fibres are the usual choice where signals need to be transmitted over short distances only, and where the bandwidth required is not large. This choice is made for cheapness: not only is the fibre cheaper than single-mode fibre, but the large number of transverse modes permits the use of an LED (operating at 850 nm or 1.3 μm) rather than a laser (usually at 1.3 μm) as the radiation source, which saves more cost. As mentioned above, the fibre is quite thick, typically with core radius 62.5 μm; partly this is to reduce intermodal dispersion, partly it is to give a large enough étendue to permit use of an LED source. Multimode operation exacts a penalty in performance, because of intermodal dispersion: the length–bandwidth product is often about 10 MHz km, compared with several GHz km for a single-mode fibre.

Multimode fibres are commonly used for applications where a few computers are linked to each other or to a main digital highway. For example, in university buildings a multimode fibre may distribute information to computers on one floor, or to one section of the building, while the entire building is linked to others by single-mode fibre. A fibre has the signals taken off and converted to electrical voltages at a 'hub', from which the signals are either re-encoded for transmission along another fibre or continue on in electrical form. The final few tens of metres from a 'hub' to a wall socket are usually provided by twisted-pair wires.

14.6 Single-mode fibres

When we need high bandwidth, we use a single-mode fibre in order to avoid intermodal dispersion. At time of writing, most long-distance traffic on optical fibres is carried by single-mode fibre at a wavelength of 1.3 μm, a frequency chosen because it minimizes material dispersion.

The practical length that an optical fibre can have depends upon how much energy loss can be tolerated. A rough guide to the loss of energy[31] in a practical fibre is 0.4 dB km^{-1} for a wavelength of 1.3 μm, or 0.2 dB km^{-1} for a wavelength of 1.55 μm. If we can afford a loss of energy by a factor 10^{-3} from one end of a fibre to the other, these figures

[31] These figures represent what is routinely achieved. They, nevertheless, represent something of a triumph of material preparation; ordinary domestic glass is noticeably absorbing in a thickness of a few centimetres.

give fibre lengths of 75 and 150 km.

If it is necessary to transmit signals through longer distances than these, we have to use 'repeaters'. A repeater detects the optical signal, converting it back into electrical voltages. It amplifies these voltages and re-modulates the signal onto a laser for transmission along the next section of fibre. This method of power restoration is inelegant—but it works. A more elegant technology is an all-optical repeater, implemented by using as amplifier an optically pumped section of erbium-doped fibre. This may cause 1.55 μm (the wavelength amplified by the erbium) to become the standard choice of wavelength in due course. In either case, the repeater or repeaters are built into the fibre-optic cable and receive electrical power via wires that are all part of the same cable structure; then the cable can be buried or cross an ocean without need for the repeaters to be accessible.

For fuller information on fibre-optic communication, the reader is referred to more specialized works, such as Senior (1992) or Gowar (1993).

Problems

Problem 14.1 (a) A step-index fibre

Figure 14.1 shows the path of a 'meridional' ray (in a plane through the fibre's axis) as it enters a step-index fibre. On a ray-optics picture, total internal reflection takes place at the interface between core and cladding provided the angle of incidence θ_i there exceeds the critical angle.

(1) Show that the greatest angle of incidence α from air is given by

$$n_{\mathrm{air}} \sin \alpha_{\max} = \sqrt{n_1^2 - n_2^2}. \tag{14.10}$$

We can identify $n_{\mathrm{air}} \sin \alpha_{\max}$ with the *numerical aperture* of the fibre, as defined in eqn (11.5).

(2) Let the fibre's core have radius a, and assume that light can be accepted into the fibre if it arrives within the area of the core. Use eqn (11.4) to show that the fibre has étendue[32]

[32] The calculation is rough, because the core area doesn't properly represent the area occupied by the field distribution. It is also rough because angle $\alpha_{\max}$ has been calculated only for the case of rays in a plane containing the fibre's axis.

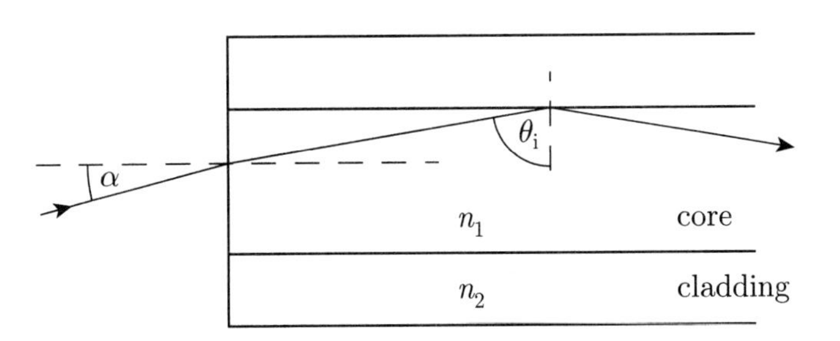

Fig. 14.1 A multimode fibre is shown from a ray-optics point of view. The inner region is the *core* (index n_1), and is surrounded by a lower-index (n_2) *cladding*. A light ray enters the core and is reflected repeatedly at the core–cladding interface. The reflection is total provided that θ_i exceeds the critical angle. The consequences for the fibre's numerical aperture, its étendue and the number of transverse modes it can support are explored in problem 14.1.

$$\text{étendue} = (\pi a^2)\,\pi(n_1^2 - n_2^2). \tag{14.11}$$

Show from eqn (11.9) that the number of transverse modes that can be launched into the fibre is

$$\text{number of transverse modes} = 2\,\frac{\pi a^2}{\lambda_{\text{vac}}^2}\,\pi(n_1^2 - n_2^2) = \frac{\mathcal{V}^2}{2}, \tag{14.12}$$

where $\mathcal{V}$ is the 'normalized frequency' defined in eqn (14.9) and a factor 2 has been inserted to allow that two polarizations may be used.[33]

(3) Calculate angle $\alpha_{\max}$ for a fibre with $n_1 = 1.53$, $n_2 = 1.51$.

(4) Find the number of transverse modes that can propagate at 850 nm in a step-index fibre whose core diameter is 62.5 μm and whose indices are 1.53, 1.51. Check against the number quoted in problem 11.8(4)(c).

(5) The condition for a step-index fibre to support only a single transverse mode is given in §14.3 as $\mathcal{V} < 2.405$. Show that then

$$2\,\frac{\pi a^2}{\lambda_{\text{vac}}^2}\,\pi(n_1^2 - n_2^2) < 2.892. \tag{14.13}$$

This is in order-of-magnitude agreement with what we should expect from eqn (14.12), which would give us the value 2.

[33] Notice the power of the method used here: it has not been necessary to pursue any complicated mathematics describing the modes within the fibre itself. It suffices to know that the number of transverse modes in the fibre must be the same as the number in air because of the conservation of mode number implied by eqns (11.9) and (11.7). As a caveat, we should note that the calculation here can be trusted only when the number of modes works out to be large—just as we trust the density of states in statistical mechanics only when the real-space volume is large.

Problem 14.2 (b) Rays in a graded-index optical fibre
Consider a fibre with a graded refractive index $n(r) = n_0(1 - \beta^2 r^2/2)$, where $r^2 = x^2 + y^2$, and $\beta r \ll 1$ for all r of interest.

(1) Show that the equation giving the path of a ray in the yz plane (so the ray is non-skew—'meridional') is

$$\mathrm{d}^2y/\mathrm{d}z^2 = -\beta^2 y,$$

provided the ray makes a small angle $\theta = \mathrm{d}y/\mathrm{d}z$ to the z-axis.[34]

(2) Show that (non-skew) rays follow sinusoidal paths along the fibre with wavelength $2\pi/\beta$.

(3) Hence show that the matrix equation, after the fashion of Chapter 7, that transforms a ray from one z to another is[35]

$$\begin{pmatrix} y \\ \theta \end{pmatrix}_{z=-l} = \begin{pmatrix} \cos\beta l & -(1/\beta)\sin\beta l \\ \beta\sin\beta l & \cos\beta l \end{pmatrix} \begin{pmatrix} y \\ \theta \end{pmatrix}_{z=0}.$$

Comment: The equations here give no indication that rays travelling within the fibre occupy any particular range of radii. For this we need a 'diffraction' picture. Compare with problem 8.6, in which a ray treatment gave the 'stability' condition for an optical cavity but no indication of the spot size for light in a 'stable' mode.

[34] *Hint*: Think about optical paths from one piece of wavefront to another.

[35] The column matrices here contain θ rather than $n\theta$. You can work with $n_0\theta$ but it doesn't simplify anything. Do not be tempted to change the lower element to $n(r)\theta = n_0(1 - \beta^2 r^2/2)\theta$. The correction (if it could ever make sense to include it) is of an order rejected early on in getting the differential equation for y.

Problem 14.3 (b) Gauss–Hermite modes in a model graded-index fibre
Consider a graded-index optical fibre in which the relative permittivity

$\varepsilon_r = \varepsilon(1-\beta^2 r^2)$, where r is the radius of cylindrical coordinates.[36,37] In a 'weakly guided approximation' the $\boldsymbol{E}$ and $\boldsymbol{H}$ fields can be considered as mostly transverse to the axis of the fibre. We can use a scalar approximation in which V stands for a suitable component of $\boldsymbol{E}$, and V obeys eqn (14.2):

$$\nabla^2 V = \frac{\varepsilon_r}{c^2}\frac{\partial^2 V}{\partial t^2} = \frac{1-\beta^2 r^2}{v^2}\frac{\partial^2 V}{\partial t^2}, \qquad \text{where} \qquad v^2 = \frac{c^2}{\varepsilon}.$$

(1) Look for a separated solution of eqn (14.2) in coordinates x, y, z:

$$V = X(x)\,Y(y)\,e^{i(kz-\omega t)}. \tag{14.14}$$

Show that X and Y obey the following equations:

$$\frac{d^2X}{dx^2} - \frac{\omega^2\beta^2}{v^2}x^2\,X = -AX, \qquad \frac{d^2Y}{dy^2} - \frac{\omega^2\beta^2}{v^2}y^2\,Y = -BY, \tag{14.15}$$

and that the propagation constant k is given by

$$k^2 = \frac{\omega^2}{v^2} - A - B.$$

(2) Show that these equations have the same form as the Schrödinger equation for a simple harmonic oscillator. Show that

$$A = (2l+1)\beta\omega/v, \qquad B = (2m+1)\beta\omega/v,$$

where l and m are integers, and so

$$k^2 = \frac{\omega^2}{v^2} - \frac{2\beta\omega}{v}(l+m+1) = \frac{\omega^2}{v^2}\left(1 - \frac{2\beta v}{\omega}(l+m+1)\right). \tag{14.16}$$

(3) Equation (14.16) shows that the different transverse modes have different values of k and, therefore, of phase velocity: there is intermodal dispersion. Interpret the second term on the right of eqn (14.16) as giving a reduction of k from ω/v for reasons paralleling those of sidenote 39 on p. 164: a beam whose amplitude is not uniform across its width necessarily has altered phase velocity.[38]

(4) Because eqns (14.15) resemble those for a quantum-mechanical harmonic oscillator, we can obtain the mode shape as a Gaussian multiplied by a Hermite polynomial (of order l in x and of order m in y) as in eqn (14.3). Show that the Gaussian part has the form $\exp(-r^2/w^2)$, where

$$w^2 = \frac{2v}{\beta\omega}. \tag{14.17}$$

(5) Show, using a criterion similar to that customary with the harmonic oscillator,[39] that the beam spreads over a range of r about equal to $r_{\max} = w\sqrt{l+m+1}$. Show that

$$\beta^2 r_{\max}^2 = \frac{2\beta v}{\omega}(l+m+1). \tag{14.18}$$

Note that the condition for this to be small is the same as requiring that in eqn (14.16) the second term on the right is small compared with the first.

[36] This expression contains the same β as that for problem 14.2 in the limit where $\beta r \ll 1$. It just happens that the present problem is tidier if we start from a simple expression for relative permittivity, rather than a simple expression for refractive index.

[37] Of course, this expression for relative permittivity can apply only for a range of r within which $\beta r \ll 1$. Outside this range there must be a cladding whose ε_r is constant. The calculations of this problem are restricted to modes whose fields remain within the range $\beta r \ll 1$.

[38] One might also 'explain' that modes with higher $l+m$ have increased phase velocity because they spread to larger values of r and so travel partly through material of lower refractive index. Since both effects act to reduce k it is hard to find a way to distinguish them. However we know of cases where k is reduced even when there is no variation of refractive index ....

[39] Figure 8.2 shows that the 'quantum-mechanical' estimate of the beam width is very sensible when l and/or m becomes large. Similar reasoning is encountered in problem 11.4(3).

(6) Show that in eqn (14.16) the ratio of the second term on the right to the first is $-(l+m+1)\lambda^2_{\text{medium}}/(\pi^2 w^2)$. Therefore, for our treatment to be valid $w \gg \lambda_{\text{med}}$: the beam inside the fibre should be many wavelengths across. Show that this in turn requires $\beta\lambda_{\text{med}} \ll 1$.

(7) The dispersion relation (14.16) has a shape that's familiar in the physics of guided waves: a k^2 that falls with increase of the transverse-mode eigenvalues.[40] If we had metal-walled waveguide, it would be appropriate to use the condition $k^2 > 0$ to set an upper limit on $(l+m+1)$. Argue that this cannot be done here, because modes with the greatest permitted values of $(l+m+1)$ would extend out to radii where ε_{r} falls to zero, and the fibre's characteristics will have departed from our graded-index model long before then.

[40] Look up the dispersion relation for waves propagating along a metal waveguide with rectangular cross-section.

(8) A multimode fibre can be made with a graded index (though it usually isn't). Deduce from part (7) that for such a fibre only the modes with small l, m can be described by our mathematics.[41] In particular, we cannot use our equations to obtain the number of propagating transverse modes. Deduce also that a single-mode fibre can be given a graded-index core but it must also have a lower-index cladding, and the electromagnetic fields must extend into the cladding.[42]

[41] In the context of optical fibres, this tells us that the present problem's model is severely limited. However, the main purpose of presenting the model is to make the connection with the 'continuum limit' of a lens chain: problem 14.4.

[42] This problem started with our writing an approximated form for the scalar wave equation. If the wave amplitude varies in the transverse direction on a distance scale of a wavelength, we have no business using this approximated wave equation: we need a full vector-field Maxwell-equations treatment. We see again that the mathematics of this problem has nothing to say to us about a single-mode fibre, or about the condition for a fibre to support just one transverse mode.

Problem 14.4 (b) Graded-index fibre as the limit of a lens chain
Consider light propagating through a lens chain where identical lenses of focal length f are separated by $2L$ in a medium of refractive index n_1. It is intended to proceed to the continuum limit where the lenses are equivalent to a continuous medium with graded refractive index.

(1) Show that the optical path through one period of the lens chain, taken on a straight path parallel to the axis but distant r from it, is

$$n_1\, 2L + (n_{\text{g}} - n_1)t - \frac{n_1 r^2}{2f}.$$

Here t is the thickness of a lens at its axis and n_{g} is the refractive index of the glass.

(2) Show from the r^2 dependence in the above optical path that in the limit the lens chain is equivalent to a fibre with refractive index $n(1-\frac{1}{2}\beta^2 r^2)$, in which

$$n = n_1 + (n_{\text{g}} - n_1)\frac{t}{2L} \qquad \text{and} \qquad \beta^2 = \frac{1}{f(2L)}\frac{n_1}{n}.$$

(3) Argue that in the continuum limit we can take $t=0$, $n=n_1$, and $\beta^2 f(2L) = 1$. Argue that the Gauss–Hermite eigenfunctions of problem 14.3 must apply also to the lens chain in the limit being taken.

(4) Problem 8.14 shows that a hemispherical cavity of length L supports a Gaussian beam whose spot size on the spherical mirror is w given by eqn (8.13). Argue that the same w will apply to a symmetrical cavity of length $2L$ and to a lens chain with spacing $2L$. So w is given by

$$w^2 = \frac{R\lambda}{\pi}\sqrt{\frac{L}{R-L}} = \frac{2f\lambda}{\pi}\sqrt{\frac{L}{2f-L}},$$

where $f = R/2$ is the focal length of each mirror of radius R and of each lens.

(5) Argue that $(2f - L)$ becomes $2f$ when we take the continuum limit so that

$$w^2 = \frac{2c}{n_1\omega}\sqrt{2Lf}.$$

(6) Put all this together and show that $w^2 = 2v/(\beta\omega)$, in agreement with eqn (14.17).

(7) [Harder] Investigate whether the diminished value of k^2 given by eqn (14.16) can be linked to the changed phase velocity incorporated in eqn (8.1)[43] in the factor $\mathrm{e}^{-\mathrm{i}(l+m+1)\alpha}$.

[43]Remember that expression (8.1) contains $\mathrm{e}^{-\mathrm{i}\alpha}$ from $1/q$ as well as the phase factor explicitly on display.

Problem 14.5 (b) Gauss–Laguerre modes
Use the description of Gauss–Laguerre modes in §14.3 to make sketches of a few of the lowest such modes, allowing both n and s to be non-zero as well as zero.[44] Check against pictures in books.

[44]*Hint*: Concentrate first on locating the nodal lines or surfaces, then fill in the bright parts in between.

Problem 14.6 (a) Material dispersion
Light travels along a fibre whose mean refractive index is n. We have

$$\text{group velocity} = \mathrm{d}\omega/\mathrm{d}k, \qquad \omega/k = c/n.$$

(1) Use these relations to show—without approximation—that[45]

$$\frac{1}{v_{\mathrm{g}}} = \frac{n}{c} + \frac{\omega}{c}\frac{\mathrm{d}n}{\mathrm{d}\omega}.$$

[45]Everyone remembers that the group velocity is $\mathrm{d}\omega/\mathrm{d}k$, so there is a temptation to think that we must differentiate with respect to k. But k is $2\pi/\lambda_{\text{medium}}$, and λ_{medium} depends upon frequency both directly and via the refractive index. The only sensible quantity to use as independent variable is frequency (or something closely related to it such as angular frequency), so I've suggested that you turn things upside down and differentiate k with respect to ω.

(2) It's common for people to use wavelength as a measure of frequency: we often say that sodium yellow light has wavelength 590 nm, rather than saying that it has frequency 5.08×10^{14} Hz. If we use wavelength for this purpose, we are giving the wavelength in vacuum (or just possibly in 'standard air'), certainly not that in the glassy medium, for the reason in sidenote 45 (this page). Show that a range of frequencies $\delta\nu = -(c/\lambda^2)\delta\lambda$ gives rise to a range of group velocities given by

$$\delta(1/v_{\mathrm{g}}) = \left(\frac{\lambda^3}{c^2}\frac{\mathrm{d}^2 n}{\mathrm{d}\lambda^2}\right)\delta\nu. \tag{14.19}$$

Equation (14.19) shows that material dispersion can be brought close to zero if we are able to choose an operating frequency at which the fibre has $\mathrm{d}^2n/\mathrm{d}\lambda^2 = 0$.

(3) Why is it the change of *group* velocity that is responsible (within material dispersion) for causing pulses to lengthen?

(4) In §14.4.2 it is stated that intermodal dispersion is caused by differences between the *phase* velocities of the transverse modes. Why is it differences of phase velocity that are harmful in this case?

15 Polarization

15.1 Introduction

In this chapter we discuss a few selected topics only. Polarization of light, including linear, circular and elliptical polarization, is usually described in elementary courses on optics, and the descriptions need not be given again here. It is assumed that the reader has also encountered **double refraction**, at least in uniaxial crystals, and seen it accounted for with Huygens wave surfaces. We, therefore, concentrate on describing the relative-permittivity tensor ε_{ij}, and seeing how it links to the elementary descriptions of double refraction.

We take the opportunity to describe just what physics can and cannot be described by means of ε_{ij}, because incorrect statements are very often found in print.

15.2 Anisotropic media

For the most part, the anisotropic media we shall deal with will be crystalline,[1] and exhibiting the phenomenon of double refraction.

When light undergoes double refraction, it splits into two waves that propagate in different ways, either with different speeds, or in different directions, or both. The two waves (in a uniaxial crystal) are called the **ordinary wave** (refracted according to Snell's law and so behaving in an 'ordinary' way) and the **extraordinary wave** (refracted differently and so outside-ordinary in its behaviour).[2]

The waves undergoing double refraction are linearly polarized in perpendicular directions. To understand the field directions, and to calculate the detailed behaviour of the ordinary and extraordinary waves, we need the relative permittivity expressed in the form of a matrix or tensor. The geometrical properties of extraordinary-wave propagation are investigated in problems 15.1 and 15.2, where it is shown that the customary Huygens-wave constructions yield correct results.

15.3 The mathematics of anisotropy

Consider a crystalline material and define within it in some way a set of Cartesian coordinates. Fields $\boldsymbol{E}$ and $\boldsymbol{D}$ can exist within the crystal, but will be related in a more complicated way than is the case for isotropic bodies.[3] Relative to these coordinates, the most general linear relation[4]

[1] Solutions of sugars in water are an exception, discussed in §15.6 and in problem 15.11.

[2] It is usual to pronounce 'extraordinary' as if it were written as two words or as 'extra-ordinary', with the first 'a' voiced, even though such a pronunciation would in other contexts be thought 'extrordinary'.

[3] There is no need to write a similarly complicated relation linking $\boldsymbol{B}$ with $\boldsymbol{H}$, because materials do not have any induced magnetization $\boldsymbol{M}$ at optical frequencies—see problem 1.4.

[4] The relation between $\boldsymbol{D}$ and $\boldsymbol{E}$ will be taken to be linear, as usual. This means we restrict attention to 'sufficiently weak' fields. There exists a whole range of phenomena described by **non-linear optics**, within which the relation between $\boldsymbol{D}$ and $\boldsymbol{E}$ is not exactly linear. However, non-linear optics is too large a topic to be dealt with adequately in the present book.

between the fields has the form

$$\left.\begin{aligned} D_1/\varepsilon_0 &= \varepsilon_{11}\,E_1 + \varepsilon_{12}\,E_2 + \varepsilon_{13}\,E_3,\\ D_2/\varepsilon_0 &= \varepsilon_{21}\,E_1 + \varepsilon_{22}\,E_2 + \varepsilon_{23}\,E_3,\\ D_3/\varepsilon_0 &= \varepsilon_{31}\,E_1 + \varepsilon_{32}\,E_2 + \varepsilon_{33}\,E_3. \end{aligned}\right\} \tag{15.1}$$

Here we write $E_1 = E_x$, $E_2 = E_y$, $E_3 = E_z$ and similarly in order to have numerical subscripts suitable for use in the equations that follow.[5] When a condensed notation is more convenient, we shall write eqns (15.1) in the 'suffix' form:

$$\frac{D_i}{\varepsilon_0} = \sum_{j=1}^{3} \varepsilon_{ij}\,E_j \tag{15.2}$$

or, using the 'summation convention' in which repeated subscripts are taken to be summed over without explicitly writing the summation sign,

$$D_i/\varepsilon_0 = \varepsilon_{ij}\,E_j. \tag{15.3}$$

It should be obvious that the three equations (15.1) can be written in a matrix form. The nine quantities ε_{ij} may then be presented as the elements of a 3×3 square matrix. However, they may also fruitfully be thought of as the elements[6] of a second-rank tensor ε_{ij}.

For the present, we shall assume that the matrix ε_{ij} is real and symmetric. Such a matrix describes a material that is doubly refracting in the usual way and is non-absorbing.[7] It is well known that a real symmetric matrix can be put into diagonal form by a suitable rotation to a new set of axes.[8] For the remainder of this chapter it will be assumed that such a choice of axes has been made. Referred to these **principal axes**, eqns (15.1) reduce to[9]

$$D_x = \varepsilon_0\,\varepsilon_{11}\,E_x, \qquad D_y = \varepsilon_0\,\varepsilon_{22}\,E_y, \qquad D_z = \varepsilon_0\,\varepsilon_{33}\,E_z. \tag{15.4}$$

Crystals are classified, so far as their optical behaviour is concerned, according to the following:

isotropic: $\varepsilon_{11} = \varepsilon_{22} = \varepsilon_{33}$,

uniaxial: $\varepsilon_{11} = \varepsilon_{22}$, ε_{33} is different,

biaxial: $\varepsilon_{11} \neq \varepsilon_{22} \neq \varepsilon_{33} \neq \varepsilon_{11}$.

In the case of uniaxial crystals, the z-axis is always chosen so that ε_{33} is made the 'different' element.

The process of finding principal axes should be familiar to the reader because of similar manoeuvres in quantum mechanics. A quick revision course is given in problem 15.3.

[5] When writing explicit expressions we shall frequently revert to writing E_x and the like where this seems more down-to-earth.

[6] The tensor has 'second rank' because its elements have two subscripts. Almost no mathematical properties of tensors will be used in the present chapter. The reader unfamiliar or uncomfortable with tensors may be reassured that the name 'second-rank tensor' can be replaced by '3×3 matrix' with no loss of insight.

[7] Section 15.4 discusses what other behaviours are possible; the meaning of 'in the usual way' will appear there.

[8] Later in the chapter it will be important whether ε_{ij} is or is not Hermitian. We, therefore, mention that a Hermitian matrix can also be put into diagonal form, but doing so requires a complex rotation which is not appropriate here.

[9] The coefficients like ε_{11} here are not the same quantities as those with the same names in eqns (15.1). No confusion should arise, as eqns (15.1) will not be used again.

15.4 The understanding of tensor ε_{ij}

In this section we give a detailed account of the physical phenomena that are, and are not, accounted for by various parts of the relative-permittivity tensor ε_{ij}. This is a topic about which remarkably diverse statements may be found in print, and opportunity is being taken here to set down what the present author believes to be correct.

We illustrate matters by writing a model tensor $\boldsymbol{\varepsilon}$ in the following form, designed to exhibit all possible behaviours for a wave travelling in the z-direction:

$$\begin{pmatrix} \varepsilon_{11} & \varepsilon_{12} & 0 \\ \varepsilon_{21} & \varepsilon_{22} & 0 \\ 0 & 0 & \varepsilon_{33} \end{pmatrix} = \begin{pmatrix} \alpha_{11} & 0 & 0 \\ 0 & \alpha_{22} & 0 \\ 0 & 0 & \alpha_{33} \end{pmatrix} + \begin{pmatrix} \mathrm{i}\beta_{11} & \mathrm{i}\beta_{12} & 0 \\ \mathrm{i}\beta_{12} & \mathrm{i}\beta_{22} & 0 \\ 0 & 0 & 0 \end{pmatrix} + \begin{pmatrix} 0 & -\mathrm{i}\gamma_{21} & 0 \\ \mathrm{i}\gamma_{21} & 0 & 0 \\ 0 & 0 & 0 \end{pmatrix} + \begin{pmatrix} 0 & \delta_{12} & 0 \\ -\delta_{12} & 0 & 0 \\ 0 & 0 & 0 \end{pmatrix}$$
$$\boldsymbol{\varepsilon} = \boldsymbol{\alpha} + \boldsymbol{\beta} + \boldsymbol{\gamma} + \boldsymbol{\delta}. \tag{15.5}$$

Here all subscripted Greek symbols α_{ij} to δ_{ij} are real, with i explicitly indicating where quantities are imaginary. The problems at the end of this chapter justify the following statements:[10]

(1) Double refraction is accounted for by a real symmetric matrix $\boldsymbol{\alpha}$. We assume that the matrix has been put into diagonal form, if necessary by imposing a rotation of axes (problem 15.3).
(2) Absorption of optical energy is absent if the matrix $\boldsymbol{\varepsilon}$ is Hermitian (problem 15.4). Thus, in particular, a real symmetric matrix like $\boldsymbol{\alpha}$ describes a non-absorbing material.
(3) Symmetric imaginary terms, as in $\boldsymbol{\beta}$, describe absorption (problem 15.7). As with $\boldsymbol{\alpha}$, this matrix can be diagonalized, and we shall make the simplifying assumption that a single rotation of axes diagonalizes both $\boldsymbol{\alpha}$ and $\boldsymbol{\beta}$.
(4) Matrix $\boldsymbol{\gamma}$ has antisymmetric imaginary elements off the diagonal. Combined with $\boldsymbol{\alpha}$ these elements contribute to a Hermitian matrix, so do not give rise to absorption. They give different phase velocities to different hands of circularly polarized light (problem 15.8).[11] An experimental consequence of such a difference is seen if plane polarized light is input, when its plane of polarization rotates on passage through the material. Such a rotation[12] can be a manifestation of Faraday rotation induced within a ferromagnetic (or otherwise magnetically ordered) material by its own internal $\boldsymbol{B}$-field.[13]
(5) Finally, tensor $\boldsymbol{\delta}$ relates to $\boldsymbol{\gamma}$ in the same way that $\boldsymbol{\beta}$ relates to $\boldsymbol{\alpha}$: it gives the absorption accompanying elliptical-wave dispersion (problem 15.9).

[10]In making this list we do not ask 'what do terms of such-a-kind mean?', because it might be that physics has no use for them. We ask 'do these terms turn out to be useful for describing a known behaviour?'

[11]Or, in the most general case, different hands of elliptical polarization.

[12]Problems 15.8, 15.10 and 15.11 exclude the more obvious possibilities of optical activity and an externally induced Faraday rotation.

[13]The reader may suspect that he is being made to contemplate a perversely obscure phenomenon. But exactly this, in thin magnetic films magnetized normal to the surface of the film, is the basis of modern high-capacity floppy discs. The sense of the internal magnetization (up or down for 0 or 1) is read optically from the rotation of the plane of polarization of light.

15.5 The Faraday effect

It is rather obvious that the Faraday effect is not accounted for within eqn (15.3), because there is a proportionality to the magnitude of an applied $\boldsymbol{B}$-field. We need to make a modification to eqn (15.3) along the lines of

$$D_i/\varepsilon_0 = \varepsilon_{ij}E_j + \eta_{ijk}B_jE_k, \tag{15.6}$$

where there is an implied summation over subscripts j and k. The B_j are the Cartesian components of the applied magnetic field.[14] In the case of the Faraday effect (rather than other magneto-optical effects),

[14]Because the right-hand side of eqn (15.6) contains a product of fields, we have here one of the phenomena included within non-linear optics. Since we do not give a systematic discussion of non-linear effects in this book, we are somewhat transgressing our usual terms of reference. However, the Faraday (and Voigt) effect is often discussed alongside the linear effects of this chapter, and we need to know about it in problem 15.8.

the **B**-field is applied in the direction of light travel (give or take a sign). Problem 15.10 investigates.

15.6 Optical activity

We give a brief account here of optical activity,[15] meaning a phenomenon in which plane-polarized light has its plane of polarization rotated as it travels through a medium. Media that are optically active include some crystals, but also some liquids such as solutions of sugars.

[15]The ideas presented here are based on Robinson (1973*b*).

The signature of optical activity is this: if the light's travel is reversed, it 'unscrews'. In this it is distinguished from the Faraday effect, where the rotation is doubled.

Optical activity is *not* accounted for within the ε tensor introduced in §15.4. Problems 15.5–15.9 investigate the possible behaviours that ε can generate, and none resembles optical activity. The γ component of ε comes closest, but reversal of the wave travel gives a doubled rotation, not an 'unscrewing'.

To account for optical activity we have to add another new term to eqn (15.3):

$$\frac{D_i}{\varepsilon_0} = \varepsilon_{ij}E_j + \zeta_{ijk}\frac{\partial E_k}{\partial x_j}, \tag{15.7}$$

with an implied summation over subscripts j and k. The electric displacement $\boldsymbol{D}$ depends upon a *gradient* of the electric field.

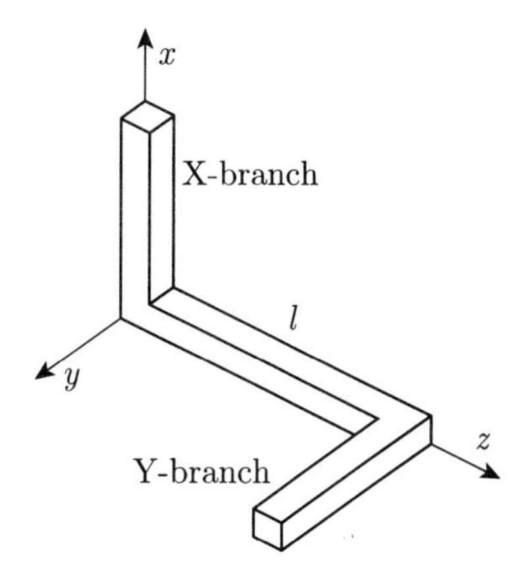

Fig. 15.1 A model for a molecule with a screw structure, showing about a quarter turn of the screw. When an $\boldsymbol{E}$-field is applied in the y-direction, it causes electrons to run along the molecule, and so induces an electric dipole moment in the x-direction. But the dipole $p_x(z)$ at z is caused by the value of $E_y(z+l)$ at $z+l$.

We may understand how it comes about that a gradient of $\boldsymbol{E}$-field is involved with the aid of Fig. 15.1. Suppose a light beam travels in the z-direction and encounters a molecule with a screw structure such as that shown in Fig. 15.1. An electric dipole moment $p_x(z)$ in the x-direction can be caused in either of two ways: by a field $E_x(z)$ in the x-direction at z, directly causing electrons to move along the X-branch of the molecule; or by field $E_y(z+l)$ in the y-direction at $z+l$, causing electrons to move along the Y-branch, and from there along the molecule into the X-branch. Field $E_y(z+l)$ is of course $E_y(z)+(\partial E_y/\partial z)l$, hence the presence of a derivative in eqn (15.7). The consequences are explored in problem 15.11.

A sugar solution is optically active and yet is also isotropic. The isotropic equivalent of eqn (15.7) is[16]

$$\boldsymbol{D}/\varepsilon_0 = \varepsilon_r\boldsymbol{E} + \zeta\,\mathrm{curl}\,\boldsymbol{E} \qquad \text{with} \qquad \boldsymbol{B}/\mu_0 = \boldsymbol{H}. \tag{15.8}$$

[16]It seems, however, to be more conventional to use a symmetrical pair of constitutive relations in the form

$$\boldsymbol{D}/\varepsilon_0 = \varepsilon_r\boldsymbol{E} + \zeta\,\mathrm{curl}\,\boldsymbol{E},$$
$$\boldsymbol{B}/\mu_0 = \mu_r\boldsymbol{H} + \zeta\,\mathrm{curl}\,\boldsymbol{H}.$$

An account, with references, may be found in Lakhtakia (1989).

Problems

Problem 15.1 (a) Field directions for o- and e-rays in a uniaxial crystal
(1) For parts (1)–(3), take the $\boldsymbol{E}$-field to lie in the xy plane (the optic axis being the z-axis). Show that it is correct, though unorthodox, to write $\boldsymbol{D} = \varepsilon_0\,\varepsilon_{11}\,\boldsymbol{E}$. Show that, when this is substituted into Maxwell's

equations, the result is a wave propagating with speed

$$\text{ordinary-wave speed} = \frac{c}{n_{\rm o}} = \frac{c}{\sqrt{\varepsilon_{11}}}. \tag{15.9}$$

(2) Show that, if the medium is uncharged, $\operatorname{div} \boldsymbol{D} = 0$. Show that the wave travels (phase velocity) in a direction $\boldsymbol{k}$ lying in a plane perpendicular to $\boldsymbol{E}$. Show that this does not exclude any direction for $\boldsymbol{k}$.

(3) Use parts (1) and (2) to show that the wave has a spherical Huygens wave surface; it must represent the ordinary wave. For this reason, $n_{\rm o} \equiv \sqrt{\varepsilon_{11}}$ is called the ordinary (or ordinary-wave) refractive index.

(4) Find double-refraction diagrams in books and confirm that the ordinary wave always has its $\boldsymbol{E}$-field perpendicular to the optic axis.

(5) Next consider a wave whose electric field $\boldsymbol{E}$ lies entirely in the z-direction. For this wave, $D_z = \varepsilon_0\, \varepsilon_{33}\, E_z$, which may be written in unconventional form as $\boldsymbol{D} = \varepsilon_0\, \varepsilon_{33}\, \boldsymbol{E}$. Substitute this into Maxwell's equations and show that it results in a wave with speed

$$\text{(special-case) extraordinary-wave speed} = \frac{c}{n_{\rm e}} = \frac{c}{\sqrt{\varepsilon_{33}}}. \tag{15.10}$$

(6) Use $\operatorname{div} \boldsymbol{D} = 0$ to show that the wave of part (5) must have its wave vector $\boldsymbol{k}$ in the xy plane. This is an example (but only an example this time) of an extraordinary wave. The quantity $n_{\rm e} \equiv \sqrt{\varepsilon_{33}}$ is the 'extraordinary refractive index'; but you have to be aware that it is only in very special cases that you can use $n_{\rm e}$ in a Snell-type refraction.

(7) For which orientation of crystal surface and incoming $\boldsymbol{k}$ *can* you use Snell's law to describe refraction of the extraordinary wave?

Problem 15.2 (b) The extraordinary wave in a uniaxial crystal
Consider an extraordinary wave travelling with its $\boldsymbol{k}$-vector making angle θ with the z-axis (the optic axis) of a uniaxial crystal. The electric and magnetic field components may be taken to have dependence

$$E_x, E_z \propto \exp\{\mathrm{i}(kz\cos\theta + kx\sin\theta - \omega t)\}.$$

(1) Show that the usual elimination of $\boldsymbol{H}$ between Maxwell's equations, to get a wave equation for $\boldsymbol{E}$, does not work now that $\boldsymbol{D} \neq \varepsilon_{\rm r}\varepsilon_0\,\boldsymbol{E}$.

(2) Substitute the given electric field into each Maxwell equation in turn. Show that it is a consistent assumption that $E_y = 0$. (E_y can exist, but it belongs to the ordinary wave.)

(3) Use $\operatorname{div} \boldsymbol{D} = 0$ to show that $\boldsymbol{D}$ is perpendicular to $\boldsymbol{k}$.

(4) Show that

$$\frac{E_z}{E_x} = -\left(\frac{\varepsilon_{11}}{\varepsilon_{33}}\right)\tan\theta. \tag{15.11}$$

(5) Show that $B_x = B_z = 0$, while

$$-\sin\theta\frac{B_y}{E_z} = \frac{k}{\omega}\left(\sin^2\theta + \frac{\varepsilon_{33}}{\varepsilon_{11}}\cos^2\theta\right) = \mu_0\,\varepsilon_0\,\varepsilon_{33}\frac{\omega}{k}. \tag{15.12}$$

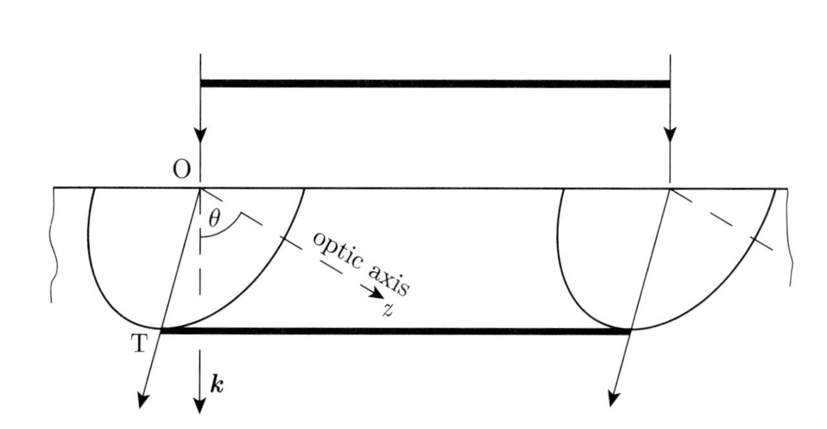

Fig. 15.2 A plane wave is incident normally on the surface of a uniaxial crystal. Double refraction may be accounted for by the Huygens-wave construction shown. An ellipsoid gives the shape of a secondary wave radiating outwards from O, and similar ellipsoids originate from all other points within the illuminated area of the surface. A transmitted wavefront is formed by drawing the surface that is the common tangent to these ellipsoids. These wavefronts are surfaces of constant phase, so the $\boldsymbol{k}$-vector and the phase velocity of the e-wave are perpendicular to those surfaces. The phase velocity is represented by the distance from O to the wavefront; it is calculated in problem 15.2.

(6) Hence show that

$$\frac{\omega^2}{c^2 k^2} = \frac{\cos^2\theta}{\varepsilon_{11}} + \frac{\sin^2\theta}{\varepsilon_{33}}. \tag{15.13}$$

(7) Show that eqn (15.13) reduces to $\omega/k = c/\sqrt{\varepsilon_{33}}$ for the case of a wave travelling in the xy plane (and with its $\boldsymbol{E}$-field in the z-direction), and to $\omega/k = c/\sqrt{\varepsilon_{11}}$ for a wave travelling along the z-axis. Check for consistency against problem 15.1.

(8) It is sometimes convenient to define a refractive index $n(\theta)$ to describe the phase speed $c/n(\theta)$ of an extraordinary wave whose *phase* velocity makes angle θ to the z-axis. Show that[17]

$$\frac{1}{n(\theta)^2} = \frac{\cos^2\theta}{n_\mathrm{o}^2} + \frac{\sin^2\theta}{n_\mathrm{e}^2}. \tag{15.14}$$

[17]This equation is used to identify the phase-matching angle for second-harmonic generation within a uniaxial crystal such as KDP (potassium dihydrogen phosphate).

(9) In elementary treatments of double refraction, we find out how an extraordinary wave propagates by drawing Huygens-wave surfaces (ellipsoidal for the e-wave, spherical for the o-wave). We now show that these geometrical constructions yield correct results. Figure 15.2 shows a Huygens-construction diagram: a plane wave is incident normally from air onto a uniaxial crystal and is transmitted as plane waves with wavefronts parallel to the surface. Argue that the phase velocity and the $\boldsymbol{k}$-vector both face vertically down the page, at angle θ to the optic axis.

(10) Show that the special cases discussed in problem 15.1 require that the ellipsoidal surface shall have proportions

$$\frac{\text{axis perpendicular to optic axis}}{\text{axis in optic-axis direction}} = \frac{c/n_\mathrm{e}}{c/n_\mathrm{o}} = \frac{n_\mathrm{o}}{n_\mathrm{e}}.$$

(11) Assume that the surface drawn in Fig 15.2 is an ellipsoid (we shall verify this by showing that the assumption yields correct results). Write down the equation for the ellipse that is the section of the ellipsoid drawn in Fig. 15.2; give the ellipsoid the size that it grows to in time δt.

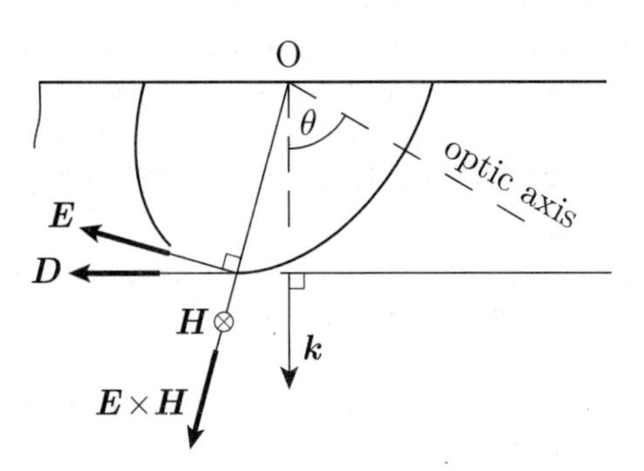

Fig. 15.3 The physical situation is the same as that in Fig. 15.2. The line down and left from O runs from the centre of the ellipsoid to the point of tangency where the ellipsoid is touched by a 'horizontal' plane (point T in Fig. 15.2, but label T is omitted here to avoid clutter). The $\boldsymbol{E}$-field is at right-angles to that line. Field $\boldsymbol{H}$ is perpendicular to the paper. The Poynting vector lies in the direction of OT: the direction of energy flow that is argued for from the Huygens method in elementary treatments.

(12) Find the coordinates of the point T where the tangent drawn meets the ellipse.

(13) Find the length of the line drawn from O perpendicular to the wavefront. Check that it agrees with the distance our wave travels in time δt at the known phase velocity.

(14) Show that the electric field $\boldsymbol{E}$ is perpendicular to the line OT.

(15) Show that the electric displacement $\boldsymbol{D}$ lies in the plane of the wavefront (part (3) of this problem).

(16) Find the direction of the Poynting vector.

(17) Confirm that the geometrical relations are summarized correctly in Fig. 15.3. Invent an argument based on interference to show independently that the energy flow is expected take place in direction OT.

(18) The phase velocity of the wave is in the direction of $\boldsymbol{k}$. What can be said about the group velocity?

Problem 15.3 (a) Diagonalization of real symmetric matrix $\boldsymbol{\alpha}$
In the run-up to eqns (15.4), it was stated that a rotation of axes can always be found that will bring a real-symmetric matrix $\boldsymbol{\alpha}$ to diagonal form. Here we rehearse the mathematics of this.

(1) In each of the three equations (15.4) taken separately, field $\boldsymbol{D}$ is parallel to field $\boldsymbol{E}$. Then, in the original axis system, we can identify the directions of the principal axes by requiring that $\boldsymbol{D}$ be parallel to $\boldsymbol{E}$, a requirement that can be met in three different ways. Argue that this requirement can be put into the form[18]

$$\begin{pmatrix} \alpha_{11} & \alpha_{12} & \alpha_{13} \\ \alpha_{21} & \alpha_{22} & \alpha_{23} \\ \alpha_{31} & \alpha_{32} & \alpha_{33} \end{pmatrix} \begin{pmatrix} E_1 \\ E_2 \\ E_3 \end{pmatrix} = \lambda \begin{pmatrix} E_1 \\ E_2 \\ E_3 \end{pmatrix}.$$

[18] The λ here is, of course, unconnected with optical wavelength.

(2) Show how this may be solved to give three eigenvalues for λ. Show that there must be three solutions for λ (not necessarily unequal).

(3) Show how three eigenvectors (E_1, E_2, E_3) can be constructed, one for each λ.

(4) For the above to make sense, the eigenvalues λ should be real. Show that for them to be real the matrix $\boldsymbol{\alpha}$ must be Hermitian.

(5) For the above to make sense, the eigenvectors must be real. Show that for them to be real the matrix $\boldsymbol{\alpha}$ must have all its elements real. Taking this with part (4), show that matrix $\boldsymbol{\alpha}$ must be real symmetric.

(6) Show that the eigenvectors belonging to different eigenvalues λ are orthogonal to each other, or can be chosen to be.[19]

(7) Show that a rotation matrix can be constructed from the eigenvectors, which carries coordinates from the old to the new axes.

[19] The phrase 'or can be chosen to be' is taken over from corresponding cases in quantum mechanics. In the event that two eigenvalues λ are equal, there are infinitely many ways of constructing their eigenvectors, and orthogonality is not guaranteed. However, it is open to us to make a choice from among the possible eigenvectors in such a way as to achieve orthogonality if we so wish—and we shall so wish.

Problem 15.4 (b) A non-absorbing medium has Hermitian ε_{ij}

(1) Show from Maxwell's equations that

$$-\boldsymbol{E}\cdot\boldsymbol{J} = \boldsymbol{E}\cdot\dot{\boldsymbol{D}} + \boldsymbol{H}\cdot\dot{\boldsymbol{B}} + \operatorname{div}(\boldsymbol{E}\times\boldsymbol{H}). \tag{15.15}$$

Notice that this equation is obtained directly from the Maxwell equations, with no use of any constitutive relations, so it must be correct in all circumstances.

(2) Consider some volume that does not contain sources of radiative energy, though energy may be absorbed within it. Energy arrives through the bounding surface[20] of the volume as $\boldsymbol{E}\times\boldsymbol{H}$; some may leave again through the bounding surface; what remains is either stored temporarily or is absorbed. Show that the rate at which energy enters the volume is

$$-\int(\boldsymbol{E}\times\boldsymbol{H})\cdot\mathrm{d}\boldsymbol{S} = \int(\boldsymbol{E}\cdot\dot{\boldsymbol{D}} + \boldsymbol{H}\cdot\dot{\boldsymbol{B}} + \boldsymbol{E}\cdot\boldsymbol{J})\,\mathrm{d}v, \tag{15.16}$$

where $\mathrm{d}\boldsymbol{S}$ is an element of surface area and $\mathrm{d}v$ is an element of volume.

[20] We are taking it as known already from standard textbook arguments that $\boldsymbol{E}\times\boldsymbol{H}$ has the interpretation of energy flow.

(3) In Chapter 1 we introduced the idea that the field quantities being used here are 'macroscopic' quantities, the result of averaging microscopic fields. In particular eqn (1.2b) shows a macroscopically averaged current density decomposed into an 'accessible' current, a polarization current and a magnetization current. This decomposition is not unique; it may be done in more than one way. In particular, 'trading' can take place between the conduction and displacement currents. When we deal with the absorption of electromagnetic waves by introducing a complex ε_{r}, we can consider that we are shifting the energy-dissipating currents from $\boldsymbol{J}_{\text{accessible}}$ into $\partial\boldsymbol{P}/\partial t$.

Argue that in eqn (15.16) we may omit the term $\boldsymbol{E}\cdot\boldsymbol{J}$ by transferring its physics into the term $\boldsymbol{E}\cdot\dot{\boldsymbol{D}}$. Within this convention, the rate of energy increase per unit volume (counting both energy stored in the fields and energy dissipated as heat) is given by $\boldsymbol{E}\cdot\dot{\boldsymbol{D}} + \boldsymbol{H}\cdot\dot{\boldsymbol{B}}$.

(4) Now apply a monochromatic wave whose field amplitudes all are written in terms of quantities[21] like $\boldsymbol{E} = \boldsymbol{E}_0\,\mathrm{e}^{\mathrm{i}(\boldsymbol{k}\cdot\boldsymbol{r}-\omega t)}$. As always with complex expressions, the actual field is given by the real part of this, so it will be $\frac{1}{2}(\boldsymbol{E}+\boldsymbol{E}^*)$ and similarly. In expressions such as $\boldsymbol{E}\cdot\dot{\boldsymbol{D}}$ it will be necessary of course to multiply real parts.[22] Show that the rate of energy increase per unit volume (the energy being either stored or

[21] The medium is taken to be linear, so the imposition of fields at angular frequency ω does not result in the generation of any other frequencies.

[22] There can be subtleties with energy when the medium supporting an electromagnetic wave is dispersive, as pointed out in problem 1.8. We avoid such difficulties here by taking the medium as non-dispersive.

dissipated) is

$$\tfrac{1}{2}(\boldsymbol{E}+\boldsymbol{E}^*)\cdot\tfrac{1}{2}(\dot{\boldsymbol{D}}+\dot{\boldsymbol{D}}^*)+\tfrac{1}{2}(\boldsymbol{H}+\boldsymbol{H}^*)\cdot\tfrac{1}{2}(\dot{\boldsymbol{B}}+\dot{\boldsymbol{B}}^*). \tag{15.17}$$

(5) Now average the rate of energy increase over a cycle of oscillation (or over a few cycles). At the end, the fields take the same values as at the beginning, so the energy they store is unchanged. Any non-zero term in the energy that has arrived must be accounted for as energy dissipated. Show that the time-average rate of dissipation of energy is

$$\tfrac{1}{4}(\boldsymbol{E}\cdot\dot{\boldsymbol{D}}^*+\boldsymbol{E}^*\cdot\dot{\boldsymbol{D}}+\boldsymbol{H}\cdot\dot{\boldsymbol{B}}^*+\boldsymbol{H}^*\cdot\dot{\boldsymbol{B}}).$$

(6) Argue that a non-magnetic material should have no magnetic contributions to the dispersion or absorption of electromagnetic waves at optical frequencies (remember problem 1.4). Then the rate of energy dissipation is

$$\begin{aligned}\text{rate of energy dissipation} &= \tfrac{1}{4}(\boldsymbol{E}\cdot\dot{\boldsymbol{D}}^*+\boldsymbol{E}^*\cdot\dot{\boldsymbol{D}})\\ &= \tfrac{1}{4}\mathrm{i}\,\omega(\boldsymbol{E}\cdot\boldsymbol{D}^*-\boldsymbol{E}^*\cdot\boldsymbol{D})\\ &= \tfrac{1}{4}\mathrm{i}\,\omega(E_i\,D_i^*-E_i^*\,D_i)\\ &= \tfrac{1}{4}\mathrm{i}\,\omega\,\varepsilon_0(E_i\,\varepsilon_{ij}^*\,E_j^*-E_i^*\,\varepsilon_{ij}\,E_j)\\ &= \tfrac{1}{4}\mathrm{i}\,\omega\,\varepsilon_0\sum_{i,j}E_i\,E_j^*(\varepsilon_{ij}^*-\varepsilon_{ji}).\end{aligned} \tag{15.18}$$

Here the summation signs for sums over indices i and j are omitted during the working but are reinstated in the final expression.

(7) Show from here that the wave undergoes no absorption if ε_{ij} is a Hermitian tensor/matrix. Conversely, if a medium absorbs, we must find the mathematical description of that absorption in non-Hermitian terms within ε_{ij} —not necessarily, as we might perhaps have expected, in imaginary contributions.

Problem 15.5 (b) Investigation of a model ε_{ij}

This is the first of a sequence of problems that investigate the behaviours introduced by components of ε_{ij} in eqn (15.5) that lie beyond real-symmetric. For this purpose let us consider a relative-permittivity tensor

$$\varepsilon_{ij}=\begin{pmatrix}\varepsilon_{11} & \varepsilon_{12} & 0\\ \varepsilon_{21} & \varepsilon_{22} & 0\\ 0 & 0 & \varepsilon_{33}\end{pmatrix}, \tag{15.19}$$

and we investigate its consequences for a plane wave travelling in the z-direction with field components proportional to $\mathrm{e}^{\mathrm{i}(kz-\omega t)}$. The matrix elements are in general complex. The constitutive relation between $\boldsymbol{B}$ and $\boldsymbol{H}$ is simply $\boldsymbol{B}=\mu_0\,\boldsymbol{H}$.

(1) Use $\operatorname{div}\boldsymbol{D}=0$ and $\operatorname{div}\boldsymbol{B}=0$ to show that field components D_z, E_z, B_z and H_z are all zero.

(2) Show that $\operatorname{div}\boldsymbol{E}=0$ and $\operatorname{div}\boldsymbol{H}=0$.

(3) The result of part (2) shows that the 'usual' elimination of $\boldsymbol{H}$ between Maxwell's equations can be performed. Show that the result is

$$\nabla^2 \boldsymbol{E} = \mu_0 \dot{\boldsymbol{J}} + \mu_0 \ddot{\boldsymbol{D}}.$$

Adopt the convention[23] introduced in problem 15.4(3) that conduction current $\boldsymbol{J}$ shall have its functions absorbed into $\boldsymbol{D}$. Show that

$$c^2 k^2 E_i = \omega^2 \varepsilon_{ij} E_j, \qquad (15.20)$$

where a summation over suffix j is implied.

[23] This convention was already assumed in writing eqn (15.19), and is part of the reason why the matrix elements may be complex.

(4) Write out eqn (15.20) in components and show that

$$\left(\varepsilon_{11} - \frac{c^2 k^2}{\omega^2}\right) E_x + \varepsilon_{12} E_y = 0, \qquad (15.21a)$$

$$\varepsilon_{21} E_x + \left(\varepsilon_{22} - \frac{c^2 k^2}{\omega^2}\right) E_y = 0. \qquad (15.21b)$$

(5) Show that these equations are consistent (meaning they are capable of having solutions other than the trivial $E_x = E_y = 0$) provided that

$$\left(\frac{ck}{\omega}\right)^4 - \left(\frac{ck}{\omega}\right)^2 (\varepsilon_{11} + \varepsilon_{22}) + (\varepsilon_{11}\varepsilon_{22} - \varepsilon_{12}\varepsilon_{21}) = 0. \qquad (15.22)$$

This dispersion equation, together with relations (15.21), can give rise to a variety of different behaviours, depending on what we assume about the tensor elements ε_{ij}.

Problem 15.6 (b) The Poynting vector and energy absorption
We continue to consider the physical situation envisaged in problem 15.5: a medium whose relative permittivity tensor is given by eqn (15.19), supporting a wave travelling in the z-direction.

Let the wave's field components all be proportional to $e^{i(kz-\omega t)}$.

(1) Show that

$$H_x = -\frac{k}{\omega\mu_0} E_y, \qquad H_y = \frac{k}{\omega\mu_0} E_x.$$

(2) The fields are to be represented by symbols $\boldsymbol{E}$, $\boldsymbol{H}$ representing complex quantities proportional to $e^{i(kz-\omega t)}$, with the usual convention that the real part is to be taken. When evaluating the Poynting vector we must multiply real expressions, so we shall write

$$\text{Poynting vector} = \tfrac{1}{2}(\boldsymbol{E} + \boldsymbol{E}^*) \times \tfrac{1}{2}(\boldsymbol{H} + \boldsymbol{H}^*).$$

Show that the Poynting vector lies in the z-direction and has time-average $\overline{N_z}$, where

$$\overline{N_z} = \frac{(k + k^*)}{4\mu_0\,\omega}\left(E_x E_x^* + E_y E_y^*\right). \qquad (15.23)$$

(3) Let the propagation constant k be written as a complex quantity

$$k = k' + \mathrm{i}k''.$$

Show from eqn (15.23) that

$$-\frac{\mathrm{d}}{\mathrm{d}z}\overline{N_z} = \frac{k'}{2\mu_0\,\omega}(E_x\,E_x^* + E_y\,E_y^*)(2k''). \qquad (15.24)$$

(4) Show that k' and k'' have the same sign if the medium absorbs.[24] Show that the absorption coefficient is $2k'' \times \mathrm{sign}(k')$.

(5) Problem 15.4 has shown that absorption is associated with non-Hermitian elements in the relative permittivity ε_{ij}. We check for consistency. Show from eqn (15.22) that

$$2\left(\frac{ck}{\omega}\right)^2 = (\varepsilon_{11} + \varepsilon_{22}) \pm \sqrt{(\varepsilon_{11} - \varepsilon_{22})^2 + 4\varepsilon_{12}\,\varepsilon_{21}}, \qquad (15.25)$$

and that this is real if ε_{ij} is Hermitian.

(6) Assume that the medium does not absorb. Equation (15.25) should have two solutions for k^2, both positive. Show from eqn (15.22) that this is so only if $(\varepsilon_{11}\,\varepsilon_{22} - \varepsilon_{12}\,\varepsilon_{21}) \geq 0$. Show that if this condition is not met, matrix ε_{ij} has a negative eigenvalue. Show that in such a case one of the two possible disturbances in the medium has evanescent form, rather than propagating as a travelling wave.[25] We shall not discuss further any medium with so extreme a behaviour.

[24] Both k' and k'' may be negative if the wave is travelling towards $-z$ and being absorbed. It is of course possible also for the wave to undergo exponential growth if the medium is arranged to amplify, as is the case within a laser. In such a case k' and k'' have opposite signs. The expression you are asked to derive covers all cases.

[25] Even then, a purely evanescent wave is one in which there is no absorption of energy, so this unusual behaviour is not in conflict with the property 'Hermitian means no absorption'.

Problem 15.7 (a) Complex diagonal elements of $\boldsymbol{\varepsilon}$: matrix $\boldsymbol{\beta}$

To isolate the effect of having imaginary components in the diagonal elements ε_{11} and ε_{22} of the relative-permittivity tensor, we take it that the diagonalization of $\boldsymbol{\alpha}$ also diagonalizes $\boldsymbol{\beta}$, so there are no off-diagonal elements β_{12}, β_{21}.

(1) Show from eqns (15.21) that field components E_x and E_y are now uncoupled: linearly polarized waves can propagate (in the z-direction) with the $\boldsymbol{E}$-field in either the x- or the y-direction. It will suffice to consider the properties of either one of these linearly polarized waves.

(2) Consider the wave with E_x as its non-zero $\boldsymbol{E}$-field component. Show that its propagation constant $k = k' + \mathrm{i}k''$ is given by

$$\varepsilon_{11} = \alpha_{11} + \mathrm{i}\beta_{11} = \frac{c^2k^2}{\omega^2} = \frac{c^2}{\omega^2}\{(k'^2 - k''^2) + 2\mathrm{i}k'k''\}. \qquad (15.26)$$

Argue that a positive imaginary part β_{11} to ε_{11} describes the behaviour of an absorbing (but otherwise unremarkable) medium.[26] Similarly for positive β_{22}.

[26] A negative β_{11} would describe a medium in which waves undergo exponential growth, such as the amplifying medium within a laser.

Problem 15.8 (b) Imaginary-antisymmetric elements of ε_{ij}: matrix $\boldsymbol{\gamma}$

We investigate the consequence of adding imaginary-antisymmetric matrix $\boldsymbol{\gamma}$ to real-symmetric $\boldsymbol{\alpha}$. The elements of $\boldsymbol{\gamma}$ are Hermitian so that ε_{ij} remains a Hermitian matrix/tensor. We know from problem 15.4 that waves propagating through a medium with such a tensor are not absorbed. We wish to find out what happens.

We rewrite eqns (15.21) as:

$$\left(\varepsilon_{11} - \frac{c^2k^2}{\omega^2}\right) E_x - \mathrm{i}\gamma_{21} E_y = 0 \qquad (15.27\mathrm{a})$$

$$\mathrm{i}\gamma_{21} E_x + \left(\varepsilon_{22} - \frac{c^2k^2}{\omega^2}\right) E_y = 0. \qquad (15.27\mathrm{b})$$

(1) Show that the consistency condition for eqns (15.27) is

$$\left(\frac{c^2k^2}{\omega^2} - \varepsilon_{11}\right)\left(\frac{c^2k^2}{\omega^2} - \varepsilon_{22}\right) - \gamma_{21}^2 = 0 \qquad (15.28\mathrm{a})$$

or, equivalently,

$$\left(\frac{ck}{\omega}\right)^4 - \left(\frac{ck}{\omega}\right)^2 (\varepsilon_{11} + \varepsilon_{22}) + (\varepsilon_{11}\,\varepsilon_{22} - \gamma_{21}^2) = 0. \qquad (15.28\mathrm{b})$$

(2) Show that there are two solutions for k^2, and that both are positive provided that $\varepsilon_{11}\varepsilon_{22} \geq \gamma_{21}^2$, which we shall assume to be the case.

(3) Let the two (real) solutions for k be written[27] as k_+ and k_- with $|k_+| > |k_-|$. Show that if $\varepsilon_{11} \geq \varepsilon_{22}$ the solutions satisfy

$$\frac{c^2k_-^2}{\omega^2} < \varepsilon_{22} \leq \varepsilon_{11} < \frac{c^2k_+^2}{\omega^2}.$$

(4) Show that the $\boldsymbol{E}$-field associated with k_+ has E_x and E_y differing in phase by 90°, and with E_y lagging E_x if $\gamma_{21} > 0$. Show that the $\boldsymbol{E}$-field traces out an ellipse at the optical frequency; that the ellipse has the x- and y-axes as its symmetry axes; and that it rotates in a positive[28] sense about Oz (from x towards y) if γ_{21} is positive. Likewise, show that the opposite property applies to the wave associated with k_-: its $\boldsymbol{E}$-field traces an ellipse in the negative sense about Oz if γ_{21} is positive.

(5) Show that the rotation senses obtained in part (4) are independent of the sign of k_+ or k_-: the $\boldsymbol{E}$-vector rotates in the same sense whether the wave travels in the positive or negative z-direction.

(6) Imagine that we input plane-polarized light travelling along the $+z$-direction. To understand how it propagates, we must decompose it into a sum of the normal-mode waves for the medium, which means making it into a linear combination of the elliptically polarized waves, one rotating positively, the other negatively. For simplicity, take the special case where the ellipses are circular. Show that the different speeds of the two waves have the consequence that the light emerges, still plane polarized, but with its plane of polarization rotated, negatively about Oz if $\gamma_{21} > 0$.

(7) Show that the rotation of the plane of polarization worked out in part (6) is *independent* of the sense of the wave's travel along the z-axis. If the wave is reversed in direction (e.g. by reflecting it from a mirror) it continues rotating in the same sense as before; it does not 'unscrew'.

[27] The point of writing $|k|$ is to allow for the possibility that the waves might travel in either sense along the z-direction, so k_+ and k_- are allowed to take either sign.

[28] The sense of rotation can be a pain in the context of circularly or elliptically polarized light. Opticians refer to light as left-circular if the $\boldsymbol{E}$-field rotates anticlockwise (with time for fixed z) as seen by an observer looking back towards the source. Radio engineers refer to radiation as having positive rotational sense if it rotates clockwise (with time for fixed z) when seen 'from behind'; the same convention as that applying to rotations in vector mathematics—clockwise when looking along the arrow. It follows that 'positive' and 'left-circular' describe the same polarization state.

Nevertheless, as shown in part (5), the sense of rotation in part (4) is neither of the above, because it is referred to the direction of the positive z axis, regardless of the direction of travel of the light.

Comment: The behaviour found in part (7) should be contrasted with that found in problems 15.11 (optical activity) and 15.10 (Faraday rotation). What we have here is clearly not optical activity, because the light does not 'unscrew' when its direction of travel is reversed. It has to be a Faraday rotation, with the $\boldsymbol{B}$-field somehow concealed in $\boldsymbol{\gamma}$. This is possible if the $\boldsymbol{B}$-field is not supplied externally but is an 'internal' field caused by ferromagnetism (or other magnetic ordering) within the optical medium itself.

Problem 15.9 (c) Real-antisymmetric off-diagonal elements of ε_{ij}: matrix $\boldsymbol{\delta}$

Analyse the effect of elements of ε_{ij} that are off-diagonal, real and antisymmetric.

Hint: Remember the Kramers–Kronig relations. It should be possible to argue that $\boldsymbol{\delta}$ describes the absorption of the elliptically polarized waves whose propagation is described by $\boldsymbol{\gamma}$.

Problem 15.10 (b) The Faraday effect

Let us investigate a simple model. Let $\boldsymbol{\varepsilon}$ have elements $\varepsilon_{11} = \varepsilon_{22} = \varepsilon$ while $\eta_{132} = -\eta_{231} = -\mathrm{i}\eta$. Show that

$$\left.\begin{aligned} D_x/\varepsilon_0 &= \varepsilon E_x - \mathrm{i}\eta B_z E_y, \\ D_y/\varepsilon_0 &= \varepsilon E_y + \mathrm{i}\eta B_z E_x. \end{aligned}\right\} \tag{15.29}$$

Follow a calculation similar to that of problems 15.5 and 15.8. Show that we obtain equations similar to eqns (15.27), except that γ_{21} is replaced by ηB_z. Show that the η terms cause radiation travelling in the z-direction to rotate its plane of polarization. Show that reversing the direction of $\boldsymbol{B}$ reverses the direction of rotation. Show that if light traverses the medium in one direction and is then reflected by a mirror to retrace its path, the rotation is doubled.

Problem 15.11 (b) Optical activity

Consider a simple model. Let tensor $\boldsymbol{\varepsilon}$ have elements $\varepsilon_{11} = \varepsilon_{22} = \varepsilon$ while tensor $\boldsymbol{\zeta}$ has $\zeta_{132} = -\zeta_{231} = -\zeta$ as its only interesting elements. Then eqn (15.7) has two interesting components for a wave travelling along the z-direction:

$$\left.\begin{aligned} \frac{D_x}{\varepsilon_0} &= \varepsilon E_x - \zeta\frac{\partial E_y}{\partial z} = \varepsilon E_x - \mathrm{i}k\zeta E_y, \\ \frac{D_y}{\varepsilon_0} &= \varepsilon E_y + \zeta\frac{\partial E_x}{\partial z} = \varepsilon E_y + \mathrm{i}k\zeta E_x. \end{aligned}\right\} \tag{15.30}$$

Follow algebra paralleling that of problems 15.5 and 15.8. Show that we obtain equations similar to eqns (15.27), except that γ_{21} is replaced by $k\zeta$. Show that the ζ terms cause radiation travelling in the z-direction to rotate its plane of polarization. Show also that the presence of k accompanying ζ causes the rotation to reverse sign when the wave travels towards $-z$, so that the plane of polarization 'unscrews'.

Two modern optical devices

16

16.1 Introduction

In this chapter we take the opportunity to give brief descriptions of two optical devices that seem to be under-represented in textbooks in relation to their importance. The CD player is familiar as a household item and as the CD-ROM reader in a computer. Such everyday equipment ought not to remain mysterious. The confocal microscope is a laboratory instrument, but one that has become the instrument of choice for a wide variety of tasks. Each of the two systems represents a sophisticated application of the principles described in earlier chapters.

16.2 Compact disc: description of the disc

The basic specification for a CD (compact disc) for storing music is given in the International Standard IEC 908 (1987), now called IEC 60908 (1999–02). There are small differences in the case of CD-ROM, and larger differences with DVD (digital versatile disc). We shall start by describing the original music CD.

Information is stored on the disc in the form of 'pits' that are described below. The pits are arranged along a single spiral track[1] running outwards from middle to edge of the disc.[2] The track starts at diameter 50 mm (preceded by a 4 mm lead-in area) and ends at at most 116 mm (plus 1 mm for 'lead-out'). The turns of the track are separated by 1.6 μm. The thickness of the disc is 1.2 mm (plus a small amount for a protective layer and label).

During the manufacturing process, a polycarbonate disc is made by injection moulding[3] to have the information impressed on one surface in the form of pits, a pit representing digit '1', and no-pit representing '0'. Each pit is 0.5 μm wide and about 100 nm deep. The information-bearing surface is aluminized to make it reflective, and the aluminium coating is covered by a protective layer and label. The completed disc is read by a laser beam *through* the thickness of the polycarbonate, so each pit presents itself to the light as a 'bump'. Ideally, all incident light is reflected whether a pit/bump is present or not, but the angular distribution of the reflected light is widened by the pit/bump, and it is this change of angular distribution that is detected by the optical system.

[1] This is in contrast to the tracks of a floppy disc or hard disc, which are concentric circles.

[2] This is in contrast to the spiral track of a black vinyl disc, which runs from the outside to the middle.

[3] There are differences in the case of writable and re-writable CDs, but all varieties must be readable using the equipment described here.

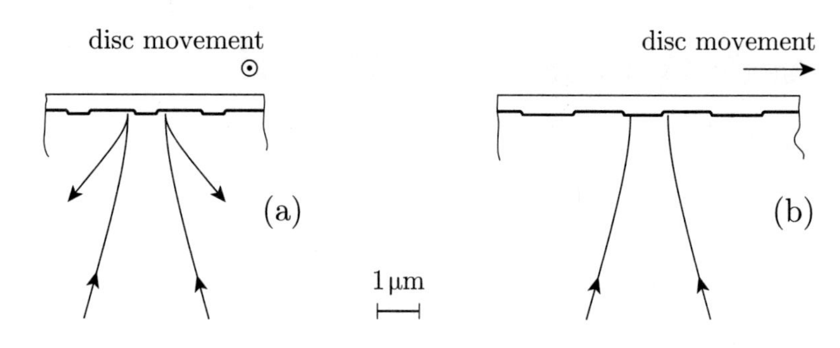

Fig. 16.1 Light from a reading laser is focused onto the reflective pits/bumps of a CD. (a) The pits/bumps have width 0.5 μm and are shown with the track movement out of the paper. Three turns of the track are shown. The light falls partly on the bumps and partly on the surrounding 'land', with the result that the reflected light is diffracted into a widened angular distribution. (b) The same situation is shown with the track movement from left to right. The pits/bumps are 0.9–3.3 μm long, so they appear larger in this section.

Figure 16.1 shows a laser beam illuminating the pits of one track.

16.3 Compact disc: the encoding scheme

While our interest here is mainly in the optical system for reading a CD, it is necessary to know a little about how the digital information is encoded before being written to the disc.

Each of two audio-frequency inputs (the left and right stereo channels) is 'sampled' at 44.1 kbit s^{-1} to 16-bit accuracy, giving a basic bit-rate of 1.4112 Mbit s^{-1}. Each 16-bit 'word' is divided into two 8-bit data bytes, and a group of 24 data bytes (192 bits) constitutes a 'frame'. The frame is treated as a unit, and is re-encoded before being written to the disc.

After re-encoding, each frame has been enlarged from 192 (data) bits to 588 (channel) bits,[4] giving a final (channel)-bit-rate of 4.3218 Mbit s^{-1}. The re-encoding achieves a number of functions, one of which is to permit error correction at the reading stage; other designed-in favourable characteristics will appear as we proceed.

[4]Here the original bits obtained by making a digital measurement of the music signal are called 'data bits'. The bits that are imprinted on the disc after re-encoding are called 'channel bits'.

Each 8-bit 'data' byte is first re-encoded as a word of 14 'channel' bits using a 'cross-interleave Reed–Solomon code' (CIRC). Between each pair of 14-channel-bit words are placed 3 'merge bits', so that the information within a frame appears as a sequence of 24×17 channel bits. The 8–14 encoding, and the merge bits, have the properties:

- at least 3, and at most 11, zeros occur in sequence, and similarly 3–11 ones occur in sequence (the encoding scheme makes this so within the 14-bit words, while the three merge bits are chosen to make it true at the joins too)[5]
- the average numbers of 0s and 1s are equal so that a.c.-coupled amplifiers can be used in the replaying electronics
- interchange of zeros and ones throughout makes no difference, so the replay circuitry does not have to preserve sign information when amplifying the digital signal.

[5]For the original data a worst-case bit sequence is 0101010, with a period of two time-slots, implying a frequency of 706 kHz. After re-encoding, the worst-case bit sequence is 000111000 with a period of six time-slots, implying a frequency of 720 kHz, hardly higher than before. This beautifully illustrates a theorem by Shannon: you can build as much error correction as you like into a long message, with negligible penalty in raised bandwidth.

The first of these properties is essential to the optical reading of the disc.

A complete frame of 588 channel bits is made up as follows:

- synchronization pattern of 24 bits (+ 3 merge bits)
- control information of 14 bits (encoded from 8) (+ 3 merge bits)
- 24 signal words of 14 bits (+ 3 merge bits after each)
- 8 error-correction words of 14 bits (+ 3 merge bits after each).[6]

Each 'channel' bit occupies a length 0.3 μm of track, and the track passes the reading optics[7] at a linear speed of $1.3\,\mathrm{m\,s^{-1}}$.

16.4 Optics of reading a compact disc

The laser used for reading a CD has wavelength 780 nm and is focused by a lens with numerical aperture 0.45. Problems 3.27 and 11.6 investigate the resolution of this optical system, and show that it is unaffected by the refractive index ($n = 1.55$) of the polycarbonate.

If light of wavelength 780 nm illuminates the focusing lens uniformly, the diffraction pattern at the focus forms an Airy disc whose FWHM (problems 3.15 and 11.9, solutions) is 0.9 μm. The length of track devoted to a (channel) bit is 0.3 μm, so a group of three ones, or a group of three zeros, is long enough to register its presence and affect the angular distribution of the scattered light. These proportions are indicated in section in Fig. 16.1(b) and in plan in Fig. 16.3. Refer to Fig. 16.1(a), which shows the reading laser beam in relation to the width of the pit/bumps. The beam illuminates a bump (encoding a '1') together with a roughly similar area of the surrounding 'land'. The bump may be thought of for now as having height $\lambda_{\text{medium}}/4 = 126\,\mu\mathrm{m}$. Then light reflected from the top of the bump is π out of phase with that reflected from the land, resulting in an interference cancellation in the light that is reflected straight back towards the source. Since all energy goes somewhere, the effect is to widen the angular distribution of the diffracted light (relative to reflection from land only as happens at a '0'). Much of the diffracted light falls outside the aperture of the lens through which it arrived, and this reduction of returned light is detected and interpreted as a '1'.

An outline of the reading optics is shown in Fig. 16.2. Light from a semiconductor laser is collimated, passed through a polarizing beam splitter, then focused onto the CD via a lens of numerical aperture 0.45. The light returned from the CD is reflected at the beam splitter into the detector which registers one of two light levels depending on the presence or absence of a pit/bump.

Real detection systems incorporate a refinement. Light from the laser is linearly polarized, and is wholly transmitted on its first encounter with the polarizing beam splitter. Next it is converted to circular polarization by a quarter-wave plate between beam splitter and focusing lens. The returned light passes a second time through the quarter-wave plate, after which it is again linearly polarized but with its plane of polarization rotated through 90°. The polarization properties of the beam splitter mean that all of this reflected light is reflected towards the detector, rather than returning towards the laser.[8]

[6] Some error correction is inherent in the 8–14-bit encoding scheme, but an algorithm makes use of the error-correction words to achieve an even lower after-correction error rate. A further error-reduction scheme is hard-wired into the system; it is the 'cross-interleave' part of the re-encoding scheme. In this, a frame of 24 8-bit data bytes has its bytes rearranged in sequence before the 8–14 re-encoding, and a complementary resequencing is applied on replay. If a continuous run of bits is unreadable, for example because of a scratch on the disc, the effect of the rearrangement is to distribute the errors thinly throughout the frame, thereby making it more likely that the error-correction system can reconstitute the original signal.

[7] Unlike the case of a vinyl-disc record player, which turns at a constant angular speed, the angular speed of a CD player is not constant but is adjusted to maintain the required linear speed, regardless of the radius at which the track is being read. The linear speed need not be very accurately constant, because information can be read off into an electronic buffer at a variable bit-rate, and then read out at a constant rate, so long as the buffer never becomes empty or full. A portable CD player uses an extra large buffer to allow for rotations of the entire apparatus affecting the reading rate.

[8] This arrangement makes efficient use of the available energy, of course, but its main function is to prevent reflected light from re-entering the laser and causing unwanted feedback effects. So far as reflection at the CD surface is concerned, there is no merit in having the light circularly polarized.

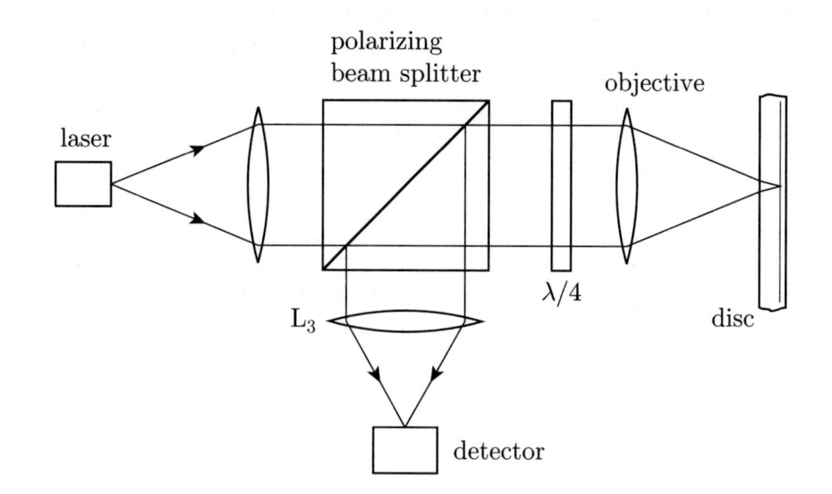

Fig. 16.2 An outline optical system for reading information from a CD. Light from the laser is focused by the objective lens onto the track inside the disc. Light reflected back from the track is captured by the objective lens and is passed via the polarizing beam splitter to the detector. When the light within the disc falls on a pit/bump, the angular distribution of the returned light is widened and the energy captured is reduced, signalling a '1'.

The laser output is linearly polarized. After two passages through the quarter-wave plate its plane of polarization is rotated through 90°, which makes it wholly take the reflected route within the polarizing beam splitter.

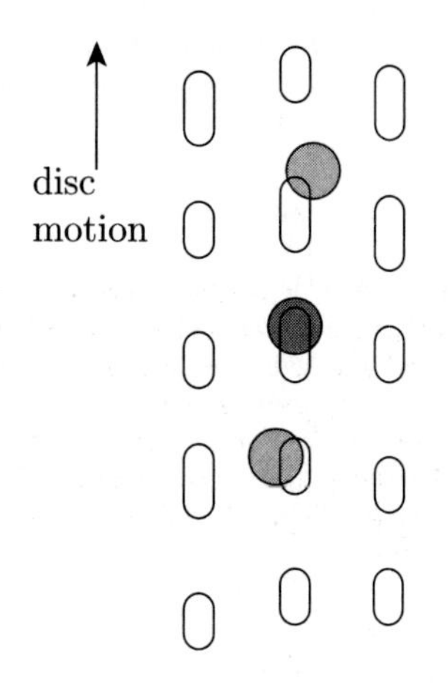

Fig. 16.3 The three-spot system for detecting and correcting tracking error. Three adjacent turns of the information track are shown, with pits/bumps and spaces that are at least 3 and at most 11 bits long. The circular area with dark shading is the most intense of the three laser spots being focused onto the track, and is the one used for reading information. The two paler spots are auxiliary spots that are the first orders of diffraction from a phase grating in the illuminating optics. If the reading light drifts away from the centre of the track, say to the left, the reflected lower auxiliary spot is less modulated by the pits/bumps than the upper, and this difference of modulation forms the basis of a feedback signal.

We have mentioned that the light reads the disc through the thickness (1.2 mm) of the disc. This means that the light beam has a diameter of 0.73 mm at the entrance surface, large enough that any scratches or other imperfections on that surface are well out of focus and can do little harm.

16.5 Feedback systems

So far, we have been concerned wholly with obtaining a detectable signal from the reflection of a light beam that is imagined to be positioned correctly over the track of pits/bumps. But the light beam will be positioned correctly only if we make it so: errors must be detected and corrected in an automatic, meaning negative-feedback, system.

Two errors must be dealt with: errors of **tracking**, where the light beam sits towards one side of the track rather than over its centre; and errors of **focus**.

16.5.1 Correction of tracking

To get an idea of the difficulty of tracking, we may note that the central hole of a CD has diameter $15\,^{+0.1}_{-0}$ mm, so if it is fitted over a 15 mm spindle the disc may be up to 50 μm off centre, which means that more than 30 different turns of the track will pass back and forth across a stationary detector during one rotation of the disc.

More than one system is used commercially for the detection of tracking error, and we shall unashamedly describe only one. Accounts of alternative methods may be found in Carriere *et al.* (2000). A phase grating is placed in the incident light path, with its rulings not quite perpendicular to the CD track, so that the ± 1 orders of diffraction are focused onto the disc to form spots that straddle the track being followed (Fig. 16.3). Additional detectors are provided in the detecting optics to collect light reflected back from each of these spots. If the reading optics drifts off towards one side of the track, the auxiliary reflection on that side suffers less a.c. modulation (from the bumps passing by) than the other because it has come more from 'land' and less from 'bumps'. Electronics can detect the difference in the size of the two a.c. signals and use that difference to impose a correction.

Although we shall not analyse its functioning, we should mention an alternative method for detecting tracking error, called the push–pull method. This uses the reflection of the main reading beam, which tends to be reflected towards one side or the other when the beam drifts away from the centre of the track it is reading. The effect is greatest if the pit/bump has height $\lambda/8$ and is zero for height $\lambda/4$. Since height $\lambda/4$ is best for reading the information signal, the push–pull method requires discs to have pits/bumps whose heights are a compromise between $\lambda/4$ and $\lambda/8$. All commercial CDs have to be made with bump heights that permit push–pull tracking to work.

16.5.2 Correction of focus

With reading optics of numerical aperture 0.45, the depth of focus is about 3 μm, while the height of the disc relative to the reading optics may vary during rotation by about 100 μm; a feedback system is required. It is again the case that more than one system is in common use, and we shall select one only for discussion.

One method for detecting focus error is to make the imaging lens (lens L_3 in Fig. 16.2) slightly astigmatic.[9] This means that what 'ought' to be a point focus is a pair of lines with a longitudinal separation, as sketched in Fig. 16.4. The photodetector is split into four quadrants

[9]The 'defect' described here is of the kind that a clinical optician calls 'astigmatism' and which he corrects by prescribing a cylindrical lens. The 'astigmatism' of sidenote 36 on p. 286 is a similar image defect (two displaced line foci), but it is produced by a different mechanism.

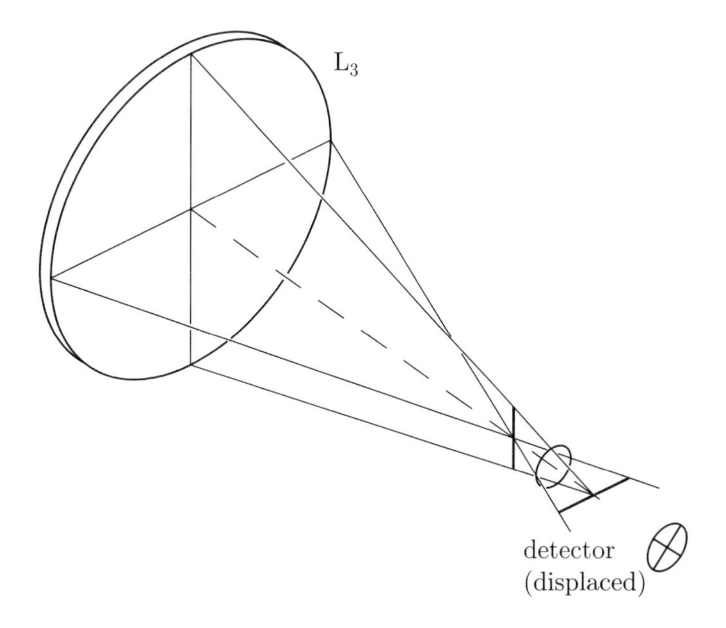

Fig. 16.4 The formation of a would-be point image by a lens possessing astigmatism. Rays passing through the horizontal plane through the middle of the lens are brought to a focus at one distance, while rays passing through the vertical plane are focused at a slightly different distance (here somewhat farther from the lens). The effect is to give a vertical 'line focus' at the closer distance, and a horizontal 'line focus' at the greater distance. In between is a 'circle of least confusion'. If a four-quadrant detector is placed at the circle of least confusion, all four quadrants will receive equal illumination. If it is placed closer, the vertical quadrants will receive more light; if farther, the horizontal quadrants will receive more.

with signals from opposite quadrants strapped together. A focus error of one sign causes the light reflected back from the CD to fall mainly on one opposite pair of quadrants; an error of the other sign illuminates the other pair more strongly. An error signal may be obtained by amplifying the difference between the photocurrents from the quadrant pairs.[10]

[10]The same detector can be used for obtaining the information signal, simply by adding the signals from all four quadrants.

16.6 CD-ROM

There is no difference optically between a CD that stores music in digital form and one that stores computer information. The 'control information' near the beginning of each frame contains a code to indicate the kind of information stored in that frame. There are some differences of detail between CD-ROM and CD because CD-ROM information is arranged into 'sectors', each with an identifying header that uniquely identifies the sector, and there is an additional layer of error correction.[11]

[11]The space devoted to the additional error correction reduces the information capacity of the disc from the 783 Mbyte available for music to 650 Mbyte.

16.7 DVD

The DVD is a development from the original CD, and is conveniently discussed by indicating how it differs from the CD. The official standard for DVD-ROM is contained in standard ECMA-267 (2001).[12]

[12]ECMA originally stood for the European Computer Manufacturers Association, but that body renamed itself in 1994 as ECMA International—European Association for Standardizing Information and Communication Systems. See http://www.ecma.ch.

A DVD (single-sided single-layer) can hold up to 4.7 Gbyte, compared with 650 Mbyte on a CD-ROM.

DVD has the refinement that information may be stored in two layers within the disc, rather than the one for CD. When two layers are used, they are spaced (in depth) by $55 \pm 15\,\mu\text{m}$. The disc capacity is increased to 8.5 Gbyte, less then double that of a single-layer disc because the information has to be stored a little more coarsely (the length devoted to a bit is increased when there are two layers, because the optical signals from one layer are potentially degraded by the other layer's out-of-focus presence). Finally, some discs are made double-sided and, if each side has two information layers, the total capacity is 17 Gbyte.

The reading wavelength for DVD is 650 nm (instead of 780 nm for CD), using a lens with numerical aperture 0.60 (instead of 0.45). The turns of a DVD track are spaced by $0.74\,\mu\text{m}$ (instead of $1.6\,\mu\text{m}$), and the length of track[13] devoted to a channel bit is $0.133\,\mu\text{m}$ (instead of $0.3\,\mu\text{m}$).

[13]The channel-bit length is $0.133\,\mu\text{m}$ when the disc has one information layer, but is increased to $0.147\,\mu\text{m}$ when the disc has two information layers read from the same side. Together with the optical enhancements (which on their own would give a factor 1.6 only), these figures (for one information layer) would scale the capacity of a CD to 3.2 Gbyte; the remaining increase to 4.7 Gbyte is mostly accounted for by more efficient encoding.

For a DVD the information is again divided into frames, but each frame contains 2048 bytes (instead of 24). Then 16 frames are grouped together into a block to which additional error correction is applied. These larger units permit the 'overhead' devoted to error correction to be reduced (as a fraction of the whole), at the expense of using more computation-intensive algorithms. An improvement also comes from encoding each byte of 8 data bits into 16 channel bits (rather than $14 + 3$; there are no merge bits as such for DVD, but there are choices that can be made during the encoding process that make the 16-bit units conform to the rules).[14]

[14]There is also a small increase in capacity from reducing the inner diameter of the data area from 50 mm to 48 mm.

The feedback system for detecting and correcting tracking errors for a DVD is different from that for a CD. A four-quadrant detector is used, and a tracking-error signal is obtained from the differences between the times at which the pit ends are detected by the four detectors. The same four-quadrant detector is also used, with an astigmatic lens, to obtain a focus-error signal, and (with all four signals added) to provide the information signal. The optical system is therefore simplified, at the cost of using more complicated electronics. Because push–pull tracking-error detection is not used, there is no need to make the pit/bump heights a compromise between $\lambda/4$ and $\lambda/8$, so a height close to $\lambda/4$ ($\sim$100 µm) is used.

16.8 The confocal microscope

The confocal microscope is an extension to optical microscopy of the scanning technique familiar in electron microscopy. Instead of forming an image all-at-once of an object, using the means described in Chapter 12, we illuminate just a single point (or rather a diffraction-limited small region) as shown in Fig. 16.5(a). Light from that same point is collected and passed to a photoelectric detector, from which a signal is stored in a computer. The sample is scanned through two dimensions (or, less often, the optical system is scanned relative to the sample), and once the scan is complete a picture can be displayed by the computer.

In the first instance, scanning is just another way of storing and displaying an image. However, once the idea of scanning has been adopted, a number of advantages can be designed into the microscope, as will appear in the discussion.

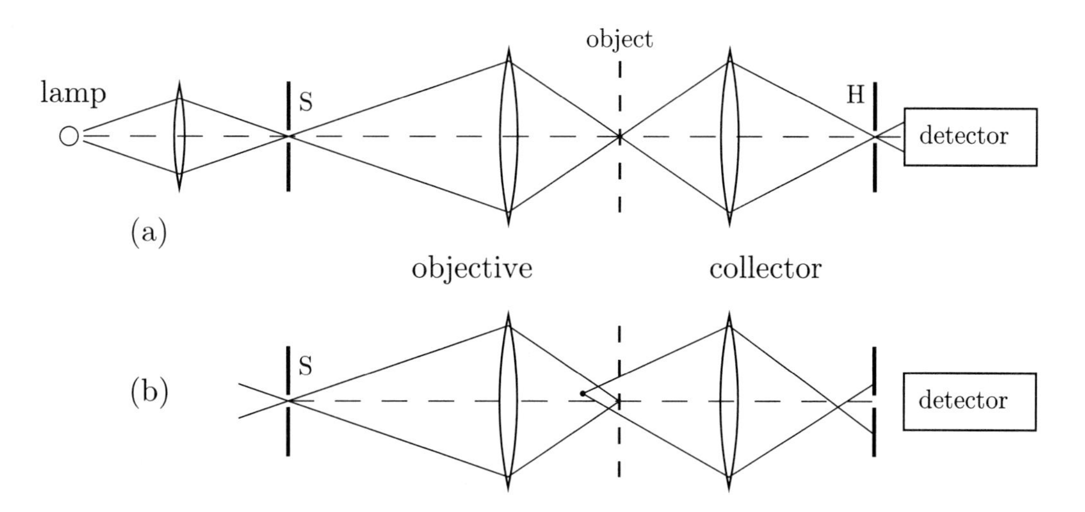

Fig. 16.5 The essentials of a confocal microscope. (a) The object is illuminated at one point by imaging point source S onto it. Light from that same point is collected and passed to the detector via a pinhole H. Holes S and H ensure that only light from the one selected point reaches the detector in any quantity. An image of an area is constructed by scanning the object and displaying the results of the scan. (b) Light scattered from a point out of the plane of interest is spread out over the aperture containing H, and only a little of it passes through H to the detector.

16.9 Confocal microscope: resolution

Refer to Fig. 16.5(a). A point in the object is being illuminated by forming an image of S onto it, and this is the point upon which attention is currently being focused. In principle, it would suffice to place a detector behind the object, without providing the collector lens or pinhole H, and an image could be formed by scanning the object as described above. Likewise, it would be possible to select a desired point for attention by using the collector lens and H, while dispensing with the objective lens and S.[15] However, the advantages of the confocal arrangement appear when we use both lenses and both pinholes.

[15]The instrument would then be in essence a conventional microscope adapted for scanning.

Consider first the resolution. The image of S, formed in the plane of best focus at the object, is a circular-aperture diffraction pattern with the form shown in Fig. 3.11. If the sensitivity to position came only from S and the objective, this would be the instrument's point spread function.[16] The same may be said by considering the image of H, by working backwards: the intensity received by H, as a function of position across the object, is a similar diffraction pattern; if only the collector and H controlled the position sensitivity, *this* would be the point spread function. When both S and H are in place, the effective point spread function is the product of the two diffraction-pattern functions. Since the product of two peaks is narrower than either peak on its own, this results in a resolution improvement[17] (over even the best microscope of Chapter 12) of about a factor 1.4.

[16]The point spread function is defined in §12.12.

[17]If the two diffraction patterns have the same size in the radial direction, their first diffraction zeros coincide, with each other and with that of their product. By the conventional measures of resolution introduced in Chapter 3, the resolution is not improved. However, this points to the roughness of those conventional measures. When we calculate the FWHM of the product it is narrower than that of each contributing peak by a factor 0.72. It cannot be doubted that this narrowing confers a benefit. See problem 16.2.

Since the two lenses, labelled objective and collector, take an equal part in defining the point upon which attention is focused, either can be considered as taking the function of a conventional microscope's objective. For this reason, the naming of the lenses is somewhat arbitrary. We have followed the nomenclature in Wilson (1990). The reason for an apparent reversal from the 'obvious' naming becomes apparent when we realise that the basic confocal microscope lends itself to a number of inventive modifications. One of these is the fluorescence microscope.

Suppose we illuminate an object with light of frequency ν_1, using the illuminating arrangement of Fig. 16.5(a). The object absorbs photons of this frequency and fluoresces at a lower frequency ν_2. A filter between object and detector selects just frequency ν_2. The lengthened wavelength of the fluorescent light means that the point spread function of the collecting optics is widened, and the resolution is now dominated by that of the illuminating optics. In such a case, it is appropriate to give tne name 'objective' to the lens that controls the resolution, even though it is 'upstream' of the object.

16.10 The confocal microscope: depth of focus

The most important characteristic of the confocal microscope is its ability to concentrate attention on a narrow range of depths within an object. These might be depths within a transparent object, or different heights of surface features on an opaque object.[18] In other words: the microscope has a very small depth of focus.

The reason for the small depth of focus may be seen from Fig. 16.5(b). In that figure, light is imagined to be scattered from a point within the cone of illumination provided by the objective lens, but away from the plane of best focus. The scattered light is imaged at a point displaced longitudinally (transversely too as drawn, but that changes during the scan) from pinhole H. The effect of the displacement is that pinhole H rejects almost all of the light from this scattering point. Out-of-focus points in the object contribute only 'dark' to the final computer-constructed image.

[18]The optical arrangement shown in Fig. 16.5 assumes the object to be transparent or part-transparent. However, it is easy to invent a modification to deal with an opaque object, along the lines of Fig. 16.2.

Problems

Problem 16.1 (a) The CD player
Check the numerical values given in the text, for consistency where they can't be checked absolutely.

Problem 16.2 (b) The resolution of a confocal microscope
Use a table of Bessel functions, or a mathematical program, to find the FWHM for the intensity of a circular-aperture diffraction pattern, and for the square of that intensity. Show that the narrowing produced by squaring is a factor 0.7178, in agreement with the statement in §16.9.

Problem 16.3 (c) The depth of focus of a confocal microscope
Investigate the depth of focus of the confocal microscope. When an object is away from the plane of best focus, its contribution to the computer-generated image is both fuzzy and reduced in intensity. Investigate both aspects.

N Notes on selected problems

Problem 1.1 Orders of magnitude

(1) Field at one Bohr radius from a proton is $5.1 \times 10^{11}\,\mathrm{V\,m^{-1}}$.

(2) Field at the surface of a proton is roughly $10^{21}\,\mathrm{V\,m^{-1}}$.

Problem 1.2 Scales of distance for averaging fields

(1) A length a with (atomic spacing) $\ll a \ll \lambda_{\text{visible}}$.

(2) The same.

(3) None. The X-rays probe distances down to an atomic spacing or less. (Averaging might be done over smaller-scale structure in the electron density, but it's not clear that this would be useful.)

(4) Opal: As (1). The interesting structure is of the order of an optical wavelength and mustn't be averaged over—but we shouldn't be doing that anyway.

(7) The atomic spacing is about an optical wavelength, so at first sight there is no scale that's suitable for averaging over. But because air molecules are in rapid motion it's sensible to describe the air first by its average density and average refractive index. Then fluctuations take place positively and negatively about that average. The fluctuations of density are comparable to the mean density itself. The density fluctuations cause the light to be scattered.[1]

[1] The blue of the sky comes from this mechanism, but from regions of higher density than that assumed here.

Problem 1.5 Wave basics

(2) The three conditions in the Answer are needed to permit the assumed $\boldsymbol{E}$ to exist as a solution of Maxwell's equations.

(3) The time-dependent Schrödinger equation is an obvious example.

Problem 1.7 The effect of having a dispersive medium

The investigation called for involves some possible surprises.

When a material is dispersive, $\boldsymbol{D}(t)$ depends upon values of $\boldsymbol{E}$ at times shortly before t as well as the value at time t itself.[2] This rather complicated physics can be handled in a remarkably simple way: eqns (1.5) can be retained provided they are understood as meaning

$$\boldsymbol{D}(\omega) = \varepsilon_{\mathrm{r}}(\omega)\,\varepsilon_0\,\boldsymbol{E}(\omega), \tag{N.1}$$

and not as relating $\boldsymbol{D}$ at time t to $\boldsymbol{E}$ at the same time t. (The corresponding equation relating $\boldsymbol{B}$ to $\boldsymbol{H}$ is, in principle, similar but need not be written because $\mu_{\mathrm{r}} = 1$ for optical frequencies.)

[2] This may be an unfamiliar idea. It is elucidated by means of a tutorial example in problem 2.22.

Wave equations (1.6) can *not* be deduced because things go wrong with the time derivatives.

A dispersive medium is necessarily also an absorbing medium—though the absorption may take place at frequencies some way from those of interest to our waves.[3] Absorption can be handled straightforwardly by working with complex amplitudes. Therefore, we work with fields given by complex expressions such as eqn (1.10) or eqn (N.2), where it is understood that the actual field is to be obtained by taking the real part. Going with these complex field expressions is a possibly-complex $\varepsilon_{\mathrm{r}}(\omega)$. All these understandings were implicit in writing eqn (N.1).

[3] This connection is exhibited in the Kramers–Kronig relations.

A field that possesses a single frequency but is otherwise general has the (complex) form

$$\boldsymbol{E} = (\text{function of } x, y, z)\, \mathrm{e}^{-\mathrm{i}\omega t}. \tag{N.2}$$

When this is substituted, together with eqn (N.1), into Maxwell's equations (1.3), we obtain replacements for the wave equations (1.6):

$$\nabla^2 \boldsymbol{E} = -\frac{\omega^2}{v^2}\boldsymbol{E}, \qquad \nabla^2 \boldsymbol{H} = -\frac{\omega^2}{v^2}\boldsymbol{H}. \tag{N.3}$$

Here v is defined as

$$v = (\mu_{\mathrm{r}}\, \mu_0\, \varepsilon_{\mathrm{r}}\, \varepsilon_0)^{-1/2} = c(\mu_{\mathrm{r}}\, \varepsilon_{\mathrm{r}})^{-1/2}, \tag{N.4}$$

where $\varepsilon_{\mathrm{r}} = \varepsilon_{\mathrm{r}}(\omega)$. This makes v mathematically similar to that given in eqn (1.7), but it is now a possibly-complex quantity and, if complex, no longer has the interpretation of phase velocity. The refractive index n is, likewise, defined by a statement identical in form to eqn (1.8), but is now frequency-dependent and possibly complex.

A trial solution of eqns (N.3) can be written in the form (1.10). Equations (1.11) remain correct when this trial solution is substituted into the new 'wave equations' (N.3). But in the list of properties following eqn (1.10), the interpretation of $\boldsymbol{k}$ is changed:

$\boldsymbol{k}$ = wave vector: direction of $\boldsymbol{k}$ is the direction of the phase velocity,
$\mathrm{Re}\, k = 2\pi/(\text{wavelength } \lambda_{\text{medium}})$,
$2\, \mathrm{Im}\, k$ = absorption coefficient,
phase velocity $= \nu\lambda_{\text{medium}} = \omega/(\mathrm{Re}\, k)$,
group velocity $= \mathrm{d}\omega/\mathrm{d}(\mathrm{Re}\, k)$, or more usefully $1/\{\mathrm{d}(\mathrm{Re}\, k)/\mathrm{d}\omega\}$.

Statement (1.12) remains correct, with complex n accompanying complex k. Indeed, the main reason for introducing a complex n is to provide an alternative expression for k for use on occasions when it is convenient to remove the factor (ω/c).

Statements about impedances remain correct, with Z complex, except that eqn (1.14) is replaced by

$$\frac{E_x}{H_y} = \frac{E_y}{-H_x} = Z \times \mathrm{sign}(\mathrm{Re}\, k). \tag{N.5}$$

In working out the intensity it must be remembered that $E_x(z,t)$ and $H_y(z,t)$ are no longer in phase with each other, since they are related by a complex Z. Also, the real parts must be taken before multiplying. We have

$$I_z = \frac{1}{T}\int_{-T/2}^{T/2}\left(\frac{E_x + E_x^*}{2}\right)\times\frac{1}{2}\left(\frac{E_x}{Z}+\frac{E_x^*}{Z^*}\right)\mathrm{d}t$$
$$= \frac{1}{4}\left(\frac{|E_x|^2}{Z}+\frac{|E_x|^2}{Z^*}\right) = \frac{|E_x|^2}{2}\,\mathrm{Re}(1/Z) = \frac{|E_x|^2}{2Z_0}\,\mathrm{Re}\,n. \qquad (\mathrm{N.6})$$

In the above, we have used the fact that E_x contains a factor $\mathrm{e}^{-\mathrm{i}\omega t}$, and have chosen T so that terms containing $\mathrm{e}^{\pm 2\mathrm{i}\omega t}$ are averaged to zero.

The scalar wave equation (1.24) is no longer obtainable, but a modification similar to eqn (N.3) replaces it. The spherical wave (eqn 1.25) is replaced by a complex amplitude

$$U(r,t) = V_0\,\frac{\mathrm{e}^{\mathrm{i}(kr-\omega t-\alpha)}}{r}. \qquad (\mathrm{N.7})$$

If all this looks like the chasing of small details, look at problem 1.8.

Problem 1.8 A paradox

The wrong statement is (a). The time-averaged energy density is $\overline{U}$, where

$$\overline{U} = \frac{1}{2}\left(\frac{\mathrm{d}(\omega\varepsilon_\mathrm{r})}{\mathrm{d}\omega}\overline{E^2} + \frac{\mathrm{d}(\omega\mu_\mathrm{r})}{\mathrm{d}\omega}\overline{H^2}\right).$$

The derivation of this equation is outlined in Landau and Lifshitz (1993), §80. To make sure everything is in order, show that the corrected statement given here is consistent with statements (b) and (c) in the problem.

Problem 1.9 Transmission line basics

(2) curl $\boldsymbol{E} \neq 0$ means that a scalar potential V obeying $\boldsymbol{E} = -\nabla V$ can't be defined. Students get so comfortable with V when learning electrostatics, and with a V that *looks like* a potential even in a.c. circuits, that they often find this hard to accept. If this causes a rethink of what you thought you understood, well, it's about time.

Problem 1.13 Basic lens optics

(3) $u = +0.8\,\mathrm{m}$; $h_\mathrm{o} = +10\,\mathrm{mm}$; $f = +0.5\,\mathrm{m}$. Hence $v = +(4/3)\,\mathrm{m}$; $h_\mathrm{i} = +(50/3)\,\mathrm{mm}$.

(4) $z_\mathrm{o} = -0.8\,\mathrm{m}$; $y_\mathrm{o} = +10\,\mathrm{mm}$. Hence $z_\mathrm{i} = +(4/3)\,\mathrm{m}$; $y_\mathrm{i} = -(50/3)\,\mathrm{mm}$.

(5) $z_\mathrm{i} = -(4/13)\,\mathrm{m}$; $y_\mathrm{i} = +(50/13)\,\mathrm{mm}$. To the left of the lens and above the axis.

(6) The meniscus lens has $f = 12\,\mathrm{m}$ both ways round.

Problem 1.17 A single-lens reflex camera

The pentaprism is 'roofed', so that light from the screen goes up, *across* (into or out of the paper), and then again follows the route drawn. There are, therefore, four reflections in all, not three.

Problem 2.2 Fourier series basics

The comment in sidenote 30 on p. 34: There are two improvements that could be made. First, it's easier to deal mathematically with continuous variables than with discrete, so it would be better—regardless of future developments—to write

$$A(\omega) = \frac{1}{\tau}\int_0^{\tau} V(t)\,\mathrm{e}^{\mathrm{i}\omega t}\,\mathrm{d}t \quad \text{with } A_p \text{ evaluated as} \quad A_p = A(p\omega_0).$$

It's no accident that this carries us another step towards eqn (2.12).

The second tidying is to cross-multiply τ, grouping (τA_p) as in problem 2.4. Then things are ready for proceeding to the limit $\tau \to \infty$.

Problem 2.4 The Fourier integral as the limit of a Fourier series

(2) If the mathematics isn't easy to feel secure with, try the following. Instead of taking a mathematician's limit $\tau \to \infty$, try letting τ be say 10^6 s (= 11 days!). Individual frequencies in the Fourier series are separated by only 10^{-6} Hz. Imagine your apparatus makes frequency measurements over 1 s, which is a very long time for optical work. A frequency interval spanning 10^{-2} Hz will contain $\sim 10^4$ terms in the series yet will certainly be unresolved. In such circumstances you should have no qualms about replacing a sum by an integral.

Problem 2.9 The Parseval theorem

(3) The Parseval theorem for complex $V(t)$ is, rather obviously,

$$\int_{-\infty}^{\infty} |V(t)|^2\,\mathrm{d}t = \int_{-\infty}^{\infty} \left|\widetilde{V}(\omega)\right|^2 \frac{\mathrm{d}\omega}{2\pi}.$$

Once this has been guessed it shouldn't be difficult to pursue the method of part (1) through to the result.

Problem 2.17 The Fourier transform for large ω

(5) A Gaussian is probably the most obvious example.

Problem 2.18 The convolution theorem

(1) The convolution theorem:

$$\begin{aligned}\widetilde{V}(\omega) &= \int_{-\infty}^{\infty} \mathrm{d}t\,\mathrm{e}^{\mathrm{i}\omega t} \int_{-\infty}^{\infty} \mathrm{d}t'\,\phi(t')\,\psi(t-t') \\ &= \int_{-\infty}^{\infty} \mathrm{d}t'\,\mathrm{e}^{\mathrm{i}\omega t'}\,\phi(t') \int_{-\infty}^{\infty} \mathrm{d}t\,\mathrm{e}^{\mathrm{i}\omega(t-t')}\,\psi(t-t').\end{aligned}$$

Now invent a new symbol for $(t - t')$ in the last integral.

Problem 2.22 Frequency dependence implies a short-term memory

(5) It certainly was necessary. Voltage expressions $v_{\text{in}}(\omega)$ and $v_{\text{out}}(\omega)$ are set up in the usual circuit-theory manner to contain $\mathrm{e}^{\mathrm{j}\omega t}$ with a single positive $\omega = \omega_0$ and a 'take the real part at the end' convention. In spite

of the notation, both are primarily functions of time t, with ω as a parameter that might later on be summed over as in step (e) of the procedure itemized in §2.5. They are, therefore, special-case expressions[4] for $U_{\rm in}(t)$ and $U_{\rm out}(t)$. By contrast, neither $\widetilde{V}_{\rm in}(\omega)$ nor $\widetilde{V}_{\rm out}(\omega)$ is a function of time t, because each is a coefficient of $e^{-i\omega t}$ (or $e^{j\omega t}$) in eqn (2.12). Moreover, the $\widetilde{V}_{\rm in}(\omega)$ of eqn (2.36) is a general function of ω, capable of describing any input waveform $V_{\rm in}(t)$ not just a single-frequency sinewave, and $\widetilde{V}_{\rm out}(\omega)$ specifies the accompanying output.

[4] Special-case because U is defined so as to handle any combination of frequencies and here we have only one.

Problem 2.25 The uncertainty principle
We remove Planck's constant by working with variables t and ω (or x and k_x would do equally well), and start from

$$0 \le I, \qquad \text{where} \qquad I \equiv \int_{-\infty}^{\infty} \left| \alpha(t - t_0)\psi(t) + \frac{d\psi}{dt} + i\omega_0\psi(t) \right|^2 dt.$$

Take $\psi(t)$ to be normalized[5] and define a normalized transform $\phi(\omega)$:

$$\psi(t) = \frac{1}{\sqrt{2\pi}} \int_{-\infty}^{\infty} \phi(\omega)\, e^{-i\omega t}\, d\omega.$$

[5] We depart from previous practice here by defining the Fourier transform to contain a coefficient $1/\sqrt{2\pi}$ in each integral. This is the convenient way to do things in a quantum-mechanical context, so that both ψ and ϕ can be normalized to 1.

Normalization means that

$$\int_{-\infty}^{\infty} \psi^*(t)\, \psi(t)\, dt = \int_{-\infty}^{\infty} \phi^*(\omega)\, \phi(\omega)\, d\omega = 1.$$

Evaluating I now gives

$$I = \alpha^2(\Delta t)^2 - \alpha + (\Delta\omega)^2. \tag{N.8}$$

Here (Δt) is the standard deviation of $|\psi|^2$ about the arbitrary origin of t at t_0. Likewise, $(\Delta\omega)$ is the standard deviation of $|\phi|^2$ about the arbitrary origin ω_0:

$$(\Delta t)^2 \equiv \int_{-\infty}^{\infty} \psi^*(t)\,(t - t_0)^2\,\psi(t)\, dt,$$

$$(\Delta\omega)^2 \equiv \int_{-\infty}^{\infty} \phi^*(\omega)\,(\omega - \omega_0)^2\,\phi(\omega)\, d\omega.$$

If I is to be non-negative for all choices of α, the quadratic expression in α in eqn (N.8) must have no real roots, and the condition for this is

$$(\Delta t)(\Delta\omega) \ge \tfrac{1}{2}. \tag{2.39}$$

Since this is true always it must be true for the special case[6] where t_0 is chosen to lie at the centroid of $|\psi|^2$ and where ω_0 is chosen to lie at the centroid of $|\phi|^2$.

[6] This choice gives Δt its smallest possible value. Remember the parallel-axis theorem for moments of inertia. Similarly $\Delta\omega$ is being given its smallest value. So the inequality (2.39) is proved to hold even after these measures have been taken to minimize the product $(\Delta t)(\Delta\omega)$.

Problem 3.4 Complementary screens
(1) Assume first that the lens is square with width D, so that the diffracted amplitude for the 'spike' is the expression given in problem 3.14(1)

with $a = b = D$. The width (centre to first zero) of the spike is $(500\,\text{mm})(500\,\text{nm})/(50\,\text{mm}) = 5\,\mu\text{m}$. The width (centre to first zero) of the diffraction pattern from the wire or slit is 2.5 mm, a width ratio of 500.

(2) One's first thought is to agree with the statement in problem 3.4 that there is plenty of room round the central spike where the only light is that diffracted from the slit/strip. However, at its centre, the spike is more intense than the slit's diffraction pattern by a factor (intensity) of $(D/a)^2 = 500^2$; things need a careful look. We use the expression for diffraction at a rectangular aperture given in problem 3.14, agreeing to observe on the Y-axis where $\phi = 0$. There the lens contributes amplitude

$$\frac{U_0\,\mathrm{e}^{\mathrm{i}kd-\mathrm{i}\omega t}}{\mathrm{i}\lambda f} D^2 \frac{\sin(\frac{1}{2}kD\sin\theta)}{\frac{1}{2}kD\sin\theta}.$$

For the slit (width a) the corresponding expression is

$$\frac{U_0\,\mathrm{e}^{\mathrm{i}kd-\mathrm{i}\omega t}}{\mathrm{i}\lambda f} Da \frac{\sin(\frac{1}{2}ka\sin\theta)}{\frac{1}{2}ka\sin\theta}.$$

These expressions have similar orders of magnitude: at any place where both numerator sines are 1 the amplitudes are equal. If this was the whole story the amplitudes sketched in Fig. 3.3 would be wrong.

To make the demonstration of Fig. 3.3 correct we have to do something different. One possibility is to try to make the central spike have less intensity in its wings, such as by making the lens circular instead of square.[7] Or we could taper the intensity of the illuminating beam to make it dimmer towards the outer edges of the lens.[8] Such measures are artificial, however, and are attempts at avoiding the issue.

The present problem has been discussing an unusual case, and applications of Babinet's principle are usually not so unfriendly. A more typical example would be the diffraction of light from many identical scatterers (lycopodium powder and blood corpuscles are often used in demonstration experiments), for which the diffraction pattern is very similar to that for light passing through a complementary array of holes. In such a case the large number of scatterers magnifies the 'wanted' part of the diffraction pattern until it swamps the wings of the central spike.

The impression we would wish to leave with the reader is that even an apparently straightforward piece of bookwork in fact needs a surprisingly critical examination.[9]

[7] This effects a reduction, but one that is barely sufficient.

[8] This idea, of apodizing the amplitude distribution across the lens, is encountered again in §12.10.

[9] I am indebted to Professor D. N. Stacey for discussion of this point.

Problem 3.7 Revision on elementary diffraction

(1) Distance r_{P} ranges through 2.5 μm and kr_{P} ranges through ~30 radians. Thus, the secondary-wave amplitudes arriving at the P of Fig. 3.2 add in a whole range of phases. It is out of the question to ignore those phase differences. By contrast, r_{P} itself, and therefore also $1/r_{\mathrm{P}}$, varies by only 50 parts per million. We make a negligible error by treating $1/r_{\mathrm{P}}$ as constant, as we then add the $\exp(\mathrm{i}kr_{\mathrm{P}})$ terms with a coefficient that is at worst in error by 50 ppm.

(4) The integral of $\{(\sin x)/x\}^2$ from 0 to π is $\mathrm{Si}(2\pi)$, where Si() is defined by Abramowitz and Stegun (1965), relation 5.2.1. The central peak contains about 90% of the energy in the whole pattern.

Problem 3.8 The proportionality constant isn't constant

(1) Use Fig. 3.4. Draw a point source at O, radiating to the right. We require all optical paths d from O to the focal plane through P_0 to be equal. Then the focal plane is a surface of constant phase: a wavefront. All rays from O to that wavefront meet the wavefront at right-angles (definition of a ray) so they are parallel to the axis. Rays travelling parallel to the axis have come from the lens's left-hand focal point.

(2) If point O does not lie at the left-hand focus of the lens, use ray optics to find its image. Light refracted through the lens forms a spherical wave that appears to diverge from (or converge towards) that image. The distance from image to P_0 is[10] $R = f^2/(f - \mathrm{OL})$. The path from image to X is $\sqrt{R^2 + X^2}$, in which we may treat X/R as small in obtaining the given expression for d. Also see a similar expression in eqn (3.26) of problem 3.23. Distance R becomes infinite, and d becomes constant, if $\mathrm{OL} = f$, in agreement with the reasoning of part (1).

[10] This may most quickly be seen by using the Newton lens formula (1.37). The reader should check the implications of R being positive or negative.

Problem 3.12 Careful English

We need to make statements about the (spatial) Fourier components being imposed on $T(x, y)$.

Problem 3.15 The 'width' of a diffraction pattern

The peak of Fig. 3.7(b) has half height for $\rho = \pm 1.3916$, giving a FWHM of 2.7831. The conventional 'width' (centre to first zero) of π is, therefore, larger than the FWHM by a factor 1.13, not a spectacular difference.

The peak of Fig. 3.11(b) has half height for $\rho = \pm 1.6163$, giving a FWHM of 3.2327. The conventional 'width' (centre to first zero) of 1.2197π exceeds the FWHM by a factor of 1.185, again not spectacular.

We might prefer to ignore the 'optics' conventions and to specify the peak widths throughout by giving the full width at half maximum height.[11] If we do, the widths are 0.8859π and 1.029π.

[11] Not recommended! There has to be a stronger reason than this to depart from an established convention.

Problem 3.16 The circular aperture

(4) The integral can be linked to the Bessel function of order 1 using relation 9.1.20 in Abramowitz and Stegun (1965), which reads after a slight change of notation:

$$\mathrm{J}_\nu(\rho) = \frac{2(\rho/2)^\nu}{\sqrt{\pi}\,\Gamma(\nu + \frac{1}{2})} \int_0^1 (1 - \xi^2)^{\nu - 1/2} \cos(\rho\xi)\, \mathrm{d}\xi.$$

In this, set $\nu = 1$. A little more rearrangement brings the integral of eqn (3.23) into the form $\pi \mathrm{J}_1(\rho)/\rho$.

(6) The limiting behaviour as $\rho \to \infty$ is given by Abramowitz and Stegun, relation 9.2.1. We have $\mathrm{J}_\nu(\rho) \to \sqrt{2/(\pi\rho)}\cos(\rho - \nu\pi/2 - \pi/4)$. The final part of this rearranges for $\nu = 1$ to $\sin(\rho - \pi/4)$, which gives the (integer + $\frac{1}{4}$) quoted in the problem.

Problem 3.17 Two pinholes
Light from *each* pinhole forms a Fraunhofer diffraction pattern centred on the lens's focal point. These patterns are superimposed centre-on-centre (did you imagine them laterally displaced from each other? if so, think again). So you get the diffraction rings characteristic of a circular aperture. The displaced pinholes also act like Young slits, so you get straight-line interference fringes too. The convolution theorem says the combined amplitude is a simple product: (amplitude from one pinhole)× (fringe pattern). In particular, the amplitude is zero when either factor is zero, so you can easily sketch the **nodal lines** where all these zeros lie (straight lines and circles). The intensity is zero there too. Once this shape is drawn right, it should be easy to imagine how the intensity maxima will look.

Problem 3.20 A grating that gives only three orders
(1) Notice that in example 3.1 we integrated each complex exponential over the full width of the grating. We were not influenced by the 'textbook' way of deriving eqn (4.1), which integrates over a single element and then sums a geometric progression to deal with the repeated elements. Of course, the 'textbook' route would get us there in the end. But the 'reasons why' are best seen by following the procedure given.

Problem 3.22 Diffraction is Fraunhofer at the image of the source
(3) Let the aperture of Fig. 3.16 have diameter a and let the ideal focus at P_0 be distant d_0 from it (the diagram marks the distance as d_P but now we shall investigate the diffracted light at distances d_P that differ somewhat from d_0). We have to apply eqn (3.19) to this case. Signs must be attached to d_s and d_P for consistency with the definitions on Fig. 3.15. Distance $d_s = -d_0$, while d_P is positive and not too far from d_0.

Because $1/d_s + 1/d_P$ may have either sign we must modify eqn (3.19):

$$\left|\frac{1}{d_s} + \frac{1}{d_P}\right| \leq \frac{1}{d_R}.$$

Thus, $1/d_P$ lies between the values $1/d_0 \pm 1/d_R$. Since the range covered by d_P will be $\ll d_0$ this rearranges to

$$d_P = d_0 \pm \lambda(d_0/a)^2,$$

which gives a focus range of $2\lambda(d_0/a)^2$.
(4) In Fig. 3.16 light arrives at the ray-optics focus within an angle a/d_0. If we move along the axis away from the focus through distance Δ, the rays that were crossing separate by $|\Delta|/(a/d_0)$. The diffraction-limited width of the image is $d_0\lambda/a$. Defocus broadens the image by the same amount as diffraction if $|\Delta|/(a/d_0) = d_0\lambda/a$ which rearranges to $2|\Delta| = 2\lambda(d_0/a)^2$, the same as the result of part (3).
(5) To set up in the laboratory, form an image of the source with the diffracting aperture absent. With a wide beam the depth of focus is

relatively small and errors of adjustment are easy to see and correct. When this is right, insert the aperture.

Problem 3.24 Fresnel conditions inside a laser cavity
Rayleigh distance = 395 mm. At half a metre, Fresnel number = 0.20.

Problem 3.25 Fresnel diffraction from a circular aperture

(5) To tackle the difficulty we need to improve on five approximations made in the problem:

- The factor $1/r_{\mathrm{P}}$ in the Kirchhoff integral has been approximated to $1/Z$.
- We have taken the areas of all Fresnel zones as being equal. However, the areas of the Fresnel zones increase at large radii, in just such a way as to cancel the decrease of $1/r_{\mathrm{P}}$ (show this).
- A decrease in the contributions from the outer zones comes from the obliquity factor, though as this never falls below $\frac{1}{2}$ (for $\theta_{\mathrm{dif}} = \pi/2$) its effect is not great.
- For large zones, the secondary waves arrive at P_0 from large angles and a scalar theory can't deal with them properly.[12]
- The incident wave can't be infinitely wide. To take the limit, take the aperture right away and consider a beam of finite width, with intensity falling quite gradually at its 'sides'. This puts a tapering 'envelope' onto the contributions of the Fresnel zones.

The last item shows that it must all work out. But, because the beam might have a variety of profiles, it's not easy to set up an algebraic demonstration. Thc point is usually discussed by drawing a phasor diagram whose curve spirals in towards the expected limit.

[12] Huygens-secondary-wave mathematics could be applied to the diffraction of a truly scalar wave; the results of problem 3.25(1)–(4) would again follow, and so would the difficulty raised in problem 3.25(5). There would, however, be no escape route via this fourth bullet point, so the vector character of the light must here be a distraction only.

Problem 3.26 Diffraction in non-optical physics

(1) Angular resolution = 3.7 mrad (milliradians). An optical telescope with similar resolution has objective diameter of 170 μm.

(2) The first diffraction zero occurs at 45° for a frequency of 1.9 kHz, so the highest usable frequency is perhaps half of this.

At 10 kHz the sound wavelength is only 33 mm, and the sound is concentrated by diffraction into a beam of angular width 0.13 rad. The cure is obvious: the sound source for high frequencies must be made smaller. But at the other extreme, at 20 Hz, the wavelength is 16.5 m; the speaker cone is too small to radiate efficiently, and we must not make it even smaller. Now you see why high-quality speaker systems divide up the frequencies into two or three ranges, giving the lowest frequencies to a large radiator (woofer), the highest to a small radiator (tweeter).

(3) Imagine that the nucleus is 'black': an encounter with a nucleus takes a 'bite' out of the neutron's wave function, the same size as the nucleus. By the Babinet principle, the downstream behaviour of the bite is the same as that of a wave that has passed through a circular hole. The bite broadens and gets shallower—it heals up—over a distance of about the Rayleigh distance.

The de Broglie wavelength for a 30 MeV neutron is 5.2 fm. Take aluminium as example target, with atomic number $A = 27$; the nuclear radius $\approx 1.2A^{1/3} = 3.6$ fm. Rayleigh distance = 2.5 fm. The downstream distance to the next nucleus is about 10^5 times this. By the time the neutron wave function reaches the next nucleus, it is again a plane wave (with slightly reduced intensity to take account of what has been eaten out of it).

(4) Relative atomic mass for arsenic = 74.9. For a kinetic energy of 100 keV the de Broglie wavelength is 1.05×10^{-14} m. This is small compared to an atomic spacing, so the arsenic atoms behave almost classically. The atomic spacing in crystalline silicon is 2.35×10^{-10} m, and the atomic diameter is about the same. The Rayleigh distance works out to be 5.26×10^{-6} m, over 20 000 atomic spacings.

Problem 3.27 Diffraction and technology: the CD

This solution ignores all details of the shape of the laser beam, treating it as providing a uniform plane wave for the lens to focus. Given the treatment in Chapter 3, the lens should focus the light down to a spot of radius $1.22 f\lambda/D$, where f is the focal length of the lens and D is its diameter. But the angles are not small here. A better estimate is that the diffraction-pattern radius is $1.22 \times \frac{1}{2}\lambda/\sin\alpha$, where $\sin\alpha$ is the 'numerical aperture' referred to in the question. This gives an Airy disc of diameter 2.1 μm. A first guess is that a CD might have adjacent turns of the track 2.1 μm apart, with bits spaced 2.1 μm apart along the track's length. Dividing this into the area of the disc gives an information capacity of 1.92 Gbit. Reading at the single-speed[13] rate of 4.32 Mbit s^{-1} would give a playing time of 7 minutes.

Fortunately manufacturers have been a good bit cleverer than we've given them credit for. First: a more careful analysis of the size of the focused laser beam, and of the acceptable cross-talk between adjacent track turns, leads to a turn spacing of only 1.6 μm. But more importantly, the length of track devoted to one bit is made only 0.3 μm. These two improvements[14] increase the capacity by a factor 9.3, giving a playing time of 69 min. (And a little pushing against permitted limits raises this further to 74 min.)

[13] This is the rate of reading for a music CD. CD-ROM readers in computers read at a multiple of this rate, at time of writing commonly of order 50.

[14] These values are discussed in §16.4.

Problem 3.29 'Impossible' polarization in diffraction

The wave equation for $\boldsymbol{E}$ (in a LIH medium) is

$$\nabla^2 \boldsymbol{E} = \frac{1}{v^2}\frac{\partial^2 \boldsymbol{E}}{\partial t^2}, \qquad \text{hence} \qquad \nabla^2 E_z = \frac{1}{v^2}\frac{\partial^2 E_z}{\partial t^2}.$$

If E_z is zero over the whole boundary of a region, it has to be zero everywhere inside too. Problem 3.29 describes a field with no E_z on the boundary, setting up a wave inside with a non-zero E_z. It won't do.

Our descriptions of diffraction mustn't be so blatantly in conflict with Maxwell, but what should be done to put things right? There *must* be a non-zero E_z on the surface spanning the diffracting aperture; this E_z could be calculated for a slit by the methods that yielded Fig. 3.20.[15]

[15] This statement is confirmed by the description of 'H-polarization' in Born and Wolf (1999), §11.4.1.

This E_z is available to 'drive' the diffracted E_z. Most of the time we can get away with thinking about the 'transverse' components of the field only, using a scalar-wave approximation, because these make the major contribution to the diffracted field; but we have to keep it in mind that this picture is too simple.

Looking back, it's not hard to find other cases where we've come too close to disbelieving Maxwell's equations. In example 3.1, imagine that the light is incident with its $\boldsymbol{E}$-field in the x-direction. The transmitted field has a non-zero $\partial E_x/\partial x$. Therefore ($\operatorname{div}\boldsymbol{E} = 0$) we must have a non-zero $\partial E_z/\partial z$, giving a 'longitudinal' component of the $\boldsymbol{E}$-field.

Problem 4.2 The multiple-element expression

(4) The amplitude has a large excursion at the ends of the range plotted, but this time it is upwards at one end and downwards at the other.

(6) Write $\frac{1}{2}\beta d = p\pi + \varepsilon$, in which ε is small over the interesting range. This permits $\sin\varepsilon \approx \varepsilon$ in the denominator, though $\sin(N\varepsilon)$ in the numerator must be left as it is. Notice the necessity for introducing ε, as $\frac{1}{2}\beta d$ itself is *not* small.

Problem 4.4 Mirror geometry for Czerny–Turner

(2) In Fig. 4.5 a direction is defined by the ray drawn from the collimating mirror to the grating. Draw a line parallel to this from the entrance slit. This is the symmetry axis of the paraboloid. From here you should be able to link this diagram with the standard one that shows rays parallel to the axis of a parabola all reflected towards its focus. (If this isn't familiar, look up the focus–directrix definition of a parabola, and use the fact that all optical paths from object to image are equal.)

Problem 4.10 A very short pulse input to a diffraction grating

(5) A harmonic of order p has (for normal incidence) $d\sin\theta = p\lambda$, so it can be identified with the pth grating order. The grating of example 3.1 gives no harmonics and no higher orders.

(7) Draw a graph like that of Fig. 4.4 for the case of an unblazed grating; a tiny piece of the graph is hinted at. Identify the values of $\sin\theta$ at which the grating maxima lie. Identify where the zeros of the envelope lie. Show that if $d = 2a$ the envelope zeros kill off all the even-numbered grating orders (except order zero). It's the same physics as the symmetrical square wave given out by an electronic generator: a symmetrical square wave has only odd-numbered harmonics.

Problem 4.11 The 'finesse' of a grating

(2) Consider normal incidence for simplicity. In first order, $\lambda = d\sin\theta$ and the angular width of the peak is $\delta\theta$, where $\lambda = Nd\,\delta(\sin\theta)$. Expressing angle θ in terms of $\sin\theta$ throughout, we have $\delta(\sin\theta) = (1/N)\sin\theta$. The number of grating orders that fit side-by-side in range $\mathrm{d}(\sin\theta)$ is

$$\frac{\mathrm{d}(\sin\theta)}{\delta(\sin\theta)} = N \times \frac{\mathrm{d}(\sin\theta)}{\sin\theta}.$$

Integrate this from the smallest $\sin\theta$ of λ_0/d to the largest $\sin\theta$ of $2\lambda_0/d$, where these define the limits of the angle range over which wavelengths ranging from λ_0 to $2\lambda_0$ can be displayed without overlapping orders.

Problem 4.15 Shortening exposure times
The cylindrical lens should form a real image of the entrance slit on the photographic film. By placing the lens fairly close to the film (and choosing a focal length that makes this possible) you can make the magnification less than 1, so the slit image is reduced in length.

Problem 5.2 Fabry–Perot basics
(3) Each of the reflecting surfaces must have high reflectivity, so cannot be a 'simple' interface between two media. It is either a thin layer of evaporated metal (aluminium or silver) or a stack of dielectric quarter-wave layers (typically nine of them). In neither case do Fresnel's equations apply to the *overall* reflection.
(6) $R = 1 - (\pi/\mathcal{F})\sqrt{R} = 1 - (\pi/30)\sqrt{R}$. The right-hand side cannot be less than $1 - (\pi/30) \approx 0.9$, so it's quite adequate to approximate $\sqrt{R}$ to 1. If you're not happy, use the R you get from this to improve the value of $\sqrt{R}$ on the right, and see what happens when you iterate.

Problem 5.3 Alternative calculation!
The equations applying to the right-hand surface are:

$$b\exp(-\mathrm{i}k_z d) = r_2 a \exp(\mathrm{i}k_z d), \qquad c = t_2 a \exp(\mathrm{i}k_z d).$$

Problem 5.4 A paradox: energy conservation
Input impedance, à la transmission lines and Chapter 6. Alternatively: there are so many waves travelling to the left within the étalon and contributing to a left-going wave 'outside' that they *can* add up to enough to cancel the first reflection. This may not 'feel right', but energy conservation does not imply amplitude conservation!

Problem 5.6 Free spectral range and resolution
(1) When the horizontal axis is δ all peaks lie at integer multiples of 2π, whatever the wavelength.

Problem 5.7 Chromatic resolving power and Q
(2) With a pulse of energy E rattling between the plates, the energy stored is just E. The energy loss is $E(1 - R)$ at each encounter with a mirror, and the encounters are spaced in time by d/c, so a coarse average rate of loss is $E(1 - R)c/d$. Then $Q \approx \omega_0 \times E \div \{E(1 - R)c/d\}$, which simplifies to $(2d/\lambda)\pi/(1 - R)$. For the reason given in the suggestion, this is as close as we'll get to the correct $(2d/\lambda)\pi\sqrt{R}/(1 - R)$.

Problem 5.14 Earthy practicalities
(1) Squeezing the seal enough to make it airtight would distort the étalon plates, and changing gas pressure would make the plates bulge.

An étalon spacer makes contact with each plate via three little pads, and the spacer is cut away between the pads. This is done for mechanical reasons, so we know precisely where plate and spacer are in contact. But given this shape, there is plenty of room for air to enter between the pads.

(2) Difficulty with making the glass optically homogeneous (free from local changes of refractive index). Section 5.7 has shown that we may well have to aim for nd varying by no more than $\lambda/120$.

Also, the costly part of making an (ordinary non-solid) étalon is getting the reflecting surfaces flat. A pair of plates can be used with several different spacers to give us several different étalons for the price of not much more than one. Each solid étalon has a single spacing only. Fine if it is to do a single job, such as discriminating modes in a laser, but too inflexible for a spectroscopic instrument.

(3) For a wavelength of 500 nm, $d \geq 2.6$ mm for air, 1.7 mm for CO_2.

(4) Assume that the gas entering the étalon box has no time to exchange heat with the box or its contents (an unrealistic worst-case assumption). Let the gas be bled in through a throttle from a reservoir at constant pressure p_{final} and temperature T_{room}.

(a) $T_{\text{final}}/T_{\text{room}} = \gamma = C_p/C_V$, the ratio of the gas's heat capacities. Then for air $\Delta T = T_{\text{final}} - T_{\text{room}} = 117\,\text{K}$.

(b) $T_{\text{final}}/T_{\text{room}} = 2\gamma/(\gamma+1)$, giving $\Delta T = 49\,\text{K}$.

If in the second case we instead assume an adiabatic compression of air we find $T_{\text{final}}/T_{\text{room}} = 2^{1-1/\gamma}$, giving $\Delta T = 64\,\text{K}$. Whichever case we take, the temperature rise is large enough to constitute a warning.

(5) About 2 degrees.

(6) Only the highly reflecting surfaces need to be flat to order $\lambda/100$. Imperfections of the surfaces of (b) and (c) will degrade the fringes, but the light meets them only once, so these surfaces need a good 'ordinary' optical quality. Strictly, this does not apply to (a), the input surface of the device. However, you have to be able to use the Fabry–Perot either way round, so this surface should be polished to the same 'ordinary' quality as the others.

Problem 5.15 The pinhole size for pressure scanning

(2) From the clues given in the problem we should equate:

$$\text{range of } \delta \text{ imposed by pinhole} = 2(\omega/c)nd(1 - \cos\theta_{\text{max}}),$$
$$\text{acceptable range of } \delta = 2\pi/\mathcal{F}.$$

(3) In the result of part (2), we can use the small-angle approximation $(1 - \cos\theta) \approx \frac{1}{2}\theta^2$. Use this to find θ_{max}. Remember that what's wanted is $2\theta_{\text{max}}$ because we work with the pinhole's diameter, not its radius.

Problem 5.17 A paradox: frequency dependences

The plates are coloured because the coatings are frequency dependent, reflecting (in the case given) more strongly in the red than the blue. Given that the coatings don't absorb, what they don't reflect they transmit; a coating that doesn't reflect much blue must transmit lots of blue.

In the blue, the poorer reflectivity means a lower finesse so the transmission peaks are wider: a larger fraction of the wavelength range is transmitted.

Problem 6.2 Design of a high-reflectance mirror

(2) The metal is a fairly good conductor (conduction current much greater than displacement current) even at optical frequencies. So the boundary condition is not so very far from $\boldsymbol{E}_{\text{tangential}} = 0$: $Z_{\text{in}} \ll Z_0$; the metal is a 'dead short'.

(3) (a) For $2p$ layers, $2p = 13.3$, rounded up to 14.
(b) For $2p$ layers, $2p = 15.1$, rounded up to 16.
(c) For $2p+1$ layers, $2p+1 = 15.9$, rounded up to 17.
(d) For $2p+1$ layers, $2p+1 = 12.4$, rounded up to 13.

(6) (a) With SiO_2 deposited first we have for 12, 13, 14, 15, 16 layers: $X = 437, 205, 1122, 526, 2882$.
(b) With TiO_2 deposited first we have for 12, 13, 14, 15, 16 layers: $X = 189, 1035, 486, 2659, 1247$.
These figures show that the best option is to use the high-index material first and last, and the worst option is the reverse: low index first and last. Even so, 'it gets worse before it gets better', so once you've started with high index you should finish with it.[16]

[16] Unless the reverse choice just happens to give the value of X you're aiming for.

(7) Calculations along the lines of part (6) show that it's now best to deposit a low-index layer first. A high-index layer is less cost-effective because its refractive index is too close to that of the substrate. The low-index layer brings the input impedance to a 'high' value (meaning above that of air); it's best now to continue towards high values. To demonstrate the principle, I've calculated X for coatings of SiO_2 and TiO_2 on a substrate with refractive index 4, with results:
SiO_2 first, 12, 13, 14, 15 layers: $X = 1149, 539, 2952, 1385$,
TiO_2 first, 12, 13, 14, 15 layers: $X = 72, 393, 185, 1010$.

(8) (a) For a Fabry–Perot coated with 7 layers: $\mathcal{F} = 83$.
(b) For a Fabry–Perot coated with 9 layers: $\mathcal{F} = 250$.
These figures refer to 'reflectivity finesse' only, and the overall performance will be degraded by other limitations, as explored in problem 5.15. Given that departures from plate flatness and the finite size of a scanning pinhole may degrade the resolution by as much as the plate reflectivity, we can hope for an overall finesse of only 80 or so when the plates are coated with nine layers.

Problem 6.3 Anti-reflection coatings

(3) Reflectivity of glass coated with a quarter-wave layer of MgF_2 is 1.26%. This is to be compared with uncoated glass which reflects about 4.3%. So it's better, but the improvement isn't very impressive.

Problem 6.5 General property of a loss-free reflector

(1) The reasoning needs to be set up with some care. The three original waves are

$$E_{\text{incident}} = E_{\text{i}} \exp(\mathrm{i}\boldsymbol{k}_{\text{i}} \cdot \boldsymbol{r} - \mathrm{i}\omega t),$$
$$E_{\text{reflected}} = r_1 E_{\text{i}} \exp(\mathrm{i}\boldsymbol{k}_{\text{r}} \cdot \boldsymbol{r} - \mathrm{i}\omega t),$$
$$E_{\text{transmitted}} = t_1 E_{\text{i}} \exp(\mathrm{i}\boldsymbol{k}_{\text{t}} \cdot \boldsymbol{r} - \mathrm{i}\omega t).$$

After reversal of t and complex conjugation these fields are changed into

$$E_{\text{i}}^* \exp(-\mathrm{i}\boldsymbol{k}_{\text{i}} \cdot \boldsymbol{r} - \mathrm{i}\omega t),$$
$$r_1^* E_{\text{i}}^* \exp(-\mathrm{i}\boldsymbol{k}_{\text{r}} \cdot \boldsymbol{r} - \mathrm{i}\omega t),$$
$$t_1^* E_{\text{i}}^* \exp(-\mathrm{i}\boldsymbol{k}_{\text{t}} \cdot \boldsymbol{r} - \mathrm{i}\omega t).$$

(2) When the second of these waves arrives from top left, a fraction t_1 is transmitted and a fraction r_1 is reflected. Similarly, when the third arrives from top right, a fraction t_2 of its amplitude is transmitted and a fraction r_2 is reflected. The two waves travelling to bottom left must add to replicate the (reversed) incident wave, and the two travelling to bottom right must add to zero. The coefficient of $\exp(-\mathrm{i}\boldsymbol{k}_{\text{i}} \cdot \boldsymbol{r} - \mathrm{i}\omega t)$ is $(r_1^* E_{\text{i}}^*)r_1 + (t_1^* E_{\text{i}}^*)t_2$, from which eqn (6.19) follows. The coefficient for the unwanted wave is $(r_1^* Ei^*)t_1 + (t_1^* E_{\text{i}}^*)r_2$, which leads to eqn (6.18).

Problem 6.6 Properties of thin-film matrices

(6) The (amplitude) reflection and transmission coefficients are given by

$$\Delta\, r_1 = n_1(A + Bn_4) - (C + Dn_4), \qquad \Delta\, t_1 = 2n_1,$$
$$\Delta\, r_2 = n_4(D + Bn_1) - (C + An_1), \qquad \Delta\, t_2 = 2n_4,$$

where Δ is the determinant of coefficients: $\Delta = n_1(A+Bn_4)+(C+Dn_4)$.

Problem 6.8 Polaroid and other absorbers

Let the absorber extend from $z = -l$ to $z = 0$ and have (complex) refractive index n_{layer}. If there are any fields at $z = 0$, they must have arrived from the right and do not penetrate to $z = -l$: the layer's input impedance at $z = -l$ is $E_x/H_y = Z_0/n_{\text{layer}}$, independent of conditions at $z = 0$. Similarly, at $z = 0$, $E_x/H_y = -Z_0/n_{\text{layer}}$, independent of what's happening at $z = -l$. The matrix mathematics must be alive and well to ensure these conditions. Notice that there may be a wave travelling towards $+z$ at $z \geq 0$ (we did not require otherwise), because with $n_{\text{layer}} \neq n_{\text{air}}$ the layer will part-reflect anything arriving from $z > 0$.

Problem 7.2 Ray matrices in special cases

(5) The 'refraction' matrix of eqn (7.2) has $B = 0$, yet does not necessarily relate object to image. So this is a counterexample to any idea that $B = 0$ forces an object–image relation as well as the reverse. Likewise, the $C = 0$ in eqn (7.1) does not force either input or output light to be composed of parallel ray bundles. The other two cases do seem to be necessary as well as sufficient.

Problem 7.3 Matrices useful for work with a lens

(4) The matrix carrying rays from z_1 to the right-hand focal plane is obtained by finding the matrix of part (3) and then requiring the A element to be zero, which yields $h_3 = n_3/P$ and

$$\begin{pmatrix} 0 & 1/P \\ -P & 1 \end{pmatrix}.$$

(6) The matrix carrying rays between the focal planes is

$$\begin{pmatrix} 0 & 1/P \\ -P & 0 \end{pmatrix}.$$

(9) A ray arrives at the thin lens distant y_1 from the axis and at angle θ_1. After refraction it leaves at $y_2 = y_1$. It passes through the lens's focal plane with $y = f\theta_1$, so its departure angle is $\theta_2 = (f\theta_1 - y_1)/f$. From here the matrix can be constructed straightforwardly.

Problem 7.4 Reflection of rays at a spherical mirror

(3) The required matrix equation relates $(y_2, n_2\theta_2)$ (at the point half way to the mirror's centre) to $(y_1, n_1\theta_1)$ (at the mirror just before reflection). The reflected ray is treated as travelling in a medium with refractive index $n_2 = -n_1$. Also, $P = -2n_1/R_s = +2n_1/R$. Then

$$\begin{pmatrix} y_2 \\ n_2\theta_2 \end{pmatrix} = \begin{pmatrix} 1 & (-\frac{1}{2}R)/n_2 \\ 0 & 1 \end{pmatrix} \begin{pmatrix} 1 & 0 \\ -P & 1 \end{pmatrix} \begin{pmatrix} y_1 \\ n_1\theta_1 \end{pmatrix},$$

which evaluates to

$$\begin{pmatrix} y_2 \\ -n_1\theta_2 \end{pmatrix} = \begin{pmatrix} 0 & R/2n_1 \\ -2n_1/R & 1 \end{pmatrix} \begin{pmatrix} y_1 \\ n_1\theta_1 \end{pmatrix}.$$

The required result now follows from working out the upper element.

Problem 7.6 Interpretation of Gaussian-beam mathematics

(11) From eqn (7.12), in the far field the (half) angular divergence of the beam is

$$\frac{w}{z - z_{\text{waist}}} = \frac{w_0}{b} = \sqrt{\frac{\lambda}{\pi|b|}},$$

so if this is much less than 1 we must have $w_0/b \ll 1$ and (equivalently) $|b| \gg \lambda$. To investigate the inequality from problem 7.5, we need to investigate values of r only up to a small multiple of w, because of the exponential fall-off of amplitude with r. So, for all z,

$$\begin{aligned} r^2 \text{ is not much more than } w(z)^2 &= \frac{w_0^2}{b^2}\left\{b^2 + (z - z_{\text{waist}})^2\right\} \\ &\ll b^2 + (z - z_{\text{waist}})^2. \end{aligned}$$

(13) It's simplest to find $1/v_{\text{g}}$ rather than v_{g}. Then

$$\frac{1}{v_{\text{g}}} = \frac{\text{d(effective } k)}{\text{d}\omega} = \frac{\text{d}(k - \text{d}\alpha/\text{d}z)}{\text{d}\omega}.$$

In this, $\tan\alpha = z/b$; how does α depend on angular frequency ω? The problem as set doesn't give enough information. Suppose that all frequency components of a beam have the same waist position z_{waist} and the same spot size w_0. Then $b = \frac{1}{2}\omega w_0^2/c$, and the ω in this gets differentiated for finding the group velocity. But, of course, a beam might be prepared in other ways.

Problem 7.7 Validity of our approximations

With U given by eqn (7.9), the scalar wave equation gives

$$\nabla^2 U - \frac{1}{v^2}\frac{\partial^2 U}{\partial t^2} = \left(2 - \frac{k^2 r^4}{4q^2} + \frac{2\mathrm{i}kr^2}{q}\right)\frac{U}{q^2}.$$

If U were an exact solution of the scalar wave equation, the right-hand side would be zero. So the non-zero right-hand side gives us an idea of how wrong U is. For locations not too far from the beam waist, $|q| \sim b$ and the interesting values of r are of order w_0. All three terms on the right are then of order U/b^2. The two derivatives being cancelled against each other on the left are of order k^2U, and the residue on the right is smaller than this by a factor $(bk)^{-2} \sim (kw_0)^{-4}$. This makes the accuracy of U appear much better than is claimed in problem 7.7. However, the comparison is overoptimistic, because the terms on the left of order k^2U are guaranteed to cancel against each other, just because the fastest-varying part of U is $\mathrm{e}^{\mathrm{i}(kz-\omega t)}$. It is the other factors in U that are of greater interest.

The better way forward is, therefore, to isolate the 'other factors' by defining $\psi(x, y, z)$ according to

$$U = U_0\,\mathrm{e}^{\mathrm{i}(kz-\omega t)}\,\psi, \qquad \text{where} \qquad \psi \approx \frac{\exp\{\mathrm{i}k(x^2+y^2)/2q\}}{q}. \tag{N.9}$$

When this is substituted into the scalar wave equation we find that

$$\frac{\partial^2\psi}{\partial x^2} + \frac{\partial^2\psi}{\partial y^2} + 2\mathrm{i}k\frac{\partial\psi}{\partial z} = -\frac{\partial^2\psi}{\partial z^2}. \tag{N.10}$$

It turns out that the Gaussian wave given by eqn (7.9), and even the Gauss–Hermite waves[17] given by eqn (8.1), are exact solutions of

[17] See problem 8.1 and its solution.

$$\frac{\partial^2\psi}{\partial x^2} + \frac{\partial^2\psi}{\partial y^2} + 2\mathrm{i}k\frac{\partial\psi}{\partial z} = 0. \tag{N.11}$$

Therefore, the error we seek to investigate comes entirely from the right-hand side of eqn (N.10). Compared with the terms on the left, $\partial^2\psi/\partial z^2$ is smaller by a factor $(kw_0)^{-2}$, as stated in the problem.

Problem 7.8 A consequence of diffraction: focus shift

(3) (a) Confocal parameter $= 0.4\,\text{mm}$, focus shift $= 3.2\,\mu\text{m}$, waist spot size $= 8.0\,\mu\text{m}$.
(b) Confocal parameter $= 2.4\,\text{m}$, focus shift $= 1.94\,\text{mm}$, waist spot size $= 0.62\,\text{mm}$.

Problem 7.11 Refraction of a Gaussian beam at a lens

(3) A ray-optics limit holds when $b \ll f$ (problem 7.8). In most cases we can assume that $b_1 \ll X$, and then $XY = f^2/(1+b_1^2/X^2) \approx f^2$. This does, of course, go wrong if we allow X to be too small.

(4) A careful set of limits needs to be taken here. First, since neither b_1 nor b_2 can be zero or infinity, $b_2 = b_1 Y/X$ means that $X = 0$ implies $Y = 0$. Then $b_2 = (f^2 - XY)/b_1 = f^2/b_1$. When $b_1 \ll f$, we start from a narrow waist and a near-ray-optics beam diverges towards the lens. If ray optics applied exactly, the refracted beam would be collimated, travelling parallel to the axis. In fact, the refracted beam has its waist at $Y = 0$, so it is collimated there rather than on exit from the lens. However, we have $b_2 = f^2/b_1 \gg f$, so the refracted beam hardly changes profile between the lens and $Y = 0$; the beam's behaviour really is close to ray-optics expectation.

(5) Yes. At its waist, the Gaussian beam has an amplitude that is a real function of x and y; and this holds only at the waist—see eqn (7.10)—where $R(z) = \infty$. The Fourier transform of a real Gaussian is another real Gaussian, so a waist should transform to another waist.

Problem 7.14 Longitudinal fields and the Poynting vector

The rectangular waveguide described in sidenote 39 on p. 164 provides a helpful precedent. The mode that is most commonly made to propagate in such a waveguide is a TE wave, having a longitudinal component of $\boldsymbol{H}$. The longitudinal $\boldsymbol{H}$ implies a 'partly-sideways' Poynting vector. The energy density in the travelling wave is 'lumpy' in the way we remember from §1.5. The sideways component of the Poynting vector contributes to the feeding of energy into and out of the lumps as the wave progresses. For a wave of angular frequency ω, the 'sideways' component of the Poynting vector varies at 2ω and has no long-term effects; but it's there.

We assume a plane-polarized Gaussian beam travelling in vacuum with field E_x given by

$$E_x = E_0 \, \mathrm{e}^{\mathrm{i}(kz-\omega t)} \frac{\exp(\mathrm{i}kr^2/2q)}{q}$$

with $q = z - \mathrm{i}b$ and $r^2 = x^2 + y^2$. From div $\boldsymbol{E} = 0$ we calculate

$$\frac{\partial E_z}{\partial z} = -E_0 \, \mathrm{e}^{\mathrm{i}(kz-\omega t)} \frac{\exp(\mathrm{i}kr^2/2q)}{q} \frac{\mathrm{i}kx}{q}.$$

In integrating this to find E_z, we should remember that there is a dependence on z through q as well as through $\mathrm{e}^{\mathrm{i}(kz-\omega t)}$. However, the variation through q is much the slower, and so we have approximately

$$E_z \approx -E_0 \, \mathrm{e}^{\mathrm{i}(kz-\omega t)} \frac{\exp(\mathrm{i}kr^2/2q)}{q} \frac{x}{q} \left(1 + \text{order } (kb)^{-1}\right).$$

The approximation here is no worse than that in E_x, since the error $(kw_0)^{-2}$ displayed in problem 7.7 is of the same order as $(kb)^{-1}$.

Components of field $\boldsymbol{B}$ can now be found from $\partial \boldsymbol{B}/\partial t = -\operatorname{curl} \boldsymbol{E}$, with the results

$$ZH_x = -\frac{xy}{q^2}E_x, \qquad ZH_y = E_x, \qquad ZH_z = -\frac{y}{q}E_x,$$

where Z is the characteristic impedance of the medium supporting the wave. From these field components, we can assemble an expression for the Poynting vector. Its time average has z-component S_z and radial component S_r, given by

$$S_z = \frac{|E_0|^2}{2Z}\frac{\exp(-2r^2/w^2)}{|q|^2}, \qquad S_r = \frac{|E_0|^2}{2Z}\frac{\exp(-2r^2/w^2)}{|q|^2}\frac{r}{R},$$

which makes the mean energy flow follow the direction of a ray. Here R is the radius of curvature of the wave's wavefront, as given by eqn (7.11).

Problem 8.1 The Gauss–Hermite wave solutions
The algebra can be heavy, and the problem is given here only so that this whole chapter does not rely on an 'it can be shown'.

First, make the substitution of eqn (N.9), defining ψ by

$$U = U_0\,e^{i(kz-\omega t)}\,\psi.$$

Follow the outline solution given above for problem 7.7 to show that, to an acceptable approximation, ψ obeys

$$\frac{\partial^2\psi}{\partial x^2} + \frac{\partial^2\psi}{\partial y^2} + 2ik\frac{\partial\psi}{\partial z} = 0. \tag{N.11}$$

Attempt to find a solution to eqn (N.11) in the form

$$\psi = \frac{\exp\{ik(x^2+y^2)/2q\}}{q}\,e^{-i(l+m)\alpha}\,F(\sqrt{2}x/w, \sqrt{2}y/w). \tag{N.12}$$

We find that F obeys the differential equation

$$\frac{\partial^2 F}{\partial\xi^2} + \frac{\partial^2 F}{\partial\eta^2} - 2\left(\xi\frac{\partial F}{\partial\xi} + \eta\frac{\partial F}{\partial\eta}\right) + 2(l+m)F = 0, \tag{N.13}$$

where $\xi = \sqrt{2}x/w$, $\eta = \sqrt{2}y/w$. The fact that x, y and z appear in eqn (N.13) only within ξ and η means that the apparently courageous form of the trial solution (N.12) is justified.

Seek a 'separated' solution in which F factorizes into

$$F(\xi,\eta) = X(\xi)\,Y(\eta).$$

Then

$$\frac{1}{X}(X'' - 2\xi X' + 2lX) = -\frac{1}{Y}(Y'' - 2\eta Y' + 2mY),$$

in which both sides are equal to some separation constant. The quantities l and m have not yet been defined, and we can exercise a choice now to make the constant zero. The result is:

$$X'' - 2\xi X' + 2lX = 0, \qquad Y'' - 2\eta Y' + 2mY = 0.$$

These have the form of Hermite's differential equation. The solution behaves acceptably only if l (in the first case, or m in the second) is an integer, and then the solution is $\mathrm{H}_l(\xi)$ or $\mathrm{H}_m(\eta)$. Backtracking through all the substitutions finally gives eqn (8.1).

We can also return to eqn (N.13) and make a separation instead in terms of functions of a scaled radius $\rho = \sqrt{2}r/w$ and the azimuth angle ϕ. After some manoeuvres we obtain a Laguerre equation with the result given in eqn (8.2) on p. 174.

Problem 8.2 The 'phase anomaly'
The question talks of the wave's phase on-axis. With odd-symmetry modes this isn't very helpful because the amplitude is zero there. Think instead about what happens off-axis. Figure 7.5 may help. The phase $(l+m+1)\alpha$ is an additional phase in a generalized version of eqn (7.14), so it's added (for example) to the phase for light travelling from top left to top right. The classic 'phase anomaly' considers what happens along ray-optics paths, so (for example) from top left to bottom right. We must take into account any phase difference between top and bottom right. An odd-symmetry mode has a minus sign here that removes the difficulty.

Problem 8.3 The condition for a low-loss cavity mode: pictorial
(1) Take the origin of z at the beam's waist. A mirror at $z_2 > 0$ must be concave to the middle if it is to fit the beam. Its radius of curvature is $R = z_2 + b^2/z_2$, so it has its centre at $z = -b^2/z_2$, to the left of the beam's waist. By a similar argument, a concave mirror to the left of the waist has its centre to the right: the centres are 'crossed over'. Then the mirror separation is less than the sum of their radii: $L < R_1 + R_2$.
(2) Let mirrors M_1 and M_2 lie at $z_1 < 0$ and $z_2 > 0$, respectively. Mirror M_2's centre lies at $z = -b^2/z_2$. If this is outside the cavity, it is to the left of M_1, so $-b^2/z_2 < z_1$. This rearranges to $b^2/|z_1| > z_2$ (care over the inequality because z_1 is negative). So M_1's centre lies to the right of M_2: both centres are outside the cavity. It is easy to reverse the inequalities and show that if one centre lies inside the cavity then both do so.
(3) The result of part (2) shows that $(1-L/R_1)$ and $(1-L/R_2)$ are either both positive or both negative; either way their product is positive. This gives one part of the required condition. For the remaining part, take $L < R_1 + R_2$ and divide by R_1R_2, which at this stage is positive. The result is $L^2/R_1R_2 < L/R_1 + L/R_2$, which is nearly there.
(4) To fit the sign convention used in the 'stability' condition, we must write $R_2 = z_2 + b^2/z_2$ but $R_1 = -(z_1 + b^2/z_1)$. Adding these gives

$$L = z_2 - z_1 = R_2 + R_1 + b^2(1/z_1 - 1/z_2) > R_1 + R_2.$$

Divide by R_1R_2, which is now negative. The negative sign reverses the inequality, after which the algebra is the same as in part (3). The quantity $(1-L/R_1)$ is positive because R_1 is negative. A simple drawing shows that $(1 - L/R_2)$ must be positive too. So the product of these brackets is positive, which gives the remaining inequality.

Problem 8.6 Ray calculation for low loss

(4) Eigenvalues $e^{\pm i\phi}$ are degenerate—belonging to the same $A+D$—so we can take linear combinations of their eigenfunctions. It's not hard to see what are the linear combinations that make the (y, α) real.

Alternatively, we can describe the propagation of the rays using a complex convention, similar to $e^{i(kz-\omega t)}$, and then the solutions we have obtained resemble the Bloch functions used to describe electron wave functions in a periodic lattice.

Problem 8.7 Use of the 'stability' chart

(2) The cavity is 'stable' when the mirror separation is less than the smaller mirror radius, 0.35 m. Separations intermediate between the values of the radii, from 0.35 m to 1 m, give an 'unstable' cavity. From 1 m to 1.35 m, the sum of the radii, the cavity is again 'stable'.[18]

[18] The numerical values are taken from an experiment in the Oxford Physics practical course, where these 'stability' conditions can be checked experimentally.

Problem 8.8 The general symmetrical cavity

(3) For propagation of the beam through one lens-chain period:

$$\begin{pmatrix} q(z_2) \\ 1 \end{pmatrix} = (\text{cm}) \begin{pmatrix} 1 & 0 \\ -1/f & 1 \end{pmatrix} \begin{pmatrix} 1 & L \\ 0 & 1 \end{pmatrix} \begin{pmatrix} q(z_1) \\ 1 \end{pmatrix}$$
$$= (\text{cm}) \begin{pmatrix} 1 & L \\ -1/f & 1 - L/f \end{pmatrix} \begin{pmatrix} q(z_1) \\ 1 \end{pmatrix}.$$

Matrix elements A, B, C, D can be read off from this, and the solution for $q(z_1)$ can be obtained by substituting them into eqn (8.5):

$$q = -\tfrac{1}{2}L - i\sqrt{Lf - L^2/4}. \qquad \text{(N.14)}$$

Requiring the quantity under the square root to be positive gives the 'stability' condition $L < 2R$.

(4) From the real part of $q(z_1)$ in eqn (N.14) we can read off the beam's waist location at the middle of the cavity.

(5) For the case $f = L/2$ we can read off from the square root in eqn (N.14) the value $b = L/2$. Had we gone back to eqn (7.11) we'd have said $2f = R = f + b^2/f$ from which $b = f = L/2$ follows as quickly.

Problem 8.9 Longitudinal modes

(2) The integer p is about 2×10^6. A cavity of length 0.5 m has longitudinal modes spaced by 300 MHz. It should be obvious that b hardly varies from one longitudinal mode to another.[19]

[19] It is tempting to use the result of problem 8.8(5) to deduce that a symmetrical confocal cavity has $b = L/2$, exactly independent of frequency. But there are so many other ways of fitting a light beam into this cavity (problem 8.11) that the conclusion is insecure.

Problem 8.11 The symmetrical confocal cavity

(1) The matrix for travel through L from z_1 to z_2 is $\begin{pmatrix} 0 & f \\ -1/f & 0 \end{pmatrix}$. Travel through $2L$ multiplies the ray's (y, θ) by two such matrices, and the product of the square matrices is $\begin{pmatrix} -1 & 0 \\ 0 & -1 \end{pmatrix}$. Multiplying out gives the required result: (y, θ) are reversed in sign. Travel through $4L$ is described by the product of two of the matrices last given, and that product is a unit matrix.

(2) A Gaussian beam, with complex radius q_1 at z_1, is transformed into a beam at z_2 with complex radius q_2, where

$$\begin{pmatrix} q_2 \\ 1 \end{pmatrix} = (\text{cm}) \begin{pmatrix} 0 & f \\ -1/f & 0 \end{pmatrix} \begin{pmatrix} q_1 \\ 1 \end{pmatrix} = (\text{cm}) \begin{pmatrix} f \\ -q_1/f \end{pmatrix}.$$

Expanding this into two equations, and dividing away (cancel me), we find $q_2 = -f^2/q_1$. Further travel through L results in $q_3 = -f^2/q_2 = q_1$.

(3) Let the first beam's waist lie at z_{w1} so that $q_1 = z_1 - z_{\text{w1}} - ib_1$. Similarly, the refracted beam's waist lies at z_{w2} so that $q_2 = z_2 - z_{\text{w2}} - ib_2$. Putting these into $q_2 = -f^2/q_1$ and separating real and imaginary parts gives us

$$z_2 - z_{\text{w2}} = f^2(z_{\text{w1}} - z_1)/|q_1|^2.$$

The mirror-cavity beam fits back over itself if $(z_2 - z_{\text{w2}}) = (z_{\text{w1}} - z_1)$, which is the case if $|q_1|^2 = f^2$. Evaluating $|q_1|^2$ gives

$$(z_1 - z_{\text{w1}})^2 = f^2 - b_1^2, \qquad \text{so} \qquad |z_1 - z_{\text{w1}}| < f;$$

thus, the beam's original waist is required to lie within the cavity, but otherwise could be anywhere along the z-axis. Once the waist location has been chosen, b_1 is determined by writing the last equation a different way round; from b_1 we can find everything else.

Problem 8.13 Practicalities with a confocal Fabry–Perot

(2) If one mirror of a confocal Fabry–Perot is tilted, its centre remains on the other mirror. We still have a confocal étalon with correct geometry, just displaced slightly. And the position and orientation of the étalon relative to the incoming light aren't critical.

(3) Insufficient versatility. The plate spacing has to equal the radius of either mirror, so only this one spacing can be used for any given pair of plates. In effect, we have a fixed free spectral range of $1/4L$.

Problem 8.14 Examples of cavity design

(1) At cavity centre, spot size = 0.22 mm. At a mirror the spot size is greater by a factor $\sqrt{2}$, giving 0.32 mm.

(3) The result of part (2) rearranges into a quadratic for R with solutions 49 m and 0.505 m. If we try to make a mirror of radius 49 m, the hollow needs to be 577 nm deep, less than a wavelength, so it can't be made to worthwhile accuracy.

(4) For spot size of 1 mm, $R - L = 5.2$ mm. For spot size of 2 mm, $R - L = 0.32$ mm.

Problem 9.6 Variations on the Young slits

The ground glass imposes a randomness on the phase of the light heading for the Young slits, but that randomness is time independent so long as we don't move the glass.

The light fields at holes Σ_1, Σ_2 have a time-independent phase difference $(\alpha_2 - \alpha_1)$, so fringes may be observed, and may be observed for all slit spacings. Even a 'slow' detector will tell us that the coherence area is as wide as the region where there's any light.

We don't know in advance what the phase of the $\boldsymbol{E}$-field will be at either of holes Σ_1 or Σ_2 because we don't know the exact shape of the ground glass. But once we've seen fringes we can predict future values of $(\alpha_2 - \alpha_1)$: no change.

If the holes Σ_1, Σ_2 were elongated into slits, the fringes would disappear. Pairs of points along the slits (corresponding to possible positions of the small holes we had before) will receive light with some random phase difference $(\alpha_2 - \alpha_1)$; but different pairs of points have different random values of $(\alpha_2 - \alpha_1)$. Where one pair of points sends 'dark' another will send 'bright', and the fringes are smeared out.

Return to the setup at the beginning of this question, with small holes Σ_1, Σ_2 and a large-area ground-glass source illuminated by a laser. We could simulate the 'incoherent' conditions of §9.4 after all: Move the ground glass so that a fresh rough surface is presented repeatedly. The coherence time τ_c of the light downstream from the ground glass will be the time taken for one piece of rough surface to be replaced by another. Thus, we simulate illumination that changes its phase[20] randomly over timescale τ_c. This is sometimes done if a laser gives out a beam that is 'too coherent' for an experiment.

[20]Notice that we do *not* introduce randomness on timescale τ_c just by using a laser with coherence time τ_c, because we've used a single transverse mode. The original randomness would be recovered (without the ground glass) only if the laser gave out a random array of transverse modes with different frequencies, those frequencies covering range $1/\tau_c$. By the time this amount of randomness has been 'designed in', we're pretty close to achieving the effects of a non-laser source.

Problem 9.8 The Michelson stellar interferometer

(5) About 100 mm. Remember the final part of problem 9.6. The fringes will be smeared if a mirror extends over more than one uniform-phase patch (dimension r_0 in sidenote 14 on p. 200).

Problem 9.9 Interference of two parallel plane waves

(4) Nothing! The light 'decides' whether to go down or to the right at the first place where amplitudes are added. That is at the half-reflecting mirror. By energy conservation, what doesn't go to the right goes down. The ground glass can't get it back. All our instincts are that randomizing the light with ground glass should destroy fringes, so it may be a surprise that it does no harm here. Things would be different if ... well, you decide.

(5) The longest path difference $l_{\max} = 3\,\text{mm}$ is smaller than the coherence length by a factor 10. Your mathematical discussion should have made repeated use of an approximation based on this.

Problem 9.10 Beats using two independent light sources

(2) The incident light has (real) electric-field amplitude $V_{\text{in}}(t)$ whose Fourier transform is $\widetilde{V}(\omega)$. The photocurrent at time t is $i(t) = D V_{\text{in}}(t)^2$ where D represents a sensitivity constant for the photodetector. Substituting for $V_{\text{in}}(t)$ with care over dummy variables gives

$$i(t) = D \int_{-\infty}^{\infty} \frac{\mathrm{d}\Omega}{2\pi}\, \widetilde{V}(\Omega)\, \mathrm{e}^{-\mathrm{i}\Omega t} \int_{-\infty}^{\infty} \frac{\mathrm{d}s}{2\pi}\, \widetilde{V}(s)\, \mathrm{e}^{-\mathrm{i}st}.$$

In the second integral set $s = \omega - \Omega$:

$$i(t) = D\int_{-\infty}^{\infty} \frac{\mathrm{d}\omega}{2\pi}\, \mathrm{e}^{-i\omega t} \int_{-\infty}^{\infty} \frac{\mathrm{d}\Omega}{2\pi} \widetilde{V}(\Omega)\, \widetilde{V}(\omega - \Omega).$$

Within this we can identify the Fourier transform $\tilde{i}(\omega)$ of $i(t)$ as

$$\tilde{i}(\omega) = D\int_{-\infty}^{\infty} \frac{\mathrm{d}\Omega}{2\pi}\, \widetilde{V}(\Omega)\, \widetilde{V}(\omega - \Omega). \tag{N.15}$$

(3) The power dissipated in resistance R by the current $i(t)$ is $R\,i(t)^2$, and the mean power dissipated within a long observation time T is

$$\frac{R}{T}\int_{-T/2}^{T/2} i(t)^2\, \mathrm{d}t = \frac{R}{T}\int_{-\infty}^{\infty} \frac{\mathrm{d}\omega}{2\pi}\, \tilde{i}(\omega)^*\, \tilde{i}(\omega) = \frac{2R}{T}\int_{0}^{\infty} \frac{\mathrm{d}\omega}{2\pi}\, \tilde{i}(\omega)^*\, \tilde{i}(\omega),$$

where we use the Parseval theorem. So the electrical power attributable to frequency range $\mathrm{d}\omega/2\pi$ is $P_{\text{elec}}(\omega)\, \mathrm{d}\omega/2\pi$, where

$$\begin{aligned} P_{\text{elec}}(\omega) &= \frac{2R}{T}\, \tilde{i}(\omega)^*\, \tilde{i}(\omega) \\ &= \frac{2RD^2}{T}\int_{-\infty}^{\infty} \frac{\mathrm{d}\Omega}{2\pi}\, \widetilde{V}(\Omega)^*\, \widetilde{V}(\omega - \Omega)^* \int_{-\infty}^{\infty} \frac{\mathrm{d}s}{2\pi}\, \widetilde{V}(s)\, \widetilde{V}(\omega - s) \\ &= \frac{2RD^2}{T}\int_{-\infty}^{\infty} \frac{\mathrm{d}\Omega}{2\pi} \int_{-\infty}^{\infty} \frac{\mathrm{d}s}{2\pi}\, \widetilde{V}(\Omega)^*\, \widetilde{V}(\omega - \Omega)^*\, \widetilde{V}(s)\, \widetilde{V}(\omega - s). \end{aligned} \tag{N.16}$$

Further than this we cannot go with just one wavetrain $V(t)$. However we are thinking of random wavetrains being emitted from some discharge lamp, and we're really interested in what to expect: ensemble-average over an ensemble of comparable wavetrains.

For any one wavetrain, $\widetilde{V}(\Omega)$ will be different from $\widetilde{V}(s)$ when frequencies Ω and s are different; there is no expected similarity between them, they are uncorrelated, and an ensemble average like $\langle \widetilde{V}(\Omega)^*\, \widetilde{V}(s)\rangle$ is zero. However, this cannot be the case if Ω and s are the same frequency, because the functions are then the same. So we ask: how close do two frequencies lie if there has to be a correlation between their $\widetilde{V}$s? The answer must be a $\Delta\nu = 1/T$ or $\Delta s = 2\pi/T$: given that observation takes place over a time-range T, a measurement can, in principle, not distinguish frequencies (ν) that are separated by less than about $1/T$. Then a product[21] of the form $\langle f(\Omega)^* f(s)\rangle$ has a peak, as a function of s, whose area we can estimate at $f(\Omega)^* f(\Omega)\,(2\pi/T)$. If it helps we can write $\langle f(\Omega)^* f(s)\rangle \approx |f(\Omega)|^2 (2\pi/T)\, \delta(s - \Omega)$.

Return to eqn (N.16). As we integrate over s, we can think that we are sliding $\widetilde{V}(s)\, \widetilde{V}(\omega - s)$ past $\widetilde{V}(\Omega)^*\, \widetilde{V}(\omega - \Omega)^*$ looking for a correlation. There is an obvious match where $s = \Omega$, but there is also another where $s = \omega - \Omega$. So the ensemble-average $P_{\text{elec}}(\omega)$ is

$$P_{\text{elec}}(\omega) = \frac{2RD^2}{T}\, \frac{2}{T}\int_{-\infty}^{\infty} \frac{\mathrm{d}\Omega}{2\pi} \Big\langle \big|\widetilde{V}(\Omega)\big|^2\, \big|\widetilde{V}(\omega - \Omega)\big|^2 \Big\rangle.$$

[21] A product like $\langle f(\Omega)\, f(s)\rangle$ ensemble-averages to zero because $f()$ has a random phase angle. So we are concerned only with terms that yield a modulus squared.

To display the physics most clearly, we divide the range of integration over Ω into three: $-\infty$ to 0, 0 to ω, and ω to ∞. The first and third pieces are equal. After a little tidying, we find

$$P_{\text{elec}}(\omega) = \frac{4RD^2}{T^2}\left\{2\int_0^\infty \frac{\mathrm{d}\Omega}{2\pi}\left\langle|\widetilde{V}(\Omega)|^2\,|\widetilde{V}(\omega+\Omega)|^2\right\rangle \right.$$
$$\left. + \int_0^\omega \frac{\mathrm{d}\Omega}{2\pi}\left\langle|\widetilde{V}(\Omega)|^2\,|\widetilde{V}(\omega-\Omega)|^2\right\rangle\right\}.$$

The two terms here show ω as made from a difference of input frequencies and from a sum, as is expected from a square-law detector. The interest here is in a radio-frequency difference between two optical frequencies, so we discard the second (the 'sum') integral. Using eqn (2.22), we reduce what remains to the required

$$P_{\text{elec}}(\omega) = 2RD^2Z^2\int_0^\infty P(\Omega)\,P(\Omega+\omega)\,\frac{\mathrm{d}\Omega}{2\pi}.$$

(4) The result here is straightforwardly obtained from that of part (3). The first two terms are 'homodyne' terms, generated by beating together frequencies originating from a single one of the input beams. The second two terms are 'heterodyne' terms, generated by beating a Fourier component from one beam against a Fourier component from the other.

Problem 9.12 Polarized and unpolarized light

(2) (a) Some ellipse. When the observation is repeated we get some other ellipse, unrelated to the first in size, shape or orientation.
(b) A mess.

(3) To mimic linearly polarized light, connect a single generator to both x and y inputs. To mimic circularly polarized light, connect a single generator to both x and y inputs, but with a 90° phase shift in one of the two signal paths. To mimic partially polarized light, use two generators, connecting one to x and a weighted sum of the two to y.

(4) (a) Over a few milliseconds only, the oscilloscope displays a circle. A repeat observation again yields a circle traced in the same direction, but with a different radius.
(b) A long-term observation shows a trace that follows a generally circular path, but with the radius varying (over timescale τ_c and longer) over quite a wide range of values. Although the polarization state is completely non-random, we still have fluctuations of intensity as foreseen in the Comment in §9.11 on p. 206. Circularly polarized light is not quite as tidy as the presentations of it usually given by physics teachers!

Problem 10.1 Properties of the autocorrelation function

(3) To derive eqn (10.15), write $V = \frac{1}{2}(U + U^*)$ and use the result of part (2).

Problem 10.2 Basic properties of $\gamma(\tau)$

(4) Statement $\mathrm{Re}\{\mathrm{e}^{\mathrm{i}\alpha}\,\gamma(\tau)\} \le 1$ is true for all values of α. Therefore α can be chosen by us to impose the tightest condition on $\gamma(\tau)$. Choose α to make $\mathrm{e}^{\mathrm{i}\alpha}\,\gamma(\tau)$ real and positive, in which case it is equal to $|\gamma(\tau)|$.

Problem 10.3 Michelson interferometer, correlation and visibility

(4) Write

$$\left\langle \left|U_1(t) - r\,\mathrm{e}^{\mathrm{i}\alpha}\,U_2(t+\tau)\right|^2 \right\rangle \ge 0,$$

and adjust both α and r.

Problem 10.5 Sinewave wave packets occurring at random times

(1) We assemble the 'innards' of the correlation function:

$$\begin{aligned}
V(t)\,V(t+\tau) &= \sum_i V_0 \cos\{\omega_0(t-t_i) + \alpha_i\} h(t-t_i) \\
&\quad \times \sum_j V_0 \cos\{\omega_0(t+\tau-t_j) + \alpha_j\} h(t+\tau-t_j) \\
&= \frac{V_0^2}{2} \sum_i \sum_j h(t-t_i)\,h(t+\tau-t_j) \\
&\quad \times \Big\{ \cos\{\omega_0(2t - t_i - t_j + \tau) + \alpha_i + \alpha_j\} \\
&\qquad + \cos\{\omega_0(t_j - t_i - \tau) + \alpha_i - \alpha_j\} \Big\}.
\end{aligned}$$

Expressions like this look much more intimidating than they are. There are two averages to be done. One is explicit in the expression (10.7) that defines the correlation function: an average over a long time T. But we also wish to know what is to be expected when we average over a distribution of wavetrains, so we are to take an ensemble average as well. It is simplest to do the ensemble-averaging first.

The first cosine ensemble-averages to zero for all i and j because $\omega_0(t_i + t_j)$ is a random phase angle. The second cosine similarly averages to zero for all terms[22] in which $j \ne i$. We are left with

$$\begin{aligned}
\big\langle V(t)\,V(t+\tau)\big\rangle &= \frac{V_0^2}{2}\cos(\omega_0\tau)\Big\langle \sum_i h(t-t_i)\,h(t+\tau-t_i)\Big\rangle \\
&= \begin{cases} \tfrac{1}{2}V_0^2\,N\cos(\omega_0\tau)\,\{\tau_{\mathrm{w}} - |\tau|\} & \text{for } |\tau| \le \tau_{\mathrm{w}} \\ 0 & \text{otherwise,} \end{cases}
\end{aligned}$$

where the factor $\{\tau_{\mathrm{w}} - |\tau|\}$ comes from the product of the hs, and is the length of time during which two top-hat envelopes overlap.

Notice that the αs cancel in the above, so it does not matter whether they are all the same or whether they are randomly distributed.

(2) The additional term resembles that of eqn (9.7) but with $(\omega - \omega_0)$ replaced by $(\omega + \omega_0)$. However, for wave packets containing several periods of oscillation within duration τ_{w}, the peak of $P(\omega)$ centred on $\omega = -\omega_0$ does not extend its 'wings' significantly into positive ωs. The term is therefore legitimately dropped, to a very good approximation.

[22] Physically, this means that there is no (stationary) interference between different wave packets, only between two pieces of the same wave packet. This result is, if anything, easier to obtain here than by arguing in words as had to be done in §9.12 and in problem 9.11(2).

(4) If the wavetrain was made by gating a continuously running sine-wave, we would have $\alpha_i = \omega_0 t_i$ for all i. Now every wave packet has a definite phase relationship with every other wave packet that it happens to overlap.[23] The correlation function contains $\cos(\omega_0\tau)$ but does not fall off with increase of $|\tau|$.

[23]The cosine containing $2\omega_0 t$ still ensemble-averages to zero because times t_i and t_j are randomly distributed.

Problem 10.6 Natural (lifetime) broadening of a spectral line

(2) When we write out an expression for $V(t)\,V(t+\tau)$, we obtain a product of cosines that is the same as that in the solution to problem 10.5. The same reasoning applies: a cosine of 'the sum' ensemble-averages to zero for all i and j because the phase $\omega_0(t_i+t_j)$ is randomly distributed; and a cosine of 'the difference' ensemble-averages to zero for all j except $j=i$. And in the process all the αs disappear so the result is independent of what assumption we make about them. We are left with

$$\big\langle V(t)\,V(t+\tau)\big\rangle = \frac{V_0^2}{2}\cos(\omega_0\tau)\exp(-\tau/2\tau_{\rm r}) \times\Big\langle \sum_i \exp\{-(t-t_i)/\tau_{\rm r}\}\,u(t-t_i)\,u(t+\tau-t_i)\Big\rangle.$$

It will suffice to evaluate things for the case where τ is positive; the product of the two step functions is then simply $u(t-t_i)$. The remaining averaging is conveniently done by integrating over t:

$$\Big\langle \sum_i \exp\{-(t-t_i)/\tau_{\rm r}\}\,u(t-t_i)\Big\rangle = \frac{1}{T}\int_{-T/2}^{T/2}\sum_i \exp\{-(t-t_i)/\tau_{\rm r}\}\,u(t-t_i)\,{\rm d}t = \sum_i \frac{1}{T}\int_{t_i}^{T/2}\exp\{-(t-t_i)/\tau_{\rm r}\}\,{\rm d}t = \sum_i \frac{1}{T}\tau_{\rm r} = N\tau_{\rm r}.$$

Finally then

$$\big\langle V(t)\,V(t+\tau)\big\rangle = \tfrac{1}{2}V_0^2\,N\tau_{\rm r}\cos(\omega_0\tau)\,\exp\{-|\tau|/2\tau_{\rm r}\},$$

in which we have used the known fact that the result must be an even function of τ.

(3) Taking the Fourier transform is not difficult. It is, however, worth taking whatever measures you can invent to streamline the integration.

Problem 10.7 Pressure (collision) broadening of a spectral line

(2) Both assumptions are wrong when there is a distribution of molecular speeds. A fast molecule bumps into something soon, so the $\tau_{\rm coll}$ in ${\rm d}t/\tau_{\rm coll}$ must be a decreasing function of molecular speed v. Conversely, a slow particle is bumped into before it has had a chance to move far, so λ in ${\rm d}x/\lambda$ must be small for small v; it is an increasing function of v.[24] A full treatment of line broadening should, in principle, allow

[24]These dependences of collision rate and free path upon speed v are worked out by Kennard (1938), §63.

for a statistical distribution of speeds, accompanied by a distribution of $\tau_{\rm coll}$ values. However, the model we are using is crude in more important ways, so this detail is not usually followed up.

Problem 10.8 More on pressure broadening

(2) The convolution can be written in the form

$$P(\omega) \propto \int_{-\infty}^{\infty} \frac{1}{(\omega' - \omega_0)^2 + a^2} \times \frac{1}{(\omega - \omega')^2 + b^2}\, \mathrm{d}\omega',$$

so the two Lorentzians being convolved are $\{(\omega - \omega_0)^2 + a^2\}^{-1}$ and $\{\omega^2 + b^2\}^{-1}$. Now

$$\text{the (inverse) transform of} \quad \frac{1}{(\omega - \omega_0)^2 + a^2} \quad \text{is} \quad \frac{\mathrm{e}^{-\mathrm{i}\omega_0\tau}\mathrm{e}^{-a|\tau|}}{2a}.$$

This[25] is exact, meaning that no term centred on $\omega = -\omega_0$ has been neglected. Using eqn (2.37) we find that the (inverse) transform of the convolution of the two Lorentzians is the product

$$2\pi\, \frac{\mathrm{e}^{-\mathrm{i}\omega_0\tau}\, \mathrm{e}^{-a|\tau|}}{2a} \times \frac{\mathrm{e}^{-b|\tau|}}{2b}.$$

[25] To derive this result 'forwards' may need contour integration; but confirming it by working from right to left is easy.

The presence of $(a + b)|\tau|$ in the exponent makes it obvious that when we transform back to functions of ω the result will have the shape of eqn (10.26). This makes the Fourier route very transparent: you can see why the result is as it is.

To do the calculation by a non-Fourier route, make the substitutions $x = (\omega - \omega_0)/a$, $z = (\omega' - \omega_0)/a$ and $\beta = b/a$. Then

$$P(\omega) \propto \frac{1}{a^3} \int_{-\infty}^{\infty} \frac{1}{z^2 + 1} \times \frac{1}{(x - z)^2 + \beta^2}\, \mathrm{d}z.$$

The integration path along the real z-axis can be continued using a large semicircle round the top half of the complex plane (or the bottom half, either will do) to form a closed contour. The integrand behaves like z^{-4} for large z so the integral taken round the semicircle is zero. The integrand has poles at $z = \pm\mathrm{i}$ and at $z = x \pm \mathrm{i}\beta$. Two of these poles lie within the contour, and the integral is $2\pi\mathrm{i} \times$ the sum of the residues.

Problem 10.9 Alternative Wiener–Khintchine theorems

(3) Although eqns (10.27) and (10.28) look very similar—and are very similar for positive ω—they are completely different for negative ω. It follows that any attempt at deriving eqn (10.28) from eqn (10.27), or the reverse, must fail; don't waste time trying it.

(4) The expression given for $U(t)$ violates the rules for constructing an analytic signal. Its Fourier transform is non-zero (though small) at negative values of ω, in the remote 'wings' of the spectral line.

The incorrect form for $U(t)$ is tempting. And that's the point of the comment about traps. If you happen to get into a tangle with this one, it can take a long time to find the mistake.

Although the form of $U(t)$ given in the problem is in principle wrong, it's 'only just' wrong. We often use such expressions, dignifying them with the name **rotating wave approximation**. We just have to remember that an approximation has been made.

Problem 10.10 Yet another Wiener–Khintchine property
Since $g(\omega - \omega_0)$ is a 'one-sided' function of frequency it makes sense to relate it to the one-sided $P(\omega)$ of eqn (10.28). Since

$$\frac{2Z_0}{n} P(\omega) = \int_{-\infty}^{\infty} \Gamma(\tau)\, \mathrm{e}^{\mathrm{i}\omega\tau}\, \mathrm{d}\tau,$$

we have the inverse:

$$\Gamma(\tau) = \int_{-\infty}^{\infty} \frac{2Z_0}{n} P(\omega)\, \mathrm{e}^{-\mathrm{i}\omega\tau} \frac{\mathrm{d}\omega}{2\pi} = \int_{0}^{\infty} \frac{2Z_0}{n} P(\omega)\, \mathrm{e}^{-\mathrm{i}\omega\tau} \frac{\mathrm{d}\omega}{2\pi},$$

where we discard negative ωs in the integration because this $P(\omega)$ is zero there. Take the special case of this where $\tau = 0$. Then

$$\Gamma(0) = \int_{0}^{\infty} \frac{2Z_0}{n} P(\omega) \frac{\mathrm{d}\omega}{2\pi}.$$

This integral can be used to evaluate the constant of proportionality linking $P(\omega)$ with $g(\omega - \omega_0)$.

Problem 10.14 Young fringes with imperfect collimation
In sidenote 49 on p. 242 it is mentioned that eqn (10.32) contains functions of time, so it ought not to contain functions of frequency such as k. We ought to be on one side or the other of the Fourier transform linking t with ω. However, the discussion is implicitly about light that contains a fairly narrow range of frequencies, perhaps one spectral line, since no mention is made of any limitation on the coherence time. Then all frequency components present in $U(x, t)$ are close together and the range of values taken by k is small in percentage terms. In such a case the use of k must be harmless.

Even so, think separately about the factors $\mathrm{e}^{\mathrm{i}kd}$ and $\mathrm{e}^{-\mathrm{i}kxX/f}/(\mathrm{i}\lambda f)$ in eqn (10.32).

Problem 10.17 Conditions for measuring $\gamma^{(2)}(\tau)$

(1) If we collect from p coherence areas we can expect fluctuations still to be present but to have variance $\propto p$ as with a random walk. The denominator of eqn (10.13) rises like p^2. So the correlation $\gamma^{(2)}(\tau) - 1$ diminishes with increase of p. It is best to use $p = 1$.

(4) The width of the coherence area is 6.27 mm, compared with an aperture in front of the detector of 5 mm.

(5) The étendue is $0.74\lambda^2$. By either comparison, we see that the detector was arranged to collect light from about one coherence area; equivalently one transverse mode. And, within that requirement, the detector was made as large as possible in order to maximize energy collection.

Problem 10.18 Overlapping wave packets in Hanbury Brown and Twiss
The number of photons per mode is 1.64×10^{-2}.

Guess the radiative lifetime as 10^{-8} s. Each wave packet is about this long, however much it's messed about by collisions.[26] Then from eqn (11.22) the average number of wave packets overlapping in one transverse mode is $(1.64 \times 10^{-2}) \times (10^{-8}) \times (8.5 \times 10^{12}) = 1394$. This applies to each of two polarizations (in the original experiment the light used was unpolarized).

[26] We assume 'soft' collisions.

Problem 10.19 The effective coherence length for intensity correlation
Intensity fluctuations are being measured by photodetectors connected to electronic equipment. If the detectors and electronics were 'infinitely fast', the electrical signals would vary on timescales down to the coherence time τ_c of the light. Errors in optical-path equality of order $c\tau_c$ would be a serious matter.

Consider now what happens if the electronics has restricted bandwidth, as was the case in the original Hanbury Brown–Twiss experiment. The electrical signals are filtered to some frequency range $\Delta\nu$, and the coherence time of those (electrical) signals is lengthened to order $1/\Delta\nu$. Compare with §9.15.4. It is *as if* the light entering the apparatus had a coherence length of $c/\Delta\nu$. Optical paths must be matched to within this distance, which may be much longer than the actual coherence length of the light arriving. For example, if the electronics responds up to 100 MHz the optical paths require matching to within 3 m, although the coherence length of incoming white light may be only a few micrometres.

Problem 10.20 Number of photons per mode in a laser beam
For 1 mW at 633 nm, the photon flux is 3.19×10^{15} photon s^{-1}. Given three modes, each of bandwidth 1 MHz, the total occupied frequency range $\delta\nu = 3$ MHz, and the number of photons per mode is 1.06×10^{9}.

Problem 10.22 A shameless model for classical chaotic light
(5) Writing out all the terms is messy:

$$\begin{aligned}
&\langle U(t)^* U(t+\tau)^* U(t+\tau)\, U(t)\rangle \\
&\quad = V_0^4 \Big\langle \sum_i \mathrm{e}^{-(t-t_i)/\tau_\mathrm{r}}\, \mathrm{e}^{-(t+\tau-t_i)/\tau_\mathrm{r}}\, u(t-t_i)\, u(t+\tau-t_i) \\
&\qquad\qquad + \sum_i \mathrm{e}^{-(t-t_i)/2\tau_\mathrm{r}}\, \mathrm{e}^{-(t+\tau-t_i)/2\tau_\mathrm{r}}\, u(t-t_i)\, u(t+\tau-t_i) \\
&\qquad\qquad \times \sum_{j\neq i} \mathrm{e}^{-(t-t_j)/2\tau_\mathrm{r}}\, \mathrm{e}^{-(t+\tau-t_j)/2\tau_\mathrm{r}}\, u(t-t_j)\, u(t+\tau-t_j) \\
&\qquad\qquad\quad + \sum_i \mathrm{e}^{-(t-t_i)/2\tau_\mathrm{r}}\, \mathrm{e}^{-(t-t_i)/2\tau_\mathrm{r}}\, u(t-t_i)^2 \\
&\qquad\qquad\quad \times \sum_{j\neq i} \mathrm{e}^{-(t+\tau-t_j)/2\tau_\mathrm{r}}\, \mathrm{e}^{-(t+\tau-t_j)/2\tau_\mathrm{r}}\, u(t+\tau-t_j)^2 \Big\rangle .
\end{aligned}$$

Here the sums over j exclude the value $j = i$, so they are taken over $(NT - 1)$ terms. However, we let $T \to \infty$ in order to take a long-time average, and then the distinction between $(NT-1)$ and NT disappears. The sums over j will, therefore, be taken as unrestricted.[27]

The equation just written looks horrendous, but that is because nothing in it has yet been simplified. There is no need to write out all the simplifying steps.[28] The three averaged sums work out to be

$$\tfrac{1}{2}V_0^4 N\tau_{\rm r}\, {\rm e}^{-|\tau|/\tau_{\rm r}} + V_0^4 (N\tau_{\rm r})^2 {\rm e}^{-|\tau|/\tau_{\rm r}} + V_0^4 (N\tau_{\rm r})^2.$$

From here to eqn (10.39) is tidying up.

[27] The reader may be unconvinced: if we ignore the restriction $j \neq i$, both double sums over i and j include all the terms with $j = i$, so it may seem as though such terms are double-counted in the two sums $\sum_i \sum_j$, and then are counted a third time in the sum over i only. In fact the double sums undercount the $j = i$ terms because the recommended ensemble-averaging procedure, integrating $\int \ldots {\rm d}t_i \ldots \int \ldots {\rm d}t_j$, treats the t_i and t_j as if they were independent random variables, and ignores the fact that the t_i and t_j are correlated ($t_j = t_i$ when $j = i$). The single sum over i is needed to remedy this deficiency.

If the reader is still unconvinced .... Change both sums over j to include $j = i$. Each double sum now includes the case $i = j = k = l$, so includes within itself the single sum. The single sum must be subtracted (instead of added) to avoid double counting. But the chance that t_j falls within ${\rm d}t_j$ must be written as

$$\left(\frac{NT-1}{T} + \delta(t_j - t_i)\right) {\rm d}t_j,$$

where the δ-function builds in the fact that $t_j = t_i$ when $j = i$. The two terms coming from the δ-function each yield the single sum, and the net result is as claimed.

[28] *Suggestion*: Work out the sums for the case where $\tau \geq 0$, because this helps with simplifying the products of $u()$ step functions. When the reduction is complete, patch in the knowledge that what is being worked out is an even function of τ, and change τ to $|\tau|$ as appropriate.

Problem 11.2 Longitudinal modes and their occupation

(3) The choice of standing waves or travelling waves is often misunderstood. It is nothing to do with boundary conditions. All we do is select a complete set of eigenfunctions with which to expand the electromagnetic field within a volume (or within a length). Such eigenfunctions could resemble $\sin kz$ or ${\rm e}^{ikz}$; Fourier's theorem tells us that both provide complete sets—and that both go wrong at the boundaries though in different ways. In the present case we have a beam incident on a surface from one side, so the appropriate choice is travelling waves. In one dimension this means giving the wave an amplitude proportional to ${\rm e}^{ikz}$ with $kL = ({\rm integer})2\pi$. The wave vector k can be positive or negative depending on the direction of travel of the wave. The number of modes associated with δk is the range of integers, so it is $(L/2\pi)\delta k = (L/c)\delta\nu$. There is a temptation to multiply by 2 because positive and negative k give the same frequency, but we don't because our wave is going in one direction only.

Problem 11.4 Étendue of a Gaussian beam

(1) The angular radius of the beam in its far field is $w(z)/z \approx w_0/b$ from eqn (7.12). Then the solid angle $\delta\Omega = \pi(w_0/b)^2$. The area from which the beam started is $\delta S = \pi w_0^2$, so the étendue is $(\pi w_0^2/b)^2 = \lambda_{\rm vac}^2$ from another application of eqn (7.12). Compare with problem 7.6(9).

(2) This exhibits a failure to meet the condition stated in sidenote 6 on p. 250, that all parts of δS must radiate into all parts of $\delta\Omega$. Within the Gaussian beam (away from its waist), light crossing an area above the axis is travelling 'uphill', so it does not go into the full far-field solid angle. We might of course worry that this objection applies equally to light crossing the waist; however our answer 'feels right' ....

(3) Consider a beam of Gauss–Hermite form (eqn 8.1), with eigenvalues l, m. For a quantum harmonic oscillator the wave functions have conventionally defined 'edges' at the limits of classical motion, which makes the wavefunctions extend $w\sqrt{2l+1}/\sqrt{2}$, $w\sqrt{2m+1}/\sqrt{2}$ either side of the origin. Use the same criterion here because it's known to work well when l and m are large. The beam at its waist fits into a rectangle $\sqrt{2}w_0\sqrt{2l+1} \times \sqrt{2}w_0\sqrt{2m+1}$. In its far field it fits into a similar rectangular shape with angular width $\sqrt{2}(w_0/b)\sqrt{2l+1} \times \sqrt{2}(w_0/b)\sqrt{2m+1}$.

The area and solid angle should not be multiplied together to give an étendue because only one transverse mode is occupied (so far).

Consider a beam that contains a random mixture of transverse modes with eigenvalues l up to maximum $l_{\max}$, m up to maximum $m_{\max}$. All modes have the same z-axis and the same waist parameter w_0. This should give a beam whose outline at the waist is not very different from $\sqrt{2}w_0\sqrt{2l_{\max}+1} \times \sqrt{2}w_0\sqrt{2m_{\max}+1}$, but now has its intensity more or less uniform within that outline. The same applies to the beam's angular spread in the far field. This beam has a properly definable étendue $(4/\pi^2)(2l_{\max}+1)(2m_{\max}+1)\lambda_{\text{vac}}^2$.

The number of transverse modes excited in the beam (assumed to have both polarizations excited) is $2(l_{\max}+1)(m_{\max}+1)$.

When this number of transverse modes excited is large, we have

$$\text{étendue} = (8/\pi^2) \times (\text{number of transverse modes}) \times \lambda_{\text{vac}}^2. \qquad \text{(N.17)}$$

Argue that the numerical coefficient in eqn (N.17) is not to be taken seriously, so the agreement with eqn (11.9) is as good as we could wish.

Problem 11.8 The number of photons per mode

(4) (a) The turns of the spiral track on a CD are 1.6 µm apart, and the length committed to one bit is 0.3 µm (independent of location on the disc). These dimensions were chosen for the industry standard in 1982, as being the smallest detail that could be reliably read by the optics available at the time, given diffraction-limited focusing. The 780 nm radiation is focused by a lens with a numerical aperture[29] of 0.45. For light occupying a single transverse mode this permits focusing to a spot of width between 0.9 and 1.1 µm depending on whether you use a diffraction-pattern calculation or use étendue.[30] For the diffraction-pattern case, the Airy disc has a FWHM of

$$1.22\frac{\lambda}{2\sin\theta_{\max}}\frac{1}{1.185} = 0.89\,\mu\text{m},$$

where the figure 1.185 is taken from the solution to problem 3.15. For the étendue route, we find the area ΔS to which light in a single transverse mode may be focused, using eqns (11.4) and (11.9):

$$\text{diameter} = \frac{2\lambda_{\text{vac}}}{\pi \times \text{numerical aperture}} = 1.10\,\mu\text{m}.$$

Clearly, even with a single transverse mode, the reading optics are only just good enough. A laser source is essential.

(b) To detect a bit reliably we need at least 3500 photons, which at 2 Gbit s^{-1} means 7×10^{12} photons per second. This is at the receiving end of a fibre. To have a useful communication range we need about 1000 times more at the sending end: 7×10^{15} photons per second, a power of 1.1 mW. This is in a single transverse mode, so an étendue of λ^2. An LED operating at 1.3 µm, if we were to try to use it, would give out radiation in a bandwidth of 1.25×10^{13} Hz;

[29] Numerical aperture is defined in eqn (11.5).

[30] Neither is adequate. The laser beam has non-uniform intensity over the area of the lens, and people designing the reader have to model this properly.

to have 7×10^{15} photons per second we would require (problem 11.2) there to be 560 photons per mode, a factor 3000 or so higher than is attainable. Yet a power of a couple of milliwatts is well within the capability of a gallium-arsenide laser.

(c) A similar calculation, using a data rate of $10\,\mathrm{Mbit\,s^{-1}}$, shows we need a power of $8\,\mu\mathrm{W}$ and about 2×10^{-3} photons per mode. An LED can supply about $40\,\mu\mathrm{W}$ into a multimode fibre, so these figures are practical.

(d) At a resolution of 600 dots per inch, each dot has a diameter of $42\,\mu\mathrm{m}$. We wish to print an A4 page (3×10^7 pixels) in about 5 s, so each pixel must be written to in $0.14\,\mu\mathrm{s}$. We have to supply a required power onto a required area, so there is a requirement on the radiance of the source. Meeting it with an LED is possible but a bit marginal, and most manufacturers use a laser.

(e) To record a hologram we need a coherence area as large as the hologram itself. The earliest holograms, as made by Gabor, were made inside a microscope, and their smallness meant that the coherence requirement wasn't too demanding. But to record onto a 100 mm square we clearly need a laser in order to have enough photons arriving in any reasonable exposure time.

Problem 11.10 Energy throughput for a grating monochromator

(4) Consider first the case where the entrance slit is wide, so that its image[31] on the photographic film is wider than the diffraction pattern there. Then, if we halve the slit width, we halve the étendue, and we halve the optical power collected, but we also halve the area of film onto which the light falls. The power per unit area remains unchanged, and so does the exposure time. In this limit there is no penalty in exposure time (and a gain in resolution) from narrowing the slit.

In the opposite limit (very narrow entrance slit), narrowing the slit further does not narrow the diffraction-pattern image on the film, and there is a penalty in exposure time similar to that encountered in the earlier parts of the problem. There's a changeover between the two regimes when $(\Delta y)_{\mathrm{slit}}$ and $f\lambda/(Nd\cos\alpha)$ are comparable.

[31] We are considering the case of a line spectrum, and with line broadening negligible. The case of a continuous spectrum is considered in problem 4.13(6).

Problem 11.13 Improving the instrumental line profile

The oscillatory behaviour has happened because the top-hat envelope of eqn (11.31) has 'sharp edges'. No realistic line profile could have produced an interferogram with such a behaviour (for-real, not just because we stopped scanning). To remove this artefact, we need to invent a new $H(\tau)$ that has a rounder shape, rather than vertical steps at $\tau = \pm 2x_{\max}/c$. We don't need any complicated optical interventions; just apply a suitable 'shading function' in the computer-generation of the Fourier transform. The manipulation of the shape of a Fourier transform here has an exact parallel in the apodizing of §12.10.

Problem 11.15 The Fourier-transform spectrometer as a whole-fringe instrument

(4) No, a whole-fringe instrument does not represent an optimum, just a standard of comparison.

(5) The aperture in front of the detector of the 'improved' instrument must have a central hole surrounded by transparent rings, the whole matched to the size of the rings in the fringe pattern. But with a Fourier-transform spectrometer, the ring fringes change size as the path difference is scanned. This makes it impossible to have a single multi-ring aperture that serves for a whole scan. In fact, the limitation represented by eqn (11.43) came about from the need to make the pinhole small enough to serve for the worst-case fringe size.

The case of a Fabry–Perot étalon is less clear-cut, because for a given étalon and wavelength the rings are of fixed size. So it might be tried, though it rarely if ever is.

In both cases, we have to ask how much improvement is possible. Suppose we prepare an aperture transmitting 10 fringes; it increases the detector power by a factor 10. But if the main source of noise is shot noise in the photocurrent, the fluctuations in photon number rise by a factor $\sqrt{10}$, so the signal/noise ratio is improved by only $\sqrt{10}$. We may well decide that the advantage to be gained does not justify the trouble.

Problem 11.19 Energy conservation in slab interference

(1) It goes 'the other way' at the beam splitter.

(2) There must be 100% reflection because there is nowhere else for the energy to go. It should be possible to show this by working out all the amplitudes and using equations from problem 6.5 to describe the 'other' surface. But the mathematics is messy.

Problem 12.6 Illumination with a condenser

(5) Any one coherence area can send diffracted light into the objective, with coherence between the light travelling in different directions. But the largest angle between rays remains $2\theta_{\max}$; this fixes the angle of arrival at the image and with it the smallest detail that can be reproduced. Increase of $\alpha_{\max}$ cannot improve this. If $\alpha_{\max}$ exceeds $\theta_{\max}$, light arriving at 'extreme' angles does not send its zeroth diffraction order into the objective, so its diffraction orders contribute in the manner of dark-ground illumination. We might expect that the only effect of the additional light is to distort relative intensities in the image (false detail?).

Problem 12.7 Worthwhile magnification for an optical microscope

(2) Consider a small area πr^2 in the object, radiating into numerical aperture $(n \sin\theta_{\max})$. The étendue is $\pi r^2 \pi (n \sin\theta_{\max})^2$, where we use eqn (11.4) and do not assume that $\theta_{\max}$ is small. In the final image, the area is magnified to $M^2 \pi r^2$, and it radiates into the eye ring of diameter d_{e} at distance D_{ldv}. The angles here are small, so the étendue is

$M^2\pi r^2\times(\pi d_{\rm e}^2/4)/D_{\rm ldv}^2$. Equating the étendues, we obtain the expression for M given in the question.

(3) Take the r in the above as $(\lambda/2)/(n\sin\theta_{\rm max})$, so it is the resolution distance in the object. The resolution distance in the final image is Mr, into which we substitute the value of M from part (2). This yields eqn (12.12).

Problem 12.9 A model for apodizing

The amplitude diffracted through angle θ to the focal plane of a lens with focal length f is

$$U(\beta)=U_0\,\mathrm{e}^{-\mathrm{i}\omega t}\,\frac{ab\,\mathrm{e}^{\mathrm{i}kd}}{2\mathrm{i}\lambda f}\,\frac{\sin(\frac{1}{2}\beta a)}{\frac{1}{2}\beta a}\,\frac{1}{1-(\beta a/2\pi)^2},$$

where $\beta=k\sin\theta$ and d is the optical path from the centre of the slit to the centre of the focal plane. Without the 'shading' function, the final fraction is replaced by 2.

The final fraction $\{1-(\beta a/2\pi)^2\}^{-1}$ has the effect of cancelling the first zeros of $\sin(\frac{1}{2}\beta a)$ either side of centre (at $\frac{1}{2}\beta a=\pm\pi$), thereby widening the central peak by a factor 2. However, the 'wings' of the diffraction pattern fall like β^{-3} for amplitude, β^{-6} for intensity, compared with β^{-1} and β^{-2} for the unshaded slit. These properties are expected for a diffracting amplitude whose 'first' discontinuity is in its second derivative.

Problem 12.11 The optical transfer function

(2) If the amplitude has a 'stripy' behaviour like $\cos\beta x$ the intensity varies like $\cos 2\beta x$. So the same detail requires an amplitude response up to space-frequency β but intensity response up to space-frequency 2β.

The observation in the question about peak widths may need elaboration. It is not hard to see that a convolution tends to widen a peak: sketch two peaks and imagine the convolution being formed by sliding one peak across the other while recording the area under their product. As a *rough* rule, the peak widths add. Examples where the width of the convolution is the sum of the widths of the contributing functions:

- The convolution of a top-hat function with itself is a triangle function with twice the original width (problem 2.20).
- The convolution of two Lorentzians (problem 10.8) is another Lorentzian whose width is the sum of the contributing widths.

But there are counterexamples. The convolution of two Gaussians is another Gaussian, with the final width the square root of the sum of squares of the contributing widths. So here the widths less-than-add.[32]

[32] This draws attention to the cautious 'almost always' in the question. Is there any case where convolution does not widen things at all? Yes there is. Suppose we do a Fraunhofer-diffraction experiment in which the diffracting aperture consists of two slits with widths a, b, one after the other and close together. The effect is like a single slit that is the narrower of a and b. Let the transmissions of the two slits be top-hat functions $H_a(x)$ and $H_b(x)$. The combined transmission can be written, perversely but correctly, as $H_a(x)\,H_b(x)$. Since this is a product, the diffraction pattern is the convolution of the patterns that the two slits would have produced if used one at a time. And the pattern is that caused by the narrower slit, unaffected by the wider slit. Therefore, the convolution of two $(\sin x)/x$ functions is the wider of the two—unwidened by convolution with the narrower. This result can of course be verified by a brute-force (contour) integration, but it is most easily obtained by the route given here.

Problem 13.1 The detail of §13.2

(8) Two reasons. One is a confusion, deliberately sown in the problem, between a $\cos^2$ variation of intensity and of amplitude. The other is the non-linear response of the photograph, which distorts the $\cos^2$ variation so as to generate harmonics.

Problem 13.7 Gabor's electron–optical microscope

(1) By scaling the hologram in proportion to the wavelength, we generate a diagram like Fig. 13.7(b) that is magnified overall, with no change of shape. This may be verified by calculation, but should be obvious.

Problem 13.8 General properties of any imaging system

(2) The transformation equations we started from could quite properly be applied to relate coordinate systems whose axes are inclined, rather than orthogonal. In this step, we reverse the signs of x_o, y_o, x_i and y_i, which can always be done formally. But we also require that the transformation looks the same afterwards as it did before. That is, the sign reversals are a **symmetry operation**. If the axes are inclined, the sign reversals do not constitute a symmetry operation, so we are here requiring that the x_o axis is at right-angles to the z_o axis and similarly.

(3) Change x_o into y_o, and y_o into $-x_o$, and similarly in the image space. This gives a symmetry operation—a 90° rotation—only if the x_o and y_o axes are at right-angles; likewise for x_i, y_i. We find $b_2 = a_1$ and $a_2 = -b_1$. After steps (2) and (3) we have input enough requirements to ensure that the axis systems in object and image spaces are fully orthogonal.[33]

We can now require that the image-space axes are oriented around z_i in such a way that points in the $x_o z_o$ plane are imaged into the $x_i z_i$ plane, which gives $a_2 = 0$.

After this the coordinate transformation has been simplified to

$$\frac{x_i}{x_o} = \frac{y_i}{y_o} = \frac{a_1}{Cz_o + 1}, \qquad z_i = \frac{c_3 z_o + d_3}{Cz_o + 1}.$$

(4) This step is mathematically straightforward, but it helps to imagine a ray diagram of a lens, together with its principal planes and focal points. We are defining the two focal lengths as distances from principal plane to focal point in just the way that is done with a thick lens.

Comment: There are implications here for problem 13.7(2). If we try to make a holographic electron microscope that resolves down to atomic dimensions, we shall need a linear magnification (lateral) of at least 10^6. Then unless something extreme has been done with f_3 and f_1 the longitudinal magnification is of order 10^{12}, giving an effective depth of focus of zero. This is sometimes offered as an explanation as to why holographic microscopy has turned out to be a dead end. However, problem 13.7(1) exhibits a method by which f_3 and f_1 could, in principle, be given values that remove the difficulty.

[33] Reasoning very similar to that pursued here can be used to investigate the possible (four-dimensional) coordinate transformations (candidates for a Lorentz transformation) between reference frames in special relativity. By demanding that 'straight lines map to straight lines', we require that an inertial frame maps onto an inertial frame. In the relativistic case we exclude the possibility that a finite point in one space can map onto infinity in the other, and that makes the transformation linear. Reasoning paralleling that of parts (2) and (3) is needed to define what is meant by orthogonal axes because the concept isn't now trivial. We end up with a Lorentz-shaped transformation in which there is a limiting velocity c and a relativistic velocity-addition theorem relating velocities in the two frames. All this without mentioning the speed of light! A very small further step identifies c with the speed of light and the reasoning is complete. The above route is long compared with the usual 'textbook' derivation of the Lorentz transformation, but is very satisfying because it shows how much *had* to happen.

Problem 14.1 A step-index fibre

(3) The acceptance angle $\alpha = 14.3°$.

(4) With a wavelength of 850 nm the number of transverse modes is 1622. For the same fibre used at a wavelength of 1.3 μm the number of transverse modes is 693. These numbers are large enough that we should be able to trust eqn (14.12).

Problem 14.3 Gauss–Hermite modes in a model graded-index fibre

(5) Identify the 'limits of classical motion' for the harmonic oscillator in the x and y directions, and show that they are $x_{\max} = w\sqrt{l+\frac{1}{2}}$ and $y_{\max} = w\sqrt{m+\frac{1}{2}}$. Then the beam shape is roughly a rectangle with half-diagonal $r_{\max}$ where $r_{\max}^2 = x_{\max}^2 + y_{\max}^2$.

This description cannot be pushed too far. For the 00 mode, it gives $r_{\max} = w$. Yet this mode is circular, rather than rectangular, and direct evaluation from e^{-r^2/w^2} gives $r_{\max} = w/\sqrt{2}$.

Problem 15.6 The Poynting vector and energy absorption

(6) When $ax^2 + bx + c = 0$ the product of the roots for x is c/a.

Problem 16.2 The resolution of a confocal microscope

The Airy function (for intensity) is $\{2J_1(\rho)/\rho\}^2$, so the point spread function (for intensity) of a confocal microscope is $\{2J_1(\rho)/\rho\}^4$. The first of these falls to $\frac{1}{2}$ for $\rho = 1.61634$, while the second falls to $\frac{1}{2}$ for $\rho = 1.16028$, smaller by a factor 0.7178.

Bibliography

Abramowitz, M. and Stegun, I. (1965). *Handbook of mathematical functions.* Dover, New York. Previously published by National Bureau of Standards in 1964.

Bleaney, B. I. and Bleaney, B. (1976). *Electricity and magnetism*, third edn. Oxford University Press, Oxford.

Born, M. and Wolf, E. (1999). *Principles of optics*, seventh edn. Cambridge University Press, Cambridge.

Boyd, G. D. and Gordon, J. P. (1961). Confocal multimode resonator for millimeter through optical wavelength masers. *Bell System Technical Journal*, **40**, 489–508.

Cagnet, M., Françon, M. and Thrierr, J. C. (1962). *Atlas of optical phenomena.* Springer-Verlag, Berlin.

Carriere, J., Narayan, R., Yeh, W.-H., Peng, C., Khulbe, P., Li, L., Anderson, R., Choi, J. and Mansuripur, M. (2000). Principles of optical disk data storage. In *Progress in optics* (ed. E. Wolf), vol. XLI, pp. 97–197. Elsevier, Amsterdam.

Davis, C. C. (1996). *Lasers and electro-optics.* Cambridge University Press, Cambridge.

ECMA-267. (2001). *120 mm* DVD*–read-only disk*, third edn. ECMA International—European association for standardizing information and communication systems, Geneva, Switzerland.

Fox, A. G. and Li, T. (1961). Resonant modes in a maser interferometer. *Bell System Technical Journal*, **40**, 453–88.

Fox, M. (2001). *Optical properties of solids.* Oxford University Press, Oxford.

Fried, D. L. and Mevers, G. E. (1974). Evaluation of r_0 for propagation down through the atmosphere. *Applied Optics*, **13**, 2620–22.

Gabor, D. (1948). A new microscopic principle. *Nature*, **161**, 777–8.

Gamow, G. (1965). *Mr Tompkins in paperback.* Cambridge University Press, Cambridge. The part cited was formerly published as *Mr Tompkins in wonderland*, CUP, 1940.

Gerrard, A. and Burch, J. M. (1975). *Introduction to matrix methods in optics.* Wiley, New York.

Glauber, R. J. (1970). Quantum theory of coherence. In *Quantum optics: proceedings of the tenth session of the Scottish Universities summer school in physics, 1969* (eds. S. M. Kay and A. Maitland), pp. 53–125. Academic Press, London.

Gowar, J. (1993). *Optical communication systems.* Prentice Hall, Hemel Hempstead.

Hajian, A. R. and Armstrong, J. T. (2001). A sharper view of the stars. *Scientific American*, **284** (March), 48–55.

Hanbury Brown, R. (1974). *The intensity interferometer: its application to astronomy.* Taylor and Francis, London.

Harburn, G., Taylor, C. A. and Welberry, T. R. (1975). *Atlas of optical transforms.* Bell, London.

Hariharan, P. and Sanders, B. C. (1996). Quantum phenomena in optical interferometry. In *Progress in optics* (ed. E. Wolf), vol. XXXVI, pp. 49–128. Elsevier, Amsterdam.

Harvey, J. E. (1979). Fourier treatment of near-field scalar diffraction theory. *American Journal of Physics*, **47**, 974–980.

Harwit, M. and Decker, A. J. (1974). Modulation techniques in spectroscopy. In *Progress in optics* (ed. E. Wolf), vol. XII, pp. 103–162. North Holland, Amsterdam.

Hecht, E. (2002). *Optics*, fourth edn. Addison-Wesley, Reading, MA.

IEC 908. (1987). *Compact disc digital audio system, ICE International Standard 908.* International Electrotechnical Commission (IEC), Geneva, Switzerland.

Jacquinot, P. (1960). New developments in interference spectroscopy. In *Reports on progress in physics*, vol. 23, pp. 267–312. Institute of Physics, London.

Jacquinot, P. and Roizen-Dossier, B. (1964). Apodising. In *Progress in optics* (ed. E. Wolf), vol. III, pp. 29–186. North Holland, Amsterdam.

Joos, G. (1934). *Theoretical physics.* Blackie, London.

Kaye, G. W. C. and Laby, T. H. (1995). *Tables of physical and chemical constants*, sixteenth edn. Longman, London.

Kennard, E. H. (1938). *Kinetic theory of gases.* McGraw-Hill, New York.

Kittel, C. (1996). *Introduction to solid state physics*, seventh edn. Wiley, New York.

Klein, M. V. and Furtak, T. E. (1986). *Optics*, second edn. Wiley, New York.

Kogelnik, H. and Rigrod, W. W. (1962). *Proc. IRE*, **50**, 220.

Koppelmann, G. (1969). Multiple-beam interference and natural modes in open resonators. In *Progress in optics* (ed. E. Wolf), vol. VII, pp. 3–66. North Holland, Amsterdam.

Lakhtakia, A., Varadan, V. K. and Varadan, V. V. (1989). *Time-harmonic electromagnetic fields in chiral media.* Springer-Verlag, Berlin. Volume 335 of series: Lecture notes in physics.

Landau, L. D. and Lifshitz, E. M. (1977). *Quantum mechanics (non-relativistic theory)*, third edn. Course of theoretical physics, vol. 3. Butterworth-Heinemann, Oxford. Formerly published by Pergamon, Oxford.

Landau, L. D. and Lifshitz, E. M. (1986). *Theory of elasticity*, third edn. Course of theoretical physics, vol. 7. Butterworth-Heinemann, Oxford. Formerly published by Pergamon, Oxford.

Landau, L. D., Lifshitz, E. M. and Pitaevskiĭ, L. P. (1993). *Electrodynamics of continuous media*, second edn. Course of theoretical physics, vol. 8. Butterworth-Heinemann, Oxford. Formerly published by Pergamon, Oxford.

Leith, E. N. and Upatnieks, J. (1962). Reconstructed wavefronts and communication theory. *J. Opt. Soc. Am.*, **52**, 1123–30.

Lipson, S. G., Lipson, H. and Tannhauser, D. S. (1995). *Optical physics*, third edn. Cambridge University Press, Cambridge.

Longhurst, R. S. (1973). *Geometrical and physical optics*, third edn. Longman, London.

Loudon, R. (2000). *The quantum theory of light*, third edn. Clarendon Press, Oxford.

Macleod, H. A. (2001). *Thin film optical filters*, third edn. Institute of Physics, Bristol.

Mandel, L. and Wolf, E. (1995). *Optical coherence and quantum optics.* Cambridge University Press, Cambridge.

Mandl, F. (1988). *Statistical physics.* Wiley, New York.

Matthews, P. T. (1974). *Introduction to quantum mechanics*, third edn. McGraw-Hill, Maidenhead.

Maystre, D. (1984). Rigorous vector theories of diffraction gratings. In *Progress in optics* (ed. E. Wolf), vol. XXI, pp. 1–67. North Holland, Amsterdam.

Nisenson, P. and Papaliolios, C. (2001). Detection of earth-like planets using apodized telescopes. *Astrophysical Journal*, **548**, L201–5.

Padgett, M., Arlt, J., Simpson, N. and Allen, L. (1996). An experiment to observe the intensity and phase structure of Laguerre–Gaussian laser modes. *American Journal of Physics*, **64**, 77–82.

Palmer, E. W., Hutley, M. C., Franks, A., Verrill, J. F. and Gale, B. (1975). Diffraction gratings. In *Reports on progress in physics*, vol. 38, part 2, pp. 975–1048. Institute of Physics, London.

Penrose, R. (1989). *The emperor's new mind.* Oxford University Press, Oxford.

Popov, E. (1993). Light diffraction by relief gratings: a macroscopic and microscopic view. In *Progress in optics* (ed. E. Wolf), vol. XXXI, pp. 139–187. Elsevier, Amsterdam.

Readhead, A. C. S. (1982). Radio astronomy by very-long-baseline interferometry. *Scientific American*, **246** (June), 38–47.

Robinson, F. N. H. (1973*a*). *Electromagnetism.* Oxford University Press, Oxford.

Robinson, F. N. H. (1973*b*). *Macroscopic electromagnetism.* Pergamon, Oxford.

Robinson, F. N. H. (1974). *Noise and fluctuations in electronic devices and circuits.* Oxford University Press, Oxford.

Rossi, B. (1965). *Optics.* Addison-Wesley, Reading, MA.

Schiff, L. I. (1968). *Quantum mechanics*, third edn. McGraw-Hill, New York.

Schmahl, G. and Rudolph, D. (1976). Holographic diffraction gratings. In *Progress in optics* (ed. E. Wolf), vol. XIV, pp. 197–244. North Holland, Amsterdam.

Senior, J. M. (1992). *Optical fiber communications: principles and practice*, second edn. Prentice Hall, New York.

Sieger, B. (1908). Die Beugung einer ebenen elektrischen Welle an einem Schirm von elliptischem Querschnitt. *Annalen der Physik*, **27**, 626–64.

Smith, H. M. (1969). *Principles of holography.* Wiley, New York.

Stacey, D. N., Stacey, V. and Malvern, A. R. (1974). A method for finding the instrumental profile of a Fabry–Perot etalon. *Journal of Physics E*, **7**, 405–8.

Stetter, F., Esselborn, R., Harder, N., Friz, M. and Tolles, P. (1976). New materials for optical thin films. *Applied Optics*, **15**, 2315–17.

Stroke, G. W. (1963). Ruling, testing and use of optical gratings for high-resolution spectroscopy. In *Progress in optics* (ed. E. Wolf), vol. II, pp. 1–72. North Holland, Amsterdam.

Stroke, G. W. (1969). *An introduction to coherent optics and holography*, second edn. Academic Press, New York.

Strom, R. G., Miley, G. K. and Oort, J. (1975). Giant radio galaxies. *Scientific American*, **233** (August), 26–35.

Tango, W. J. and Twiss, R. Q. (1980). Michelson stellar interferometry. In *Progress in optics* (ed. E. Wolf), vol. XVII, pp. 239–77. North Holland, Amsterdam.

Thorne, A. P. (1988). *Spectrophysics*, second edn. Chapman and Hall, London.

Walther, H. (1962). Das Kernquadrupolmoment des Mn^{55}. *Zeitschrift für Physik*, **170**, 507–25.

Walton, A. J. (1986). The Abbe theory of imaging: an alternative derivation of the resolution limit. *European Journal of Physics*, **7**, 62–3.

Wharmby, D. O. (1997). Radiation and light production. In *Lamps and lighting* (eds. J. R. Coaton and A. M. Marsden), fourth edn., pp. 95–118. Arnold, London.

Wilson, T. (1990). Confocal microscopy. In *Confocal microscopy* (ed. T. Wilson), pp. 1–64. Academic Press, London.

Woodgate, G. K. (1980). *Elementary atomic structure*, second edn. Oxford University Press, Oxford.

Yariv, A. (1997). *Optical electronics in modern communications*, fifth edn. Oxford University Press, Oxford.

Index